Tom Richmond
**General Topology**

## Also of Interest

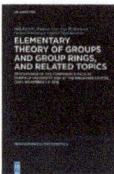

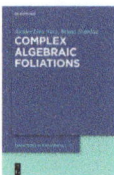

Tom Richmond

# General Topology

—

An Introduction

**DE GRUYTER**

**Mathematics Subject Classification 2010**
Primary: 54-01, 54A05; Secondary: 54A10, 54D10, 54D15, 54E15, 54F05

**Author**
Prof. Tom Richmond
Department of Mathematics
Western Kentucky University
1906 College Heights Blvd.
Bowling Green, KY 42101
USA
tom.richmond@wku.edu

ISBN 978-3-11-068656-2
e-ISBN (PDF) 978-3-11-068657-9
e-ISBN (EPUB) 978-3-11-068672-2

**Library of Congress Control Number: 2020935444**

**Bibliographic information published by the Deutsche Nationalbibliothek**
The Deutsche Nationalbibliothek lists this publication in the Deutsche Nationalbibliografie;
detailed bibliographic data are available on the Internet at http://dnb.dnb.de.

© 2020 Walter de Gruyter GmbH, Berlin/Boston
Cover image: Ocelia_MG / iStock / Getty Images Plus
Typesetting: VTeX UAB, Lithuania
Printing and binding: CPI books GmbH, Leck

www.degruyter.com

# Preface

Topology is the study of continuous deformations. It is an abstract geometry which is not concerned with properties such as area, length, or shape, which may change as an object is continuously deformed, but rather with more fundamental properties such as connectedness. As the study of continuous deformations, topology includes the study of continuity, limits, and convergence. Thus, topology may be viewed as the formal study of nearness.

While topological concepts can be identified in much earlier work, the development of topology as a separate discipline was strongly influenced by the push in the mid-1800s by Karl Weierstrass, Georg Friedrich Bernhard Riemann, and others to put mathematics on a solid formal foundation, and later that century, by Georg Cantor and Richard Dedekind. The foundations of the specific areas of general topology presented in this book were started in the early 1900s by Felix Hausdorff and Maurice Fréchet and continued by Pavel Alexandroff, Pavel Urysohn, and Nicholas Bourbaki, among many others.

Topology has developed into a broad and useful area encompassing far too much to be included in any single book. This book provides a solid introduction to basic concepts and applications of general topology and lays the foundation for continued study in topology. Much of early topology was developed to put analysis on a solid foundation, and introductory topology is often presented as a handmaiden of analysis and an introduction to algebraic topology. A distinguishing feature of this text is the additional emphasis on the connections between topology and order, which are fundamental to applications in computer science. Many of the techniques introduced in the latter half of the book are part of the growing area of *asymmetric topology*, which addresses topological spaces in which a point $x$ having a neighborhood disjoint from $y$ does not imply that $y$ has a neighborhood disjoint from $x$ (spaces which are not $T_1$), or distance functions (quasi-metrics) $q$ which may allow the distance from $x$ to $y$ to be different from the distance from $y$ to $x$.

After the preliminaries in Chapter 0, Chapters 1–7 contain a classical introduction to general topology, motivated and developed from intuitive ideas. Boundary points are used to introduce open and closed sets. The quotient topology is developed as the topology of saturated open sets relative to an equivalence relation. Before continuity is defined, homeomorphisms are introduced as bijections between the points and open sets of two topological spaces. Most of this material should be covered in an introductory topology course. The topics after the Chapter 7 provide some traditional and some modern applications of introductory general topology.

An equivalence relation on a set $X$ is essentially equality of some specified attribute of the elements of $X$. A partial order on some attribute of the elements of $X$ gives a *quasiorder* on the elements of $X$. That is, a partial order on the equivalence classes from an equivalence relation on $X$ gives a quasiorder on $X$. Such constructions

https://doi.org/10.1515/9783110686579-201

clearly fall in the area of discrete mathematics, and with the traditional focus of topology applied to the continuous mathematics used in analysis, it may be surprising to find that this construction yields a topology on $X$. Indeed, if $X$ is finite, every topology arises from a quasiorder, so the study of topology on finite sets may be rephrased as the study of quasiorders. This is done in Chapter 8. Computers only deal with finite sets: $\pi$ can only be represented by a finite rational decimal approximation, and line segments contain only finitely many points (pixels). Convergence, continuity, and nearness in computer applications thus require topologies on finite sets. The discrete techniques needed for these applications are a recurring theme of this book.

After viewing topologies as (quasi-)orders in Chapter 8, Chapter 9 investigates some classical results about orders on collections of topologies. Chapter 10 considers topologies on ordered sets. Further connections between topology and order are given in Chapters 11 and 12, particularly in the sections on quasi-metrics, quasi-uniformities, and partial metrics. Distance functions, or *metrics*, are extremely useful in defining nearness, convergence, and topologies. Typically distance functions are assumed to satisfy properties of the familiar Euclidean distances between points in the real line or in the Euclidean plane. Variations of distance functions which do not behave exactly as the Euclidean distance function occur in many natural ways and are the topic of much research. For example, the distance from $a$ to $b$ may differ from the distance from $b$ to $a$ when driving in a city with one-way streets. Or, motivated by loss of resolution situations, the distance from one pixel to itself (or from New York to New York) may be greater than zero. Some variations of distance functions are introduced in Chapter 11. Chapter 12 introduces uniform spaces, which provide a global approach to specifying nearness to points. Chapter 13 formally returns to the classical topological idea of continuous deformation applied not only to points, but to curves or sets in the plane. This chapter lays the groundwork for further study in algebraic topology and is motivated by the question of whether two disjoint planar sets moving continuously may go from being disjoint to having overlapping interiors without first having their boundaries touch.

Another important part of this book is the assortment of exercises. There are over 740 exercises designed to reinforce the concepts, illustrate further applications, and suggest areas for continued investigation.

**Course outline**

For an introductory one-semester course at the upper undergraduate or beginning graduate level, a traditional course may start with a day or two of review in Chapter 0, then cover at least Sections 1.1 through 7.2, omitting Sections 1.4.1 (Neighborhood bases), 1.6 (Limit points), 5.2 (Compactness in metric spaces), and 6.3 (The cocountable topology). Additional topics from Chapters 8, 11, or 13 may be added as time permits. For one-semester courses with more advanced students, Chapters 1–7 may be covered entirely, with additional topics selected from the remaining chapters. All sections may be covered in a two-semester course. Though the connections between Sections 8.1–8.3, 11.2, and 12.4 provide an elegant motivating cohesion for the topics of these chapters, the chapters after Chapter 7 may be largely presented independently. Section 9.3 depends on Section 8.1, Theorem 11.2.12 depends on Section 8.3, Section 12.4 depends on 11.1, and Section 12.5 depends on Sections 8.3 and 11.2.

The selection of topics reflects my personal interests, which have been greatly influenced by many students and colleagues. In particular, the works of Hans-Peter Künzi and his collaborators and students have been a major influence, as well as the works of Darrell Kent, Ralph Kopperman, and Marcel Erné. Thanks to Filiz Yıldız, Anneliese Schauerte, Hans-Peter Künzi, and many students for their comments on drafts of this book. Their comments have greatly improved the text, but I am solely responsible for any remaining errors.

Tom Richmond

# Contents

# 0 Preliminaries

Most of the concepts of this chapter should be familiar to the reader. We present them here briefly as a refresher and to clarify the terminology. More details can be found, for example, in [41].

## 0.1 Real numbers

Mathematics involves logic and patterns, often in quantitative settings requiring numbers. The *natural numbers* $\mathbb{N} = \{1, 2, 3, 4, 5, \ldots\}$ are those that occur naturally in counting, and are sometimes called the *counting numbers*. Historically, zero came much later than the counting numbers. Numbers were used to count things. If you had no things, you would not need to count, and would not need a number to quantify nothing to count. The natural numbers together with zero form the set of *whole numbers*. Negative numbers came into widespread usage even later. The whole numbers together with their negatives form the set of *integers*, denoted $\mathbb{Z} = \{\ldots, -4, -3, -2, -1, 0, 1, 2, 3, 4, \ldots\}$. The symbol $\mathbb{Z}$ comes from the German word for numbers, *Zahlen*. Ratios of integers give us the set of *rational numbers* $\mathbb{Q} = \{\frac{a}{b} : a, b \in \mathbb{Z}, b \neq 0\}$. The symbol $\mathbb{Q}$ is used to signify quotients of integers. The Greeks were very adept at using positive integers and ratios of positive integers. They were not happy to discover that some numbers they encountered—such as the length of the diagonal of a unit square—were not ratios of integers. Such numbers are *irrational numbers*. If we visualize the set of *real numbers* $\mathbb{R}$ as all the distances on a number line, then the irrational numbers are the real numbers which are not rational.

We list some well-known facts about rational and irrational numbers.

**Theorem 0.1.1.**
(a) *If $n \in \mathbb{N}$, then $\sqrt{n}$ is either an integer or irrational.*
(b) *The product or quotient of two nonzero rational numbers is rational.*
(c) *The product or quotient of a nonzero rational number and an irrational number is irrational.*
(d) *The product or quotient of two irrational numbers may be rational or irrational.*
(e) *Every nonempty open interval $(a, b)$ contains rational numbers and irrational numbers.*

*Proof.* Suppose $n \in \mathbb{N}$. (a) can be rephrased as: if $\sqrt{n}$ is rational, then it is an integer. Suppose $\sqrt{n} = \frac{a}{b}$ where $a, b \in \mathbb{N}$. Squaring and clearing the denominators gives $nb^2 = a^2$. Since $nb^2$ is a perfect square, all of its prime factors appear with even multiplicities, and thus all of the prime factors of $n$ appear with even multiplicities, so $n$ is a perfect square and $\sqrt{n} \in \mathbb{N}$.

https://doi.org/10.1515/9783110686579-001

(c) Suppose to the contrary that $(\frac{a}{b})\alpha = \frac{c}{d}$, where $a, b, c, d$ are nonzero integers and $\alpha$ is irrational. Solving for $\alpha$ gives $\alpha = \frac{bc}{ad}$, contrary to $\alpha$ being irrational. Similar arguments apply for the other half of (c) and (b).

(d) $\sqrt{2}\sqrt{2}$ and $\frac{\sqrt{2}}{\sqrt{2}}$ are rational; $\sqrt{6}\sqrt{3} = \sqrt{18}$ and $\frac{\sqrt{6}}{\sqrt{3}} = \sqrt{2}$ are irrational.

(e) Given an open interval $(a, b)$ with $a < b$, let $n$ be a positive integer with $1/n$ less than the length $b - a$ of $(a, b)$. Now all multiples of $\frac{1}{n}$ are rational, and some multiple $\frac{m}{n}$ must fall in $(a, b)$, or else $(a, b)$ falls between two consecutive multiples of $\frac{1}{n}$. This cannot happen: two consecutive multiples of $\frac{1}{n}$ form an interval of length $\frac{1}{n}$, which cannot contain the interval $(a, b)$ of length greater than $\frac{1}{n}$. By choosing $n \in \mathbb{N}$ large enough to make $\frac{\pi}{n} < b - a$, a similar argument shows that some multiple of $\frac{\pi}{n}$ must fall in $(a, b)$, and all multiples of $\frac{\pi}{n}$ are irrational. □

## 0.2 Sets

Suppose we have several objects under consideration, and we wish to specify some list of these objects for further consideration. The objects under consideration make up the *universal set* $U$, and a well-define list of selected objects gives a *set* $A$ in $U$. The objects included in a set $A$ are the *elements* of the set $A$. If $a$ is an element of $A$, we write $a \in A$. If $A$ and $B$ are two sets in $U$ and every element of $A$ is an element of $B$, then $A$ is a *subset* of $B$ and $B$ is a *superset* of $A$, denoted by $A \subseteq B$ or $B \supseteq A$. If $A \subseteq B$ and $A \neq B$, then $A$ is a *proper subset* of $B$, denoted $A \subset B$. The set with no elements is called the *empty set*, denoted $\emptyset$ or $\{\}$. A set $\{a\}$ with a single element is called a *singleton set*. The *union* of two sets $A$ and $B$ is $A \cup B = \{x : x \in A \text{ or } x \in B\}$. The *intersection* of $A$ and $B$ is $A \cap B = \{x : x \in A \text{ and } x \in B\}$. The *complement* of $A$ in the universal set $U$ is $A^C = U - A = \{x \in U : x \notin A\}$. If $A$ and $B$ are sets, the *set difference* $A - B$ is $\{x : x \in A \text{ and } x \notin B\} = A \cap B^C = A \cap (U - B)$. The set difference $A - B$ is also called the complement of $B$ in $A$. If $A \subseteq B \subseteq U$, then $U - B \subseteq U - A$; that is, a smaller set has a larger complement. Two sets $A$ and $B$ are *disjoint* if $A \cap B = \emptyset$. The number of elements in a finite set $A$ is the *cardinality* of $A$, denoted $|A|$.

A *collection* is a set whose elements are themselves sets. Elements of sets are denoted by lower case letters, sets are denoted by upper case letters, and collections are denoted by upper case script letters. The collection of all subsets of $U$ is the *power set* of $U$, denoted $\mathcal{P}(U)$. For example, if $U = \{1, 2, 3\}$, then $\mathcal{P}(U) = \{\emptyset, \{1\}, \{2\}, \{3\}, \{1, 2\}, \{1, 3\}, \{2, 3\}, \{1, 2, 3\}\}$. If $U$ has $n$ elements, then the power set $\mathcal{P}(U)$ has $2^n$ elements—that is, $U$ has $2^n$ subsets. This follows since any subset of $U$ is obtained by making a binary decision for each of the $n$ elements to include the element or exclude it. Note that $A \subseteq U$ is equivalent to $A \in \mathcal{P}(U)$.

**Example 0.2.1.** For $U = \{1, 2, 3, 4\}$, let $\mathcal{D}$ be the collection of all subsets of $U$ which contain an odd number of elements, and let $\mathcal{E}$ be the collection of all subsets of $U$

which contain an even number of elements. That is,

$$\mathcal{D} = \{A \in \mathcal{P}(U) : |A| \text{ is odd}\} = \{A \subseteq U : |A| \text{ is odd}\}, \quad \text{and}$$
$$\mathcal{E} = \{A \in \mathcal{P}(U) : |A| \text{ is even}\} = \{A \subseteq U : |A| \text{ is even}\}.$$

Thus,

$$\mathcal{D} = \{\{1\}, \{2\}, \{3\}, \{4\}, \{1, 2, 3\}, \{1, 2, 4\}, \{1, 3, 4\}, \{2, 3, 4\}\} \quad \text{and}$$
$$\mathcal{E} = \{\emptyset, \{1, 2\}, \{1, 3\}, \{1, 4\}, \{2, 3\}, \{2, 4\}, \{3, 4\}, \{1, 2, 3, 4\}\}.$$

Now $\mathcal{D}$ and $\mathcal{E}$ are disjoint subcollections of $\mathcal{P}(U)$, and $\mathcal{D} \cup \mathcal{E} = \mathcal{P}(U)$. Observe that

$$2 \notin \mathcal{D},$$
$$\{2\} \in \mathcal{D},$$
$$\{2\} \not\subseteq \mathcal{D},$$
$$\{\{2\}\} \subseteq \mathcal{D},$$
$$\{\{2\}, \{4\}, \{1, 3, 4\}\} \subseteq \mathcal{D},$$
$$\mathcal{D} \subseteq \mathcal{P}(U),$$
$$\mathcal{D} \in \mathcal{P}(\mathcal{P}(U)),$$
$$\{\mathcal{D}, \mathcal{E}\} \subseteq \mathcal{P}(\mathcal{P}(U)),$$
$$\{\mathcal{D}, \mathcal{E}\} \in \mathcal{P}(\mathcal{P}(\mathcal{P}(U))),$$
$$\mathcal{D} - \{\{3\}, \{1, 3\}, \{2, 4\}\} = \mathcal{D} - \{\{3\}\},$$
$$\mathcal{D} \cup \{\{1, 2\}, \{1, 2, 3\}, \{2, 3, 4\}\} = \mathcal{D} \cup \{\{1, 2\}\}.$$

Often we will need to give each element of a collection $\mathcal{C}$ a label, and we may write $\mathcal{C} = \{A_i : i \in I\} = \{A_i\}_{i \in I}$. Here, the subscript $i$ is an *index*, the set $I$ is the *index set*, and $\mathcal{C}$ is thus an *indexed collection*. Operations on indexed collections are defined as expected. If $\mathcal{C} = \{A_i : i \in I\}$, then

$$\bigcup \mathcal{C} = \bigcup_{i \in I} A_i = \{x : x \in A_i \text{ for some } i \in I\},$$
$$\bigcap \mathcal{C} = \bigcap_{i \in I} A_i = \{x : x \in A_i \text{ for every } i \in I\}.$$

If the index set $I$ is finite, then the intersection $\bigcap_{i \in I} A_i$ is called a *finite intersection*. That is, a finite intersection is an intersection of a finite collection of sets. Similarly, the adjectives in *infinite intersections, countable intersections*, and *uncountable intersections* refer to the cardinality of the index set. If no restriction whatsoever is given on the index set, we may refer to an *arbitrary intersection*. In practice, *arbitrary intersection* should suggest intersecting a finite or infinite collection. Similar terminology applies to unions.

Occasionally, we may consider a collection of subsets of a universal set $U$ indexed by the empty set. In this case, we take

$$\bigcap_{i \in \emptyset} A_i = U \quad \text{and} \quad \bigcup_{i \in \emptyset} A_i = \emptyset.$$

Intuitively, the more sets you intersect, the smaller the intersection gets, so the fewer sets you intersect, the larger the intersection gets, so intersecting no sets gives the largest possible set $U$. Similarly, the fewer sets you union, the smaller the result, so unioning no sets gives the smallest possible set, $\emptyset$.

A collection $C = \{A_i : i \in I\}$ is *mutually disjoint* if for $i, j \in I$, $A_i \neq A_j$ implies $A_i \cap A_j = \emptyset$, and is *nested* if for every $i, j \in I$, either $A_i \subseteq A_j$ or $A_j \subseteq A_i$.

Rules for taking complements of unions or intersections are named for the British mathematician Augustus De Morgan (1806–1871). *De Morgan's Laws* state that the complement of an intersection is the union of the complements, and the complement of a union is the intersection of the complements. That is,

$$U - \bigcap_{i \in I} A_i = \bigcup_{i \in I} (U - A_i) \quad \text{and} \quad U - \bigcup_{i \in I} A_i = \bigcap_{i \in I} (U - A_i).$$

Another often-used property involves containments. If $A_i \subseteq B_i$ for each $i \in I$, then $\bigcup_{i \in I} A_i \subseteq \bigcup_{i \in I} B_i$ and $\bigcap_{i \in I} A_i \subseteq \bigcap_{i \in I} B_i$.

Given two sets $A$ and $B$, their *Cartesian product* is the set $A \times B = \{(a, b) : a \in A, b \in B\}$ of all ordered pairs whose first coordinate is an element of $A$ and whose second coordinate is an element of $B$. If $\{A_i\}_{i \in I}$ is a collection of sets, the Cartesian product $\prod_{i \in I} A_i$ of the collection consists of all the vectors $(x_i)_{i \in I}$ where $x_i \in A_i$ for each $i \in I$.

It may not be easy to recognize when a set is infinite or finite. Mathematicians do not know whether the set of twin primes (that is, primes which differ by two, such as 17 and 19) is finite or infinite. However, the concepts of finite and infinite are still well-defined. One must be precise when discussing infinite sets, paying careful attention to what the adjective "infinite" quantifies. It is incorrect to say $C = \{\{n, n + 1\}\}_{n \in \mathbb{N}}$ has infinite elements. Every element of $C$ has form $\{n, n + 1\}$, so every element of $C$ is finite. We could properly say that $C$ has infinitely many elements.

## 0.3 Quantifiers and logic

A statement is a sentence which is either true or false. If $S$ and $T$ are statements, the implication "$S$ implies $T$" will be denoted $S \Rightarrow T$. The implication $S \Rightarrow T$ is true if, whenever $S$ is true, $T$ is also true. If $S \Rightarrow T$, then we say $S$ implies $T$, $T$ is implied by $S$, $S$ is a *sufficient condition* for $T$, or $T$ is a *necessary condition* for $S$. In words, $S \Rightarrow T$ may appear as "if $S$ then $T$", "$T$ if $S$", or "$S$ only if $T$".

If $S \Rightarrow T$ and $T \Rightarrow S$, then we write $S \iff T$ and say $S$ and $T$ are *logically equivalent statements* (or simply *equivalent*), or $S$ occurs *if and only if* (or *iff*) $T$ occurs. In this case $S$ is a *necessary and sufficient condition* for $T$.

The *negation* of a statement $S$, denoted $\sim S$, is the statement "not S". That is, $S$ is true if and only if $\sim S$ is false.

The *converse* of an implication $S \Rightarrow T$ is the implication $T \Rightarrow S$. If an implication is true, its converse may be true or false, as illustrated in Example 0.3.1.

The *contrapositive* of an implication $S \Rightarrow T$ is the implication $\sim T \Rightarrow \sim S$. An implication and its contrapositive are logically equivalent: either both are true or both are false.

**Example 0.3.1.** The implications below illustrate that the converse of a true implication may be true or false. Also note that an implication and its contrapositive have the same truth value.

| | | | | |
|---|---|---|---|---|
| Implication: | $x = -2$ | $\Rightarrow$ | $x^2 = 4.$ | True |
| Converse: | $x^2 = 4$ | $\Rightarrow$ | $x = -2.$ | False |
| Contrapositive: | $x^2 \neq 4$ | $\Rightarrow$ | $x \neq -2.$ | True |
| Implication: | $x = 3$ | $\Rightarrow$ | $x + 1 = 4.$ | True |
| Converse: | $x + 1 = 4$ | $\Rightarrow$ | $x = 3.$ | True |
| Contrapositive: | $x + 1 \neq 4$ | $\Rightarrow$ | $x \neq 3.$ | True |
| Implication: | $\sin x = 0$ | $\Rightarrow$ | $x = 0.$ | False |
| Converse: | $x = 0$ | $\Rightarrow$ | $\sin x = 0.$ | True |
| Contrapositive: | $x \neq 0$ | $\Rightarrow$ | $\sin x \neq 0.$ | False |

The *conjunction* $S \wedge T$ of statements $S$ and $T$ is the statement "S and T". The *disjunction* $S \vee T$ of statements $S$ and $T$ is the statement "S or T". The following facts are also called *De Morgan's Laws*:

$$\sim (S \wedge T) = \sim S \vee \sim T \quad \text{and}$$
$$\sim (S \vee T) = \sim S \wedge \sim T.$$

The only way an implication $S \Rightarrow T$ may fail is if $S$ occurs but $T$ does not. Thus, $\sim (S \Rightarrow T) = S \wedge \sim T$.

The universal quantifier $\forall$ means "for every". The existential quantifier $\exists$ means "there exists". For example, $\bigcap_{i \in I} A_i = \{x : \forall i \in I, x \in A_i\}$ and $\bigcup_{i \in I} A_i = \{x : \exists i \in I \text{ such that } x \in A_i\}$.

The only way the statement "there exists $x$ for which $S$ is true" could fail is if for every $x$, $S$ is false. Similarly, the statement "for every $x$, $S$ is true" fails if and only if there exists an $x$ for which $S$ fails. That is,

$$\sim (\forall x, S) = \exists x \text{ such that } \sim S \quad \text{and}$$
$$\sim (\exists x \text{ such that } S) = \forall x, \sim S.$$

**Example 0.3.2.** Some statements and their negations are given.

$S$: $\forall x \in \mathbb{R}, x^2 - 1 > 0$,

$\sim S$: $\exists x \in \mathbb{R}$ such that $x^2 - 1 \not> 0$,

$T$: $\exists x \in \mathbb{N}$ such that $x^2 = x + 3$,

$\sim T$: $\forall x \in \mathbb{N}, x^2 \neq x + 3$,

$P$: $\forall y \in [0,1]\ \exists x \in [0, 2\pi)$ with $\sin x = y$,

$\sim P$: $\exists y \in [0,1]$ such that $\sim (\exists x \in [0, 2\pi)$ with $\sin x = y)$
$= (\exists y \in [0,1]$ such that $\forall x \in [0, 2\pi), \sin x \neq y)$,

$Q$: $\exists y \in \mathbb{R}$ such that $\forall x \in \mathbb{R}, \frac{y}{x} = 0$,

$\sim Q$: $\forall y \in \mathbb{R} \sim (\forall x \in \mathbb{R}, \frac{y}{x} = 0)$
$= (\forall y \in \mathbb{R}\ \exists x \in \mathbb{R}$ such that $\frac{y}{x} \neq 0)$.

We note that $S$, $T$, and $Q$ are false and $P$ is true. Had the statement $T$ started with $\exists x \in \mathbb{R}$, then it would have been true. Statement $Q$ fails precisely for one point $x = 0$.

## 0.4 Functions

If $X$ and $Y$ are sets, a *function* from $X$ to $Y$ is a rule that assigns to each element of $X$ a unique element of $Y$. To denote that $f$ is a function from $X$ to $Y$, we write $f : X \to Y$. The set $X$ is the *domain* and the set $Y$ is the *codomain* of the function. If $f : X \to Y$ is a function and $x \in X$, then the unique element of $Y$ which $f$ assigns to $x$ is denoted $f(x)$ and is called the *value of $f$ at $x$* or the *image of $x$ under $f$*. While $f$ must assign a value to each point $x$ of its domain, not every point of the codomain must be used as an output value. The set $f(X) = \{f(x) : x \in X\} \subseteq Y$ is called the *image* or *range* of $f$. A function $f : X \to Y$ is *onto* or *surjective* if $f(X) = Y$, that is, if for every $y \in Y$, there exists $x \in X$ with $f(x) = y$. A function $f : X \to Y$ is *one-to-one* or *injective* if $w \neq x \Rightarrow f(w) \neq f(x)$, or equivalently, $f(w) = f(x) \Rightarrow w = x$. Note that $f : X \to Y$ fails to be one-to-one if there exist two distinct elements $w \neq x$ in $X$ which map to the same one element $f(w) = f(x)$ in $Y$. A one-to-one onto function may be called a *bijection*. If $f : X \to Y$ is a bijection, then, for any $y \in Y$, there exists $x \in X$ with $f(x) = y$ since $f$ is onto, and there is a unique $x \in X$ with $f(x) = y$ since $f$ is one-to-one. Thus, $f : X \to Y$ is a bijection if and only if there exists a function $f^{-1} : Y \to X$ called the *inverse function of $f$* with $f^{-1}(y) = x$ if and only if $f(x) = y$.

Functions are rules denoted by $f$ or by the rule of assignment $y = f(x)$; technically, $f(x)$ represents the value of a function and not the function itself. By widespread abuse of notation, "the function $f(x)$" is sometimes used as a shortened version of "the function $y = f(x)$". Of course, it would be even shorter to simply say "the function $f$", but often it is important to know the independent variable (that is, the "input" variable).

Suppose $f : X \to Y$ is a function, $A \subseteq X$, and $B \subseteq Y$. The set $f(A) = \{f(a) : a \in A\}$ is the *image of $A$ under $f$*. The set $f^{-1}(B) = \{x \in X : f(x) \in B\}$ is the *inverse image of $B$ under $f$*. Note that the inverse image of a set $B$ is defined whether or not $f$ has an inverse function. In this notation, $f^{-1}$ denotes the inverse relation, which may or may not be a function.

For example, consider $f : \mathbb{R} \to \mathbb{R}$ defined by $f(x) = \sin x$. The image of $\{\frac{n\pi}{2} : n \in \mathbb{Z}\}$ under $f$ is $\{f(\frac{n\pi}{2}) : n \in \mathbb{Z}\} = \{0, 1, -1\}$. The inverse image of $\{0\}$ under $f$ is $f^{-1}(\{0\}) = \{x \in \mathbb{R} : f(x) \in \{0\}\} = \{n\pi : n \in \mathbb{Z}\}$. Likewise, $f^{-1}([0, \infty)) = \bigcup_{n\in\mathbb{Z}}[2n\pi, (2n+1)\pi]$.

The behavior of images and inverse images of unions, intersections, and complements is described in the next theorem.

**Theorem 0.4.1.** *Suppose $f : X \to Y$ is a function.*
(a) $f(A \cup B) = f(A) \cup f(B)$ *for all $A, B \subseteq X$.*
(b) $f(A \cap B) \subseteq f(A) \cap f(B)$ *for all $A, B \subseteq X$ and $f(A \cap B) = f(A) \cap f(B)$ for all $A, B \subseteq X$ if and only if $f$ is one-to-one.*
(c) $f(A - B) \supseteq f(A) - f(B)$ *for all $A, B \subseteq X$ and $f(A - B) = f(A) - f(B)$ for all $A, B \subseteq X$ if and only if $f$ is one-to-one.*
(d) $f^{-1}(C \cup D) = f^{-1}(C) \cup f^{-1}(D)$ *for all $C, D \subseteq Y$.*
(e) $f^{-1}(C \cap D) = f^{-1}(C) \cap f^{-1}(D)$ *for all $C, D \subseteq Y$.*
(f) $f^{-1}(C - D) = f^{-1}(C) - f^{-1}(D)$ *for all $C, D \subseteq Y$.*

The proof is left as an exercise.

If $f : A \to B$ and $g : B \to C$ are functions, then the *composition* of $f$ and $g$ is $g \circ f : A \to C$ defined by $(g \circ f)(a) = g(f(a))$ for all $a \in A$.

A *sequence* in $A$ is a function $f : \mathbb{N} \to A$. The $n$th term of the sequence is $f(n)$. If we write $f(n) = a_n$, then we may specify the sequence by $(f(n))_{n\in\mathbb{N}} = (a_n)_{n\in\mathbb{N}}$, or simply as $(a_n)$.

If $X$ and $Y$ are subsets of $\mathbb{R}$, a function $f : X \to Y$ is *increasing* if $a \leq b$ in $X$ implies $f(a) \leq f(b)$ in $Y$, and is *strictly increasing* if $a < b$ in $X$ implies $f(a) < f(b)$ in $Y$. *Decreasing* and *strictly decreasing* functions are defined dually.

Intuitively, a subsequence of a sequence $(a_n)_{n\in\mathbb{N}}$ is a sequence obtained by (possibly) omitting some of the terms of $(a_n)_{n\in\mathbb{N}}$. For example, for $(a_n)_{n\in\mathbb{N}} = (1, 4, 9, 16, 25, 36, \ldots) = (n^2)_{n\in\mathbb{N}}$, we have $(b_n)_{n\in\mathbb{N}} = (4, 25, 64, 121, \ldots) = (a_2, a_5, a_8, a_{11}, \ldots)$ is a subsequence of $(a_n)_{n\in\mathbb{N}}$. The $n$th term of $(b_n)_{n\in\mathbb{N}}$ is the $(3n - 1)$st term of $(a_n)_{n\in\mathbb{N}}$. That is, for every $n \in \mathbb{N}$, $b_n = a_{3n-1} = a_{\sigma(n)}$ where $\sigma : \mathbb{N} \to \mathbb{N}$ is the strictly increasing function $\sigma(n) = 3n - 1$. In general, $(b_n)_{n\in\mathbb{N}}$ is a subsequence of $(a_n)_{n\in\mathbb{N}}$ if and only if $(b_n)_{n\in\mathbb{N}} = (a_{\sigma(n)})_{n\in\mathbb{N}}$ for some strictly increasing function $\sigma : \mathbb{N} \to \mathbb{N}$. The sequence $(\sigma(n))_{n\in\mathbb{N}}$ tells which terms of $(a_n)_{n\in\mathbb{N}}$ to retain in the subsequence $(b_n)_{n\in\mathbb{N}}$.

## 0.5 Countable and uncountable sets

To count the objects in a finite set, we set up a one-to-one correspondence (that is, a bijection) between the objects and a set $\{1, 2, 3, \ldots, n\}$. We say sets $A$ and $B$ have the same cardinality, denoted $|A| = |B|$, if there exists a bijection from $A$ to $B$. A set $A$ is *countably infinite* it has the same cardinality as the set $\mathbb{N}$ of counting numbers. That is, $A$ is countably infinite if and only if there exists a bijection $f : \mathbb{N} \to A$. Thus, $A$ is countably infinite if and only if the elements of $A$ can be listed as a sequence $(a_n)_{n \in \mathbb{N}}$ of distinct terms. A set is *countable* if and only if it is finite or countably infinite. A set is *uncountable* if it is not countable.

For example, the set $\{1, 4, 9, 16, 25, \ldots\} = \{n^2 : n \in \mathbb{N}\}$ is countably infinite since $f : \mathbb{N} \to \{1, 4, 9, 16, 25, \ldots\}$ defined by $f(n) = n^2$ is a bijection.

**Theorem 0.5.1.**

(a) *Every subset $A$ of a countable set $B$ is countable.*

(b) *If $f : A \to B$ is one-to-one and $B$ is countable, then $A$ is countable.*

(c) *If $f : A \to B$ is onto and $A$ is countable, then $B$ is countable.*

(d) *If $A$ and $B$ are countably infinite sets, then $A \cup B$ is countably infinite.*

(e) *If $A$ and $B$ are countably infinite sets, then $A \times B$ is countably infinite.*

(f) *If $\{A_n\}_{n \in \mathbb{N}}$ is a countable collection of countably infinite sets $A_n$, then the union $\bigcup_{n \in \mathbb{N}} A_n$ is countably infinite.*

*Proof.* (a) Since every countable set $B$ is in one-to-one correspondence with a subset of $\mathbb{N}$, it suffices to show that every subset of (a subset of) $\mathbb{N}$ is countable. If $A$ is finite, clearly $A$ is countable, so suppose $A$ is an infinite subset of $\mathbb{N}$. Define $f : \mathbb{N} \to A$ by $f(1) = $ the smallest element of $A$, $f(2) = $ the smallest element of $A - \{f(1)\}$, and in general $f(n) = $ the smallest element of $A - \{f(1), \ldots, f(n-1)\}$. Since $f$ is a bijection, $A$ is countable.

(b) follows from (a) since $A$ has the same cardinality as $f(A) \subseteq B$ if $f$ is one-to-one.

(c) Suppose $f : A \to B$ is onto and $A$ is countable. Since $f$ is onto, for every $b \in B$, there exists (at least one) $a_b \in f^{-1}(\{b\})$. Thus, we may define a function $g : B \to A$ by taking $g(b)$ to be one element $a_b \in f^{-1}(b)$. This function is one-to-one from $B$ into the countable set $A$, so by (b), $B$ is countable.

(d) Suppose $A$ and $B$ are countably infinite. Then there exist bijections $f : \mathbb{N} \to A$ and $g : \mathbb{N} \to B$. Define $h : \mathbb{N} \to A \cup B$ by $h(2n) = f(n)$ and $h(2n - 1) = g(n)$. Now $h$ is clearly onto, so by (c), $A \cup B$ is countable.

(e) Suppose $A = \{a_n\}_{n \in \mathbb{N}}$ and $B = \{b_n\}_{n \in \mathbb{N}}$ are countably infinite. List the elements $(a_i, b_j)$ of $A \times B$ in a rectangular array as shown at the top of Figure 0.1. Define $f(1), f(2), f(3), \ldots$ in the diagonal pattern suggested in the second part of Figure 0.1. This will provide a bijection between $\mathbb{N}$ and $A \times B$. The details are left to the reader.

(f) is similar to (e). List all the elements of $\bigcup_{n \in \mathbb{N}} A_n$ in an array with the elements $a_{n,1}, a_{n,2}, a_{n,3}, \ldots$ of $A_n$ on the $n$th row. Define $f(1), f(2), f(3), \ldots$ in the diagonal pattern as suggested in Figure 0.1. This function will not be one-to-one if the sets $A_n$ are not mutually disjoint, but by (c), the union will be countable. $\quad\square$

**Theorem 0.5.2.** *The set $\mathbb{Q}$ of rational numbers is countable. The set $\mathbb{R}$ of reals is uncountable.*

$$A \times B$$

| $(a_1, b_1)$ | $(a_1, b_2)$ | $(a_1, b_3)$ | $(a_1, b_4)$ | $\cdots$ |
| $(a_2, b_1)$ | $(a_2, b_2)$ | $(a_2, b_3)$ | $(a_2, b_4)$ | $\cdots$ |
| $(a_3, b_1)$ | $(a_3, b_2)$ | $(a_3, b_3)$ | $(a_3, b_4)$ | $\cdots$ |
| $(a_4, b_1)$ | $(a_4, b_2)$ | $(a_4, b_3)$ | $(a_4, b_4)$ | $\cdots$ |
| $\vdots$ | $\vdots$ | $\vdots$ | $\vdots$ | $\ddots$ |

$f(1)$

$f(2), f(3)$

$f(4), f(5), f(6)$

$f(7), f(8), f(9), f(10)$

| $(a_1, b_1)$ | $(a_1, b_2)$ | $(a_1, b_3)$ | $(a_1, b_4)$ | $\cdots$ |
| $(a_2, b_1)$ | $(a_2, b_2)$ | $(a_2, b_3)$ | $(a_2, b_4)$ | $\cdots$ |
| $(a_3, b_1)$ | $(a_3, b_2)$ | $(a_3, b_3)$ | $(a_3, b_4)$ | $\cdots$ |
| $(a_4, b_1)$ | $(a_4, b_2)$ | $(a_4, b_3)$ | $(a_4, b_4)$ | $\cdots$ |
| $\vdots$ | $\vdots$ | $\vdots$ | $\vdots$ | $\ddots$ |

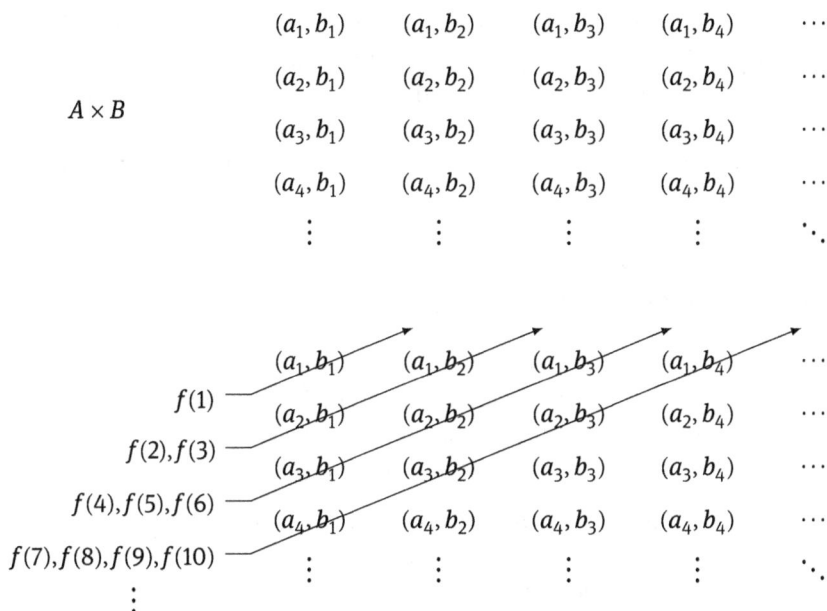

**Figure 0.1:** The procedure for enumerating the elements of $A \times B$.

*Proof.* For $n \in \mathbb{N}$, the set $Q_n = \{\frac{a}{n} : a \in \mathbb{Z}\}$ is a countable set: it is indexed by $\mathbb{Z}$, which is the union of the countable sets $\mathbb{N} \cup \{0\}$ and $-\mathbb{N}$. Now $\mathbb{Q} = \bigcup_{n \in \mathbb{N}} Q_n$, and by Theorem 0.5.1(f), $\mathbb{Q}$ is countable.

To see that $\mathbb{R}$ is uncountable, we will show that $[0, 1)$ is uncountable. Suppose to the contrary that $[0, 1)$ is a countable set $\{r_i\}_{i \in \mathbb{N}}$ indexed by $\mathbb{N}$. Express each of these real numbers $r_i$ in its decimal representation, avoiding repeating nines. (So, $\frac{1}{2}$ is represented as $0.5\overline{0}$, and not as $0.4\overline{9}$.) Let $d_{i,j}$ be the $j$th digit in the decimal expansion of $r_i$, as suggested in Figure 0.2.

$r_1 = 0.d_{1,1}\, d_{1,2}\, d_{1,3}\, d_{1,4} \cdots$
$r_2 = 0.d_{2,1}\, d_{2,2}\, d_{2,3}\, d_{2,4} \cdots$
$r_3 = 0.d_{3,1}\, d_{3,2}\, d_{3,3}\, d_{3,4} \cdots$
$r_4 = 0.d_{4,1}\, d_{4,2}\, d_{4,3}\, d_{4,4} \cdots$

$\vdots$

**Figure 0.2:** An alleged enumeration of the real numbers in $[0, 1)$.

We will find a contradiction by producing a real number $x \in [0, 1)$ which is not in the countable list $\{r_1, r_2, r_3, \dots\}$. We will give the decimal expansion of $x$. Let $x_j$ represent the $j$th digit of the decimal expansion of $x$, and take $x_j$ to be any digit different from

$0, d_{j,j}$, and 9. Thus, the decimal expansion of $x$ differs from that of $r_j$ in the $j$th digit. Because we have avoided zeros and nines in the decimal expansion of $x$, this means that $x \neq r_j$ for any $j \in \mathbb{N}$, giving the desired contradiction. (Note that the decimal numbers $0.5\overline{0}$ and $0.4\overline{9}$ are equal even though they differ in every digit of their decimal expansions. Avoiding zeros and nines in the decimal expansion of $x$ prevents this potential problem.) □

## 0.6 Equivalence relations and partitions

If $A$ and $B$ are sets, a *relation from A to B* is a subset $R$ of $A \times B$. If $(a, b) \in R$, we say $a$ is related to $b$ by the relation $R$, denoted $aRb$. Thus, a relation from $A$ to $B$ is just a set of ordered pairs telling which elements of $A$ are related to which elements of $B$.

A relation from a set $A$ to the same set $A$ is called a *relation on A*.

**Definition 0.6.1.** An *equivalence relation R* on set $A$ is a relation which is
*reflexive:* $aRa$ for every $a \in A$,
*symmetric:* $aRb \Rightarrow bRa$ for every $a, b \in A$, and
*transitive:* $aRb$ and $bRc$ imply $aRc$ for every $a, b, c \in A$.

The prototype for an equivalence relation on a set $A$ is equality. For every $a, b,$ $c \in A$, clearly $a = a$, if $a = b$ then $b = a$, and if $a = b$ and $b = c$, then $a = c$, so equality is an equivalence relation. In general, an equivalence relation on a set $A$ will be equality of some attribute of the elements of $A$. For example, a prospector may use an equivalence relation on gold nuggets defined by taking two gold nuggets to be equivalent if and only if their weights are equal.

Given an equivalence relation, it is not always easy to recognize it as equality of some attribute. For example, define a relation $\approx$ on $\mathbb{N}$ by taking $a \approx b$ if and only if $ab$ is a perfect square. It is easy to verify that $\approx$ is an equivalence relation, but it may not be immediately obvious what attributes of $a$ and $b$ are equal if $a \approx b$. In fact, $a \approx b$ if and only if their *square-free parts* are equal. Any natural number $a$ may be factored into the largest factor $m^2$ that is a perfect square and a square-free factor $n$ which only has prime factors of multiplicity one. For example, the square-free part of $a = 2^3 3^6 5^9 7^3 = (2 \cdot 3^3 5^4 7)^2 (2 \cdot 5 \cdot 7)$ is $2 \cdot 5 \cdot 7$. Now if $ab$ is a perfect square, then the prime factorization of $b$ must have odd multiplicities on precisely the same prime factors as $a$, namely $2, 5,$ and $7$. Thus, the square-free parts of $a$ and $b$ must be equal.

A *partition* of a set $A$ is a collection $\mathcal{P} = \{B_i\}_{i \in I}$ of nonempty, mutually disjoint subsets of $A$ whose union is $A$. The elements $B_i$ of a partition are called the *blocks* of the partition. If $\mathcal{P} = \{B_i\}_{i \in I}$ is a partition of set $A$ and some of the blocks $B_i$ are further partitioned into "subblocks", the resulting collection of subblocks is also a partition of $A$ called a *refinement* of $\mathcal{P}$, or a *finer partition* than $\mathcal{P}$. Formally, if $\mathcal{P}$ and $\mathcal{R}$ are partitions of set $A$, $\mathcal{R}$ is a refinement of $\mathcal{P}$ (or $\mathcal{R}$ is finer than $\mathcal{P}$, or $\mathcal{P}$ is coarser than $\mathcal{R}$) if for every block $B_R \in \mathcal{R}$, there exists $B_P \in \mathcal{P}$ with $B_R \subseteq B_P$. If partition $\mathcal{R}$ is finer than

$\mathcal{P}$ and $\mathcal{R} \neq \mathcal{P}$, then we say $\mathcal{R}$ is *strictly finer* than $\mathcal{R}$ (or $\mathcal{P}$ is *strictly coarser* than $\mathcal{R}$). Subsets of $A$ which are unions of blocks of a partition of $A$ are said to be *saturated* with respect to the partition. Thus, partition $\mathcal{R}$ is finer than partition $\mathcal{P}$ if and only if each block of $\mathcal{P}$ is $\mathcal{R}$-saturated.

**Example 0.6.2.** Pixelate the square $S = [0, 5) \times [0, 5)$ into the set of pixels $\mathcal{P} = \{[m, m+1) \times [n, n+1) : m, n \in \{0, 1, 2, 3, 4\}\}$, as suggested in Figure 0.3. Now $\mathcal{P}$ is a partition of $S$. A saturated set and a non-saturated set relative to $\mathcal{P}$ are also shown there. The collections $\mathcal{V} = \{[m, m+1) \times [0, 5) : m \in \{0, 1, 2, 3, 4\}\}$ and $\mathcal{H} = \{[0, 5) \times [n, n+1) : n \in \{0, 1, 2, 3, 4\}\}$ partition $S$ into vertical and horizontal strips, respectively. Since each block of $\mathcal{V}$ is $\mathcal{P}$-saturated, $\mathcal{P}$ is a refinement of $\mathcal{V}$, and similarly, $\mathcal{P}$ is a refinement of $\mathcal{H}$. Note that $\mathcal{V}$ is neither finer nor coarser than $\mathcal{H}$: these two partitions are non-comparable.

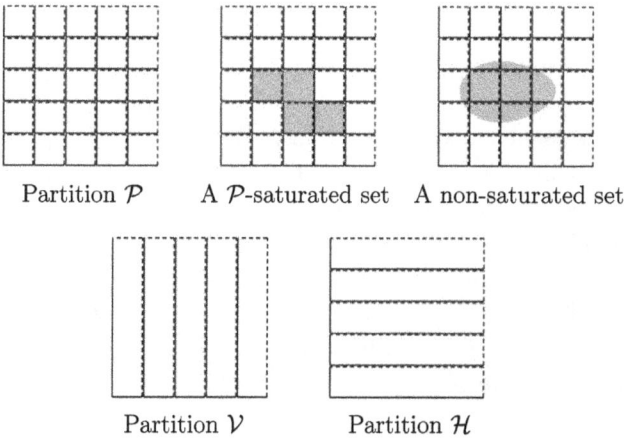

Partition $\mathcal{P}$     A $\mathcal{P}$-saturated set   A non-saturated set

Partition $\mathcal{V}$        Partition $\mathcal{H}$

**Figure 0.3:** Partitions and Saturated Sets. $\mathcal{P}$ is finer than $\mathcal{V}$ and $\mathcal{H}$.

If $\approx$ is an equivalence relation on $A$ and $a \in A$, the *equivalence class of $a$* is the set $[a] = \{x \in A : x \approx a\}$ of all elements of $A$ which are equivalent to $a$.

The following result shows the fundamental connection between equivalence relations and partitions.

**Theorem 0.6.3.** *Every partition $\mathcal{P}$ of a set $A$ determines an equivalence relation $\approx_{\mathcal{P}}$ on $A$ defined by $a \approx_{\mathcal{P}} b$ if and only if there exists a block $B_i \in \mathcal{P}$ with $\{a, b\} \subseteq B_i$.*

*Every equivalence relation $\approx$ on set $A$ determines a partition $\mathcal{P}_{\approx} = \{[a] : a \in A\}$ whose blocks are the equivalence classes.*

*Furthermore, $\mathcal{P}_{\approx_{\mathcal{P}}} = \mathcal{P}$ and $\approx_{\mathcal{P}_{\approx}} = \approx$.*

It is easy to transfer terminology about partitions into the language of equivalence relations. For example, if $\approx$ is an equivalence relation on set $A$, we say a set $B \subseteq A$ is *saturated* with respect to $\approx$ if $x \in B$ implies $[x] \subseteq B$.

## 0.7 Partial orders

Order relations are used to determine preference. In any optimization problem, an order relation is required to determine which solutions are better than which.

**Definition 0.7.1.** A *partial order* on set $A$ is a relation $\leq$ which is
*reflexive:* $a \leq a$ for every $a \in A$,
*antisymmetric:* $a \leq b$ and $b \leq a$ imply $a = b$ for all $a, b \in A$, and
*transitive:* $a \leq b$ and $b \leq c$ imply $a \leq c$ for every $a, b, c \in A$.

A set $A$ with a partial order $\leq$ is called a *partially ordered set*, or *poset*, denoted $(A, \leq)$.

A *total order* (or *linear order*) on set $A$ is a partial order $\leq$ on $A$ with the additional property that, for every $a, b \in A$, either $a \leq b$ or $b \leq a$.

Note that equivalence relations and partial orders are both special kinds of reflexive, transitive relations. Reflexive, transitive relations are called *quasiorders* and are studied in Chapter 8.

Partial orders allow us to define monotone functions. A function $f : (X, \leq_X) \to (Y, \leq_Y)$ is *increasing* if $a \leq b$ in $X$ implies $f(a) \leq f(b)$ in $Y$. It is *strictly increasing* if $a < b$ in $X$ implies $f(a) < f(b)$ in $Y$. *Decreasing* and *strictly decreasing* functions are defined dually. A function is *(strictly) monotone* if it is either (strictly) increasing or (strictly) decreasing. Increasing functions are sometimes called *order preserving* functions, and decreasing functions are called *order reversing* functions. As a special case, we see that a sequence $(a_n)_{n \in \mathbb{N}}$ of real numbers is increasing if $i \leq j$ in $\mathbb{N}$ implies $a_i \leq a_j$ in $\mathbb{R}$.

If $(A, \leq_A)$ and $(Z, \leq_Z)$ are posets, we may define a partial order $\leq$ on the Cartesian product $A \times Z$ by taking

$$(a, y) \leq (b, z) \quad \text{if and only if} \quad a \leq_A b \text{ and } y \leq_Z z.$$

This order on $A \times Z$ is called the *product order* (or *coordinate-wise order*).

If $(A, \leq)$ is a poset, an element $a \in A$ is *maximum* if $a \geq x$ for all $x \in A$, and is *maximal* if $x \in A$ and $x \geq a$ implies $x = a$. Thus, $a$ is maximum (or largest) if it is larger than every element and is maximal if there is no element strictly larger than it. In totally ordered sets, these concepts agree. *Minimum elements* and *minimal elements* are defined dually. The distinction between maximal and maximum is illustrated in the following example.

**Example 0.7.2.** Consider the poset $(\mathcal{P}(\{1, 2, 3\}) - \{\{1, 2, 3\}\}, \subseteq)$ of all proper subsets of $\{1, 2, 3\}$, ordered by set inclusion. Finite posets such as this may be represented by a *Hasse diagram* (see Figure 0.4) showing each point and upward line segments from each point leading to the larger points.

The sets $\{1, 2\}$, $\{1, 3\}$, and $\{2, 3\}$ are maximal, since there are no larger proper subsets of $\{1, 2, 3\}$. However, these maximal sets are not maximum since none is larger than

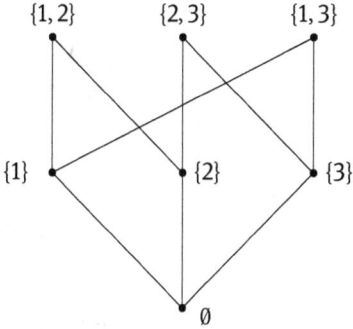

**Figure 0.4:** The proper subsets of {1, 2, 3}, ordered by inclusion.

every element of the poset. There is a minimum element, namely ∅, which is contained in (below) every other element of the poset. Furthermore, there is no element below ∅, so ∅ is also minimal.

In the Hasse diagram shown in Figure 0.4, ∅ is below {1, 2}, but there is no line segment directly connecting these two points of the diagram. A direct line segment is not needed (and should not be included in the Hasse diagram) since there is a point {1} strictly between ∅ and {1, 2}. The segments from ∅ to {1} and from {1} to {1, 2} show ∅ ⊆ {1} ⊆ {1, 2}, and by transitivity, we can determine ∅ ⊆ {1, 2} from the diagram without a direct segment connecting them. In a Hasse diagram, a line from $a$ up to $b$ should be drawn if and only if $a < b$ and there is no point strictly between $a$ and $b$. In this case, we say $a$ is *covered by* $b$ or $b$ *covers* $a$. Formally, we say $b$ covers $a$ in a poset $(X, \leq)$ if $a < b$ and whenever $a \leq x \leq b$, either $x = a$ or $x = b$.

If $B$ is a subset of a poset $(A, \leq)$, an element $a \in A$ is an *upper bound* of $B$ if $a \geq b$ for every $b \in B$. If it exists, the least element in the set of upper bounds of $B$ is called the *least upper bound* of $B$ or the *supremum* of $B$, denoted lub $B$ = sup $B$. *Lower bounds* of a set $B$ are defined dually, as are the *greatest lower bound* or *infimum* of $B$, denoted glb $B$ = inf $B$. A set $B$ is *bounded above (below)* if it has an upper (lower) bound, and is *bounded* if it has both an upper and a lower bound. A set $B$ in a poset is *convex* (or, for emphasis, *order convex*) if $a, b \in B$ and $a \leq x \leq b$ imply $x \in B$. You may have encountered the definition of convexity of a set $B$ in the plane or a vector space. The definition of convexity in any setting is as follows. $B$ is convex if whenever $a$ and $b$ are in $B$ and $x$ is between $a$ and $b$, then $x$ must be in $B$. The variability lies in how "between" is defined. A proper subset of a poset $A$ is an *interval* if it has one of the following forms:

$$[a, b] = \{x \in A : a \leq x \leq b\} \qquad [a, b) = \{x \in A : a \leq x < b\}$$
$$(a, b] = \{x \in A : a < x \leq b\} \qquad (a, b) = \{x \in A : a < x < b\}$$
$$[a, \rightarrow) = \{x \in A : x \geq a\} \qquad (a, \rightarrow) = \{x \in A : x > a\}$$
$$(\leftarrow, b] = \{x \in A : x \leq b\} \qquad (\leftarrow, b) = \{x \in A : x < b\}$$

If $A$ is a totally ordered set, $A = (\leftarrow, \rightarrow)$ is also be considered to be an interval. If $b < a$, then $[a, b] = \{x \in A : a \le x$ and $x \le b\} = \emptyset$. From the definition, we see that $[a, b]$ is defined to be the smallest convex set containing $a$ and $b$. Intervals of any form in a poset are always convex. The converse does not hold; a poset may have a convex set which is not an interval. In any totally ordered set, such as the real line $\mathbb{R}$ with the usual order, $I$ is an interval if and only if $I$ is convex. That is, an interval in a totally ordered set may be defined to be any set $I$ with the property that if $a, b \in I$ and $a \le x \le b$, then $x \in I$.

**Example 0.7.3.** Give each of the following subsets of $\mathbb{R} \times \mathbb{R}$ the product order. Note that $(x, y) \le (z, w)$ if and only if $(x, y)$ is south and west of $(z, w)$. The sets are depicted in Figure 0.5.

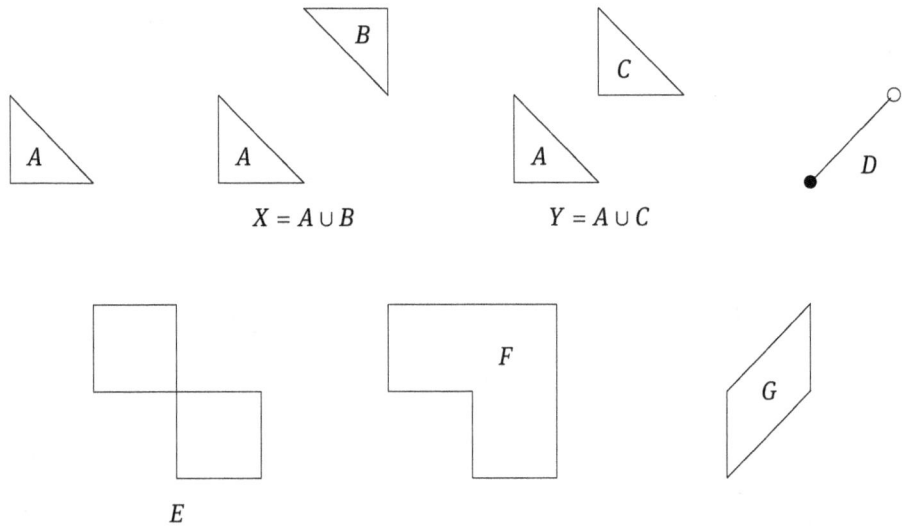

$$X = A \cup B \qquad Y = A \cup C$$

**Figure 0.5:** Subsets of the plane with the product order. Note that $E$ and $F$ are (order) convex, but $D$ and $G$ are not.

$A = \{(x, y) \in [0, 1] \times [0, 1] : y \le 1 - x\}$ has no maximum element. For each $x \in [0, 1]$, the point $(x, 1 - x)$ is maximal. The minimum point of $A$ is $(0, 0)$, and any minimum point is also necessarily minimal.

Let $B = \{(x, y) \in [1, 2] \times [1, 2] : y \ge 3 - x\}$ and $X = A \cup B$. In $X$, every point of $B$ is an upper bound of $A$, but $A$ has no least upper bound. For every $x \in [1, 2]$, the point $(x, 3 - x)$ is a minimal upper bound of $A$. The point $(0, 0)$ is the greatest lower bound (infimum) of $A$.

Let $C = \{(x, y) \in [1, 2] \times [1, 2] : y \le 3 - x\}$ and $Y = A \cup C$. In $Y$, $\inf A = (0, 0) \in A$ and $\sup A = (1, 1) \notin A$. $Y$ has no maximum element, but infinitely many maximal elements.

Let $D = \{(x, x) : x \in [0, 1)\}$. $D$ has no maximum element and no maximal elements. $D$ is not convex in $\mathbb{R}^2$: $(0, 0) \le (0.3, 0.5) \le (0.8, 0.8)$ and both $(0, 0)$ and $(0.8, 0.8)$ are in $D$, but $(0.3, 0.5)$ is not in $D$.

Let $E = ([0, 1] \times [1, 2]) \cup ([1, 2] \times [0, 1])$. $E$ is convex in $\mathbb{R}^2$, but is not an interval. However, $E$ is the union of the two intervals $[(0, 1), (1, 2)]$ and $[(1, 0), (2, 1)]$.

Let $F = ([0, 2] \times [0, 2]) - ([0, 1) \times [0, 1))$. The maximum point of $F$ is $(2, 2)$, and $F$ has two minimal points $(1, 0)$ and $(0, 1)$. $F$ is convex in $\mathbb{R}^2$.

Let $G = \{(x, y) \in [0, 1] \times \mathbb{R} : x \le y \le x + 1\}$. The point $(0, 0)$ is the minimum point of $G$ and the infimum of $G$, and $(1, 2)$ is the maximum point and supremum of $G$. $G$ is not convex in $\mathbb{R}^2$: $(0, 0)$ and $(1, 1)$ are both in $G$ and $(0, 0) \le (0.5, 0) \le (1, 1)$, but $(0.5, 0) \notin G$.

Note that in the space $X = A \cup B$ of the previous example, $A$ was a nonempty set bounded above which had no supremum. Such an example cannot occur in the real line $\mathbb{R}$ with the usual order. This is called the *completeness* property of the real line, which we will assume without proof.

**Axiom 0.7.4** (Completeness of $\mathbb{R}$). *Every nonempty set of real numbers which is bounded above has a supremum in $\mathbb{R}$. Equivalently, every increasing sequence of real numbers bounded above has a limit. Dually, every nonempty set of real numbers which is bounded below has an infimum in $\mathbb{R}$, and every decreasing sequence of real numbers bounded below has a limit.*

The completeness of the reals is one way to say that there are no holes in the real line. The set $X = [0, 1) \cup (1, 2]$ contains a nonempty set $A = [0, 1)$ bounded above which has no supremum, since there is no least element among the upper bounds $(1, 2]$ of $A$. Thus, $X$ is not complete; there is something missing at the upper end of $[0, 1)$, namely lub $A$.

## Exercises

1. Find the union, or state that the notation is not defined.
   (a) $\{0, 1, 2\} \cup 3$
   (b) $\{0, 1, 2\} \cup \{3\}$
   (c) $\{0, 1, 2\} \cup \{\{3\}\}$
   (d) $\{\{0\}, \{1\}, \{2\}\} \cup 3$
   (e) $\{\{0\}, \{1\}, \{2\}\} \cup \{3\}$
   (f) $\{\{0\}, \{1\}, \{2\}\} \cup \{\{3\}\}$
2. True or false: $\{A, B\} \cup \emptyset = \{A, B\} \cup \{\emptyset\}$. Justify your answer.
3. Fill in the blank to make a true statement. Is your answer unique?
   (a) $\{\{1\}, \{2, 3\}\} \cup \underline{\hspace{1cm}} = \{\{1\}, \{2, 3\}\}$
   (b) $\{\{1\}, \{2, 3\}\} \cup \underline{\hspace{1cm}} = \{\{1\}, \{2, 3\}, \emptyset\}$
4. For each of the following, find all solutions $A \subseteq \{1, 2, 3, 4\}$.
   (a) $\{1, 2\} \cup A = \{1, 2, 4\}$
   (b) $\{1, 2, 4\} - A = \{1, 2\}$
5. For each of the following, find the number of solutions for $C \subseteq \mathcal{P}(\{1, 2, 3, 4\})$, and give three solutions.
   (a) $\{\{1\}, \{2, 3\}\} \cup C = \{\{1\}, \{2, 3\}, \{2, 3, 4\}\}$
   (b) $\{\{1\}, \{2, 3\}, \{2, 3, 4\}\} - C = \{\{1\}, \{2, 3\}\}$

6. Let $W_k = \{(x, y) \in \mathbb{R}^2 : x > 0, 0 \leq y < kx\}$ and $S_n = [\frac{1}{n}, 2 + \frac{1}{n}) \times (\frac{1}{n}, 3 - \frac{1}{n}]$. Sketch the sets indicated.

   (a) $W_2$
   (b) $S_3$
   (c) $S_2 \cap W_1$
   (d) $S_2 \cup W_1$
   (e) $S_1 \cup S_2$

   (f) $S_1 \cap S_3$
   (g) $S_2 - W_2$
   (h) $W_2 - S_2$
   (i) $W_2 - W_1$

   (j) $S_3 - S_1$
   (k) $\bigcup \{W_k : k \in \mathbb{N}\}$
   (l) $\bigcup \{S_n : n \in \mathbb{N}\}$
   (m) $\bigcap \{S_n : n \in \mathbb{N}\}$

7. Determine whether the statements are true or false. For those that are false, explain and correct the errors.

   (a) The set $\mathbb{N}$ has infinite points.
   (b) $\{\mathbb{N}\}$ is an infinite set.
   (c) $\{\mathbb{N}, \mathbb{R}\}$ has infinite elements.
   (d) $\{\mathbb{N}, \mathbb{R}\}$ has infinitely many elements.

8. Give a proof or counterexample: $\{A_i\}_{i \in I}$ is a collection of mutually disjoint sets if and only if for every pair $i \neq j$ in $I$, $A_i$ and $A_j$ are disjoint.

9. Suppose $\{A_i\}_{i \in I}$ and $\{B_i\}_{i \in I}$ are collections of subsets of set $X$, and $a \in X$. Fill in the blanks below with the appropriate symbol $\subseteq, \supseteq$, or $=$. Provide counterexamples for any inclusions that fail.

   (a) $\bigcap_{i \in I}(A_i \cup B_i)$ _____ $(\bigcap_{i \in I} A_i) \cup (\bigcap_{i \in I} B_i)$.
   (b) $\bigcap_{i \in I}(\{a\} \cup B_i)$ _____ $\{a\} \cup (\bigcap_{i \in I} B_i)$.

10. Consider the collection $\mathcal{C} = \{(-\infty, -a) \cup (a, \infty) : a \geq 0\} \cup \{(-\infty, -a] \cup [a, \infty) : a \geq 0\}$.

    (a) Find $\bigcap \mathcal{C}$ and $\bigcup \mathcal{C}$.
    (b) Is $\mathcal{C}$ nested? Is it a collection of mutually disjoint sets?
    (c) Which of the following are members of $\mathcal{C}$? $A = (3, \infty)$, $B = \emptyset$, $C = \mathbb{R}$, $D = (-\infty, -2) \cup (2, \infty)$, $E = \mathbb{R} - (-3, 3)$, $F = [\pi, \infty)$

11. For each implication below, give its converse and contrapositive. Determine the truth value of each.

    (a) $n \in \mathbb{N} \Rightarrow 4(n + \frac{1}{2})^2 \in \mathbb{N}$.
    (b) $xy \in \mathbb{N} \Rightarrow x \in \mathbb{N}$ and $y \in \mathbb{N}$.
    (c) $x \in \mathbb{R} \Rightarrow x^2 \neq -1$.

12. For each statement below, give its negation. Determine the truth value of each.

    (a) $\exists y \in \mathbb{R}$ such that $\forall x \in \mathbb{R}$, $xy = 0$.
    (b) $\forall x \in (0, \infty) \ \exists n \in \mathbb{N}$ such that $\frac{1}{n} < x$.
    (c) $\exists y \in (0, \infty)$ such that $\forall x \in (0, \infty)$, $xy = 1$.
    (d) $\forall y \in (0, \infty) \ \exists x \in (0, \infty)$ such that $xy = 1$.
    (e) $\forall x \in \mathbb{R} \ \exists (n, a) \in \mathbb{N} \times [0, \frac{1}{2}]$ such that $x = n + a$ or $x = n - a$.

13. For each statement below, give its negation. Determine the truth value of each.

    (a) $\forall n \in \mathbb{N}$, $n$ is odd or $n + 1$ is odd.
    (b) $\exists n \in \mathbb{N}$ such that $\forall m \in \mathbb{N}$, $n \geq m$.
    (c) $\forall m \in \mathbb{N}, \exists n \in \mathbb{N}$ such that $n > m$.
    (d) $\exists m, M \in \mathbb{R}$ such that $\forall x \in \mathbb{R}$, $m \leq \sin x \leq M$.
    (e) $\forall n \in \mathbb{N}$ and $\forall x \in (0, 0.5), \exists m \in \mathbb{N}$ such that $m = n + 2x$.

14. Complete the sentences:
    (a) $f : A \to B$ is onto if and only if for every $b \in B, f^{-1}(\{b\}) \ldots$
    (b) $f : A \to B$ is one-to-one if and only if for every $b \in B, f^{-1}(\{b\}) \ldots$
15. Prove parts (a), (c), and (e) of Theorem 0.4.1.
16. Prove parts (b), (d), and (f) of Theorem 0.4.1.
17. Suppose $f : X \to X$ is a function. Prove:
    (a) For all $U \subseteq X, U \subseteq f^{-1}(f(U))$, and equality holds if and only if $f$ is one-to-one.
    (b) For all $V \subseteq X, f(f^{-1}(V)) \subseteq V$, and equality holds if and only if $f$ is onto.
18. Suppose $f : A \to B$ and $g : B \to C$ are functions. Show
    (a) $g \circ f$ is one-to-one $\Rightarrow f$ is one-to-one.
    (b) $g \circ f$ is onto $\Rightarrow g$ is onto.
19. Suppose $f : A \to B$ is a function. Recall that there exists a function $g : B \to A$ such that $f \circ g = id_B$ and $g \circ f = id_A$ if and only if $f$ is one-to-one and onto. Such a function $g$ is called the inverse of $f$.
    (a) Show that there exists a function $g : B \to A$ such that $f \circ g = id_B$ if and only if $f$ is onto. Such a function $g$ is called a *right inverse* of $f$.
    (b) Show that there exists a function $h : B \to A$ such that $h \circ f = id_A$ if and only if $f$ is one-to-one. Such a function $g$ is called a *left inverse* of $f$.
    (c) If $g$ is a right inverse of $f$ and $h$ is a left inverse of $f$, show that $g = h = f^{-1}$.
20. Consider the sequence $(a_n)_{n \in \mathbb{N}} = (1, 2, 3, 1, 2, 3, 1, 2, 3, 1, 2, 3, \ldots)$. Determine whether the sequences $(b_n)_{n \in \mathbb{N}}$ below are subsequences of $(a_n)_{n \in \mathbb{N}}$. If so, identify a strictly increasing function $\sigma : \mathbb{N} \to \mathbb{N}$ such that $b_n = a_{\sigma(n)}$. If not, justify your answer.
    (a) $(1, 2, 3, 1, 2, 3, 1, 2, 3, \ldots)$
    (b) $(3, 1, 2, 3, 1, 2, 3, 1, 2, 3, \ldots)$
    (c) $(1, 1, 1, 1, 1, \ldots)$
    (d) $(3, 2, 1, 3, 2, 1, 3, 2, 1, \ldots)$
    (e) $(1, 4, 1, 1, 1, 1, 1, \ldots)$
21. Consider the sequence $(a_n)_{n \in \mathbb{N}} = (1, 2, 3, 2, 3, 4, 3, 4, 5, 4, 5, 6, 5, 6, 7, \ldots)$. Determine whether the sequences $(b_n)_{n \in \mathbb{N}}$ below are subsequences of $(a_n)_{n \in \mathbb{N}}$. If so, identify a strictly increasing function $\sigma : \mathbb{N} \to \mathbb{N}$ such that $b_n = a_{\sigma(n)}$. If not, justify your answer.
    (a) $(1, 2, 3, 4, 5, 6, 7, 8, \ldots)$
    (b) $(3, 3, 2, 4, 4, 3, 5, 5, 4, 6, 6, 5, \ldots)$
    (c) $(1, 1, 1, 1, 1, \ldots)$
    (d) $(3, 2, 1, 4, 5, 6, 9, 8, 7, \ldots)$
    (e) $(3, 2, 5, 4, 7, 6, 9, 8, \ldots)$
22. Consider the sequence $(a_n)_{n \in \mathbb{N}} = (1, 2, 1, 2, 3, 1, 2, 3, 4, 1, 2, 3, 4, 5, \ldots)$. Notice that $a_n = 1$ if and only if $n$ belongs to the sequence $(t_n)_{n \in \mathbb{N}} = (1, 3, 6, 10, 15, 21, \ldots)$ of triangular numbers, where $t_n = 1 + 2 + \cdots + n = n(n+1)/2$. Show that every sequence in $\mathbb{N}$ is a subsequence of $(a_n)_{n \in \mathbb{N}}$.
23. Determine whether the following sets are countable or uncountable.

(a) For $n \in \mathbb{N}$, $C_n = \{A \subseteq \{1, 2, 3, \ldots, n\} : A \text{ is finite}\}$

(b) $C = \{A \subseteq \mathbb{N} : A \text{ is finite}\}$

(c) $D = \{A \subseteq \mathbb{R} : A \text{ is finite}\}$

(d) Any mutually disjoint collection $\mathcal{F} = \{(a_i, b_i)\}_{i \in I}$ of open intervals in $\mathbb{R}$. (Hint: Consider Theorem 0.1.1(e).)

24. (a) Show that if $f : A \to B$ is onto and $B$ is uncountable, then $A$ is uncountable.

(b) For each $i \in \mathbb{N}$, let $D_i$ be the countable set $\{0, 1, 2, 3, \ldots, 9\}$. Consider the map $f : \prod_{i \in \mathbb{N}} D_i \to [0, 1]$ defined by $f((d_i)_{i \in \mathbb{N}}) = 0.d_1 d_2 d_3 \cdots$, the number whose $i$th decimal digit is $d_i$. Show that $f$ is onto.

(c) Show that a countable product of countable sets need not be countable.

25. A real number $a$ is an *algebraic number* if and only if $a$ is a zero of a polynomial $p$ with integer coefficients. A real number is *transcendental* if it is not algebraic. Show that the set of algebraic numbers is countable and the set of transcendental numbers is uncountable.

26. Define an equivalence relation $\sim$ on $\mathbb{R}$ by $x \sim y$ if and only if $x^2 = y^2$.

(a) Find a saturated set (relative to this equivalence relation) with four elements.

(b) Find a saturated set with five elements.

(c) Find a countably infinite saturated set.

(d) Find an interval of positive length which is saturated.

27. Let $\mathcal{V}, \mathcal{H}$, and $\mathcal{P}$ be the partitions of Example 0.6.2. How many saturated sets does each have? Give a general result about the number of saturated sets with respect to a finite partition.

28. Note that the partition $\mathcal{P}$ of Example 0.6.2 of $[0, 5)^2$ arises from the equivalence relation $(a, b) \sim (x, y)$ if and only if $(\lfloor a \rfloor, \lfloor b \rfloor) = (\lfloor x \rfloor, \lfloor y \rfloor)$ where $\lfloor z \rfloor$ is the greatest integer less than or equal to $z$. The function $f(z) = \lfloor z \rfloor$ is called the *floor function* of $z$. Use the floor function to describe equivalence relations on $[0, 5)^2$ which generate the partitions $\mathcal{V}$ and $\mathcal{H}$ given in Example 0.6.2.

29. Consider the partition $\mathcal{P} = \{\mathbb{D}, \mathbb{E}\}$ of $\mathbb{Z}$ where $\mathbb{D} = \{2n+1 : n \in \mathbb{Z}\}$ is the set of odd integers and $\mathbb{E} = \{2n : n \in \mathbb{Z}\}$ is the set of even integers. Give two partitions of $\mathbb{Z}$ finer than $\mathcal{P}$. How many partitions of $\mathbb{Z}$ are strictly coarser than $\mathcal{P}$? Justify your answer.

30. Define a relation $\approx$ on $\mathbb{N}$ by taking $a \approx b$ if and only if $ab$ is a perfect square. Show that $\approx$ is an equivalence relation.

31. A relation $R$ on a nonempty set $A$ is *irreflexive* if for every $a \in A$, $(a, a) \notin R$.

(a) Exhibit an example to prove that irreflexive is different from "not reflexive".

(b) Does either of "irreflexive" or "not reflexive" imply the other?

(c) If $A$ is a finite set, show that the number of reflexive relations on $A$ is equal to the number of irreflexive relations on $A$.

32. Suppose $\mathcal{P}$ and $\mathcal{R}$ are partitions of $A$. Recall that $\mathcal{R}$ is a refinement of $\mathcal{P}$ if for every block $B_R \in \mathcal{R}$, there exists $B_P \in \mathcal{P}$ with $B_R \subseteq B_P$. Show that this is equivalent to: for every block $B_P$ of $\mathcal{P}$, there exists a subcollection $\{B_i\}_{i \in I}$ of $\mathcal{R}$ which partitions $B_P$.

33. Define a relation $\trianglelefteq$ on the power set $\mathcal{P}(\{1, 2, 3, 4\})$ by $A \trianglelefteq B$ if and only if the sum of the elements of $A$ (listed without repetition) is less than or equal to the sum of the elements of $B$ (listed without repetition).
    (a) Which of the defining properties of a partial order does $\trianglelefteq$ satisfy?
    (b) Let $\mathcal{D}$ be the collection of subsets of $\{1, 2, 3, 4\}$ which contain an odd number of elements, as in Example 0.2.1. Which of the defining properties of a partial order does $\trianglelefteq$ satisfy when restricted to $\mathcal{D}$?

34. Rephrase the statements of Exercise 13(b) and 13(c) in terms of maximal and maximum elements.

35. Give a proof or counterexample for each statement below about a partially ordered set $(X, \leq)$.
    (a) If $m$ is minimum in $X$, then $m$ is minimal.
    (b) If $b$ is a lower bound of $A \subseteq X$ and $c < b$, then $c$ is a lower bound of $A$.
    (c) If $X$ has a maximum element, then it is unique.
    (d) If $A \subseteq B \subseteq X$, then $\inf A \leq \inf B$.

36. Suppose $C \subseteq X$ is convex and $a$ is maximal in $C$. Must $C - \{a\}$ be convex? Give a proof or counterexample.

37. Consider the collection $C = \{\{1\}, \{2\}, \{1, 2\}, \{1, 3\}, \{2, 3\}, \{1, 3, 4\}, \{5, 6\}\}$ ordered by set inclusion $\subseteq$.
    (a) Draw the Hasse diagram for $(C, \subseteq)$.
    (b) Find all maximum, maximal, minimum, and minimal elements.
    (c) Find a nonconvex set in $(C, \subseteq)$.

38. Order each collection below by set inclusion. Discuss the maximum, maximal, minimum, and minimal elements.
    (a) $\{[0, x] \subseteq \mathbb{R} : x \in [0, 1)\}$
    (b) $\{[x, x + 1] \subseteq \mathbb{R} : x \in [0, 1)\}$

39. Consider the poset $(\mathcal{P}(U), \subseteq)$. Suppose $A_i \in \mathcal{P}(U)$ for every $i \in I$. What are $\sup\{A_i : i \in I\}$ and $\inf\{A_i : i \in I\}$?

40. Shown are three subsets $X$, $Y$, and $Z$ of $\mathbb{R}^2$ with the product order, each of which contains $A = \{(x, y) \in \mathbb{R}^2 : x \geq 0, y \geq 0, y \leq 1 - x\}$. Find the set of upper bounds of $A$ and discuss the supremum of $A$ in each poset $X$, $Y$, and $Z$.

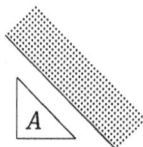

$X = A$             $Y$             $Z$

(a) $X = A$
(b) $Y = A \cup ([1, 3] \times [2, 3]) \cup ([2, 3] \times [1, 3])$
(c) $Z = A \cup \{(x, y) \in \mathbb{R}^2 : y \geq 1.5 - x\}$

# 1 Topologies

## 1.1 The Euclidean metric

A *metric* on a set $X$ is a distance function, giving the distance $d(x, y)$ between two points $x, y \in X$. The familiar Euclidean distance formula between points in the plane $\mathbb{R}^2$ is the prototype of a metric, and the properties we expect from Euclidean distances are the basis for our definition of a metric. We will see that there are many other ways to measure distances between points in $\mathbb{R}^2$, and metrics may be applied to sets other than $\mathbb{R}^n$. For example, if you misspell a word, your word processor spell-check application may present suggestions based on the distance between the "word" you typed and a word in its dictionary. In Chapter 6, we will introduce two metrics to measure distances between continuous functions.

**Definition 1.1.1.** A *metric* on a set $X$ is a function $d : X \times X \to \mathbb{R}$ satisfying
(a)  $d(x, y) \geq 0$ for all $x, y \in X$ (nonnegativity);
(b)  $d(x, y) = 0$ if and only if $x = y$;
(c)  $d(x, y) = d(y, x)$ for all $x, y \in X$ (symmetry);
(d)  $d(x, y) \leq d(x, z) + d(z, y)$ for all $x, y, z \in X$ (triangle inequality).

A *metric space* is a pair $(X, d)$ where $X$ is a set and $d$ is a metric on $X$.

The first three defining conditions of a metric are easily understood: distances are never negative, two points are equal if and only if the distance between them is zero, and the distance from $x$ to $y$ always equals the distance from $y$ to $x$. The triangle inequality is easily illustrated using our usual concept of distance in the Euclidean plane. If $x, y$, and $z$ are three points in the plane, then $d(x, y)$ is the length of the straight line segment between $x$ and $y$. Since the shortest distance between two points in the Euclidean plane follows a straight line, $d(x, y)$ is less than or equal to the length of the polygonal path from $x$ to $z$ and on to $y$. Thus, the triangle inequality says that the length of one side of a triangle is less than or equal to the sum of the lengths of the other two sides, as suggested in Figure 1.1

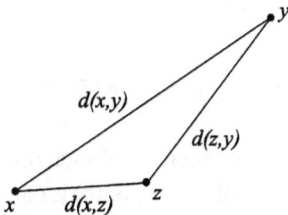

**Figure 1.1:** The triangle inequality: $d(x, y) \leq d(x, z) + d(z, y)$.

https://doi.org/10.1515/9783110686579-002

Vectors in $\mathbb{R}^n$ will be denoted by $x = (x_1, x_2, \ldots, x_n)$. We are most familiar with the Euclidean metric $d : \mathbb{R}^n \times \mathbb{R}^n \to [0, \infty)$ given by

$$d(x,y) = \sqrt{(x - y)^2} = |x - y| \quad \text{for } x, y \in \mathbb{R},$$

$$d(x,y) = \sqrt{(x_1 - y_1)^2 + (x_2 - y_2)^2} \quad \text{for } x, y \in \mathbb{R}^2,$$

$$d(x,y) = \sqrt{(x_1 - y_1)^2 + (x_2 - y_2)^2 + (x_3 - y_3)^2} \quad \text{for } x, y \in \mathbb{R}^3,$$

$$d(x,y) = \sqrt{(x_1 - y_1)^2 + (x_2 - y_2)^2 + \cdots + (x_n - y_n)^2} \quad \text{for } x, y \in \mathbb{R}^n.$$

The sets $\mathbb{R}$ and $\mathbb{R}^2$ with the Euclidean metric are called the *Euclidean line* and the *Euclidean plane*, and in general, $\mathbb{R}^n$ with the Euclidean metric is called the *$n$-dimensional Euclidean space*. While it is intuitively clear that the Euclidean metric on $\mathbb{R}^n$ satisfies the triangle inequality, a formal proof of this is not trivial. One such proof is outlined in Exercise 15.

If $d$ is a metric on $X$ and $Y \subseteq X$, then since $d$ tells us the distance between any two points in $X$, it tells us the distance between any two points in $Y \subseteq X$. It is easy to show that this inherited distance on $Y$ obeys all the conditions defining a metric. Formally, if $d : X \times X \to \mathbb{R}$ is a metric on $X$ and $Y \subseteq X$, then the restriction $d|_{Y \times Y}$ of $d$ to $Y \times Y$ is a metric on $Y$. Thus, we may speak of a subset of $\mathbb{R}^n$ with the Euclidean metric.

A metric allows us to determine which points are near a given point $x$. This nearness can be characterized by the "balls" centered at $x$.

**Definition 1.1.2.** Suppose $(X, d)$ is a metric space. For $x \in X$ and $\varepsilon > 0$, the ball centered at $x$ with radius $\varepsilon$, or simply the $\varepsilon$-ball centered at $x$, is

$$B(x, \varepsilon) = \{y \in X : d(x,y) < \varepsilon\}.$$

Note that our definition of $B(x, \varepsilon)$ requires $\varepsilon > 0$. This convention guarantees that $x \in B(x, \varepsilon)$, so every ball is nonempty.

In the Euclidean line $\mathbb{R}$, the $\varepsilon$-ball centered at $x$ consists of all points on the real line within $\varepsilon$ of $x$, and thus $B(x, \varepsilon)$ is the open interval $(x - \varepsilon, x + \varepsilon)$. For example, $B(5, 0.2) = (4.8, 5.2)$. In the Euclidean plane $\mathbb{R}^2$, the ball $B(x, \varepsilon)$ is the disk centered at $x$ with radius $\varepsilon$, not including the boundary of the disk. In the 3-dimensional Euclidean space $\mathbb{R}^3$, the ball $B(x, \varepsilon)$ is the solid inside the sphere centered at $x$ with radius $\varepsilon$, not including the boundary sphere.

The balls are used to define important concepts such as convergence of sequences and continuity of functions.

**Definition 1.1.3.** A *tail* of sequence $(x_n)_{n \in \mathbb{N}}$ is a subsequence $(x_n)_{n \geq k}$ obtained from the original by deleting the initial terms $x_1, x_2, \ldots, x_{k-1}$.

A sequence $(x_n)_{n \in \mathbb{N}}$ is *eventually in a set $S$* if there exists an integer $k \in \mathbb{N}$ such that $x_n \in S$ for all $n \geq k$. That is, $(x_n)_{n \in \mathbb{N}}$ is eventually in $S$ if some tail of the sequence is contained in $S$. A sequence $(x_n)_{n \in \mathbb{N}}$ is *eventually constant* if it is eventually in a singleton set $\{a\}$.

A sequence $(x_n)_{n \in \mathbb{N}}$ is *frequently in a set S* if for any $k \in \mathbb{N}$, there exists $n > k$ such that $x_n \in S$.

Clearly if $(x_n)_{n \in \mathbb{N}}$ is eventually in a set $S$, it is frequently in $S$.

**Example 1.1.4.**

(a) In the Euclidean line $\mathbb{R}$, the sequence $(\frac{(-1)^n}{n})_{n \in \mathbb{N}}$ is eventually in $B(-1, 1.02) = (-2.02, 0.02)$, since for $n \geq k = 51$, $x_n = \frac{(-1)^n}{n} \in (-2.02, 0.02)$. However, the sequence is not eventually in $B(2, 2) = (0, 4)$ since no tail of the sequence is contained in $B(2, 2)$. This sequence is eventually in $B(0, \varepsilon) = (-\varepsilon, \varepsilon)$ for any $\varepsilon > 0$.

(b) In the Euclidean line $\mathbb{R}$, the sequence $(z_n)_{n \in \mathbb{N}} = (0, 5, 0, 5, 0, 5, 0, 5, \ldots)$ is frequently in $B(0, .3)$ and frequently in $B(5, .02)$, but is not eventually in either of these balls. It is eventually in the ball $B(2, 7) = (-5, 9)$ and eventually in the set $S = \{0, \pi, \sqrt{7}, 5, 12.6\}$.

**Definition 1.1.5.** A sequence $(x_n)_{n \in \mathbb{N}}$ in a metric space $(X, d)$ *converges to* $a \in X$ if for every $\varepsilon > 0$, $(x_n)_{n \in \mathbb{N}}$ is eventually in $B(a, \varepsilon)$. If $(x_n)_{n \in \mathbb{N}}$ converges to $a$, we say $a$ is the *limit* of the sequence, and write $\lim_{n \to \infty} x_n = a$ or $x_n \to a$.

Referring to the previous example, in the Euclidean line $\mathbb{R}$, the sequence $(\frac{(-1)^n}{n})_{n \in \mathbb{N}}$ is eventually in every $\varepsilon$-ball around 0, so the sequence converges to 0. The sequence $(z_n)_{n \in \mathbb{N}} = (0, 5, 0, 5, 0, 5, 0, 5, \ldots)$ does not converge to any limit. To see this, suppose to the contrary that $\lim_{n \to \infty} x_n = a$. Consider the ball $B(a, 1) = (a - 1, a + 1)$, an interval of length 2. Now this ball may contain 0 or 5, but cannot contain both. Thus, the sequence is not eventually in $B(a, 1)$, contrary to the definition of $\lim_{n \to \infty} x_n = a$. The sequence $(0, 5, 0, 5, 0, 5, \ldots)$ is frequently in every $\varepsilon$-ball $B(5, \varepsilon)$ around 5. This occurs since there is a subsequence (consisting of the even-indexed terms) which converges to 5. This idea is investigated in Exercise 4.

Our next result is that a sequence in Euclidean space cannot converge to two different limits. Our familiarity with limits of sequences from calculus, which is based on the Euclidean metric, makes this result not surprising. The surprise will come later when we examine topological spaces in which a sequence might converge to more than one point.

**Theorem 1.1.6.** *If a sequence* $(x_n)_{n \in \mathbb{N}}$ *in a metric space* $(X, d)$ *converges, then the limit is unique.*

*Proof.* Suppose $(x_n)_{n \in \mathbb{N}}$ is a sequence converging to $a$ in a metric space $(X, d)$, and $b \neq a$. We will show that $(x_n)_{n \in \mathbb{N}}$ does not converge to $b$. Since $b \neq a$, $\varepsilon = \frac{1}{2}d(a, b)$ is greater than zero. First, we will show that, as Figure 1.2 suggests, $B(a, \varepsilon)$ and $B(b, \varepsilon)$ are disjoint. Suppose to the contrary that there exists $x \in B(a, \varepsilon) \cap B(b, \varepsilon)$. Then $d(a, x) < \varepsilon$ and $d(x, b) < \varepsilon$, and adding gives $d(a, x) + d(x, b) < 2\varepsilon = d(a, b)$, contrary to the triangle inequality. So, $B(a, \varepsilon)$ and $B(b, \varepsilon)$ are disjoint.

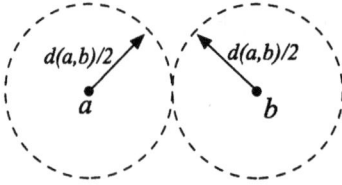

**Figure 1.2:** Potential distinct limits $a$ and $b$ of a sequence in a metric space.

Since $(x_n)_{n\in\mathbb{N}}$ converges to $a$, it is eventually in $B(a,\varepsilon)$, which is disjoint from $B(b,\varepsilon)$. Thus, no tail of the sequence is in $B(b,\varepsilon)$, so the sequence does not converge to $b$. □

The following property of metric spaces will be used frequently.

**Theorem 1.1.7.** *If $(X,d)$ is a metric space, $x \in X$, and $y \in B(x,\varepsilon)$, then there exists $\varepsilon_y > 0$ such that $B(y,\varepsilon_y) \subseteq B(x,\varepsilon)$.*

*Proof.* Figure 1.3 suggests the appropriate value of $\varepsilon_y$.

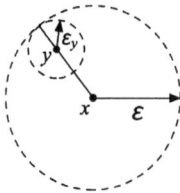

**Figure 1.3:** If $B(x,\varepsilon)$ contains $y$, then it contains a ball $B(y,\varepsilon_y)$ around $y$.

Under the hypotheses, take $\varepsilon_y = \varepsilon - d(x,y)$. Since $y \in B(x,\varepsilon)$, $\varepsilon_y > 0$. Now $z \in B(y,\varepsilon_y)$ implies $d(y,z) < \varepsilon_y = \varepsilon - d(x,y)$. Adding $d(x,y)$ to both sides of the equation and applying the triangle inequality, we have $d(x,z) \leq d(x,y) + d(y,z) < \varepsilon$, so $z \in B(x,\varepsilon)$. Since $z$ was an arbitrary point of $B(y,\varepsilon_y)$, we have $B(y,\varepsilon_y) \subseteq B(x,\varepsilon)$. □

Now we present another metric on $\mathbb{R}^2$ to illustrate that, while our intuition from Euclidean geometry will often provide the correct motivation for metric arguments, we may need to be careful in the details.

**Example 1.1.8** (Sup-metric). Define $m : \mathbb{R}^2 \times \mathbb{R}^2 \to \mathbb{R}$ by

$$m((x_1,y_1),(x_2,y_2)) = \sup\{|x_1 - x_2|, |y_1 - y_2|\}.$$

Thus, the distance between two points in the plane is the larger of the horizontal distance $|x_1 - x_2|$ separating them and the vertical distance $|y_1 - y_2|$ separating them. For example, to get from $(-2,3)$ to $(5,1)$, we rise by $y_2 - y_1 = -2$ and run by $x_2 - x_1 = 7$, so in the supremum metric, $m((-2,3),(5,1)) = \sup\{|-2|, |7|\} = 7$.

We will confirm that the function $m$ is indeed a metric. As the supremum of two nonnegative numbers, $m((x_1, y_1), (x_2, y_2))$ is always nonnegative. Also, $m((x_1, y_1), (x_2, y_2)) = \sup\{|x_1 - x_2|, |y_1 - y_2|\} = 0$ if and only if both of the nonnegative numbers $|x_1 - x_2|$ and $|y_1 - y_2|$ are zero, which happens if and only if $(x_1, y_1) = (x_2, y_2)$. Symmetry follows since $m((x_1, y_1), (x_2, y_2)) = \sup\{|x_1 - x_2|, |y_1 - y_2|\} = \sup\{|x_2 - x_1|, |y_2 - y_1|\} = m((x_2, y_2), (x_1, y_1))$ for all $(x_1, y_1), (x_2, y_2) \in \mathbb{R}^2$. It only remains to check the triangle inequality. Since the sup-metric is defined in terms of the Euclidean metric $d(x_1, x_2) = |x_1 - x_2|$ on $\mathbb{R}$, it is not surprising that the triangle inequality for the sup-metric will be based on that for the Euclidean metric on $\mathbb{R}$. For any $x_1, x_2, x_3, y_1, y_2, y_3 \in \mathbb{R}$, we have

$$|x_1 - x_3| \leq |x_1 - x_2| + |x_2 - x_3|$$
$$\leq \sup\{|x_1 - x_2|, |y_1 - y_2|\} + \sup\{|x_2 - x_3|, |y_2 - y_3|\}$$
$$= m((x_1, y_1), (x_2, y_2)) + m((x_2, y_2), (x_3, y_3)).$$

Similarly, $|y_1 - y_3| \leq m((x_1, y_1), (x_2, y_2)) + m((x_2, y_2), (x_3, y_3))$, so $m((x_1, y_1), (x_3, y_3)) = \sup\{|x_1 - x_3|, |y_1 - y_3|\} \leq m((x_1, y_1), (x_2, y_2)) + m((x_2, y_2), (x_3, y_3))$. This shows that $m$ satisfies the triangle inequality.

In $\mathbb{R}^2$ with the sup-metric, $B((0, 0), \varepsilon) = \{(x, y) \in \mathbb{R}^2 : \sup\{|x - 0|, |y - 0|\} = \sup\{|x|, |y|\} < \varepsilon\}$, which is the set of all points in the plane such that the vertical distance $|y|$ and horizontal distance $|x|$ from the origin are both less than $\varepsilon$. This is the square $(-\varepsilon, \varepsilon) \times (-\varepsilon, \varepsilon)$. Similarly, in this metric, the ball $B((a, b), \varepsilon)$ is the square region $(a - \varepsilon, a + \varepsilon) \times (b - \varepsilon, b + \varepsilon)$, as depicted in Figure 1.4.

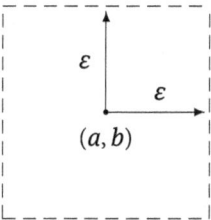

**Figure 1.4:** The sup-metric ball $B((a, b), \varepsilon)$.

## Exercises

1.  Suppose $d : X \times X \to [0, \infty)$ is a function with $d(x, x) = 0$ for all $x \in X$. Show that if $x, y, z \in X$ are not all distinct, then $d(x, y) + d(y, z) \geq d(x, z)$. Thus, in confirming that $d$ is a metric, if the other conditions are verified first, one need only confirm the triangle inequality for *distinct* points $x, y, z \in X$.
2.  In a metric space $(X, d)$, show that if $0 < \delta < \varepsilon$, then $B(x, \delta) \subseteq B(x, \varepsilon)$.
3.  In a metric space $(X, d)$, show that if $x \in B(y, \alpha) \cap B(z, \beta)$, then there exists $\varepsilon > 0$ s. t. $B(x, \varepsilon) \subseteq B(y, \alpha) \cap B(z, \beta)$.

4. Show that a sequence $(x_n)_{n\in\mathbb{N}}$ in Euclidean space is frequently in every ball $B(a,\varepsilon)$ centered at $a$ if and only if there is a subsequence $(x_{\sigma(n)})$ converging to $a$.

5. **(Taxicab metric)** The *taxicab metric* (or *Manhattan metric*) on $\mathbb{R}^2$ is given by $d((x,y),(a,b)) = |x-a| + |y-b|$. Thus, the distance from one point to another is the sum of the vertical and horizontal (Euclidean) distances between them. Show that the taxicab metric really is a metric.

6. In the taxicab metric defined in Exercise 5, draw the balls $B((0,0),2)$ and $B((2,3),1)$.

7. Suppose $d : X \times X \to \mathbb{R}$ is a metric on a set $X$ and $k$ is a positive real number. Show that $kd : X \times X \to \mathbb{R}$ is also a metric on $X$. If $d$ measures Euclidean distances in units of meters, what value of $k$ would give a metric $kd$ which measures Euclidean distances in millimeters?

8. In the metric spaces described below, find all points $x \in X$ such that $B(x,15) = B(5,15)$. Carefully justify that you have found all answers.
   (a) $X = \mathbb{R}$ with the Euclidean metric.
   (b) $X = \{1,5,10,50\} \subseteq \mathbb{R}$ with the Euclidean metric.

9. Under what conditions does the intersection $B(x,\varepsilon) \cap B(y,\delta)$ of two balls equal a ball in
   (a) the Euclidean metric on $\mathbb{R}$?
   (b) the Euclidean metric on $\mathbb{R}^2$?
   (c) the sup-metric on $\mathbb{R}^2$?

10. Describe all points $x,y,z$ in the Euclidean plane for which equality holds in the triangle inequality, that is, for which $d(x,y) + d(y,z) = d(x,z)$.

11. In the Euclidean plane, show that $a = b$ if and only if $B(a,1) = B(b,1)$.

12. Suppose $d : X \times X \to \mathbb{R}$ is a metric on $X$. Show that $f : X \times X \to \mathbb{R}$ defined by $f(x,y) = \sqrt{d(x,y)}$ is also a metric on $X$.

13. Let $f : \mathbb{R} \to \mathbb{R}$ be any one-to-one function. Define $d_f : \mathbb{R} \times \mathbb{R} \to \mathbb{R}$ by $d_f(x,y) = |f(x) - f(y)|$ for all $x,y \in \mathbb{R}$.
   (a) Show that $d_f$ is a metric on $\mathbb{R}$.
   (b) For $f(x) = x$, find the $d_f$-ball $B(0,1)$ centered at 0 with radius 1.
   (c) For $g(x) = \arctan x$, find the $d_g$-ball $B(0,1)$.
   (d) For $h(x) = x^{-1}$ for $x \neq 0$ and $h(0) = 0$, find the $d_h$-ball $B(0,1)$.

14. Let $d_f, d_g$ and $d_h$ be the metrics on $\mathbb{R}$ defined in Exercise 13. Discuss the convergence of the sequences $(n)_{n\in\mathbb{N}}$ and $(\frac{1}{n})_{n\in\mathbb{N}}$ in each of the metric spaces $(\mathbb{R}, d_f)$, $(\mathbb{R}, d_g)$, and $(\mathbb{R}, d_h)$.

15. For $x,y \in \mathbb{R}^n$, let $x \cdot y = x_1 y_1 + x_2 y_2 + \cdots + x_n y_n$ and $\|x\| = \sqrt{x \cdot x}$. Then the Euclidean metric $d$ on $\mathbb{R}^n$ is given by $d(x,y) = \|x - y\|$. To prove that $d$ satisfies the triangle inequality, it suffices to show $\|x + y\| \leq \|x\| + \|y\|$ for any $x,y \in \mathbb{R}$. Complete the steps below to do this.
   (a) Given $x,y \in \mathbb{R}^n$, let $z = \|y\|x - \|x\|y$ and show that $z \cdot z \geq 0$ implies $x \cdot y \leq \|x\|\|y\|$. This is the *Cauchy–Schwarz inequality*.
   (b) Show that $0 \leq \|x + y\| = (x + y) \cdot (x + y) \leq (\|x\| + \|y\|)^2$, so $\|x + y\| \leq \|x\| + \|y\|$.

## 1.2 Open sets in metric spaces

In the Euclidean line $\mathbb{R}$ familiar from calculus, an interval $(a, b)$ which contains neither of its endpoints is called an open interval and an interval $[a, b]$ which contains both of its endpoints is called a closed interval. Similarly, an open set in Euclidean space $\mathbb{R}^n$ is a set which contains none of its boundary points and a closed set is one which contains all of its boundary points. See Figure 1.5. These definitions of open and closed sets hold not only in Euclidean space, but in all metric spaces (and, indeed, in all topological spaces). Thus, we need a formal definition of boundary points.

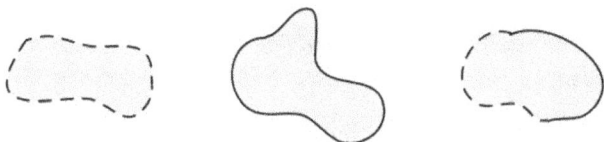

**Figure 1.5:** An open set, a closed set, and a set that is neither.

**Definition 1.2.1.** Suppose $A$ is a subset of a metric space $(X, d)$. A *boundary point of $A$* is a point $x \in X$ such that every ball around $x$ (with arbitrary radius $\varepsilon > 0$) intersects both $A$ and $X - A$. The set of all boundary points of $A$ is called the *boundary* of $A$ and is denoted $\partial A$.

We note that an immediate consequence of the definition is that $\partial A = \partial(X - A)$. Now we may formally define open and closed sets in metric spaces.

**Definition 1.2.2.** A set $A$ in a metric space $(X, d)$ is *open* (or, for emphasis, *d-open*) if it contains none of its boundary points, and is *closed* (or *d-closed*) if it contains all of its boundary points.

In the case $d$ is the Euclidean metric on $\mathbb{R}^n$, $d$-open and $d$-closed sets may be called *Euclidean open* and *Euclidean closed* sets.

**Example 1.2.3.** Consider the following sets in Euclidean space.
(a) For $A = (0, 1] \subseteq \mathbb{R}$, $\partial A = \{0, 1\}$. For every $\varepsilon > 0$, $B(0, \varepsilon) = (-\varepsilon, \varepsilon)$ intersects both $A = (0, 1]$ and $(-\infty, 0] \subseteq \mathbb{R} - A$, so $0 \in \partial A$, and similarly, $1 \in \partial A$. Furthermore, for $x \notin \{0, 1\}$, $x$ is not a boundary point: for $\varepsilon = \min\{|x|, |1 - x|\}$, $B(x, \varepsilon)$ is either completely contained in $A$ or completely contained in $\mathbb{R} - A$. Since $A$ contains some, but not all, of its boundary points, $A$ is neither open nor closed.
(b) For $A = (0, 1] \cup \{2\}$, we have $\partial A = \{0, 1, 2\}$, and $A$ is neither open nor closed.
(c) For $A = (3, \infty) \subseteq \mathbb{R}$, we have $\partial A = \{3\}$, and $\partial A \cap A = \emptyset$, so $A = (3, \infty)$ is open.
(d) For $A = \mathbb{R}$ in the Euclidean space $\mathbb{R}$, the complement of $A$ is empty, so no ball around any point $x$ intersects the complement of $A$. Thus, we have $\partial \mathbb{R} = \emptyset$. Now $\mathbb{R}$

contains none of its boundary points (since it has none), but also contains all of its boundary points since $\partial\mathbb{R} = \emptyset \subseteq \mathbb{R}$. Thus, $\mathbb{R}$ is both open and closed.

(e) For the set $\mathbb{Q}$ of rational numbers, $\partial\mathbb{Q} = \mathbb{R}$. For every real number $x$ and every $\varepsilon > 0$, $B(x, \varepsilon) = (x - \varepsilon, x + \varepsilon)$ contains some rational numbers and some irrational numbers, so $x \in \partial\mathbb{Q}$. Since $\partial\mathbb{Q} \not\subseteq \mathbb{Q}$, the set $\mathbb{Q}$ is not closed. Since $\partial\mathbb{Q} \cap \mathbb{Q} \neq \emptyset$, the set $\mathbb{Q}$ is not open.

(f) The set $A = (0, 1) \times \mathbb{R} \subseteq \mathbb{R}^2$ has boundary $\partial A = \{(x, y) : x = 0 \text{ or } x = 1\}$. Since $A$ contains none of its boundary points, $A$ is open. Furthermore, since $A$ does not contain all of its boundary points, $A$ is not closed.

The open sets and closed sets are fundamentally related by the following theorem.

**Theorem 1.2.4.** *In a metric space $(X, d)$, $A$ is open if and only if $X - A$ is closed; $A$ is closed if and only if $X - A$ is open.*

*Proof.* Recalling that the boundary of $A$ equals the boundary of its complement, $B$ is open if and only if $B \cap \partial B = \emptyset$, and $B$ is closed if and only if $\partial B \subseteq B$, we note the following equivalences in a metric space $X$:

$$
\begin{aligned}
A \text{ is open} &\iff A \cap \partial A = \emptyset \\
&\iff \partial A \subseteq X - A \\
&\iff \partial(X - A) \subseteq X - A \\
&\iff X - A \text{ is closed.}
\end{aligned}
$$

Thus, $A$ is open if and only if its complement is closed. Replacing $A$ by $X - A$ shows that $A$ is closed if and only if its complement is open. $\square$

We will focus on the open sets in this section. The closed sets and open sets are the extremes—they contain either all or none of their boundary points. Most sets would fall somewhere between these extremes, including some but not all of their boundary points. That is, most sets would be neither open nor closed. In particular, notice that "not open" does not imply closed, and "not closed" does not imply open. The third set shown in Figure 1.5 is simultaneously not open and not closed.

**Theorem 1.2.5.** *$U$ is open in a metric space $(X, d)$ if and only if for every $x \in U$, there is a ball $B(x, \varepsilon)$ contained in $U$.*
  *That is, $U$ is open if and only if*

$$\forall x \in U, \ \exists \varepsilon > 0 \quad \text{such that} \quad B(x, \varepsilon) \subseteq U.$$

*Proof.* ($\Rightarrow$) Suppose $U$ is open in $X$. Then

$$x \in U \Rightarrow x \notin \partial U$$

$$\Rightarrow \sim (\text{every ball around } x \text{ intersects both } U \text{ and } X - U)$$

$\Rightarrow$ there exists a $B(x, \varepsilon)$ which does not intersect both $U$ and $X - U$

$\Rightarrow$ there exists a $B(x, \varepsilon)$ which does not intersect $X - U$

(since $x \in U$, every $B(x, \varepsilon)$ must intersect $U$)

$\Rightarrow$ there exists a $B(x, \varepsilon) \subseteq U$.

($\Leftarrow$) Suppose that for every $x \in U$ there exists $\varepsilon_x > 0$ such that $B(x, \varepsilon_x) \subseteq U$. Now for every $x \in U, B(x, \varepsilon_x)$ does not intersect $X - U$, so $x \notin \partial U$. Thus, $U$ contains none of its boundary points, so $U$ is open. $\qquad\square$

This theorem is suggested in Figure 1.6.

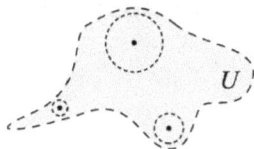

**Figure 1.6:** $U$ is open if and only if it contains balls centered at each of its points.

**Corollary 1.2.6.** *In a metric space, every ball $B(x, \varepsilon)$ is open.*

*Proof.* Apply Theorems 1.2.5 and 1.1.7. $\qquad\square$

**Corollary 1.2.7.** *$U$ is open in a metric space $(X, d)$ if and only if $U$ is a union of balls.*

*Proof.* By "union", we mean an arbitrary union, allowing unions of infinite collections, finite collections, or even the empty collection of balls. The union of an empty collection is $\emptyset$, so $U = \emptyset$ is a union of balls.

If $U$ is a nonempty open set in $X$, by Theorem 1.2.5, for each $x \in U$, there exists $\varepsilon_x$ with $\{x\} \subseteq B(x, \varepsilon_x) \subseteq U$. Taking the union over all $x \in U$ gives

$$U = \bigcup_{x \in U} \{x\} \subseteq \bigcup_{x \in U} B(x, \varepsilon_x) \subseteq \bigcup_{x \in U} U = U,$$

and thus $U$ is a union $\bigcup_{x \in U} B(x, \varepsilon_x)$ of balls.

Conversely, suppose $U$ is a union $\bigcup_{i \in I} B(x_i, \varepsilon_i)$ of balls and $x \in U$. Then there exists $j \in I$ such that $x \in B(x_j, \varepsilon_j)$. By Theorem 1.1.7, there is a ball $B(x, \varepsilon)$ centered at $x$ which is contained in $B(x_j, \varepsilon_j) \subseteq U$. Now by Theorem 1.2.5, $U$ is open in $X$. $\qquad\square$

**Definition 1.2.8.** An *open neighborhood* of $x$ is an open set containing $x$. A *neighborhood* of $x$ is any set which contains an open neighborhood of $x$.

Some introductory textbooks in topology define all neighborhoods to be open. Because arbitrary neighborhoods of $x$ are defined in terms of open neighborhoods, in many applications this is adequate; there are enough open neighborhoods to define nearness. However, in applications involving nesting of neighborhoods and supersets of neighborhoods, the more general definition is helpful.

The collection of all neighborhoods of $x$ will be used to quantify nearness to $x$. The nicest neighborhoods of $x$ in a metric space are the balls centered at $x$. Indeed, Theorem 1.2.5 shows that any neighborhood of $x$ contains a ball centered at $x$. The definitions of boundary and sequential convergence given above in terms of balls can in fact be restated using "every neighborhood" instead of "every ball." So, even if we do not have a metric to generate balls, these concepts can be equally well utilized if we only know which sets are open—and consequently, which sets are neighborhoods of any given point. A *topology* on a set $X$ will be a collection of subsets of $X$ which serve as the open sets. The theorem below lists fundamental properties of open sets in a metric space which we will wish any topology—that is, any collection of open sets—to satisfy.

**Theorem 1.2.9.** *In a metric space $(X, d)$:*
(a) *$\emptyset$ and $X$ are open.*
(b) *Finite intersections of open sets are open.*
(c) *Arbitrary unions of open sets are open.*

*Proof.* (a) $\emptyset$ and $X$ have no boundary points, and thus are open.

(b) Suppose $U_1, U_2, \ldots, U_n$ are open in $X$ and $x \in U_1 \cap U_2 \cap \cdots \cap U_n$. By Theorem 1.2.5, there exist $\varepsilon_1, \varepsilon_2, \ldots, \varepsilon_n$ such that $B(x, \varepsilon_i) \subseteq U_i$ for $i = 1, \ldots, n$. Take $\varepsilon = \min\{\varepsilon_1, \varepsilon_2, \ldots, \varepsilon_n\}$. Then $\varepsilon > 0$ and $B(x, \varepsilon) \subseteq U_1 \cap U_2 \cap \cdots \cap U_n$. Thus, Theorem 1.2.5 implies $U_1 \cap U_2 \cap \cdots \cap U_n$ is open.

(c) By Corollary 1.2.7, a union of open sets in $X$ is a union of unions of balls, which is again a union of balls, and is thus open. □

Note that arbitrary intersections of open sets need not be open: If $U_n = (0, 1 + \frac{1}{n})$ in the Euclidean line $\mathbb{R}$, then $U_n$ is open for each $n \in \mathbb{N}$, but the intersection $\bigcap_{n \in \mathbb{N}} U_n = (0, 1]$ is not open.

## Exercises

1. The proof of Theorem 1.2.9(b) cannot be modified to show that arbitrary intersections of open sets are open. Why not? Which step of the proof fails to generalize to arbitrary intersections?

2. Recall that in a metric space $(X, d)$, $x$ is a boundary point of $A$ if and only if every ball $B(x, \varepsilon)$ centered at $x$ intersects both $A$ and $X - A$. Show that this statement is equivalent: In a metric space $(X, d)$, $x$ is a boundary point of $A$ if and only if every neighborhood $U$ of $x$ intersects both $A$ and $X - A$.

3. For each given subset of the Euclidean plane $\mathbb{R}^2$, (a) determine the boundary points, (b) determine whether the set is open, (c) write each set that is open as a union of balls, and (d) determine whether the set is closed.

$A = \mathbb{R}^2$

$B = (-1, 1) \times \mathbb{R}$

$C = \{(x, y) \in \mathbb{R}^2 : x > 0\}$

$D = \{(x, y) \in \mathbb{R}^2 : y = x^2\}$

$E = \{(x, y) \in \mathbb{R}^2 : x \geq 0, y > 0\}$

$F = \{(x, y) \in \mathbb{R}^2 : y < x\}$

4.  From the sets given in Exercise 3, determine all that are
    (a) a neighborhood of $(3, 2)$;
    (b) a neighborhood of $(1, 1)$;
    (c) a neighborhood of $(0, 1)$;
    (d) a neighborhood of $(-2, 0)$;
    (e) a neighborhood of $(0, -1)$.

5.  Some open sets $U$ in the Euclidean plane $\mathbb{R}^2$ are given below. For every point $(x, y) \in U$, find an $\varepsilon > 0$ such that $B((x, y), \varepsilon) \subseteq U$.
    (a) $U = B((0, 0), 1) \cup B((1, 1), 2)$
    (b) $U = (-1, 1) \times (-1, 1)$
    (c) $U = \{(x, y) \in \mathbb{R}^2 : y < x\}$

6.  Sketch each given subset of the Euclidean plane $\mathbb{R}^2$ and determine whether it is
    (a) open, (b) closed, (c) not open, (d) not closed, and (e) a neighborhood of $(1, 0)$.

    $A = \bigcap_{n=1}^5 B((0, 0), 1 + \frac{1}{n})$

    $B = \bigcap_{n=1}^\infty B((0, 0), 1 + \frac{1}{n})$

    $C = \bigcap_{n=1}^3 B((n, 0), \frac{1}{3})$

    $D = \bigcup_{n=1}^3 B((n, 0), \frac{1}{3})$

    $E = \bigcup_{n=1}^\infty [\frac{1}{n}, 3 - \frac{1}{n}] \times [\frac{1}{n}, 3 - \frac{1}{n}]$

    $F = \bigcup_{n=1}^\infty [0, 3 - \frac{1}{n}] \times [0, 3 - \frac{1}{n}]$

7.  Suppose $X$ is a metric space.
    (a) Show that $X - \{a\}$ is open for each $a \in X$.
    (b) Show that every subset $A$ of $X$ is an intersection of open sets.

8.  Suppose $A$ is a subset of a metric space $X$. Show that $x \in \partial A$ if and only if there exists a sequence $(a_n)_{n \in \mathbb{N}}$ in $A$ converging to $x$ and a sequence $(b_n)_{n \in \mathbb{N}}$ in $X - A$ converging to $x$.

## 1.3 Topologies

In a metric space, open sets were defined in terms of boundaries of sets, which were defined in terms of $\varepsilon$-balls, which were defined using a distance function. Thus, open sets were defined in terms of the distance function. But there are ways to determine nearness—which is the goal of topology—which cannot be measured by a metric. All that is needed is a suitable collection of open sets. The open sets may then be used to define neighborhoods, convergence, and continuity, among other topological con-

cepts. The characteristic properties of open sets allowing the quantification of nearness are those given for open sets in a metric space in Theorem 1.2.9. This is the basis
for the definition below.

**Definition 1.3.1.** A *topology* on a nonempty set $X$ is a collection $\mathcal{T}$ of subsets of $X$,
called the *open sets*, satisfying:

(a) $\emptyset \in \mathcal{T}$ and $X \in \mathcal{T}$.

(b) If $U_1, U_2, \ldots, U_n \in \mathcal{T}$, then $\bigcap_{i=1}^{n} U_i \in \mathcal{T}$ ($\mathcal{T}$ *is closed under finite intersections*).

(c) If $I$ is an arbitrary index set and $U_i \in \mathcal{T}$ for all $i \in I$, then $\bigcup_{i \in I} U_i \in \mathcal{T}$ ($\mathcal{T}$ *is closed under arbitrary unions*).

A *topological space* is a pair $(X, \mathcal{T})$ where $X$ is a nonempty set and $\mathcal{T}$ is a topology on $X$.

If the topology $\mathcal{T}$ is understood, we may refer to "the topological space $X$" just
naming the underlying set of the topological space $(X, \mathcal{T})$. If we wish to emphasize the
topology, an element of a topology $\mathcal{T}$ may be called a $\mathcal{T}$-open set.

The condition (b) that a topology $\mathcal{T}$ must be closed under finite intersections can
be replaced by the condition (b') that if $U_1, U_2 \in \mathcal{T}$, then $U_1 \cap U_2 \in \mathcal{T}$ (that is, $\mathcal{T}$ is
closed under *binary intersections*). Clearly (b) implies (b'). Suppose (b') holds and
$U_1, U_2, \ldots, U_n \in \mathcal{T}$. Proceeding inductively, if $U_1 \cap \cdots \cap U_k \in \mathcal{T}$, then, by (b'), $(U_1 \cap \cdots \cap U_k) \cap U_{k+1} \in \mathcal{T}$. Thus, $\mathcal{T}$ is closed under finite intersections if and only if it is
closed under binary intersections.

If $(X, d)$ is a metric space, then the collection of $d$-open sets is a topology on $X$
called the *metric topology generated by $d$*. In the special case of $\mathbb{R}^n$ with the Euclidean
metric, the resulting topology is the *Euclidean topology* $\mathcal{T}_\mathcal{E}$, and $(\mathbb{R}^n, \mathcal{T}_\mathcal{E})$ is called the
*n-dimensional Euclidean space*.

While a metric on $X$ gave us one way to generate open sets, the concept of a topology is that we start by specifying the open sets and use them to characterize nearness, convergence, and continuity. Metric spaces are concrete examples of topological
spaces which are easy to visualize, but not every topological space is a metric space.

We note that the open sets in a metric space were all generated by a nicer collection
of open sets, the balls, in the sense that every open set was a union of balls. Such a
collection of open sets will be called a *basis* for the topology. Bases will be studied in
the next section.

Now, we will give the natural extensions of some metric space concepts to topological spaces.

**Definition 1.3.2.** If $(X, \mathcal{T})$ is a topological space and $x \in X$, an *open neighborhood of $x$*
is an open set containing $x$. A *neighborhood of $x$* is a set containing an open neighborhood of $x$.

To emphasize the topology, a neighborhood of $x$ in $(X, \mathcal{T})$ may be called a $\mathcal{T}$-*neighborhood of $x$*. (As noted earlier, some elementary textbooks initially require all neighborhoods to be open, which is adequate for an elementary development.)

Definition 1.1.5 of convergence of a sequence in a metric space carries over easily to convergence in an arbitrary topological space.

**Definition 1.3.3.** A sequence $(x_n)_{n\in\mathbb{N}}$ in a topological space $X$ *converges to* $a \in X$ if and only if $(x_n)_{n\in\mathbb{N}}$ is eventually in every neighborhood of $a$. That is, $\lim_{n\to\infty} x_n = a$ if and only if for any neighborhood $U$ of $x$, there exists $k \in \mathbb{N}$ such that $n \geq k$ implies $x_n \in U$.

Checking sequential convergence to a point $a$ requires checking something for every neighborhood of $a$, with the idea that as the neighborhoods get "smaller and smaller" the sequence still eventually gets in the neighborhoods. The formal way to quantify the "smaller and smaller" concept is to require the condition to hold for every possible neighborhood.

We note that "every neighborhood" in Definition 1.3.3 could be replaced by "every open neighborhood". Clearly if $(x_n)_{n\in\mathbb{N}}$ is in every neighborhood of $a$, then it is in every open neighborhood of $a$. Conversely, since every neighborhood of $a$ contains an open neighborhood of $a$, if $(x_n)_{n\in\mathbb{N}}$ is in every open neighborhood of $a$, it must be in every neighborhood of $a$.

The next examples illustrate some topologies and how they may be used to quantify nearness.

**Example 1.3.4** (The discrete topology). If $X$ is any nonempty set, the discrete topology on $X$ is $\mathcal{T}_D = \mathcal{P}(X) = \{U : U \subseteq X\}$. Every subset of $X$ is open in the discrete topology (so it is easily confirmed that $\mathcal{T}_D$ is in fact a topology). Any topology is a collection of subsets of $X$; the discrete topology is the collection of all subsets of $X$ and thus is the largest possible topology on $X$. In the discrete topology on $X$, each singleton set $\{x\}$ is open and is therefore a neighborhood of $x$. This suggests that $x$ is not near any other points. If a sequence $(x_n)_{n\in\mathbb{N}}$ converges to $a$, then the sequence is eventually in every neighborhood of $a$, and thus is eventually in the neighborhood $\{a\}$. Thus, the only convergent sequences are those which are eventually constant.

The discrete topology models discrete situations. To model the Euclidean plane by a finite set of pixels on a computer screen, we might stipulate that an open set in the plane should illuminate an open set of pixels on the screen. If the point $(0,0)$ is centered in a pixel occupying a square $[-r,r] \times [-r,r]$, then the open Euclidean ball $B((0,0),r)$ is contained in $[-r,r] \times [-r,r]$ and thus would illuminate the single pixel, so this single pixel should be open. Similarly, every pixel should be open, and we have the discrete topology on the set of pixels. This natural model is in many ways inadequate. A better way to model pixels is suggested in Example 1.4.6, which incorporates the boundaries between pixels.

The discrete topology on a set $X$ is *metrizable*, that is, it arises as the metric topology for some metric on $X$. Indeed, there are many metrics which generate the discrete topology. The most common one is the *discrete metric*

$$d(x,y) = \begin{cases} 0 & \text{if } x = y, \\ 1 & \text{if } x \neq y. \end{cases}$$

The discrete metric is used in instances where exact precision is needed, such as in comparing an input 'password' to the actual password. If the entered password is not exact, then you will not be permitted partial access based on entering something "almost correct." Any entry other than the exact password has distance 1 to the password, which is not considered close.

**Example 1.3.5** (The indiscrete topology). The indiscrete topology on $X$ is $\mathcal{T}_I = \{\emptyset, X\}$. Thus, the only open sets in the indiscrete topology are those required by the definition of a topology. If $x \in X$, the only neighborhood of $x$ is the whole space $X$. Every point $x$ is close to every other point in the space. In the indiscrete topology on $\mathbb{R}$, consider the sequence $(0, 1, 0, 1, 0, 1, 0, 1, \ldots)$. The sequence is eventually in (in fact, always in) every neighborhood of 7, and thus $(0, 1, 0, 1, 0, 1, 0, 1, \ldots)$ converges to 7. There was nothing special about 7, or the sequence $(0, 1, 0, 1, 0, 1, 0, 1, \ldots)$. Every sequence $(x_n)_{n \in \mathbb{N}}$ in $\mathbb{R}$ with the indiscrete topology converges to every point $a \in \mathbb{R}$. If $X$ has more than one point, every sequence converges to more than one limit, so by Theorem 1.1.6, $(X, \mathcal{T}_I)$ is not metrizable.

The discrete topology and the indiscrete topology are the extremes. In the discrete topology, every set is open, each $\{x\}$ is a neighborhood of $x$, no point is close to any other, and the only convergent sequences are the eventually constant sequences. In the indiscrete topology, every point is close to every other, and every sequence converges to every point.

**Example 1.3.6.** Consider the set $X = \{a, b, c\}$ and the collections of subsets $\mathcal{T}_1 = \{\emptyset, \{a, b\}, \{b\}, \{b, c\}\}$ and $\mathcal{T}_2 = \{\emptyset, \{a, b\}, \{b, c\}, X\}$ as shown below.

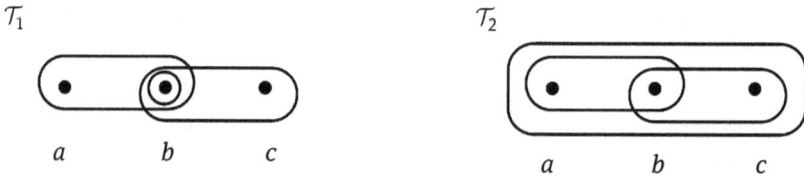

$\mathcal{T}_1$ is not a topology on $X$ since $X \notin \mathcal{T}_1$. The collection $\mathcal{T}_2$ is not a topology on $X$ since it is not closed under finite intersections: $\{a, b\}$ and $\{b, c\}$ are in $\mathcal{T}_2$ but their intersection $\{b\}$ is not.

Consider the set $Y = \{a, b, c, d\}$ and the collections of subsets $\mathcal{T}_3 = \{\emptyset, \{a\}, \{c\}, \{c, d\}, Y\}$ and $\mathcal{T}_4 = \{\emptyset, \{a, b, c\}, \{c\}, \{c, d\}, Y\}$ as shown below.

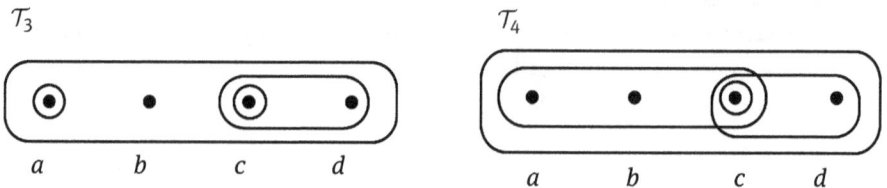

$\mathcal{T}_3$ is not a topology on $Y$ since it is not closed under arbitrary unions. The sets $\{a\}$ and $\{c\}$ are in $\mathcal{T}_3$, but their union $\{a, c\}$ is not.

The collection $\mathcal{T}_4$ is a topology on $Y$. The $\mathcal{T}_4$-neighborhoods of $a$ are $\{a, b, c\}$ and $\{a, b, c, d\}$. The constant sequence $(b, b, b, b, b, \ldots)$ is eventually in (in fact, entirely in) every neighborhood of $b$ and in every neighborhood of $a$. Thus, the constant sequence $(b, b, b, b, b, \ldots)$ converges to $a$ and to $b$. The sequence $(c, d, c, d, c, d, c, d, \ldots)$ is eventually in both open neighborhoods $\{c, d\}$ and $\{a, b, c, d\}$ of $d$ and thus $(c, d, c, d, c, d, c, d, \ldots)$ converges to $d$. This sequence is not eventually in the neighborhood $\{c\}$ of $c$. Thus, $(c, d, c, d, c, d, c, d, \ldots)$ does not converge to $c$.

**Example 1.3.7** (Single transmitter topology). Let $X = [-1, 1]$ and let $\mathcal{T}$ be the collection of subsets of $X$ satisfying

$$0 \in U \implies (-1, 1) \subseteq U.$$

Thus, $\mathcal{T}$ contains any subset of $X$ which does not contain 0, and the supersets of $(-1, 1)$, all of which do contain 0.

We will confirm that $\mathcal{T}$ is a topology on $X$. Clearly $\emptyset$ and $X$ are in $\mathcal{T}$. Suppose $U_1, \ldots, U_n$ are in $\mathcal{T}$. If $0 \in U_1 \cap \cdots \cap U_n$ then $0 \in U_k$ for all $k \in \{1, \ldots, n\}$, so $(-1, 1) \subseteq U_k$ for all $k \in \{1, \ldots, n\}$, and thus $(-1, 1) \subseteq U_1 \cap \cdots \cap U_n$. Thus, $0 \in U_1 \cap \cdots \cap U_n$ implies $(-1, 1) \subseteq U_1 \cap \cdots \cap U_n$, which is the definition of the finite intersection $U_1 \cap \cdots \cap U_n$ being open. Finally, suppose $U_i \in \mathcal{T}$ for all $i \in I$, where $I$ is an arbitrary index set. If $0 \in \bigcup_{i \in I} U_i$, then there exists $k \in I$ such that $0 \in U_k$. Since $U_k$ is open, we have $(-1, 1) \subseteq U_k \subseteq \bigcup_{i \in I} U_i$. Thus, $0 \in \bigcup_{i \in I} U_i$ implies $(-1, 1) \subseteq \bigcup_{i \in I} U_i$, so the arbitrary union $\bigcup_{i \in I} U_i$ is open. Thus, $\mathcal{T}$ is a topology.

In this topology, if $x \neq 0$, then $\{x\}$ is a neighborhood of $x$, so points different from $x$ are not close to $x$. However, every neighborhood of 0 contains $(-1, 1)$, and thus 0 is close to all the points in $(-1, 1)$. For example, this topology may represent the transmission reach from any point in $X = [-1, 1]$ if there is a single functioning transmitter at 0 which reaches every point in $(-1, 1)$. Transmitters at every other point $x \neq 0$ are not functioning, so "transmissions" from $x \neq 0$ do not reach any other point.

The sequence $(.2, .2, .2, .2, .2, \ldots)$ converges to $.2$ and to 0, and to no other points. Clearly the sequence is in every neighborhood of $.2$, since any neighborhood of $.2$ must contain $.2$. Also, $.2$ is in every neighborhood of 0 since every neighborhood of 0 contains $(-1, 1)$. Furthermore, if $a \notin \{0, .2\}$, then the sequence is not eventually in the neighborhood $\{a\}$ of $a$, so the sequence does not converge to any $a \notin \{0, .2\}$.

In the indiscrete topology on $\mathbb{R}$, $\mathcal{T}_4$ of Example 1.3.6, and the single transmitter topology on $[-1, 1]$, we saw sequences which converged to more than one limit. Theorem 1.1.6 showed that in a metric space no sequence can have more than one limit. Examining the proof of Theorem 1.1.6, the crucial property which prevented a sequence in a metric space from converging to distinct points $a$ and $b$ was that $a$ and $b$ had disjoint neighborhoods. To ensure that no sequence ever has two distinct limits, we could impose the condition that every pair of distinct points $a$ and $b$ possess disjoint neighborhoods, that is, every pair of distinct points $a$ and $b$ are "separated" by disjoint neighborhoods. This separation property is named for Felix Hausdorff (1868–1942). It is also called the $T_2$ property, short for the original German terminology *Trennungsax-iom 2*, which means "separation axiom 2".

**Definition 1.3.8.** A topological space $(X, \mathcal{T})$ satisfies the *Hausdorff property* (or the *$T_2$-property*) if for every $a, b \in X$ with $a \neq b$, there exist $N_a, N_b \in \mathcal{T}$ with $a \in N_a$, $b \in N_b$, and $N_a \cap N_b = \emptyset$. (See Figure 1.7.) If $X$ satisfies the Hausdorff property, we say $X$ is a *Hausdorff space*, or is a $T_2$ *space*.

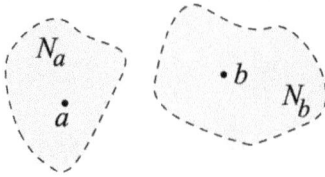

**Figure 1.7:** The Hausdorff property: distinct points have disjoint neighborhoods.

We have already noted the first benefit of this property, stated below. Its proof is similar to that of Theorem 1.1.6 and is left to the exercises.

**Theorem 1.3.9.** *If $(X, \mathcal{T})$ is a Hausdorff topological space, then every convergent sequence in $X$ has a unique limit.*

The indiscrete topology on $\mathbb{R}$ is clearly not Hausdorff. Given $a \neq b$ in $\mathbb{R}$, the only neighborhood of $a$ or of $b$ is $\mathbb{R}$, so distinct points do no have disjoint neighborhoods. This is manifested by sequences with more than one limit. In the single transmitter topology, $a = 0$ and $b = .2$ do not have disjoint neighborhoods, so this topology is not Hausdorff. Since any metric topology is Hausdorff, the indiscrete topology on $\mathbb{R}$ and the single transmitter topology are not metric topologies for any metric. That is, these topologies are not *metrizable*.

In Section 6.3 we will see that the converse of Theorem 1.3.9 holds under certain conditions, but not in general.

**Example 1.3.10.** Let $X$ be the interval $(0, 1)$.

The collection $\mathcal{T}_1 = \{(0, b) : 0 \leq b \leq 1\} \cup \{(a, 1) : 0 \leq a \leq 1\}$ of intervals having 0 or 1 as an endpoint is not a topology on $X$. It contains $\emptyset = (0, b)$ for $b = 0$ and $X = (0, b)$ for $b = 1$, but is neither closed under arbitrary unions nor finite intersections. The sets $(0, \frac{1}{3})$ and $(\frac{2}{3}, 1)$ are in $\mathcal{T}_1$ but their union is not. The sets $(0, \frac{2}{3})$ and $(\frac{1}{3}, 1)$ are in $\mathcal{T}_1$ but their intersection is not.

The collection $\mathcal{T}_2 = \{(0, b) : \frac{2}{3} < b \leq 1\} \cup \{(a, 1) : 0 \leq a < \frac{1}{3}\}$ is not a topology on $X$. It contains $X = (0, 1)$, and is closed under arbitrary unions. Indeed, arbitrary unions of a collection of sets in $\mathcal{T}_2$ all of the form $(0, b)$ or all of the form $(a, 1)$ are again of that form and are in $\mathcal{T}_2$, and if a collection of sets from $\mathcal{T}_2$ contains a set of form $(0, b)$ and a set of form $(a, 1)$, then since $a < \frac{1}{3} < \frac{2}{3} < b$, the union is $(0, 1) = X \in T_2$. However, $T_2$ is not a topology since it does not contain $\emptyset$ and is not closed under finite intersections. The sets $(0, \frac{2}{3})$ and $(\frac{1}{3}, 1)$ are in $T_1$ but their intersection is not.

The collection $T_3 = \{(0, b) : \frac{1}{2} < b < 1\} \cup \{(a, 1) : 0 < a < \frac{1}{2}\} \cup \{(a, b) : 0 \leq a < \frac{1}{2} < b \leq 1\} \cup \{\emptyset\}$ is a topology on $X$. The verification is omitted.

**Example 1.3.11** (The cofinite topology). The *cofinite topology* (or *finite complement topology*) on a set $X$ consists of the empty set together with all subsets of $X$ which have a finite complement. That is $T_c = \{\emptyset\} \cup \{U \subseteq X : X - U \text{ is finite}\}$. The details of confirming that this is a topology are left to the exercises.

## Exercises

1. Some collections of subsets $T_i$ of $\{1, 2, 3, 4\}$ are shown below. Assume that each collection contains the empty set. Which are topologies? For all that fail to be a topology, state each condition of the definition which does not hold.

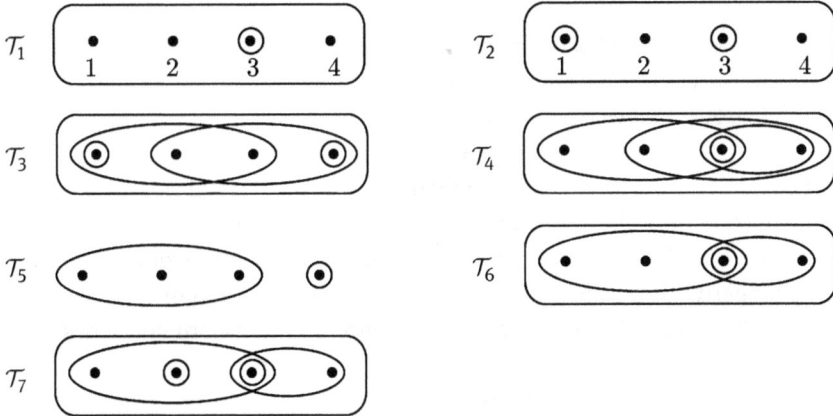

2. **(Right ray topology)** The right ray topology on $\mathbb{R}$ is the collection $\{(a, \infty) : a \in \mathbb{R}\} \cup \{\emptyset, \mathbb{R}\}$.
   (a) Verify that this is a topology.
   (b) Find all limits of the sequence $(2, 3, 2, 3, 2, 3, 2, 3, \ldots)$ in this topological space.
   (c) Is this topology Hausdorff? Justify your answer.
   (d) Is the collection $T = \{[a, \infty) : a \in \mathbb{R}\} \cup \{\emptyset, \mathbb{R}\}$ a topology on $\mathbb{R}$? Justify your answer.

3. Which of the following collections are topologies on $X = [0, 1]$? For those that are not topologies, state every condition of the definition which does not hold.
   (a) $T_1 = \{A \subseteq X : A \text{ is finite}\}$
   (b) $T_2 = \{A \subseteq X : \{0, \frac{1}{2}, 1\} \subseteq A\} \cup \{\emptyset\}$
   (c) $T_3 = \{[0, b] : \frac{1}{2} < b \leq 1\} \cup \{\emptyset\}$
   (d) $T_4 = \{[0, b) : \frac{1}{2} < b < 1\} \cup \{(a, 1] : 0 \leq a < \frac{1}{2}\} \cup \{(a, b) : 0 \leq a < \frac{1}{2} < b \leq 1\} \cup \{\emptyset, X\}$

4. For $k \geq 0$, let $W_k$ be the wedge $\{(x,y) \in \mathbb{R}^2 : x > 0, 0 \leq y < kx\}$. For each set $X_i$ and collection $\mathcal{T}$ of subsets of $X_i$ below, determine whether $\mathcal{T}$ is a topology on $X_i$. For those that are not topologies, identify every condition of the definition which does not hold.

   (a) $X_1 = \{(x,y) \in \mathbb{R}^2 : x > 0, y \geq 0\}$ and $\mathcal{T} = \{\emptyset, W_1, W_2, X_1\}$.

   (b) $X_1 = \{(x,y) \in \mathbb{R}^2 : x > 0, y \geq 0\}$ and $\mathcal{T} = \{\emptyset, W_2, W_3, W_3 - W_1, X_1\}$.

   (c) $X_1 = \{(x,y) \in \mathbb{R}^2 : x > 0, y \geq 0\}$ and $\mathcal{T} = \{\emptyset, X_1\} \cup \{W_{1-\frac{1}{n}} : n \in \mathbb{N}\}$.

   (d) $X_1 = \{(x,y) \in \mathbb{R}^2 : x > 0, y \geq 0\}$ and $\mathcal{T} = \{W_k : k > 0\}$.

   (e) $X_2 = \{(x,y) \in \mathbb{R}^2 : x > 0, y > 0\}$ and $\mathcal{T} = \{W_k \cap X_2 : k \geq 0\}$.

   (f) $X_3 = \{(x,y) \in \mathbb{R}^2 : x > 0, 0 < y < x\}$ and $\mathcal{T} = \{W_k \cap X_3 : 0 \leq k \leq 1\}$.

5. Suppose $U$ is a subset of a topological space $X$. Show that $U$ is open if and only if for every $x \in U$, there exists a neighborhood of $x$ contained in $U$.

6. (a) Show that a point $x$ in a topological space $X$ has a smallest neighborhood $N_x$ if and only if $\bigcap\{U : U$ is a neighborhood of $x\}$ is a neighborhood of $x$.

   (b) Which of the following topological spaces contain a point with a smallest neighborhood?

   i. $\mathbb{R}$ with the Euclidean topology.

   ii. $\mathbb{R}$ with the discrete topology.

   iii. $\mathbb{R}$ with the cofinite topology.

   iv. The space $(Y, \mathcal{T}_4)$ given in Example 1.3.6.

7. Consider this property $P$ which a topological space $X$ might satisfy: For any three distinct points $a, b, c \in X$, there exist mutually disjoint neighborhoods $N_a, N_b$, and $N_c$ of $a, b$, and $c$. Does the $T_2$ property imply property $P$? Does property $P$ imply the $T_2$ property? Give a proof or counterexample for each question.

8. Prove Theorem 1.3.9: A convergent sequence in a Hausdorff space has a unique limit.

9. Show that every metric space is Hausdorff.

10. Verify that the cofinite topology is indeed a topology.

11. Show that the cofinite topology on $\mathbb{R}$ is not Hausdorff by:

    (a) exhibiting distinct points $a$ and $b$ which do not have disjoint neighborhoods;

    (b) exhibiting a sequence which has more than one limit.

12. Show that the discrete topology is the only Hausdorff topology on a finite set $X = \{x_1, \ldots, x_n\}$.

13. Describe all nonempty sets $X$ for which the indiscrete topology on $X$ is Hausdorff.

14. Describe all nonempty sets $X$ for which the cofinite topology on $X$ is Hausdorff.

15. **(Particular point topology)** Let $X$ be any nonempty set and $a \in X$. The *particular point topology on X determined by a* is $P_a = \{U \subseteq X : a \in U\} \cup \{\emptyset\}$. Show that this is a topology.

16. **(Excluded point topology)** Let $X$ be any nonempty set and $a \in X$. The *excluded point topology on X determined by a* is $E_a = \{U \subseteq X : a \notin U\} \cup \{X\}$. Show that this is a topology.

17. What can be said about the set $X$ if
    (a) the particular point topology $P_a$ on $X$ equals the excluded point topology $E_a$?
    (b) the particular point topology $P_a$ on $X$ equals the discrete topology?
    (c) the particular point topology $P_a$ on $X$ equals the indiscrete topology?
    (d) the cofinite topology on $X$ equals the discrete topology?
    (e) the discrete topology on $X$ has 64 open sets?
    (f) the particular point topology is Hausdorff?
18. (Nested collections)
    (a) For $n \in \mathbb{N}$, let $I_n = [0, n] \subseteq \mathbb{R}$. Show that $\{I_n : n \in \mathbb{N}\} \cup \{\emptyset, [0, \infty)\}$ is a topology on $X = [0, \infty)$.
    (b) Suppose $U_1 \subseteq U_2 \subseteq U_3 \subseteq \cdots \subseteq U_n \subseteq X$. Show that the finite nested collection $\{\emptyset, U_1, U_2, \ldots, U_n, X\}$ is a topology on $X$.
    (c) Is every infinite nested collection of subsets of $X = \mathbb{R}$ which contains $\emptyset$ and $\mathbb{R}$ a topology? Give a proof or counterexample.
19. Suppose $X$ is a finite set and $|X| = n$. Since every topology on $X$ is a collection of subsets of $X$, the number of collections of subsets of $X$ will be an upper bound for the number of topologies on $X$. How many subsets does $X$ have? How many collections of subsets of $X$ are there?
20. A topology $\mathcal{T}$ on $X$ is called a *finite topology* on $X$ if the collection $\mathcal{T}$ of open sets is finite.
    (a) Which of these topologies on $\mathbb{R}$ are finite topologies?
    $\mathcal{T}_1$ = the discrete topology
    $\mathcal{T}_2$ = the indiscrete topology
    $\mathcal{T}_3$ = the Euclidean topology
    $\mathcal{T}_4 = \{\emptyset, \mathbb{Z}, \mathbb{R}\}$
    $\mathcal{T}_5$ = the right ray topology = $\{(a, \infty) : a \in \mathbb{R}\} \cup \{\emptyset, \mathbb{R}\}$.
    (b) If $\mathcal{T}$ is a finite topology on $X$ and $a \in X$, show that $a$ has a smallest neighborhood.
    (c) In the Euclidean topology on $\mathbb{R}$, does 0 have a smallest neighborhood?
21. Suppose $X$ is a set, $A \subseteq X$, and $p, q$ are distinct points in $X$.
    (a) Show that $\{\emptyset, A, X\}$ is a topology on $X$. (Note that the only topology strictly contained in this topology is the indiscrete topology.)
    (b) Show that $\mathcal{T}_{\{p,q\}} = \{U \subseteq X : p \in U \text{ or } \{p, q\} \cap U = \emptyset\}$ is a topology on $X$. (We will see in Section 9.3 that the only topology which strictly contains this topology is the discrete topology.)

## 1.4 Basis for a topology

We saw that all the open sets in a metric space were generated by the balls in two ways:
    A set $U$ is open if and only if it is a union of balls (Corollary 1.2.7).

A set $U$ is open if and only if for every $x \in U$, there exists a ball $B(x, \varepsilon)$ which is contained in $U$. (Theorem 1.2.5).

A nice collection of open sets which can be used in these ways to generate the topology is called a *basis* for the topology.

**Definition 1.4.1.** If $\mathcal{T}$ is a topology on $X$, a *basis* for $\mathcal{T}$ is a collection $\mathcal{B} \subseteq \mathcal{T}$ of open subsets of $X$ such that every open set is a union of elements of $\mathcal{B}$. The open sets in $\mathcal{B}$ are called *basic open sets with respect to* $\mathcal{B}$, and if $B \in \mathcal{B}$ and $x \in B$, we say $B$ is a *basic neighborhood* of $x$ with respect to $\mathcal{B}$. If $\mathcal{B}$ is a basis for $\mathcal{T}$, we say $\mathcal{T}$ is *generated* by $\mathcal{B}$, and write $\mathcal{T} = \mathcal{T}_{\mathcal{B}}$ or $\mathcal{T} = [\mathcal{B}]$.

When we refer to an open set, we must know which topology is being used. When referring to a basic open set, we must also know which basis is being used. When the topology $\mathcal{T}$ and basis $\mathcal{B}$ are understood, we may simply say "open" or "basic open" instead of the more precise "$\mathcal{T}$-open" or "basic open with respect to $\mathcal{B}$".

Recall that the union of an empty collection is $\emptyset$, so the open set $\emptyset$ will always be realized as a union (namely, the empty union) of elements of a basis $\mathcal{B}$. To avoid the case of the empty union, we could equivalently define a basis for a topology $\mathcal{T}$ to be a collection $\mathcal{B} \subseteq \mathcal{T}$ such that every *nonempty* open set is a union of elements of $\mathcal{B}$.

To reiterate our motivating example, if $(X, d)$ is a metric space, the collection $\mathcal{B} = \{B(x, \varepsilon) : x \in X, \varepsilon > 0\}$ of all balls is a basis for the metric topology on $X$ generated by the metric $d$.

Every topology $\mathcal{T}$ has a basis: $\mathcal{B} = \mathcal{T}$ is a basis for $\mathcal{T}$. A topology may have many different bases, as the following example illustrates.

**Example 1.4.2.** Let $\mathcal{T}_{\mathcal{E}}$ be the Euclidean topology on $\mathbb{R}$. Consider the collections

$$\mathcal{B}_1 = \{(x - \varepsilon, x + \varepsilon) : x \in \mathbb{R}, \varepsilon > 0\},$$
$$\mathcal{B}_2 = \mathcal{T}_{\mathcal{E}},$$
$$\mathcal{B}_3 = \left\{ \left(x - \frac{1}{n}, x + \frac{1}{n}\right) : x \in \mathbb{R}, n \in \mathbb{N} \right\}, \quad \text{and}$$
$$\mathcal{B}_4 = \{(x - \varepsilon, x + \varepsilon) - \{x\} : x \in \mathbb{R}, \varepsilon > 0\}.$$

Each is a collection of open sets. $\mathcal{B}_1$ is the collection of balls in the Euclidean metric, and thus is a basis for $\mathcal{T}_{\mathcal{E}}$. Clearly every open set in $\mathcal{T}_{\mathcal{E}}$ is a union of sets in $\mathcal{T}_{\mathcal{E}}$, so $\mathcal{B}_2$ is a basis for $\mathcal{T}_{\mathcal{E}}$.

Note that every $\varepsilon$-ball $(x - \varepsilon, x + \varepsilon)$ is a union of $\frac{1}{n}$-balls: for each $z \in (x - \varepsilon, x + \varepsilon)$, there exists a $\frac{1}{n}$-ball $B_z$ centered at $z$ and contained in $(x - \varepsilon, x + \varepsilon)$, and $(x - \varepsilon, x + \varepsilon) = \bigcup \{B_z : z \in (x - \varepsilon, x + \varepsilon)\}$. Thus, every open set is a union of $\frac{1}{n}$-balls, so $\mathcal{B}_3$ is a basis.

The elements of $\mathcal{B}_4$ are balls with their center points removed; such sets are called *deleted balls*. Any $\varepsilon$-ball $(x - \varepsilon, x + \varepsilon)$ is the union of the two deleted balls $((x - \varepsilon, x + \varepsilon) - \{x\}) \cup ((y - \frac{\varepsilon}{2}, y + \frac{\varepsilon}{2}) - \{y\})$ where $y = x + \frac{\varepsilon}{4}$, so every open set is a union of deleted balls, and thus $\mathcal{B}_4$ is a basis.

The balls in a metric space are the prototype of a basis. We used the property of balls given in Corollary 1.2.7 to form our definition. Now we will also show that the property of balls in a metric space given in Theorem 1.2.5 carries over to any basis for any topological space.

**Theorem 1.4.3.** *Suppose $\mathcal{T}$ is a topology on set $X$ and $\mathcal{B} \subseteq \mathcal{T}$ is a collection of open sets. The following are equivalent.*
(a) *$\mathcal{B}$ is a basis for $\mathcal{T}$.*
(b) *$U$ is open if and only if $U$ is a union of elements of $\mathcal{B}$.*
(c) *$U$ is open if and only if for every $x \in U$, there exists a basic open set $B \in \mathcal{B}$ which contains $x$ and is contained in $U$.*

We note that (c) could be formally written as:
(c') $U \in \mathcal{T}$ if and only if $\forall x \in U \ \exists B \in \mathcal{B}$ such that $x \in B$ and $B \subseteq U$.

*Proof.* The definition of $\mathcal{B}$ being a basis for $\mathcal{T}$ was that $U$ is open only if $U$ is a union of elements of $\mathcal{B}$. The converse also holds: $U$ is open if $U$ is a union of elements of $\mathcal{B}$. Indeed, since $\mathcal{B} \subseteq \mathcal{T}$, any union of elements of $\mathcal{B}$ is a union of open sets, which is open. Thus, (a) and (b) are equivalent.

To see (b) and (c) are equivalent, it suffices to show that (b2) $U$ is a union of elements of $\mathcal{B}$ is equivalent to (c2) for every $x \in U$, there exists a basic open set $B \in \mathcal{B}$ which contains $x$ and is contained in $U$. Suppose (b2). Express $U$ as the union $\bigcup\{B_i : i \in I\}$ of basic open sets $B_i \in \mathcal{B}$. For any $x \in U$, there exists $j \in I$ such that $x \in B_j$. Furthermore, $B_j \subseteq \bigcup\{B_i : i \in I\} = U$, proving (c2). Now suppose (c2). Then, for every $x \in U$, there exists a basic open set $B_x \in \mathcal{B}$ with $\{x\} \subseteq B_x \subseteq U$. Taking the union over all $x \in U$ gives $U = \bigcup_{x \in U}\{x\} \subseteq \bigcup_{x \in U} B_x \subseteq \bigcup_{x \in U} U = U$, so $U$ is a union of basis elements, proving (b2). $\square$

Note that Theorem 1.4.3(c) may be interpreted as saying every neighborhood $U$ of $x$ contains a *basic neighborhood of $x$*, that is, a basic open set $B \in \mathcal{B}$ which contains $x$. In the discrete topology $\mathcal{T}_D$ on $X$, every point $x$ has a smallest neighborhood $\{x\}$. Since $\{x\}$ must contain a basic neighborhood of $x$, it follows that $\{x\}$ must be a basic open set in any basis $\mathcal{B}$ for $\mathcal{T}_D$. Furthermore, since every set is a union of singleton sets, the collection $\mathcal{B} = \{\{x\} : x \in X\}$ of singleton sets in $X$ is a basis for the discrete topology. Similarly, if a point $x$ in a topological space $(X, \mathcal{T})$ has a smallest neighborhood $N_x$, then $N_x \in \mathcal{B}$ for any basis $\mathcal{B}$ for $\mathcal{T}$. If every point $x \in X$ has a smallest neighborhood $N_x$, then $\{N_x : x \in X\}$ is a basis for $\mathcal{T}$ by the equivalence of (a) and (c) in Theorem 1.4.3.

Recall the single transmitter topology $\mathcal{T} = \{U \subseteq [-1, 1] : 0 \in U \Rightarrow (-1, 1) \subseteq U\}$ on $X = [-1, 1]$ given in Example 1.3.7. In this topology, if $x \neq 0$, then $\{x\}$ is open, and $U$ is a neighborhood of $0$ if and only if $(-1, 1) \subseteq U$. Thus, every point has a smallest neighborhood, and the collection of these smallest neighborhoods $\{\{x\} : x \in [-1, 1], x \neq 0\} \cup \{(-1, 1)\}$ is a basis for $\mathcal{T}$.

Of course, generally a point will not have a smallest neighborhood. In the Euclidean topology on $\mathbb{R}$, $B(x, \frac{1}{n}) = (x - \frac{1}{n}, x + \frac{1}{n})$ is a neighborhood of $x$ for every $n \in \mathbb{N}$. If there were a smallest neighborhood $N_x$ of $x$, it would have to contain $x$ and be contained in each $B(x, \frac{1}{n})$, giving $\{x\} \subseteq N_x \subseteq \bigcap\{B(x, \frac{1}{n}) : n \in \mathbb{N}\} = \{x\}$, so $N_x = \{x\}$. But this is a contradiction, since $\{x\}$ is not a neighborhood of $x$ in the Euclidean topology.

We will see several topological concepts such as convergence of sequences or boundary points of a set, which are defined in terms of arbitrary neighborhoods of a point, with an unwritten emphasis on "no matter how small". Since any neighborhood of $x$ contains a basic neighborhood of $x$, such properties may be characterized or confirmed by using only the basic neighborhoods of the point.

In our discussion so far, we have started with a given topology $\mathcal{T}$ and considered bases generating that topology. We may ask, if we start with a collection of subsets of $X$, when might it be a basis for some (as yet unknown) topology on $X$? Suppose that $\mathcal{B}$ is a basis for a topology $\mathcal{T}$ on $X$. Since $X$ is an open set, condition (b) of Theorem 1.4.3 implies that $\bigcup \mathcal{B} = X$. If $B_1, B_2 \in \mathcal{B}$, then $B_1$ and $B_2$ are both open, so $B_1 \cap B_2$ is open. If $x \in B_1 \cap B_2$, then the neighborhood $B_1 \cap B_2$ of $x$ must contain a basic neighborhood $B_3$ of $x$. That is, there exists $B_3 \in \mathcal{B}$ with $x \in B_3$ and $B_3 \subseteq B_1 \cap B_2$, as suggested in Figure 1.8. In fact, these properties are all that is needed to guarantee that a collection $\mathcal{B}$ is a basis for some topology on $X$.

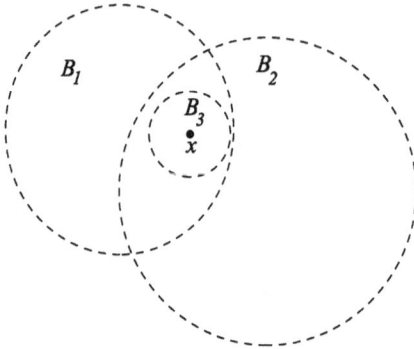

**Figure 1.8:** If $B_1, B_2 \in \mathcal{B}$ and $x \in B_1 \cap B_2$, then there exists $B_3 \in \mathcal{B}$ with $x \in B_3 \subseteq B_1 \cap B_2$.

**Theorem 1.4.4.** *A collection $\mathcal{B}$ of subsets of $X$ is a basis for some topology on $X$ if and only if*

(a) $\bigcup \mathcal{B} = X$, *and*

(b) *if $B_1, B_2 \in \mathcal{B}$ and $x \in B_1 \cap B_2$, then there exists $B_3 \in \mathcal{B}$ with $x \in B_3 \subseteq B_1 \cap B_2$.*

*Proof.* If $\mathcal{B}$ is a basis for some topology on $X$, the discussion before the statement of the theorem shows that (a) and (b) are satisfied.

Suppose $\mathcal{B}$ satisfies (a) and (b) above. If $\mathcal{B}$ is to be a basis for a topology, then that topology must consist of all unions of elements of $\mathcal{B}$. Let $\mathcal{T}_{\mathcal{B}}$ be the collection of all unions of elements of $\mathcal{B}$. We will show that $\mathcal{T}_{\mathcal{B}}$ is a topology. By condition (a) above, $X \in \mathcal{T}_{\mathcal{B}}$, and $\emptyset \in \mathcal{T}_{\mathcal{B}}$ since $\emptyset$ is the union of the empty collection of elements of $\mathcal{B}$. An arbitrary union of elements of $\mathcal{T}_{\mathcal{B}}$ is a union of unions of elements of $\mathcal{B}$, which is just a larger union of elements of $\mathcal{B}$, so $\mathcal{T}_{\mathcal{B}}$ is closed under the formation of arbitrary unions. For finite intersections, suppose $U_1, U_2 \in \mathcal{T}_{\mathcal{B}}$. If $U_1 \cap U_2 = \emptyset$, then $U_1 \cap U_2 \in \mathcal{T}_{\mathcal{B}}$. Otherwise, suppose $x \in U_1 \cap U_2$. Since $U_1$ is a union of sets in $\mathcal{B}$, there exists $B_1 \in \mathcal{B}$ with $\{x\} \subseteq B_1 \subseteq U_1$. Similarly, there exists $B_2 \in \mathcal{B}$ with $\{x\} \subseteq B_2 \subseteq U_2$. By condition (b), there exists $B_x \in \mathcal{B}$ with $\{x\} \subseteq B_x \subseteq B_1 \cap B_2 \subseteq U_1 \cap U_2$. Now taking the union over all $x \in U_1 \cap U_2$ shows that

$$U_1 \cap U_2 = \bigcup_{x \in U_1 \cap U_2} \{x\} \subseteq \bigcup_{x \in U_1 \cap U_2} B_x \subseteq \bigcup_{x \in U_1 \cap U_2} U_1 \cap U_2 = U_1 \cap U_2,$$

so $U_1 \cap U_2 = \bigcup_{x \in U_1 \cap U_2} B_x$ is a union of elements of $\mathcal{B}$ and thus is in $\mathcal{T}_{\mathcal{B}}$. Having shown the result for intersections of two open sets, the result follows for finite intersections inductively. □

**Example 1.4.5.** Let $X = \{1, 2, 3, 4\}$, and consider the collections $\mathcal{A} = \{\{1, 2\}, \{2\}, \{2, 3\}\}$, $\mathcal{B} = \{\{1, 2\}, \{2\}, \{2, 3\}, \{4\}\}$, and $\mathcal{C} = \{\{1, 2, 3\}, \{2\}, \{2, 3, 4\}\}$, as depicted in Figure 1.9.

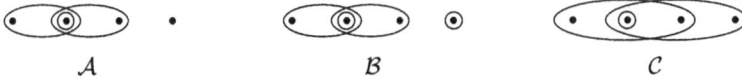

**Figure 1.9:** Which collections are bases?

$\mathcal{A}$ is not a basis for a topology on $X$ since $\bigcup \mathcal{A} \neq X$. $\mathcal{A}$ defines no neighborhood of the point 4.

The union of $\mathcal{B}$ is $X$. To check the second defining condition of a basis for $\mathcal{B}$, we must verify an implication for every $x \in B_1 \cap B_2$, where $B_1$ and $B_2$ are arbitrary members of $\mathcal{B}$. Furthermore, the implication is always true if $B_1 = B_2$ (for then we may take $B_3 = B_1 = B_2$), so we only need to verify the implication for distinct sets $B_1, B_2 \in \mathcal{B}$ with nonempty intersection. There is only one such pair in $\mathcal{B}$, and their intersection contains only one point, namely $2 \in \{1, 2\} \cap \{2, 3\}$. Now $B_3 = \{2\} \in \mathcal{B}$ contains 2 and is contained in $\{1, 2\} \cap \{2, 3\}$, so $\mathcal{B}$ is a basis for a topology on $X$. The topology generated by $\mathcal{B}$ is $\mathcal{T}_{\mathcal{B}} = \mathcal{B} \cup \{\emptyset, \{1, 2, 3\}, \{1, 2, 4\}, \{2, 4\}, \{2, 3, 4\}, X\}$.

$\mathcal{C}$ is not a basis for a topology on $X$. Indeed, $\{1, 2, 3\}, \{2, 3, 4\} \in \mathcal{C}$ and $3 \in \{1, 2, 3\} \cap \{2, 3, 4\}$, but there is no $C_3 \in \mathcal{C}$ with $3 \in C_3 \subseteq \{1, 2, 3\} \cap \{2, 3, 4\}$.

Many topologies are defined in terms of a basis, as the next two examples illustrate.

**Example 1.4.6 (The digital line).** Consider a line of pixels. As discussed after Example 1.3.4, each pixel should be open since it is illuminated by an open set contained in the pixel. Considering only the pixels, this gives the discrete topology, which does not adequately model closeness of pixels. To remedy this, we will also consider the boundaries between pixels. Suppose the pixels correspond to the odd integers, with the even integers representing the boundaries between adjacent pixels, as suggested in Figure 1.10. Each boundary between pixels is close to its two neighboring pixels. This suggests the topology $\mathcal{T}_{DL}$ on $\mathbb{Z}$ whose basis is shown in Figure 1.10. With this topology, $\mathbb{Z}$ is called the digital line. The basis for $\mathcal{T}_{DL}$ is $\mathcal{B} = \{\{n\} : n \text{ is an odd integer}\} \cup \{\{n-1, n, n+1\} : n \text{ is an even integer}\}$.

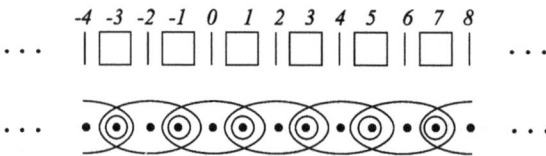

**Figure 1.10:** The digital line topology.

**Example 1.4.7** (Partition topologies). Let $\mathcal{P}$ be a partition of a set $X$. Then $\bigcup \mathcal{P} = X$, and because the blocks of $\mathcal{P}$ are mutually disjoint, $x \in P_1 \cap P_2$ for $P_1, P_2 \in \mathcal{P}$ implies $P_1 = P_2$, so with $P_3 = P_1$, we have $P_3 \in \mathcal{P}$ and $x \in P_3 \subseteq P_1 \cap P_2$. Thus, $\mathcal{P}$ is basis for a topology on $X$. The topology $\mathcal{T}_{\mathcal{P}}$ generated by a partition is called a *partition topology*.

Recall that a partition $\mathcal{R}$ is a refinement of partition $\mathcal{P}$ if $\mathcal{R}$ is obtained by partitioning the blocks of $\mathcal{P}$, as illustrated in Figure 1.11. If $\mathcal{R}$ is a refinement of $\mathcal{P}$, then $\mathcal{R}$ has smaller blocks than $\mathcal{P}$, and thus $\mathcal{T}_{\mathcal{R}}$ has smaller neighborhoods than $\mathcal{T}_{\mathcal{P}}$. This suggests the terminology of one topology being finer than another.

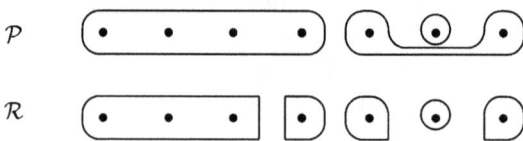

**Figure 1.11:** Partition $\mathcal{R}$ is a refinement of partition $\mathcal{P}$. Basis $\mathcal{R}$ generates a finer topology than basis $\mathcal{P}$.

**Definition 1.4.8.** If $\mathcal{F}$ and $\mathcal{C}$ are topologies on the same set $X$, then we say $\mathcal{F}$ is *finer* than $\mathcal{C}$ (or $\mathcal{C}$ is *coarser* than $\mathcal{F}$) if and only if $\mathcal{C} \subseteq \mathcal{F}$. Thus, $\mathcal{F}$ is finer than $\mathcal{C}$ if $\mathcal{F}$ contains all the $\mathcal{C}$-open sets and possibly more.

**Example 1.4.9** (Lower limit topology). The *lower limit topology* (also called the *Sorgenfrey topology*) on $\mathbb{R}$ is the topology $\mathcal{T}_l$ generated by the basis $\mathcal{B} = \{[a, b) \subseteq \mathbb{R} : a < b\}$. $\mathbb{R}$ with the lower limit topology is denoted $\mathbb{R}_l$. It is easy to check that $\mathcal{B}$ is a basis for a topology. Taking unions of basis elements will produce sets such as $[0, 1) \cup [2, 3)$, $\bigcup_{n>7}[5, n) = [5, \infty)$, and even $\bigcup\{[5 + \frac{1}{n}, 8) : n \in \mathbb{N}\} = (5, 8)$. Similarly, every interval $(a, b)$ with $a < b$ is open in the lower limit topology. Thus, every union of intervals of form $(a, b)$ is open in the lower limit topology. That is, any set open in the Euclidean topology $\mathcal{T}_\mathcal{E}$ is open in the lower limit topology. This proves that $\mathcal{T}_\mathcal{E} \subseteq \mathcal{T}_l$, so the lower limit topology is finer than the Euclidean topology on $\mathbb{R}$. The lower limit topology models situations in which you need "close, but no less than". For example, if you need a log to cross from one side of a gorge to the opposite side 4 meters away, logs of length $l \in [4, b)$ will be of interest.

By defining a topology on a set $X$, we specify which sets are open, and these open sets are then used to define neighborhoods and sequential convergence. We cannot answer the question "Does the sequence $(\frac{1}{n})_{n=1}^\infty$ converge in $\mathbb{R}$?" until we know which topology on $\mathbb{R}$ we are using. In the Euclidean topology familiar from calculus, this sequence converges to 0. In the indiscrete topology, this sequence converges to every real number $r$. In the discrete topology, this sequence does not converge. In the lower limit topology, the sequence converges to 0, but $(\frac{-1}{n})_{n=1}^\infty$ does not converge: If $a < 0$, then $(\frac{-1}{n})_{n=1}^\infty$ is eventually out of the neighborhood $[a, a/2)$ of $a$, and if $a \geq 0$, the sequence is never in the neighborhood $[a, a + 1)$ of $a$. The lower limit topology captures the usual convergence we expect from the Euclidean topology for decreasing sequences, but not for arbitrary sequences.

*Terminology:* If $\mathcal{S} \subseteq \mathcal{T}$ is a collection of open sets, by an $\mathcal{S}$-neighborhood of $x$ we mean a neighborhood of $x$ which is an element of $\mathcal{S}$. In particular, if $\mathcal{B}$ is a basis for $\mathcal{T}$, a $\mathcal{B}$-neighborhood of $x$ will be a basic open set from $\mathcal{B}$ containing $x$.

Since a basis determines the topology, we should be able to tell when one topology is finer than another by comparing the bases. The finer topology should have smaller basic neighborhoods at each point. This result is formally stated now.

**Theorem 1.4.10.** *Suppose $\mathcal{B}_\mathcal{F}$ and $\mathcal{B}_\mathcal{C}$ are bases for topologies $\mathcal{F}$ and $\mathcal{C}$ on $X$. Then $\mathcal{F}$ is finer than $\mathcal{C}$ (and $\mathcal{C}$ is coarser than $\mathcal{F}$) if and only if for every $x \in X$, every $\mathcal{B}_\mathcal{C}$-neighborhood of $x$ contains a $\mathcal{B}_\mathcal{F}$-neighborhood of $x$.*

*Proof.* Suppose $\mathcal{F}$ is finer than $\mathcal{C}$, $x \in X$, and $B_\mathcal{C} \in \mathcal{B}_\mathcal{C}$ with $x \in B_\mathcal{C}$. Since $B_\mathcal{C} \in \mathcal{B}_\mathcal{C} \subseteq \mathcal{C} \subseteq \mathcal{F}$, $B_\mathcal{C}$ is an $\mathcal{F}$-neighborhood of $x$, so by Theorem 1.4.3(c), $B_\mathcal{C}$ must contain a $\mathcal{B}_\mathcal{F}$-basic neighborhood of $x$, as needed.

Conversely, suppose that, for every $x \in X$ and every $B_\mathcal{C} \in \mathcal{B}_\mathcal{C}$ with $x \in B_\mathcal{C}$, there exists $B_\mathcal{F} \in \mathcal{B}_\mathcal{F}$ with $x \in B_\mathcal{F} \subseteq B_\mathcal{C}$. To show $\mathcal{C} \subseteq \mathcal{F}$, suppose $U \in \mathcal{C}$. Now for each $x \in U$, there exists $B_\mathcal{C} \in \mathcal{B}_\mathcal{C}$ with $x \in B_\mathcal{C} \subseteq U$. By our hypothesis, there exists $B_x \in \mathcal{B}_\mathcal{F}$ with $x \in B_x \subseteq B_\mathcal{C} \subseteq U$. Taking the union over all $x \in U$, we find that $U = \bigcup\{B_x : x \in U\}$, and as a union of basic open sets in $\mathcal{B}_\mathcal{F} \subseteq \mathcal{F}$, $U \in \mathcal{F}$, as needed. □

We have seen that the lower limit topology on $\mathbb{R}$ is finer than the Euclidean topology. The theorem above allows us to confirm this as follows: Since every basic Euclidean neighborhood $(x - \varepsilon, x + \varepsilon)$ of an arbitrary $x \in \mathbb{R}$ contains a basic lower limit neighborhood $[x, x + \varepsilon)$, the lower limit topology is finer than the Euclidean topology.

For another example, consider $\mathbb{R}^2$ in the metric topology $\mathcal{T}_\varepsilon$ from the Euclidean metric and in the metric topology $\mathcal{T}_{\text{sup}}$ from the sup-metric. The balls form a basis for any metric topology. In the Euclidean metric the balls $B_\varepsilon(x, \varepsilon)$ are circular regions, but the sup-metric produces square balls $B_{\text{sup}}(x, \varepsilon)$. It is easy to see that $B_\varepsilon(x, \varepsilon) \subseteq B_{\text{sup}}(x, \varepsilon)$, so by the previous theorem, the Euclidean metric topology is finer than the sup-metric topology. However, every round $\varepsilon$-ball $B_\varepsilon(x, \varepsilon)$ contains a square $\frac{\varepsilon}{2}$-ball $B_{\text{sup}}(x, \frac{\varepsilon}{2})$ as shown in Figure 1.12, so the sup-metric topology is finer than the Euclidean metric topology. Thus, $\mathcal{T}_{\text{sup}} \subseteq \mathcal{T}_\varepsilon \subseteq \mathcal{T}_{\text{sup}}$, so the topologies are the same. A subset of the plane is a union of open disks if and only if it is a union of open squares. This shows that different metrics may generate the same topology.

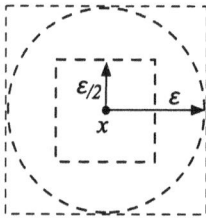

**Figure 1.12:** $B_{\text{sup}}(x, \frac{\varepsilon}{2}) \subseteq B_\varepsilon(x, \varepsilon) \subseteq B_{\text{sup}}(x, \varepsilon)$, so $\mathcal{T}_{\text{sup}} \subseteq \mathcal{T}_\varepsilon \subseteq \mathcal{T}_{\text{sup}}$.

The prototype of a basis for a topological space are the open intervals $(a, b)$ (that is, the balls) in $\mathbb{R}$, giving the Euclidean topology. The open intervals $(a, b)$ may be generated from open rays $(a, \infty)$ and $(-\infty, b)$ by taking intersections. The collection $\mathcal{S} = \{(a, \infty) : a \in \mathbb{R}\} \cup \{(-\infty, b) : b \in \mathbb{R}\}$ is not a basis, since, for example, $S_1 = (-1, \infty)$ and $S_2 = (-\infty, 1)$ are in $\mathcal{S}$ and $0 \in S_1 \cap S_2$, but there is no $S_3 \in \mathcal{S}$ with $S_3 \subseteq S_1 \cap S_2$. The collection $\mathcal{S}$ of open rays in the Euclidean topology on $\mathbb{R}$ is the prototype of a *subbasis* for a topology.

**Definition 1.4.11.** A collection $\mathcal{S}$ of subsets of $X$ is a *subbasis* for a topology $\mathcal{T}$ on $X$ if the collection $\mathcal{B}_\mathcal{S} = \{S_1 \cap \cdots \cap S_n : n \in \mathbb{N}, S_1, \ldots, S_n \in \mathcal{S}\}$ of finite intersections of elements of $\mathcal{S}$ is a basis for $\mathcal{T}$.

If $\mathcal{S}$ is a subbasis for $\mathcal{T}$, then the open sets in $\mathcal{T}$ are generated as arbitrary unions of finite intersections of elements of $\mathcal{S}$. The result below tells us when a collection is a subbasis for some topology.

**Theorem 1.4.12.** *A collection $\mathcal{S}$ of subsets of a set $X$ is a subbasis for a topology on $X$ if and only if $\bigcup \mathcal{S} = X$.*

The proof is left as an exercise.

### 1.4.1 Neighborhood bases

The Euclidean topology on $\mathbb{R}^2$ has a basis of balls. Then, a basic neighborhood of $a \in \mathbb{R}^2$ is any ball containing $a$, whether the ball is centered at $a$ or not. But, every ball $B(x, \varepsilon)$ containing $a$ contains a ball $B(a, \delta)$ centered at $a$. Thus, if we are checking some condition which should hold for every neighborhood of $a$, then it would suffice to check the condition for every ball *centered at* $a$. The collection $\{B(a, \varepsilon) : \varepsilon > 0\}$ of balls centered at $a$ in the Euclidean topology is the prototype for a *neighborhood base* at a point $a$.

**Definition 1.4.13.** If $(X, \mathcal{T})$ is a topological space, a *neighborhood base* at $a \in X$ is a collection $\mathcal{N}(a)$ of neighborhoods of $a$ such that every neighborhood of $a$ contains a neighborhood of $a$ from $\mathcal{N}(a)$. That is, a neighborhood base at $a \in X$ is a collection $\mathcal{N}(a) \subseteq \mathcal{T}$ such that, for any neighborhood $U$ of $a$, there exists $N \in \mathcal{N}(a)$ with $a \in N \subseteq U$.

Every point of a topological space $(X, \mathcal{T})$ has a neighborhood base: taking $\mathcal{N}(a)$ to be the collection of all basic neighborhoods of $a$ from any basis $\mathcal{B}$ for $\mathcal{T}$ gives a neighborhood base. In particular, all neighborhoods of $a$ form a neighborhood base at $a$, as do all open neighborhoods of $a$. As with bases, the usefulness of this concept will be in choosing neighborhood bases which simplify matters.

To show that a sequence $(x_n)$ converges to $x$ in Euclidean space, from the definition, we should check that $(x_n)$ is eventually in every open set containing $x$. However, it suffices to check that $(x_n)$ is eventually in every open ball $B(y, \varepsilon)$ containing $x$ (that is, every neighborhood from the standard basis), and furthermore, it suffices to check that $(x_n)$ is eventually in every open ball $B(x, \varepsilon)$ centered at $x$. That is, we need only check the neighborhoods from a neighborhood base at $x$ to determine whether $(x_n)$ converges to $x$.

The result below shows that we may check when one topology is finer than another by comparing the neighborhood bases at every point.

**Theorem 1.4.14.** *Suppose $\mathcal{F}$ and $\mathcal{C}$ are topologies on $X$, $\mathcal{B}_\mathcal{F}$ is a basis for $\mathcal{F}$, $\mathcal{B}_\mathcal{C}$ is a basis for $\mathcal{C}$, and for each $x \in X$, $\mathcal{N}_\mathcal{F}(x)$ is a neighborhood base for $x$ in $\mathcal{F}$ and $\mathcal{N}_\mathcal{C}(x)$ is a neighborhood base for $x$ in $\mathcal{C}$. Then the following are equivalent.*
(a) *$\mathcal{F}$ is finer than $\mathcal{C}$.*
(b) *For every $x \in X$, every $\mathcal{C}$-neighborhood of $x$ contains a $\mathcal{F}$-neighborhood of $x$.*
(c) *For every $x \in X$, every $\mathcal{B}_\mathcal{C}$-neighborhood of $x$ contains a $\mathcal{B}_\mathcal{F}$-neighborhood of $x$.*
(d) *For every $x \in X$, every $\mathcal{N}_\mathcal{C}(x)$-neighborhood of $x$ contains a $\mathcal{N}_\mathcal{F}(x)$-neighborhood of $x$.*
(e) *For every $x \in X$, every $\mathcal{C}$-neighborhood of $x$ contains a $\mathcal{B}_\mathcal{F}(x)$-neighborhood of $x$.*

*Proof.* The equivalence of (a) and (c) is Theorem 1.4.10. The equivalences (c) $\Longleftrightarrow$ (b) and (b) $\Longleftrightarrow$ (e) follow since every neighborhood of $x$ contains a basic neighborhood of $x$, and (c) $\Longleftrightarrow$ (d) follows since every neighborhood of $x$ contains a neighborhood of $x$ from a neighborhood base at $x$. $\qquad\square$

**Definition 1.4.15.** A topological space $X$ is *first countable* if every $x \in X$ has a countable neighborhood base.

Every metric space $M$ is first countable since every point $x \in M$ has a countable neighborhood base $\{B(x, \frac{1}{n}) : n \in \mathbb{N}\}$.

**Example 1.4.16.** $\mathbb{R}$ with the cofinite topology is not first countable. Suppose to the contrary that it is first countable, and $\mathcal{N}(0) = \{B_i : i \in \mathbb{N}\}$ is a countable neighborhood base at 0. For every $y \neq 0$, $\mathbb{R} - \{y\}$ is a neighborhood of 0, so there exists $B_y \in \mathcal{N}(0)$ with $\{0\} \subseteq B_y \subseteq \mathbb{R} - \{y\}$. Taking the intersection over all $y \neq 0$, it follows that

$$\{0\} \subseteq \bigcap \mathcal{N}(0) = \bigcap\{B_i : i \in \mathbb{N}\} \subseteq \bigcap_{y \neq 0} B_y \subseteq \{0\}.$$

Thus, $\mathbb{R} - \{0\} = \mathbb{R} - \bigcap\{B_i : i \in \mathbb{N}\} = \bigcup\{\mathbb{R} - B_i : i \in \mathbb{N}\}$, which is a countable union of finite sets, and thus is countable. But if $\mathbb{R} - \{0\}$ is countable, then $\mathbb{R}$ is countable, and this contradiction shows that there is not a countable neighborhood base at 0, so the cofinite topology is not first countable.

## Exercises

1. Which collections below are a basis for a topology on $\mathbb{R}$? Justify your answers.
   (a) $\mathcal{B}_1 = \{[a, b] : a < b\}$
   (b) $\mathcal{B}_2 = \{(a, b) \cup (c, d) : a < b < c < d\}$
   (c) $\mathcal{B}_3 = \{(a, a + 2) : a \in \mathbb{R}\}$
   (d) $\mathcal{B}_4 = \{A \subseteq \mathbb{R} : 0 \in A \text{ or } 1 \in A\}$
   (e) $\mathcal{B}_5 = \{A \subseteq \mathbb{R} : 0 \notin A \text{ or } 1 \notin A\}$
   (f) $\mathcal{B}_6 = \{A \subseteq \mathbb{R} : 0 \notin A \text{ and } 1 \in A\}$

2. Of the collections depicted in Exercise 1 of Section 1.3, which are bases for a topology on $\{1, 2, 3, 4\}$?

3. Suppose $(X, \mathcal{T})$ is a topological space and $\mathcal{B}$ is a basis for $\mathcal{T}$. Prove that $(x_n)_{n=1}^{\infty}$ converges to $a$ in $X$ if and only if the sequence is eventually in every *basic* neighborhood $B$ of $a$.

4. Which collections below are bases for the right ray topology $\mathcal{T} = \{(a, \infty) : a \in \mathbb{R}\} \cup \{\emptyset, \mathbb{R}\}$ on $\mathbb{R}$? Justify your answer.
   (a) $\{(a, \infty) : a \in \mathbb{R}\}$
   (b) $\{(a, \infty) : a \in \mathbb{Q}\}$
   (c) $\{(a, \infty) : a \in \mathbb{N}\}$
   (d) $\{(a, \infty) : a \in \mathbb{R}\} \cup \{(0, 1)\}$

5. Let $\mathcal{B} = \{[x, x + \varepsilon) : x \in \mathbb{Q}, \varepsilon > 0\} \cup \{(x - \varepsilon, x] : x \in \mathbb{R} - \mathbb{Q}, \varepsilon > 0\}$.
   (a) Show that $\mathcal{B}$ is a basis for a topology on $\mathbb{R}$.
   (b) Is $\mathcal{T}_{\mathcal{B}}$ finer, coarser, or incomparable to the Euclidean topology $\mathcal{T}_{\mathcal{E}}$?

(c) Discuss the convergence of these sequences in $(\mathbb{R}, \mathcal{T}_B)$: $(\pi/n)_{n=1}^{\infty}$, $(-\pi/n)_{n=1}^{\infty}$, $(\pi + 1/n)_{n=1}^{\infty}$, $(\pi - 1/n)_{n=1}^{\infty}$.

6. **(Multiples topology on $\mathbb{N}$)** For $n \in \mathbb{N}$, let $M_n = \{kn : k \in \mathbb{N}\}$ be the set of positive multiples of $n$.

   (a) Show that $\mathcal{B} = \{M_n : n \in \mathbb{N}\}$ is a basis for a topology on $\mathbb{N}$. This topology is called the *multiples topology on $\mathbb{N}$*.

   (b) In the multiples topology, give six distinct neighborhoods of 10.

   (c) In the multiples topology, does every $k \in \mathbb{N}$ have a smallest neighborhood? Explain.

   (d) Prove or disprove: The multiples topology on $\mathbb{N}$ is Hausdorff.

7. In the multiples topology on $\mathbb{N}$ (see Exercise 6), find the following or explain why they do not exist.

   (a) All limits of the sequence $(6, 12, 18, 24, 30, 36, 42, 48, \ldots) = (6n)_{n=1}^{\infty}$.

   (b) All limits of the sequence $(2, 3, 2, 3, 2, 3, 2, 3, 2, 3, \ldots)$.

   (c) All limits of the sequence $(2, 4, 2, 4, 2, 4, 2, 4, 2, 4, \ldots)$.

   (d) All limits of the sequence of digits to the right of the decimal in the decimal expansion of $\frac{13579}{99999}$.

   (e) A sequence with no limit.

8. Find a characterization of all constant sequences in $\mathbb{N}$ with the multiples topology (see Exercise 6) which converge to exactly three limits. Prove that your characterization is true.

9. Suppose $C$ is a collection of subsets of $X$ which is closed under finite intersections and has $\bigcup C = X$.

   (a) Give a proof or counterexample: $C$ must be a topology.

   (b) Give a proof or counterexample: $C$ must be a basis for a topology.

10. Give a proof or counterexample: If $\mathcal{B}$ is a basis for a topology, then $\mathcal{B}$ is closed under finite intersections.

11. Suppose $\mathcal{B}$ is a basis for topology $\mathcal{T}$ on $X$, and $C$ is any larger collection of open sets. (That is, suppose $\mathcal{B} \subseteq C \subseteq \mathcal{T}$.) Show that $C$ is also a basis for $\mathcal{T}$.

12. Suppose $\mathcal{B}$ is a basis for a topology on $X$, and $C$ is any larger collection of subsets of $X$, that is, $\mathcal{B} \subseteq C \subseteq P(X)$. Must $C$ be a basis for some topology on $X$? Give a proof or counterexample. (Compare to Exercise 11.)

13. Among all the topologies on a set $X$, is there always a finest topology? Is there always a coarsest one?

14. On $X = [-1, 1]$, consider the discrete topology $\mathcal{T}_D$ and the single transmitter topology $\mathcal{T}$. Carefully justify which (if either) of these topologies is finer than the other.

15. On $X = \mathbb{R}^2$, consider the metric topology $\mathcal{T}_{\text{taxi}}$ generated by the taxicab metric of Exercise 5 of Section 1.1 and the Euclidean topology $\mathcal{T}_\mathcal{E}$. Carefully justify which (if either) of these topologies is finer than the other.

16. Discuss sequential convergence in the digital line. Find the limits of all constant sequences. Describe all convergent sequences which are not eventually constant.

17. If $d : X \times X \to \mathbb{R}$ is a metric on $X$ and $k$ is a positive real number, then $kd : X \times X \to \mathbb{R}$ is also a metric on $X$. If $\mathcal{T}_d$ and $\mathcal{T}_{kd}$ are the metric topologies generated by these metrics, carefully justify which (if either) of these topologies is finer than the other.

18. Consider $X = \{1, 2, 3, 4\}$ with the topology generated by the basis $\{\{1, 2, 3\}, \{3\}, \{4\}\}$.
    (a) To which point or points of $X$, if any, does the sequence $(2, 3, 2, 3, 2, 3, 2, 3, \ldots)$ converge?
    (b) To which point or points of $X$, if any, does the sequence $(1, 1, 1, 1, 1, \ldots)$ converge?
    (c) Find a sequence in $X$ which does not converge.
    (d) Find a convergent sequence in $X$ which has a unique limit.

19. Can an increasing sequence $(x_n)_{n=1}^{\infty}$ of real numbers converge in the lower limit topology? What about a strictly increasing sequence?

20. Find a sequence $(y_n)_{n=1}^{\infty}$ which is not decreasing and not eventually decreasing but converges to 0 in the lower limit topology.

21. Let $X = [0, 1]$, $\mathcal{B}_1 = \{(a, b) : 0 \le a < b \le 1\}$, $\mathcal{B}_2 = \{\{0\} \cup (a, 1) : 0 \le a < 1\}$, and $\mathcal{B}_3 = \{(0, b) \cup \{1\} : 0 < b \le 1\}$.
    (a) Show that $\mathcal{B} = \mathcal{B}_1 \cup \mathcal{B}_2 \cup \mathcal{B}_3$ is a basis for a topology on $X$.
    (b) Is $\mathcal{T}_{\mathcal{B}}$ finer, coarser, neither, or equal to the Euclidean topology $\mathcal{T}_{\mathcal{E}}$ on $X$, which has a basis $\{B(x, \varepsilon) \cap X : x \in X, \varepsilon > 0\}$.
    (c) Find the limit of the sequence $(\frac{1}{n})_{n=1}^{\infty}$ in $(X, \mathcal{T}_{\mathcal{B}})$ and in $(X, \mathcal{T}_{\mathcal{E}})$.
    (d) Find a nonconstant sequence in $(X, \mathcal{T}_{\mathcal{B}})$ which converges to 0.
    (e) Give a geometric discussion of how $(X, \mathcal{T}_{\mathcal{B}})$ and $(X, \mathcal{T}_{\mathcal{E}})$ are related.

22. Prove: If $(x_n)_{n=1}^{\infty}$ converges to $a$ in a topological space $(X, \mathcal{T})$ and $\mathcal{T}_C$ is a topology on $X$ coarser than $\mathcal{T}$, then $(x_n)_{n=1}^{\infty}$ converges to $a$ in $(X, \mathcal{T}_C)$.

**Exercises 23–27** refer to these topologies. The **bow-tie topology** $\mathcal{T}_{BT}$ on $\mathbb{R}^2$ has a basis consisting of sets of form

$$BT((a, b), \varepsilon) = \{(x, y) : |y - b| < |x - a| < \varepsilon\} \cup \{(a, b)\},$$

and the **deleted radius topology** $\mathcal{T}_{DR}$ on $\mathbb{R}^2$ has a basis consisting of sets of form

$$DR((a, b), \varepsilon) = (B_{\mathcal{E}}((a, b), \varepsilon) - \{(x, y) : x = a\}) \cup \{(a, b)\},$$

where $B_{\mathcal{E}}(x, \varepsilon)$ is the Euclidean metric ball, as suggested in Figure 1.13.

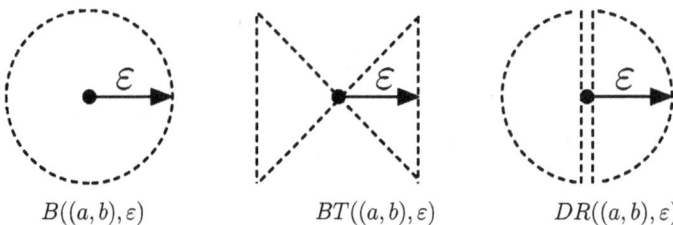

$B((a, b), \varepsilon)$ $\qquad\qquad$ $BT((a, b), \varepsilon)$ $\qquad\qquad$ $DR((a, b), \varepsilon)$

**Figure 1.13:** Basic open sets in the Euclidean, bow-tie, and deleted radius topologies.

23. Which of the topologies $\mathcal{T}_\mathcal{E}, \mathcal{T}_{BT}$, and $\mathcal{T}_{DR}$ are finer than which? If one of these topologies is finer than another, prove it. If one is not finer than another, prove it.

24. Verify that the basis indicated for the bow-tie topology is indeed a basis for a topology.

25. Verify that the basis indicated for the deleted radius topology is indeed a basis for a topology.

26. Consider the bow-tie, deleted radius, and Euclidean topologies on $\mathbb{R}^2$. Give examples of the sequences in $\mathbb{R}^2$ described or show that no such sequence can exist.

    (a) A sequence $(x_n)_{n=1}^\infty$ which converges in the Euclidean topology but not in the bow-tie topology.
    (b) A sequence $(x_n)_{n=1}^\infty$ which converges in the deleted radius topology but not in the bow-tie topology.
    (c) A sequence $(x_n)_{n=1}^\infty$ which converges in the bow-tie topology but not in the deleted radius topology.
    (d) A sequence $(x_n)_{n=1}^\infty$ which converges to two distinct limits in the bow-tie topology.
    (e) A sequence $(x_n)_{n=1}^\infty$ which is not eventually constant and converges to the same limit in all three topologies.
    (f) A sequence $(x_n)_{n=1}^\infty$ which converges in one of the three topologies and converges to a different limit in another of the three topologies.
    (g) A sequence $(x_n)_{n=1}^\infty$ which diverges in all three topologies.

27. Let $A = DR((0,0),5) \cup (\{0\} \times [-1,1]) \subseteq \mathbb{R}^2$, where $DR((0,0),5)$ is a basic open set in the deleted radius topology. Determine whether $A$ is open in (a) the Euclidean topology, (b) the deleted radius topology, and (c) the bow-tie topology. For each of these topologies in which $A$ is open, write $A$ as a union of basis elements from that topology.

28. Show that if two Hausdorff topologies on $X$ are comparable (that is, one is finer than the other) and $(x_n)_{n=1}^\infty$ is a sequence in $X$, then $(x_n)_{n=1}^\infty$ cannot converge to different limits in the different topologies. Why are the assumptions of Hausdorff and comparability necessary?

29. Prove Theorem 1.4.12 by showing that if $\mathcal{S}$ is a collection of subsets of $X$ with $\bigcup \mathcal{S} = X$, then $\mathcal{B}_\mathcal{S} = \{S_1 \cap \cdots \cap S_n : n \in \mathbb{N}, S_1, \ldots, S_n \in \mathcal{S}\}$ is a basis for a topology.

30. Determine which of the following collections of subsets of $\mathbb{R}$ is a subbasis for a topology on $\mathbb{R}$. For those that are subbases, give the associated basis and topology.

    (a) $\mathcal{S}_1 = \{(-\infty, -3), (0, \infty)\}$
    (b) $\mathcal{S}_2 = \{(-\infty, 3), (0, \infty)\}$
    (c) $\mathcal{S}_3 = \{(-\infty, 3), (0, \infty), (1, 5)\}$
    (d) $\mathcal{S}_4 = \{(-\infty, -3), (0, \infty), (-4, 4)\}$

## 1.5 Closure and interior

In our earlier discussion of metric spaces, we defined the boundary points of a set $A$ to be those points $x$ such that every ball $B(x, \varepsilon)$ centered at $x$ laps into $A$ and laps out of $A$. Then we defined a set to be open if it contains none of its boundary points, and closed if it contains all of its boundary points. We immediately noted that, in the metric space setting, a set is open if and only if its complement is closed. In this section, we will extend these concepts to topological spaces.

A topology on $X$ is a collection of subsets of $X$ which we call the open sets. We could, in a natural extension, define boundary points in terms of neighborhoods (taking the place of balls in our Euclidean metric approach), then define closed sets in terms of boundary points. However, because closed sets play a more fundamental role than boundary points, here we will define closed sets directly, without reference to boundary points.

**Definition 1.5.1.** Suppose $X$ is a topological space. A subset $F$ of $X$ is *closed* if and only if its complement $X - F$ is open.

An extremely important fact to recognize immediately is that closed does not mean "not open". A set may be open, closed, neither, or both. In the Euclidean topology on $\mathbb{R}$, the set $[0, 1)$ is not open and not closed, showing that "not open" is different from "closed" and "not closed" is different from "open". The sets $\emptyset$ and $\mathbb{R}$ are both open and closed.

Open sets are frequently denoted by the letters $U$ or $V$, from the German word *Umgebung* and the French word *voisinage* for "neighborhood". Closed sets are frequently denoted by the letter $F$, from the French word *fermé* for "closed".

By taking complements, statements about open sets can be transformed into statements about closed sets. For example, we know that if $U_1, \ldots, U_n$ are open in $X$, then the finite intersection $U_1 \cap \cdots \cap U_n$ is open. That is, if $X - U_1, \ldots, X - U_n$ are closed in $X$, then $X - (U_1 \cap \cdots \cap U_n)$ is closed, and applying De Morgan's law we have $(X - U_1) \cup \cdots \cup (X - U_n)$ is closed. That is, if $F_1, \ldots, F_n$ are closed, then $F_1 \cup \cdots \cup F_n$ is closed, proving that finite unions of closed sets are closed.

In short, taking complements toggles open sets to closed sets, unions to intersections, and intersections to unions. In a similar manner, the statement that arbitrary unions of open sets are open toggles to the fact that arbitrary intersections of closed sets are closed. Thus, the collections $\mathcal{T}$ of open sets and $\mathcal{F}$ of closed sets are each closed under finite unions and finite intersections, but generally only one ($\mathcal{T}$) is closed under arbitrary unions and only one ($\mathcal{F}$) is closed under arbitrary intersections. To help remember which is which, keep in mind the counterexamples from the Euclidean topology on $\mathbb{R}$ of the infinite intersection of open sets $\bigcap_{n=1}^{\infty} (\frac{-1}{n}, 1 + \frac{1}{n}) = [0, 1]$ which is not open, and the infinite union of closed sets $\bigcup_{n=2}^{\infty} [\frac{1}{n}, 1 - \frac{1}{n}] = (0, 1)$ which is not closed.

In the Euclidean space $\mathbb{R}$, the set $A = [0,1) \cup (3,5) \cup \{7\}$ is not closed, but we may obtain a closed set by adding some points. We could add all the points of $[1,3] \cup [5,6]$ to get the closed set $[1,6] \cup \{7\}$, but we could have obtained a closed set by adding fewer points. The minimal way we could obtain a closed set by adding points would be to add $\{1,3,5\}$ to $A$ to get $[0,1] \cup [3,5] \cup \{7\}$, which is the smallest closed set containing $A$. There is no smallest open set containing $A$, since there is no smallest neighborhood of 7. However, there is a largest open set contained in $A$, namely $(0,1) \cup (3,5)$. This discussion prompts our next definition.

**Definition 1.5.2.** Suppose $A$ is a subset of a topological space $X$. The *closure of $A$*, denoted cl $A$, is the intersection of all closed sets containing $A$. The *interior of $A$*, denoted int $A$, is the union of all open sets contained in $A$.

Note that because arbitrary intersections of closed sets are closed, the closure of $A$ is a closed set which contains $A$. Furthermore, cl $A$ is the smallest closed set which contains $A$, for if $F$ is any closed set containing $A$, then cl $A$ is the intersection of $F$ with the other closed sets containing $A$, so cl $A \subseteq F$. Similarly, the interior of $A$ is open, and is the largest open set contained in $A$.

We always have int $A \subseteq A \subseteq$ cl $A$, and these inclusions could be equality. Note that $A$ is closed if and only if $A =$ cl $A$. (If $A =$ cl $A$, then $A$ is closed since cl $A$ is a closed set. Conversely, if $A$ is closed, then $A$ is the smallest closed set containing $A$, so $A =$ cl $A$.) Similarly, $A$ is open if and only if $A =$ int $A$. Thus, int $A =$ cl $A$ if and only if $A =$ int $A =$ cl $A$ is both open and closed.

Furthermore, if $A \subseteq B$, then cl $A \subseteq$ cl $B$ and int $A \subseteq$ int $B$.

**Example 1.5.3.** Let $X = \{a,b,c,d\}$ with the topology $\mathcal{T} = \{\emptyset, \{a,b\}, \{b\}, \{b,c,d\}, X\}$, as shown on the left in Figure 1.14. Since the interior of set $A$ is the largest open set contained in $A$, we see that int$\{b,c,d\} = \{b,c,d\}$, int$\{b,c\} = \{b\}$, and int$\{c,d\} = \emptyset$. The closure of a set $A$ is the smallest closed set containing $A$, and to recognize these for this simple example, we may draw all the closed sets, as seen on the right of Figure 1.14. Now we can recognize that cl$\{b,c,d\} = X$, cl$\{c,d\} = \{c,d\}$, and cl$\{c\} = \{c,d\}$.

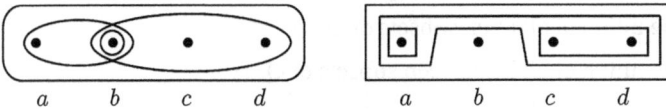

**Figure 1.14:** The $\mathcal{T}$-open sets and the $\mathcal{T}$-closed sets.

The following theorem provides one of the most powerful characterizations of closure and interior.

**Theorem 1.5.4.** *Suppose $X$ is a topological space and $A \subseteq X$.*
(a) *$x \in$ cl $A$ if and only if every neighborhood of $x$ intersects $A$.*
(b) *$x \in$ int $A$ if and only if there exists a neighborhood of $x$ contained in $A$.*

*Proof.* For (a), we will show the equivalent statement that $x \notin \mathrm{cl}\, A$ if and only if there exists a neighborhood of $x$ which does not intersect $A$. If $x \notin \mathrm{cl}\, A$, then $X - \mathrm{cl}\, A$ is a neighborhood of $x$ which does not intersect $A$. Conversely, if there exists a neighborhood $N$ of $x$ which does not intersect $A$, let $N'$ be an open neighborhood of $x$ contained in $N$. Now $X - N'$ is a closed set containing $A$ but not containing $x$. Thus, the smallest closed set containing $A$ does not contain $x$, so $x \notin \mathrm{cl}\, A$.

For (b), if $x \in \mathrm{int}\, A$, then $\mathrm{int}\, A$ is a neighborhood of $x$ contained in $A$. Conversely, if there exists a neighborhood $U$ of $x$ contained in $A$, there exists an open neighborhood $U'$ of $x$ with $U' \subseteq U \subseteq A$. Since $\mathrm{int}\, A$ is the union of all open sets contained in $A$, $x \in U'$ implies $x \in \mathrm{int}\, A$. $\qquad\square$

Since every open set is a union of basic open sets, it should be no surprise that the results hold if we only consider basic neighborhoods.

**Corollary 1.5.5.** *Suppose $X$ is a topological space with a specified basis and $A \subseteq X$.*
*(a) $x \in \mathrm{cl}\, A$ if and only if every basic neighborhood of $x$ intersects $A$.*
*(b) $x \in \mathrm{int}\, A$ if and only if there exists a basic neighborhood of $x$ contained in $A$.*

*Proof.* For (a), if every neighborhood of $x$ intersects $A$ then, in particular, every basic neighborhood of $x$ intersects $A$. Conversely, since every arbitrary neighborhood of $x$ contains a basic neighborhood of $x$, if all the basic neighborhoods of $x$ intersects $A$, then so must every arbitrary neighborhood. The proof of (b) is similar. $\qquad\square$

**Corollary 1.5.6.** *If $X$ is a topological space and $A \subseteq X$, then $X - \mathrm{cl}\, A = \mathrm{int}(X - A)$.*

*Proof.* Applying both parts of Theorem 1.5.4, we see that the following are equivalent:
(a) $x \in X - \mathrm{cl}\, A$. (b) there exists a neighborhood $U$ of $x$ which does not intersect $A$. (c) there exists a neighborhood $U$ of $x$ with $U \subseteq X - A$. (d) $x \in \mathrm{int}(X - A)$. $\qquad\square$

**Example 1.5.7.** In $\mathbb{R}$ with the Euclidean topology, $\mathrm{int}\, \emptyset = \emptyset = \mathrm{cl}\, \emptyset$, $\mathrm{int}\, \mathbb{R} = \mathbb{R} = \mathrm{cl}\, \mathbb{R}$, $\mathrm{int}((1,2] \cup [3,4)) = (1,2) \cup (3,4)$, and $\mathrm{cl}((1,2] \cup [3,4)) = [1,2] \cup [3,4]$. For the set $\mathbb{Q}$ of rational numbers,

$$\mathrm{int}\, \mathbb{Q} = \text{the union of all open subsets of } \mathbb{Q}$$
$$= \text{the union of all basic open subsets of } \mathbb{Q}$$
$$= \text{the union of all intervals } (x - \varepsilon, x + \varepsilon) \text{ contained in } \mathbb{Q}$$
$$= \emptyset.$$

Furthermore, $x \in \mathrm{cl}\, \mathbb{Q}$ if and only if every basic neighborhood $(x - \varepsilon, x + \varepsilon)$ of $x$ intersects $\mathbb{Q}$. Since every interval $(x - \varepsilon, x + \varepsilon)$ contains rational points, every real number $x$ is in the closure of $\mathbb{Q}$. That is, $\mathrm{cl}\, \mathbb{Q} = \mathbb{R}$. This is sometimes stated by saying $\mathbb{Q}$ is *dense* in $\mathbb{R}$.

**Definition 1.5.8.** In a topological space $X$, a set $D \subseteq X$ is *dense* in $X$ if $\mathrm{cl}\, D = X$.

Thus, $D$ is dense in $X$ if and only if every nonempty (basic) open set in $X$ intersects $D$. Since every open interval in $\mathbb{R}$ intersects the irrationals, the irrationals are dense in $\mathbb{R}$ with the Euclidean topology.

**Example 1.5.9.** Consider the right ray topology $\mathcal{T} = \{\emptyset, \mathbb{R}\} \cup \{(a, \infty) : a \in \mathbb{R}\}$ on $\mathbb{R}$. The closed sets are $\mathbb{R}, \emptyset$, and rays of form $(-\infty, a]$. The set $A = (0, 1)$ contains no nonempty open set $(a, \infty)$, so the largest open set it contains is $\text{int}(0, 1) = \emptyset$. The smallest closed set $(-\infty, a]$ containing $A$ is $\text{cl}\, A = (-\infty, 1]$. The set $B = \{1, 2, 3\} \cup [\pi, \infty)$ does contain nonempty open sets of form $(a, \infty)$; the largest one it contains is $\text{int}\, B = (\pi, \infty)$. Since $B$ is not contained in any closed set of form $(-\infty, a]$, the smallest closed set containing $B$ is $\text{cl}\, B = \mathbb{R}$. The set $\mathbb{Q}$ contains no nonempty open sets $(a, \infty)$ and is contained in no proper closed set $(-\infty, a]$, so $\text{int}\, \mathbb{Q} = \emptyset$ and $\text{cl}\, \mathbb{Q} = \mathbb{R}$. Thus, the sets $B$ and $\mathbb{Q}$ are dense in $\mathbb{R}$ with the right ray topology.

**Example 1.5.10.** Consider the cofinite topology on $\mathbb{R}$. Recall that $U$ is open if and only if $\mathbb{R} - U$ is finite or $U = \emptyset$. Here, the open sets are defined in terms of their complements, the closed sets. Thus, in the cofinite topology, $F$ is closed if and only if $F$ is finite or $F = \mathbb{R}$. If $A$ is infinite, the only closed set containing $A$ is $\mathbb{R}$, so $\text{cl}\, A = \mathbb{R}$. Thus, any infinite subset of $\mathbb{R}$ is dense in $\mathbb{R}$ with the cofinite topology.

For interiors, $\text{int}(0, 7] = \emptyset$ since the interval $(0, 7]$ contains no set whose complement is finite. Since $\text{int}\, A$ is an open set contained in $A$, if $\text{int}\, A$ is nonempty, then $A$ contains a cofinite set and thus $A$ is cofinite. That is, in $\mathbb{R}$ with the cofinite topology, the only sets with nonempty interiors are the nonempty open sets.

The points of metric spaces, and in particular Euclidean spaces $\mathbb{R}^n$, have countable neighborhood bases $\{B(x, \frac{1}{n}) : n \in \mathbb{N}\}$. This allows us to characterize the closure of a set in terms of sequences.

**Theorem 1.5.11.** *If $(X, d)$ is a metric space and $A \subseteq X$, then $x \in \text{cl}\, A$ if and only if there exists a sequence in $A$ converging to $x$.*

*Proof.* Suppose $(X, d)$ is a metric space, $A \subseteq X$, and $x \in \text{cl}\, A$. Then, for any $n \in \mathbb{N}$, the basic neighborhood $B(x, \frac{1}{n})$ intersects $A$, so we may choose $x_n \in A \cap B(x, \frac{1}{n})$. Now the sequence $(x_n)_{n=1}^{\infty}$ is a sequence in $A$ and converges to $x$, for every neighborhood $U$ of $x$ contains a basic neighborhood $B(x, \frac{1}{k})$, and $x_n \in U$ for all $n \geq k$. Conversely, suppose there is a sequence in $A$ converging to $x$. Then every neighborhood of $x$ contains some tail of that sequence, and thus contains points of $A$. Thus, $x \in \text{cl}\, A$. $\square$

Closure may be thought of as a function $\text{cl} : \mathcal{P}(X) \to \mathcal{P}(X)$ which maps a subset $A \subseteq X$ to its closure $\text{cl}\, A \subseteq X$. Union and intersection are operations on $\mathcal{P}(X)$. If $S$ is a set and $*$ an operation on $S$, recall the algebraic terminology that a function $\varphi : S \to S$ *preserves the operation* $*$ if $\varphi(a * b) = \varphi(a) * \varphi(b)$ for all $a, b \in S$. The closure function does not preserve the operation of intersection. That is, in general, $\text{cl}(A \cap B) \neq \text{cl}\, A \cap \text{cl}\, B$. For example, in the Euclidean line $\mathbb{R}$, with $A = (0, 1)$ and $B = (1, 2)$, note that $\text{cl}(A \cap B) = \text{cl}\, \emptyset = \emptyset$, which does not equal $\text{cl}\, A \cap \text{cl}\, B = [0, 1] \cap [1, 2] = \{1\}$. While the closure function does not preserve the operation of intersection, it does preserve the operation of union. The interior function int : $\mathcal{P}(X) \to \mathcal{P}(X)$ which maps a subset $A \subseteq X$ to int $A \subseteq X$ preserves intersections. The interactions between closure and interior with unions and intersections are given below. The proofs are left to the exercises.

**Theorem 1.5.12.** *Suppose $X$ is a topological space, and $A, B \subseteq X$.*
(a) $\mathrm{cl}(A \cap B) \subseteq \mathrm{cl}\, A \cap \mathrm{cl}\, B.$
(b) $\mathrm{cl}(A \cup B) = \mathrm{cl}\, A \cup \mathrm{cl}\, B.$
(c) $\mathrm{int}(A \cap B) = \mathrm{int}\, A \cap \mathrm{int}\, B.$
(d) $\mathrm{int}(A \cup B) \supseteq \mathrm{int}\, A \cup \mathrm{int}\, B.$

### 1.5.1 Boundary points

In a metric space, we defined $x$ to be a boundary point of $A$ if and only if every ball centered at $x$ intersects both $A$ and $X - A$. This suggests the definition of boundary points in an arbitrary topological space.

**Definition 1.5.13.** If $A$ is a subset of a topological space $X$, then $x \in X$ is a *boundary point of $A$* if and only if every neighborhood of $x$ intersects both $A$ and $X - A$. The set of all the boundary points of $A$ is called the *boundary of $A$*, denoted $\partial A$.

Again we note that an equivalent definition would be that $x \in X$ is a boundary point of $A$ if and only if every *basic* neighborhood of $x$ with respect to a given basis (and in particular, every *open* neighborhood) intersects both $A$ and $X - A$. Clearly if every neighborhood intersects $A$ and $X - A$, then every basic neighborhood does. If every basic neighborhood intersects $A$ and $X - A$, then every neighborhood does, since every neighborhood of $x$ contains a basic neighborhood of $x$.

From Definition 1.5.13, Theorem 1.5.4, and Theorem 1.5.11, we have the following corollary.

**Corollary 1.5.14.** *If $A$ is a subset of a topological space $X$, then $\partial A = \mathrm{cl}\, A \cap \mathrm{cl}(X - A)$. Furthermore, if $X$ is a metric space, then $x \in \partial A$ if and only if there exist sequences $(a_n)_{n=1}^{\infty}$ in $A$ and $(b_n)_{n=1}^{\infty}$ in $X - A$ which both converge to $x$.*

The following result confirms our expectations.

**Theorem 1.5.15.** *Suppose $A$ is a subset of a topological space $X$.*
(a) *$A$ is open if and only if it contains none of its boundary points.*
(b) *$A$ is closed if and only it contains all of its boundary points.*

*Proof.* (a) Suppose $A$ is open and $b \in \partial A$. Then every neighborhood of $b$ intersects $X - A$, so no neighborhood of $b$ is contained in the open set $A$. Thus, $b \notin A$. Conversely, suppose that $A$ contains none of its boundary points, and $a \in A$. Since $a$ is not a boundary point of $A$, there exists a neighborhood $N$ of $a$ which does not intersect both $A$ and

$X - A$. Because $N$ contains $a \in A$, $N$ intersects $A$ and thus $N$ does not intersect $X - A$. That is, $N \subseteq A$. We have shown that, for every $a \in A$, there exists a neighborhood $N$ of $a$ with $N \subseteq A$, which shows that $A$ is open.

(b) Suppose $A$ is closed and $b \in \partial A$. Then every neighborhood of $b$ intersects $A$, so $b \in \operatorname{cl} A = A$. Thus, $A$ contains all of its boundary points. Conversely, suppose $A$ contains all of its boundary points. If $x \in \operatorname{cl} A$, then either there exists a neighborhood of $x$ contained in $A$, in which case $x \in A$, or every neighborhood of $x$ intersects $X - A$, in which case $x \in \partial A \subseteq A$. Thus, $\operatorname{cl} A \subseteq A$, so $A = \operatorname{cl} A$ is closed. □

The following theorem is similar. Its proof is left to the exercises.

**Theorem 1.5.16.** *Suppose $A$ is a subset of a topological space $X$.*
(a) $\operatorname{int} A = A - \partial A$.
(b) $\operatorname{cl} A = \operatorname{int} A \cup \partial A$.
(c) $\partial A = \operatorname{cl} A - \operatorname{int} A$.

**Example 1.5.17.** For each $n \in \mathbb{Z}$, suppose a metal rod fills the interval $(n, n + 1)$, and the rods are connected by insulated couplings at each $n \in \mathbb{Z}$. For the transfer of heat or electricity, the usual Euclidean nearness applies between the couplings, but if an electrical current is applied to a coupling at $x = n \in \mathbb{Z}$, the current does not transfer to any other points. This situation is modeled by the topology $\mathcal{T}$ on $\mathbb{R}$ having basis $\mathcal{B} = \{(a, b) \subseteq \mathbb{R} : a < b\} \cup \{\{n\} : n \in \mathbb{Z}\}$. (Another basis for this topology would be $\{(a, b) \subseteq \mathbb{R} : \exists n \in \mathbb{Z} \text{ with } (a, b) \subseteq (n, n + 1)\} \cup \{\{n\} : n \in \mathbb{Z}\}$.)

In this topology, $A = [0, 1] = \{0\} \cup (0, 1) \cup \{1\}$ is a union of basis elements, and therefore is open. The complement of $A$ can also be expressed as a union of open sets, so $A$ is also closed. Thus, $A$ must contain all of its boundary points and none of its boundary points, so $\partial A = \partial[0, 1] = \emptyset$. For example, $0 \notin \partial[0, 1]$ since the neighborhood $\{0\}$ of $0$ does not intersect $X - A = \mathbb{R} - [0, 1]$. Formally, since $A$ is open and closed, $A = \operatorname{int} A = \operatorname{cl} A$, so Theorem 1.5.16 tells us $\partial A = \operatorname{cl} A - \operatorname{int} A = \emptyset$.

For $B = [0, \pi)$, we have $\partial B = \partial[0, \pi) = \{\pi\}$. Applying Theorem 1.5.16, we see $\operatorname{int}[0, \pi) = [0, \pi) - \partial[0, \pi) = [0, \pi) - \{\pi\} = [0, \pi)$, so $[0, \pi)$ is open. Also, $\operatorname{cl}[0, \pi) = [0, \pi) \cup \partial[0, \pi) = [0, \pi) \cup \{\pi\} = [0, \pi]$.

In general, in this topology, the boundary of an interval $I$ of form $(a, b)$, $[a, b)$, $(a, b]$, or $[a, b]$ is $\partial I = \{a, b\} - \mathbb{Z}$. Thus, the closure $\operatorname{cl} I = I \cup \partial I$ of such an interval keeps the included endpoints but only adds non-integer endpoints. The interior $\operatorname{int} I = I - \partial I$ only removes the non-integer endpoints.

## Exercises

1. Use De Morgan's laws to prove that arbitrary intersections of closed sets are closed.
2. Find a set that is neither open nor closed in the cofinite topology, showing that "not open" does not imply closed and "not closed" does not imply open.

3. Suppose $A \subseteq B$. Carefully justify that $\operatorname{cl} A \subseteq \operatorname{cl} B$ and $\operatorname{int} A \subseteq \operatorname{int} B$.
4. Show that $U$ is a neighborhood of $x$ if and only if $x \in \operatorname{int} U$.
5. In the proof of Theorem 1.5.4, we state "If $x \notin \operatorname{cl} A$, then $X - \operatorname{cl} A$ is a neighborhood of $x$ which does not intersect $A$." Carefully justify this statement.
6. Show that both parts of Corollary 1.5.5 fail if the basis and basic neighborhoods are replaced by a subbasis and subbasic neighborhoods.
7. Sketch the subsets of the Euclidean plane $\mathbb{R}^2$ below and find the interior, closure, and boundary of each. $B_\varepsilon((x,y),\varepsilon)$ represents a ball in the Euclidean metric on $\mathbb{R}^2$.
   (a) $A = [0,1) \times (0,2)$
   (b) $B = B_\varepsilon((0,0),1) \cup ([-2,2] \times \{0\})$
   (c) $C = (\mathbb{Q} \cup (0,1)) \times (\mathbb{Q} \cup (0,1))$
   (d) $D = \{(x,y) \in \mathbb{R}^2 : y = \sin x\}$
   (e) $E = \{(\frac{1}{n}, \frac{1}{n}) \in \mathbb{R}^2 : n \in \mathbb{N}\}$
   (f) $F = \bigcup_{x\in[0,2]} B_\varepsilon((x,0),1)$
8. In the topology on $X = \{1,2,3,4,5\}$ whose basis is shown, find the interior, closure, and boundary of each set below.
   (a) $\{1,2,3\}$      (b) $\{4,5\}$      (c) $\{3,4\}$      (d) $\{5\}$

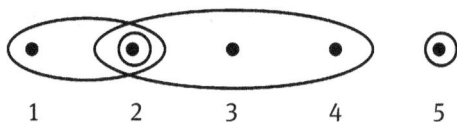

9. In the topology on $X = \{1,2,3,4,5\}$ whose basis is shown, find the interior, closure, and boundary of each set below.
   (a) $\{1,2,3\}$      (b) $\{4,5\}$      (c) $\{3,4\}$      (d) $\{5\}$

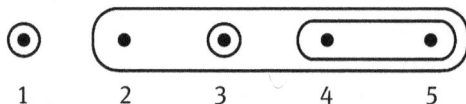

10. In the digital line, find the interior, closure, and boundary of the following sets.
    (a) $A = \{1,2,3\}$
    (b) $B = \{2,3,4\}$
    (c) $\mathbb{D} = \{2n+1 : n \in \mathbb{Z}\}$
    (d) $\mathbb{E} = \{2n : n \in \mathbb{Z}\}$
11. Give $\mathbb{N}$ the *multiples topology* having basis $\mathcal{B} = \{M_n : n \in \mathbb{N}\}$, where for $n \in \mathbb{N}$, $M_n = \{kn : k \in \mathbb{N}\}$ is the set of positive multiples of $n$.
    (a) Find the interior, closure, and boundary of $A = \{3,4,5\}$.
    (b) Find the interior, closure, and boundary of $P = \{p \in \mathbb{N} : p \text{ is prime}\}$.
12. In $\mathbb{R}$ with the right ray topology $\mathcal{T} = \{\emptyset, \mathbb{R}\} \cup \{(a,\infty) : a \in \mathbb{R}\}$, find the interior, closure, and boundary of the sets below.

(a) $A = (-\infty, 0) \cup [1, \infty)$

(b) $B = (-\infty, 2)$

(c) $C = \mathbb{N}$

13. Consider $\mathbb{R}$ with the right ray topology $\mathcal{T} = \{\emptyset, \mathbb{R}\} \cup \{(a, \infty) : a \in \mathbb{R}\}$.

(a) Give necessary and sufficient conditions for a subset to be dense.

(b) Give necessary and sufficient conditions for a subset to have empty interior.

14. Consider $\mathbb{R}$ with the right ray topology $\mathcal{T} = \{\emptyset, \mathbb{R}\} \cup \{(a, \infty) : a \in \mathbb{R}\}$.

(a) Give a proof or counterexample: If $\operatorname{cl} A \neq \mathbb{R}$ then $\operatorname{int} A = \emptyset$.

(b) Give a proof or counterexample: If $\operatorname{int} A = \emptyset$ then $\operatorname{cl} A \neq \mathbb{R}$.

15. In $\mathbb{R}$ with the cofinite topology, find the interior, closure, and boundary of the sets $A = \{1, 2, \pi\}$ and $B = \mathbb{N}$.

16. Let $X = [-1, 1]$ have the single transmitter topology $\mathcal{T} = \{U \subseteq X : 0 \in U \Rightarrow (-1, 1) \subseteq U\}$. Find the interior, closure, and boundary of the sets $A = \{1\}$, $B = [0.2, 0.3)$, and $C = \{0\}$.

17. Let $X = [-1, 1]$ have the single transmitter topology $\mathcal{T} = \{U \subseteq X : 0 \in U \Rightarrow (-1, 1) \subseteq U\}$.

(a) Show that a nonzero point $x$ is never in the boundary of any set, so $\partial A \subseteq \{0\}$ for every $A \subseteq X$.

(b) Find all sets $A \subseteq X$ which have empty boundary.

18. Show that if $A \subseteq B \subseteq \operatorname{cl} A$, then $\operatorname{cl} B = \operatorname{cl} A$.

19. In $\mathbb{R}$ with the Euclidean topology, let $S = \mathbb{Q} \cap (-1, 1)$.

(a) Show that $\partial(\partial S) \neq \partial S$. (Note: In any topological space $X$, closure and interior are *idempotent* operators; that is, $\operatorname{cl}(\operatorname{cl} A) = \operatorname{cl} A$ and $\operatorname{int}(\operatorname{int} A) = \operatorname{int} A$ for every $A \subseteq X$. This shows that the boundary operator is not idempotent.)

(b) For the set $A = [0, 1)$ in the Euclidean line (or $A = [0, 1) \times (0, 2]$ in the Euclidean plane), if $\operatorname{int} A \subseteq B \subseteq \operatorname{cl} A$, then $A$ and $B$ have the same interior, closure, and boundary. This need not be true in general. With $S$ as above, find a set $B$ in $\mathbb{R}$ with $\operatorname{int} S \subseteq B \subseteq \operatorname{cl} S$ whose interior, closure, and boundary are all distinct from the interior, closure, and boundary of $S$.

20. Prove: If $x$ is any point in a Hausdorff space $X$, then $\{x\}$ is closed.

21. Show that if $D$ is dense in $X$ and $D \subseteq E \subseteq X$, then $E$ is dense in $X$.

22. A topological space $X$ is *resolvable* if it has two disjoint dense subsets. Consider the Euclidean, discrete, indiscrete, cofinite, and right ray topologies on $\mathbb{R}$. Which of these topologies make $\mathbb{R}$ a resolvable space?

23. Provide a proof or counterexample: If $A \subseteq B$, then $\partial A \subseteq \partial B$.

24. (a) Find two sets $A$ and $B$ for which $\partial(A \cap B)$ is a proper subset of $\partial A \cap \partial B$.

(b) Find two sets $C$ and $D$ for which $\partial C \cap \partial D$ is a proper subset of $\partial(C \cap D)$.

25. Prove Theorem 1.5.12(a): For any sets $A$ and $B$ in a topological space, $\operatorname{cl}(A \cap B) \subseteq \operatorname{cl} A \cap \operatorname{cl} B$. Give an example where equality fails.

26. Prove Theorem 1.5.12(b): For any sets $A$ and $B$ in a topological space, $\operatorname{cl}(A \cup B) = \operatorname{cl} A \cup \operatorname{cl} B$. (Hint: Use the contrapositive for one direction.)

27. Prove Theorem 1.5.12(c): For any sets $A$ and $B$ in a topological space, $\text{int}(A \cap B) = \text{int}\,A \cap \text{int}\,B$.

28. Prove Theorem 1.5.12(d): For any sets $A$ and $B$ in a topological space, $\text{int}(A \cup B) \supseteq \text{int}\,A \cup \text{int}\,B$. Give an example where equality fails.

29. Prove Theorem 1.5.16(a): For a subset $A$ of a topological space $X$, $\text{int}\,A = A - \partial A$.

30. Prove Theorem 1.5.16(b): For a subset $A$ of a topological space $X$, $\text{cl}\,A = \text{int}\,A \cup \partial A$.

31. Prove Theorem 1.5.16(c): For a subset $A$ of a topological space $X$, $\partial A = \text{cl}\,A - \text{int}\,A$. Then show that $\partial A$ is always a closed set.

32. Suppose $A$ is a subset of the Euclidean line $\mathbb{R}$ and $\inf A$ exists. Show that $\inf A \in \partial A$ (and hence, $\inf A \in \text{cl}\,A$).

33. Determine which statements below are true. Prove those that are true and give counterexamples for those that are false.
    (a) $\partial A = \partial(\text{cl}\,A)$
    (b) $\text{cl}(X - A) = \text{cl}(X - \text{cl}\,A)$
    (c) $\partial(\text{cl}\,A) \subseteq \partial A$

34. Suppose $X$ is a topological space, $a \in X$, $\{a\}$ is not open, and $U$ is a neighborhood of $a$. Show that $a \in \partial(U - \{a\})$, and thus $a \in \text{cl}(U - \{a\})$.

35. Consider the topology on $\mathbb{R}$ having basis $\{(a, b) \subseteq \mathbb{R} : \exists n \in \mathbb{Z} \text{ with } (a, b) \subseteq (n, n + 1)\} \cup \{\{n\} : n \in \mathbb{Z}\}$, discussed in Example 1.5.17. Give a metric $d$ on $\mathbb{R}$ which generates this topology, and prove that $d$ is indeed a metric.

36. Give the appropriate topology which models the situation described.
    (a) A packaging facility can set its equipment to pack rice in bags to be sold as "$x$ pound" bags, for any $x \in [.25, 15.25]$. Bags are close to $x$-pounds in the usual sense, except that underweight bags cannot be shipped, and thus are not considered close to the correct weight.
    (b) Nearness to a point in the plane is determined by remote imaging from a camera mounted on a north–south cable positioned at that point. The mounting prevents a camera angle of due north or due south.
    (c) An app checks your password and allows access based on the closeness of your input to the correct password. Describe the topology used to measure closeness.
    (d) An app asks the user to "press any key to proceed". Describe the topology used to measure closeness of the entry to the correct response.
    (e) At a security gate of Bill Gates' compound, a nonempty set of people are permitted to enter if and only if they are accompanied by Bill Gates.
    (f) A game has nine levels, which must be completed sequentially. Describe the topology on the set of possible fully-completed levels attainable.
    (g) A building has eight rooms, each with one light which illuminates the entire room and only that room. On the set of points in the building, describe the topology of all possible sets which may be simultaneously illuminated.

37. (The infinitude of primes) In 1955, H. Fürstenberg [18] gave the topological proof outlined below that there are infinitely many primes. (See [35] for other proper-

ties of this topology.) For integers $a$ and $d$ with $d \neq 0$, the *arithmetic progression* containing $a$ with common difference $d$ is

$$AP(a, d) = \{a + nd : n \in \mathbb{Z}\} = \{m \in \mathbb{Z} : m \equiv a \bmod d\}.$$

(a) Show that $\mathcal{B} = \{AP(a, d) : a, d \in \mathbb{Z}, d \neq 0\}$ is a basis for a topology $\mathcal{T}$ on $\mathbb{Z}$.
(b) Show that

$$\mathbb{Z} \setminus \{1, -1\} = \bigcup_{p \text{ a prime}} AP(p, p).$$

(c) Show that if $p$ is prime, $AP(p, p)$ is $\mathcal{T}$-closed.
(d) Show that if there are only finitely many primes $p$, then (b) implies $\{1, -1\}$ is a $\mathcal{T}$-open set. Explain why this is a contradiction.

## 1.6 Limit points

Consider the set $A = \{\frac{1}{n} : n \in \mathbb{N}\} \cup [2, 3) \cup \{4\}$ in the Euclidean line $\mathbb{R}$. The closure of $A$ is $\operatorname{cl} A = A \cup \{0, 3\}$. Since $4 \in \operatorname{cl} A$, every neighborhood of 4 intersects $A$, but some just barely intersect $A$. For example, $B(4, 0.1) \cap A = (3.9, 4.1) \cap A = \{4\}$. Similarly, for any given $n \in \mathbb{N}$, every neighborhood of $\frac{1}{n}$ intersects $A$ (so $\frac{1}{n} \in \operatorname{cl} A$), but the neighborhoods of $\frac{1}{n}$ which are contained in $(\frac{1}{n+1}, \frac{1}{n-1})$ intersect $A$ only in the one point $\frac{1}{n}$. Other points, such as $x = 0$, $x = 2.5$, or $x = 3$ have the property that not only does every neighborhood of $x$ intersect $A$, but every neighborhood of $x$ intersects $A$ in a point other than $x$ itself. These are the *limit points of $A$*.

**Definition 1.6.1.** Suppose $X$ is a topological space and $A \subseteq X$. A *limit point of $A$* is a point $x \in X$ such that every neighborhood of $x$ intersects $A$ in a point other than $x$. The set of limit points of a set $A$ is called the *derived set* of $A$, denoted $A'$.

More formally, we could say $x$ is a limit point of $A$ if and only if for every neighborhood $N$ of $x$, $(N \cap A) - \{x\} \neq \emptyset$. Now $(N \cap A) - \{x\} = (N - \{x\}) \cap A$. A set of the form $N - \{x\}$ where $N$ is a neighborhood of $x$ is called *deleted neighborhood* (or *punctured neighborhood)* of $x$. With this terminology, we see that $x$ is a limit point of $A$ if and only if every deleted neighborhood of $x$ intersects $A$. Also, since $(N \cap A) - \{x\} = N \cap (A - \{x\})$, we see that $x$ is a limit point of $A$ if and only if $x \in \operatorname{cl}(A - \{x\})$. Limit points are also called *accumulation points* or *cluster points*.

One of the motivations for limit points is the fundamental result that the closure of a set is obtained by adding the limit points.

**Theorem 1.6.2.** *If $X$ is a topological space and $A \subseteq X$, then $\operatorname{cl} A = A \cup A'$.*

*Proof.* Clearly $A \subseteq \operatorname{cl} A$, and if $x \in A'$, then every (deleted) neighborhood of $x$ intersects $A$, so $x \in \operatorname{cl} A$. Thus $A \cup A' \subseteq \operatorname{cl} A$. To show $\operatorname{cl} A \subseteq A \cup A'$, we need only show that cl $A - A \subseteq A'$. If $x \in$ cl $A - A$, then every neighborhood of $x$ intersects $A$ and $x$ is not in this intersection, so every neighborhood of $x$ intersects $A$ in a point other than $x$. This says $x \in A'$, as needed.  □

The previous theorem says cl $A = A \cup A'$, and Theorem 1.5.16 says cl $A = A \cup \partial A$. Unfortunately, there is little correlation between limit points and boundary points. For $A = (0,1) \cup \{2\}$ in the Euclidean line, the boundary $\partial A$ of $A$ is the three point set $\{0,1,2\}$, while the derived set $A'$ is the infinite set $[0,1]$. In particular, a limit point such as $\frac{1}{2}$ need not be a boundary point, and a boundary point such as 2 need not be a limit point.

Limit points are related to limits of sequences, at least for metric spaces, as we see next. (More generally, limit points are related to limits, but there are non-metric spaces in which convergence cannot always be determined by sequences—the countable indexing of a sequence may be inadequate in such spaces.)

**Theorem 1.6.3.** *In a metric space $(X,d)$, $x$ is a limit point of $A \subseteq X$ if and only if $x$ is the limit of a sequence $(a_i)_{i=1}^{\infty}$ of distinct points in $A$.*

One may ask whether a limit point of the set of limit points of $A$ is a limit point of $A$. That is, is $A'' \subseteq A'$? This need not be true, but is true in certain spaces, including all Hausdorff spaces. These concepts are investigated in Exercises 9 and 10.

## Exercises

1. Prove Theorem 1.6.3.
2. In the finite topological space $X$ of Exercise 8 of Section 1.5, find the derived set of each subset of $X$ given there.
3. In the finite topological space $X$ of Exercise 9 of Section 1.5, find the derived set of each subset of $X$ given there.
4. Consider $\mathbb{R}$ with the cofinite topology. Suppose $A \subseteq \mathbb{R}$ is finite and $B \subseteq \mathbb{R}$ is infinite. Find $A'$ and $B'$.
5. Consider the digital line $\mathbb{Z}$.
   (a) Describe all sets $A$ which have 2 as a limit point.
   (b) Describe all sets $A$ which have 3 as a limit point.
   (c) Find the set of all limit points of $\mathbb{D} = \{2n + 1 : n \in \mathbb{Z}\}$ and $\mathbb{E} = \{2n : n \in \mathbb{Z}\}$.
6. Show that if the singleton set $\{x\}$ is open in $X$ (that is, $x$ is an *isolated point* of $X$), then $x \notin A'$ for any $A \subseteq X$.
7. Suppose $X$ is a topological space and $A, B \subseteq X$. Prove the following. Compare the results to the analogous statements about closures.
   (a) $\emptyset' = \emptyset$.
   (b) If $A \subseteq B$, then $A' \subseteq B'$.
   (c) $(A \cap B)' \subseteq A' \cap B'$, and equality may fail.
   (d) $(A \cup B)' \subseteq A' \cup B'$.

8. (a) Show that if $X$ is a Hausdorff space and $a \in X$, then $\{a\}' = \emptyset$.
   (b) Find a set $A$ in a topological space $X$ with $A'' \neq A'$, proving that unlike the closure operator, the derived set operator is not idempotent.

9. Suppose $X$ is a topological space with the property that every singleton set $\{x\}$ in $X$ is closed. (Hausdorff spaces, including metric spaces, have this property. See Exercise 20 of Section 1.5.)
   (a) Show that $x$ is a limit point of $A \subseteq X$ if and only if every neighborhood of $x$ intersects $A$ in infinitely many points.
   (b) If $A \subseteq X$, show that $A'' \subseteq A'$.
   (c) Show that $A'$ is closed for every $A \subseteq X$.

10. Let $X = \{1, 2, 3\}$ with the indiscrete topology.
    (a) Find the derived set $A'$ for each of the eight subsets of $X$.
    (b) Find a set $A \subseteq X$ with $A'' \not\subseteq A'$.
    (c) For which subsets $A$ is $A'$ closed?

# 2 New topologies from existing topologies

## 2.1 Subspaces

If we know how to add real numbers then, since $\mathbb{Q} \subseteq \mathbb{R}$, we know how to add rational numbers. The operation of addition on $\mathbb{Q}$ is "inherited" from the larger set $\mathbb{R}$. Similarly, if we know a topology on a set $X$ and $Y \subseteq X$, then $Y$ should inherit a topological structure from $X$. The inherited topology is the *subspace topology* on $Y$.

**Definition 2.1.1.** Suppose $(X, \mathcal{T})$ is a topological space and $Y \subseteq X$. The *subspace topology* on $Y$ is

$$\mathcal{T}_Y = \{U \cap Y : U \in \mathcal{T}\}.$$

Thus, the open sets in the subspace $Y \subseteq X$ are open sets in $X$ restricted to (that is, intersected with) $Y$. The subspace topology on $Y \subseteq X$ is also called the topology inherited from $X$. By saying $Y$ is a *subspace* of $X$, we imply that $Y$ carries the subspace topology inherited from $X$.

It is easy to show that $\emptyset, Y \in \mathcal{T}_Y$, $\bigcap_{i=1}^{n}(U_i \cap Y) = (\bigcap_{i=1}^{n} U_i) \cap Y$, and for arbitrary index sets $I$, $\bigcup_{i \in I}(U_i \cap Y) = (\bigcup_{i \in I} U_i) \cap Y$, showing that $\mathcal{T}_Y$ is indeed a topology.

As always, in determining whether a set is open or not, the answer depends on the topological space $(X, \mathcal{T})$ involved. This is easy to overlook in dealing with subspaces. The interval $[0, 1)$ is not open in $\mathbb{R}$ with the Euclidean topology, but $[0, 1)$ is open in the subspace $Y = [0, \infty)$ of $\mathbb{R}$ with the Euclidean topology, since $[0, 1) = (-1, 1) \cap Y$ is realized as the restriction of an open set $(-1, 1)$ in $\mathbb{R}$ to $Y$. If $[0, 1)$ is open in $Y$, then it contains none of its boundary points. The point $0$ is not a boundary point of $[0, 1)$ in the subspace $Y$ since, for example, $[0, \frac{1}{4}) = (-\frac{1}{4}, \frac{1}{4}) \cap Y$ is a neighborhood of $0$ which does not intersect $Y - [0, 1)$.

The following theorem tells us that closed sets in the subspace $Y$ are also recognizable from the topology on $X$ in a natural way.

**Theorem 2.1.2.** *Suppose $Y$ is a subspace of $X$. Then $C$ is closed in the subspace topology on $Y$ if and only if $C$ has the form $F \cap Y$ for some set $F$ closed in $X$.*

*Proof.* Figure 2.1 suggests the proof, which formally follows from the equivalence of the following statements.

$C$ is closed in the subspace $Y$.
$C = Y - (U \cap Y)$ where $U$ is open in $X$.
$C = Y - U = Y \cap (X - U)$ where $U$ is open in $X$.
$C = Y \cap F$ where $F = X - U$ is closed in $X$. $\qquad\qquad\square$

If $Y$ is a subspace of $X$, then the open sets in $Y$ are determined by the open sets in $X$, which are determined by the basic open sets from any basis $\mathcal{B}$. Thus, the open sets in the subspace $Y$ are determined by the basic open sets in $X$. The theorem below makes this relation precise.

https://doi.org/10.1515/9783110686579-003

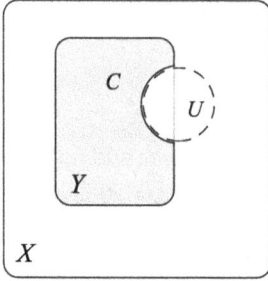

**Figure 2.1:** $C = Y - (U \cap Y)$ if and only if $C = (X - U) \cap Y$.

**Theorem 2.1.3.** *If $\mathcal{B}$ is a basis for a topology $\mathcal{T}$ on set $X$ and $Y \subseteq X$, then*

$$\mathcal{B}_Y = \{B \cap Y : B \in \mathcal{B}\}$$

*is a basis for the subspace topology $\mathcal{T}_Y$ on $Y$.*

*Proof.* Recall that $\mathcal{B}_Y$ is a basis for $\mathcal{T}_Y$ if and only if $\mathcal{B}_Y$ is a collection of $\mathcal{T}_Y$-open sets such that every set open in $\mathcal{T}_Y$ is a union of sets in $\mathcal{B}_Y$. Clearly $\mathcal{B}_Y$ is a collection of $\mathcal{T}_Y$-open sets. If $U_Y$ is open in $\mathcal{T}_Y$, then $U_Y = U \cap Y$ for some $U \in \mathcal{T}$. Since $\mathcal{B}$ is a basis for $\mathcal{T}$, we may write $U$ as a union $\bigcup_{i \in I} B_i$ of basis elements $B_i \in \mathcal{B}$. Now $U_Y = U \cap Y = (\bigcup_{i \in I} B_i) \cap Y = \bigcup_{i \in I} (B_i \cap Y)$, which is a union of elements from $\mathcal{B}_Y$. □

**Example 2.1.4.** Consider the plane $\mathbb{R}^2$ with the deleted radius topology, having a basis of sets of form $\mathrm{DR}((a, b), \varepsilon) = (B_{\mathcal{E}}((a, b), \varepsilon) - \{(x, y) : x = a\}) \cup \{(a, b)\}$, where $B_{\mathcal{E}}((a, b), \varepsilon)$ is the Euclidean $\varepsilon$-ball centered at $(x, y)$. We may view any line in $\mathbb{R}^2$ as a copy of $\mathbb{R}$ embedded in the plane, and as a subset of the plane, the line inherits the subspace topology. If the line is not vertical, the basic neighborhoods in the deleted radius topology intersect the line in an open interval $(a, b)$ or the union $(a, b) \cup (b, c)$ of two open intervals sharing an endpoint, as suggested on the left in Figure 2.2. These form a basis for the Euclidean topology on the line, so the subspace topology for a non-vertical line is the Euclidean topology. If the line is vertical, the basic deleted radius neighborhoods intersect the line in singletons and open intervals, as suggested on the right in Figure 2.2, and these sets form a basis for the subspace topology. Since every singleton is a basic open set, the subspace topology for a vertical line is the discrete topology.

The results below follow immediately from the definitions.

**Theorem 2.1.5.** *Suppose $X$ is a topological space and $A \subseteq Y \subseteq X$.*
(a) *If $A$ is open in the subspace $Y$ and $Y$ is open in $X$, then $A$ is open in $X$.*
(b) *If $A$ is closed in the subspace $Y$ and $Y$ is closed in $X$, then $A$ is closed in $X$.*

*Proof.* Suppose $A$ is open in the subspace topology on $Y$ and $Y$ is open in $X$. From the definition of the subspace topology, $A = U \cap Y$ for some $U$ open in $X$. Now $U$ and $Y$ are both open in $X$, so $U \cap Y = A$ is open in $X$. A similar argument shows (b). □

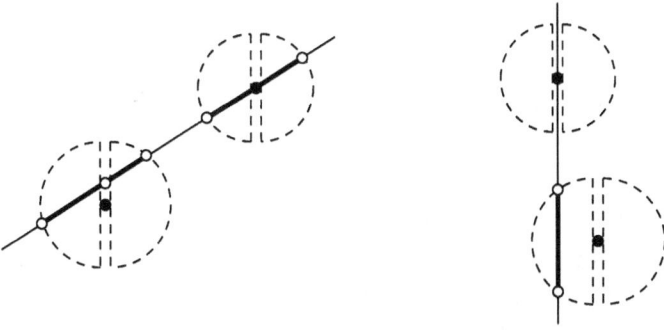

**Figure 2.2:** Basic open sets for non-vertical and vertical lines as a subspace of the plane with the deleted radius topology.

While an open set in a subspace $Y$ of $X$ is obtained by intersecting an open set of $X$ with $Y$, the same cannot be said for the interior of a set. Consider $X = \mathbb{R}$ with the Euclidean topology and $Y = [0, \infty)$. The set $A = [0, 1) = (-1, 1) \cap Y$ is open in $Y$ and thus $\text{int}_Y A = A$. However, $\text{int}_X(A) \cap Y = \text{int}_{\mathbb{R}}([0, 1)) \cap [0, \infty) = (0, 1) \cap [0, \infty) = (0, 1) \neq A = \text{int}_Y A$. Furthermore, $\partial_Y[0, 1) = \{1\} \neq \partial_X([0, 1)) \cap Y = \{0, 1\}$. The following theorem shows the containments which do hold.

**Theorem 2.1.6.** *Suppose $Y$ is a subspace of $X$ and $A \subseteq Y$.*
(a) $\text{int}_X(A) \cap Y \subseteq \text{int}_Y A$.
(b) $\text{cl}_X(A) \cap Y = \text{cl}_Y A$.
(c) $\partial_X(A) \cap Y \supseteq \partial_Y A$.

*Proof.* (a) Since $\text{int}_X A \subseteq A \subseteq Y$, $\text{int}_X(A) = \text{int}_X(A) \cap Y$ is an open set in $Y$ contained in $A$, and thus $\text{int}_X(A) \cap Y \subseteq \text{int}_Y A$ since $\text{int}_Y A$ is the largest open set in $Y$ contained in $A$.

(b) $\text{cl}_X(A) \cap Y$ is a closed set in $Y$ containing $A$, so $\text{cl}_Y A \subseteq \text{cl}_X(A) \cap Y$ since $\text{cl}_Y A$ is the smallest such closed set. Conversely, if $y \notin \text{cl}_Y A$, then there exists an open neighborhood $U_Y$ of $y$ in $Y$ which does not intersect $A \subseteq Y$. Now $U_Y = U \cap Y$ for some open set $U$ in $X$, and $U \cap A = U \cap (A \cap Y) = (U \cap Y) \cap A = U_Y \cap A = \emptyset$, so $U$ is a neighborhood of $y$ in $X$ which does not intersect $A$, so $y \notin \text{cl}_X A$, and thus $y \notin \text{cl}_X(A) \cap Y$. Thus $y \in \text{cl}_Y A$ if and only if $y \in \text{cl}_X(A) \cap Y$.

(c) follows from the equations

$$\begin{aligned}
\partial_Y A &= \text{cl}_Y A - \text{int}_Y A \\
&= (\text{cl}_X(A) \cap Y) - \text{int}_Y A \quad \text{(by (b))} \\
&= \text{cl}_X A \cap (Y - \text{int}_Y A) \\
&\subseteq \text{cl}_X A \cap (Y - \text{int}_X A) \quad \text{(by (a))} \\
&= (\text{cl}_X A - \text{int}_X A) \cap Y \\
&= \partial_X(A) \cap Y.
\end{aligned}$$

$\square$

Subspaces allow us to easily define the concept of an *isolated point* of a set $A$ in a topological space $X$.

**Definition 2.1.7.** If $X$ is a topological space and $A \subseteq X$, then $x \in A$ is an isolated point of $A$ if $\{x\}$ is open in the subspace topology on $A$. In particular, if $A = X$, $x$ is an isolated point of $X$ if $\{x\}$ is open.

Thus, a point $x$ of $(X, \mathcal{T})$ is an isolated point of $A$ if there exists a neighborhood $N \in \mathcal{T}$ of $x$ such that $N \cap A = \{x\}$. For example, consider the subset $A = \{2\} \cup [3, 4) \cup \{5\} \cup (7, \infty)$ of the real line. If $\mathbb{R}$ is given the Euclidean topology, then 2 and 5 are the only isolated points of $A$. If $\mathbb{R}$ is given the discrete topology, then every point of $A$ is isolated. If $X = \mathbb{R}$ with the Euclidean topology, then $X$ has no isolated points, but if $X = \mathbb{R}$ with the discrete topology, every point of $X$ is isolated.

## Exercises

1.  Suppose $\mathcal{T}$ is a topology on $X$ and $Y \subseteq X$. Show that $\emptyset, Y \in \mathcal{T}_Y$, $\bigcap_{i=1}^{n}(U_i \cap Y) = (\bigcap_{i=1}^{n} U_i) \cap Y$, and for arbitrary index sets $I$, $\bigcup_{i \in I}(U_i \cap Y) = (\bigcup_{i \in I} U_i) \cap Y$, confirming that the subspace topology $\mathcal{T}_Y$ really is a topology.
2.  Suppose $(X, \mathcal{T})$ is a topological space and $Z \subseteq Y \subseteq X$. Now $Z$ may be given the subspace topology $\mathcal{T}_Z$ it inherits from $X$ or the subspace topology $\mathcal{T}_{Y_Z}$ it inherits from the subspace $(Y, \mathcal{T}_Y)$ of $X$. Show that these topologies on $Z$ are equal.
3.  Let $X$ be the real line with the Euclidean topology, and $Y = [0, 5) \cup \{10\} \cup [12, 15)$. For each set below, determine whether it is open, closed, neither, or both in the subspace $Y$.
    $A = [0, 5)$
    $B = \{10, 12\}$
    $C = (1, 2)$
    $D = \{10\} \cup [12, 13)$
    $E = \{10\} \cup [12, 13]$
    $F = \{10\} \cup [13, 14)$
    $G = [0, 5) \cup \{10\} \cup [12, 15)$
4.  Describe the subspace topology $\mathbb{N}$ inherits (a) from $\mathbb{R}$ with the Euclidean topology, (b) from $\mathbb{R}$ with the cofinite topology, (c) from $\mathbb{R}$ with the discrete topology, and (d) from $\mathbb{R}$ with the indiscrete topology.
5.  Give $\mathbb{Z}$ the digital line topology. Describe the subspace topology on the set $\mathbb{D}$ of odd integers and on the set $\mathbb{E}$ of even integers.
6.  For any natural number $n > 2$, let $P_n = \{\frac{p+1}{2} \in \mathbb{N} : p \text{ is prime and } p \geq n\}$ with the subspace topology from the digital line.
    (a) Is $P_{13}$ Hausdorff? Justify your answer.
    (b) What is the twin primes conjecture? Rephrase it in terms of $P_n$ being Hausdorff.
7.  Prove that every subspace of a Hausdorff topological space is Hausdorff.

8. Find a topology on $X = \{1, 2, 3, 4, 5\}$ which is not Hausdorff, but has the property that every subspace of $X$ which does not contain 5 is Hausdorff.

9. Find a finite topological space $X$ with the property that $X$ is not Hausdorff, but every proper subspace of $X$ is Hausdorff.

10. Consider the plane $\mathbb{R}^2$ with the bow-tie topology, defined above Figure 1.13.
    (a) What topology does a horizontal line inherit? Which other lines inherit the same topology?
    (b) What topology does a vertical line inherit? Which other lines inherit the same topology?

11. Shown is a basis for a topology on $X = \{a, b, c, d\}$.

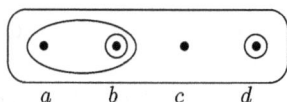

    (a) Which points of $X$ are isolated?
    (b) Which one-point subsets of $X$ are closed?
    (c) List all subspaces $Y$ of $X$ in which $a$ is an isolated point.
    (d) List all nonempty subspaces $Y$ of $X$ which are Hausdorff.

12. Suppose $A$ is a subset of topological space $X$ and $a$ is an isolated point of $A$. Show that any sequence in $A$ which converges to $a$ is eventually constant.

13. A set $A$ is *regular closed* if $A = \mathrm{cl}(\mathrm{int}\, A)$.
    (a) Among the closed subsets $A = \mathbb{N}^2$, $B = [0, 1]^2 \cup (\{0\} \times [-2, 2])$, and $C = \{(x, y) \in \mathbb{R}^2 : x^2 + y^2 \le 4\}$ of $\mathbb{R}^2$ with the Euclidean topology, which are regular closed sets?
    (b) If $X = \mathbb{R}^2$ with the Euclidean topology, show that if a subset $A \subseteq X$ has an isolated point, then $A$ is not regular closed.
    (c) Give an example to show that the converse of (b) does not hold.
    (d) Give an example to show that the result of (b) does not always hold if $X$ is a subspace of $\mathbb{R}^2$ with the Euclidean topology.

14. If $\mathcal{S}$ is a subbasis for a topology $\mathcal{T}$ on set $X$ and $Y \subseteq X$, is $\mathcal{S}_Y = \{S \cap Y : S \in \mathcal{S}\}$ a subbasis for the subspace topology $\mathcal{T}_Y$ on $Y$? Justify your answer. (Compare Theorem 2.1.3.)

## 2.2 Finite products of spaces

Given topologies on sets $X$ and $Y$, we should be able to define a topology on the Cartesian product $X \times Y$. From our experience with the Euclidean space $\mathbb{R}^2 = \mathbb{R} \times \mathbb{R}$, we would expect a point $(x, y)$ to be close to a point $(a, b)$ if and only if it is close in both coordinates. That is, a (basic) neighborhood of $(a, b)$ should consist of points $(x, y)$ where $x$ is in a (basic) neighborhood of $a$ and $y$ is in a (basic) neighborhood of $b$. This prompts our definition of the product topology.

**Definition 2.2.1.** For each $i = 1, \ldots, n$, suppose $(X_i, \mathcal{T}_i)$ is a topological space and $\mathcal{B}_i$ is any basis for $\mathcal{T}_i$. The *product topology* on $X_1 \times \cdots \times X_n$ is the topology generated by the basis

$$\mathcal{B} = \{B_1 \times \cdots \times B_n : B_i \in \mathcal{B}_i \text{ for every } i = 1, \ldots, n\}.$$

It is routine to check that $\mathcal{B}$ is a basis for a topology on the product $X_1 \times \cdots \times X_n$. Since the product topology was defined in terms of arbitrary bases $\mathcal{B}_i$ for the topologies $\mathcal{T}_i$, to see that the product topology is well-defined, we should also confirm that every collection of bases $\mathcal{B}_i$ will generate the same topology. That is, we will not get a different "product topology" by choosing a different set of bases for the coordinate factors. For each $i = 1, \ldots n$, suppose $\mathcal{B}_i$ and $\mathcal{C}_i$ are bases for the topology $\mathcal{T}_i$ on $X_i$. Let $\mathcal{B} = \{B_1 \times \cdots \times B_n : B_i \in \mathcal{B}_i \text{ for every } i = 1, \ldots, n\}$ and $\mathcal{C} = \{C_1 \times \cdots \times C_n : C_i \in \mathcal{C}_i \text{ for every } i = 1, \ldots, n\}$ be bases for $\mathcal{T}_\mathcal{B}$ and $\mathcal{T}_\mathcal{C}$, respectively. If $B_1 \times \cdots \times B_n$ is a $\mathcal{B}$-neighborhood of $(x_1, \ldots, x_n)$, then each $B_i$ is a $\mathcal{T}_i$-neighborhood of $x_i$, and since $\mathcal{C}_i$ is a basis for $\mathcal{T}_i$, there exists $C_i \in \mathcal{C}_i$ with $x \in C_i$ and $C_i \subseteq B_i$. Now $C_1 \times \cdots \times C_n$ is a $\mathcal{T}_\mathcal{C}$ neighborhood of $(x_1, \ldots, x_n)$ contained in the $\mathcal{T}_\mathcal{B}$ neighborhood $B_1 \times \cdots \times B_n$ of $(x_1, \ldots, x_n)$, so $\mathcal{T}_\mathcal{C}$ is finer than $\mathcal{T}_\mathcal{B}$. Interchanging $\mathcal{B}$ and $\mathcal{C}$ shows that $\mathcal{T}_\mathcal{B}$ is also finer than $\mathcal{T}_\mathcal{C}$, so $\mathcal{T}_\mathcal{C} = \mathcal{T}_\mathcal{B}$. Thus, the product topology is unique, regardless of the bases $\mathcal{B}_i$ or $\mathcal{C}_i$ used on the coordinates.

Since the choice of basis for the topology $\mathcal{T}_i$ on the factor $X_i$ is irrelevant, we may choose $\mathcal{B}_i = \mathcal{T}_i$ for each $i = 1, \ldots, n$. This shows that the product topology on $(X_1, \mathcal{T}_1) \times \cdots \times (X_n, \mathcal{T}_n)$ has basis

$$\{U_1 \times \cdots \times U_n : U_i \in \mathcal{T}_i \text{ for every } i = 1, \ldots, n\}.$$

In short, a basis for the product topology has the form

$$(\text{open set}) \times (\text{open set}) \times \cdots \times (\text{open set}),$$

or indeed,

$$(\text{basic open set}) \times (\text{basic open set}) \times \cdots \times (\text{basic open set}).$$

The product topology of two copies of the Euclidean line $(\mathbb{R}, \mathcal{T}_\mathcal{E})$ would have a basis of sets of form $(x - \varepsilon, x + \varepsilon) \times (y - \varepsilon, y + \varepsilon)$. Such sets are just the square balls generated by the sup metric on $\mathbb{R}^2$. However, we have seen that the sup metric topology is precisely the Euclidean topology on $\mathbb{R}^2$.

The example of the product $(\mathbb{R}, \mathcal{T}_\mathcal{E}) \times (\mathbb{R}, \mathcal{T}_\mathcal{E})$ also points out that the basis for the product topology given in Definition 2.2.1 is, in general, not a topology. The standard basis on each factor generates open squares in $\mathbb{R}^2$. An open circular disk (that is, a Euclidean metric ball) is a union of such open squares, but is not itself the product of two (basic) open sets.

Cartesian coordinates allows us to visualize basic open sets in $\mathbb{R}^3$ by drawing the three mutually perpendicular copies of $\mathbb{R}$ as the coordinate axes. Since we have only three spatial dimensions, we cannot represent $\mathbb{R}^4$ with mutually perpendicular axes. To allow the representation of higher dimensional products of $\mathbb{R}$, we may draw our axes parallel. Figure 2.3 shows a basic open set in the Cartesian coordinate system and the same set in a parallel coordinate system, and it shows the representation of a basic open set in the product topology on $\mathbb{R}^4$.

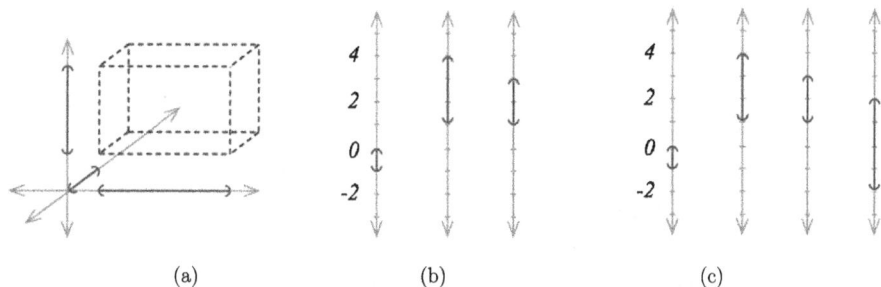

**Figure 2.3:** (a) $(-1, 0) \times (1, 4) \times (1, 3)$ in perpendicular coordinates. (b) $(-1, 0) \times (1, 4) \times (1, 3)$ in parallel coordinates. (c) $(-1, 0) \times (1, 4) \times (1, 3) \times (-2, 2)$ in parallel coordinates.

**Example 2.2.2** (The digital plane). Computer graphics try to represent sets in the Euclidean plane, which may contain infinitely many points, by a finite set of square pixels. Since there are tiny Euclidean open sets which illuminate a single pixel, every pixel should be open. When applied just to the set of pixels, we get the discrete topology, in which every set is open. To better model which pixels are near, we will introduce "boundaries" between pixels. Every pixel will have a horizontal boundary line above and below and a vertical boundary line on its left and right, each shared by two adjacent pixels. Furthermore, every pixel has four corner boundary points, each shared by four pixels. To model this, we may place the pixels on the points $(2n + 1, 2k + 1) \in \mathbb{Z}^2$. Then the vertical boundary lines fall at points $(2n, 2k + 1) \in \mathbb{Z}^2$, the horizontal boundary lines fall at points $(2n + 1, 2k)$, and the corner boundary points fall at $(2n, 2k) \in \mathbb{Z}^2$. Each object has a smallest neighborhood, as represented in Figure 2.4. Every pixel is open. The smallest neighborhood of a boundary line consists of that line and the two adjacent pixels. The smallest neighborhood of a corner point contains the corner point, four adjacent pixels, and four adjacent boundary lines. $\mathbb{Z}^2$ with this topology is called the *digital plane*.

Notice that the basis of smallest neighborhoods of the points of the digital plane $\mathbb{Z}^2$ are all rectangular, of form $B_1 \times B_2$ where $B_1$ and $B_2$ are of form $\{2n - 1, 2n, 2n + 1\}$ or $\{2k + 1\}$. Thus, the digital plane topology on $\mathbb{Z}^2$ has a basis of form $\{B_1 \times B_2 : B_1, B_2 \in \mathcal{B}\}$ where $\mathcal{B}$ is the basis for the digital line topology on $\mathbb{Z}$. This shows that the digital plane is the product of two copies of the digital line, given the product topology.

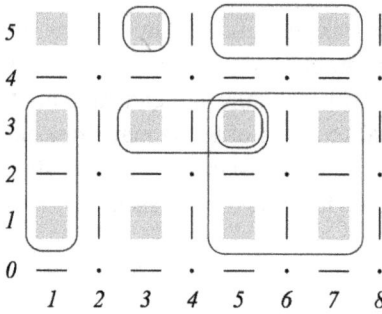

**Figure 2.4:** The digital plane and some basic open sets.

**Example 2.2.3.** Let $S$ be the two-element set $\{0, 1\}$ with the indiscrete topology, and let $T$ be the same set $\{0, 1\}$ with the discrete topology. Give $\mathbb{R}$ the Euclidean topology.

Since $\{0, 1\}$ is the only nonempty open set in $S$, the basic open sets in the product topology for $\mathbb{R} \times S$ are of form $(a, b) \times \{0, 1\}$, as shown in Figure 2.5(a). The open sets are unions of such basic open sets, so the space $\mathbb{R} \times S$ consists of two mirror image copies of $\mathbb{R}$. This product is not Hausdorff: For any $x \in \mathbb{R}$, $(x, 0) \neq (x, 1)$ but every neighborhood of $(x, 0)$ contains $(x, 1)$.

Since $T = \{0, 1\}$ with the discrete topology has a basis $\{\{0\}, \{1\}\}$ consisting of the singletons in $T$, basic open sets in the product topology for $\mathbb{R} \times T$ are of form $(a, b) \times \{0\}$ or $(c, d) \times \{1\}$, as shown in Figure 2.5(b).

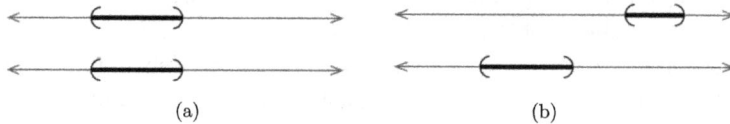

**Figure 2.5:** (a) A basic open set in $\mathbb{R} \times S$. (b) Two basic open sets in $\mathbb{R} \times T$.

The underlying sets for $\mathbb{R} \times S$ and $\mathbb{R} \times T$ are equal, and consist of two copies $\mathbb{R} \times \{0\}$ and $\mathbb{R} \times \{1\}$ of the real line. In $\mathbb{R} \times S$, in effect we have one copy of the Euclidean line and a carbon copy, mirror image of it. The two copies are inseparably linked. An open set in $\mathbb{R} \times S$ has the form $U \times \{0, 1\}$ where $U$ is open in $\mathbb{R}$, and thus consists of two copies $U \times \{0\}$ and $U \times \{1\}$ of the set $U \subseteq \mathbb{R}$. In $\mathbb{R} \times T$ we have two copies of the Euclidean line which are not linked in any way. An open set in $\mathbb{R} \times T$ consists of a Euclidean open set in the first copy of $\mathbb{R}$ and an independent Euclidean open set in the second copy of $\mathbb{R}$.

If $X$ and $Y$ are disjoint sets with topologies $\mathcal{T}_X$ and $\mathcal{T}_Y$, respectively, then a natural topology on the union $X \cup Y$ is the *union topology* generated by the basis $\mathcal{T}_X \cup \mathcal{T}_Y$. Thus, open sets in the union have the form $U_X \cup U_Y$ where $U_X$ is open in $X$ and $U_Y$ is open in $Y$ (including the possibility that $U_X, U_Y$, or both are empty).

The previous example of the product of the Euclidean line $\mathbb{R}$ with the discrete space $T = \{0,1\}$ gave two independent copies of $\mathbb{R}$, with open sets consisting of an open set from each copy. Thus, this was the union topology on two "disjoint" copies of $\mathbb{R}$, that is, on the *disjoint union* of $\mathbb{R}$ with $\mathbb{R}$. To distinguish the two copies of $\mathbb{R}$, the product construction essentially indexed them, obtaining disjoint copies $\mathbb{R} \times \{0\}$ and $\mathbb{R} \times \{1\}$. The same process works for combining any two topological spaces $(X_1, T_1)$ and $(X_2, T_2)$. We relabel each $x \in X_1$ as $(x, 1)$ and relabel each $x \in X_2$ as $(x, 2)$. Now if $X_1$ and $X_2$ are not disjoint, each $x \in X_1 \cap X_2$ appears as $(x, 1)$ when viewed as an element of $X_1$ and as $(x, 2)$ when viewed as an element of $X_2$. This technical indexing trick allows us to treat the sets $X_1$ and $X_2$ as being disjoint, and is the basis of our formal definition.

**Definition 2.2.4.** Suppose $(X_1, T_1)$ and $(X_2, T_2)$ are topological spaces. Their disjoint union, denoted $X_1 \mathbin{\mathring{\cup}} X_2$, is the set $(X_1 \times \{1\}) \cup (X_2 \times \{2\})$ with the topology $\{(U_1 \times \{1\}) \cup (U_2 \times \{2\}) : U_1 \in T_1, U_2 \in T_2\}$.

We note that the definition of the disjoint union of topological spaces can be extended in a natural way to arbitrary unions of topological spaces.

**Example 2.2.5.** Consider the coordinate axes $A_x = \{(x, 0) \in \mathbb{R}^2 : x \in \mathbb{R}\}$ and $A_y = \{(0, y) \in \mathbb{R}^2 : y \in \mathbb{R}\}$ in $\mathbb{R}^2$. If $A_x$ and $A_y$ each have the subspace topology, $B = \{(x, 0) \in A_x : |x| < 1\}$ is open in $A_x \mathbin{\mathring{\cup}} A_y$. Notice that the definition of $B \subseteq A_x$ implies $B$ only contains the copy of $(0,0)$ on the $x$-axis, and not the copy on the $y$-axis, and $(0,0) \in A_x$ has neighborhoods which do not contain any points of $A_y$ in the disjoint union. However, in the subspace of $A_x \cup A_y$ of $\mathbb{R}^2$, $B$ is not open, since $(0,0) \in A_x$ is also in $A_y$, and $B$ contains no neighborhood of $(0,0)$. See Figure 2.6.

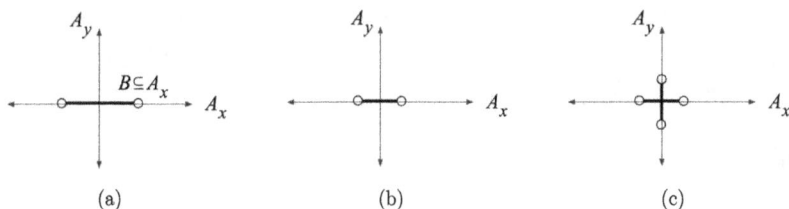

**Figure 2.6:** (a) $B \subseteq A_x$ is open in $A_x \mathbin{\mathring{\cup}} A_y$. (b) A neighborhood of $(0,0) \in A_x$ in $A_x \mathbin{\mathring{\cup}} A_y$. (c) A neighborhood of $(0,0)$ in $A_x \cup A_y$.

## Exercises

1. Verify that the collection $\mathcal{B}$ given in Definition 2.2.1 of the product topology really is a basis for a topology on $X_1 \times \cdots \times X_n$.

2. Prove that $(x_n, y_n) \to (x, y)$ in the product space $X \times Y$ if and only if $x_n \to x$ in $X$ and $y_n \to y$ in $Y$.

3. If $A$ is closed in $X$ and $B$ is closed in $Y$, show that $A \times B$ is closed in the product space $X \times Y$.

4. Carefully describe the open sets in the product of the Euclidean line $\mathbb{R}$ with the two-point set $\{0, 1\}$ with the nested topology $\mathcal{T} = \{\emptyset, \{0\}, \{0, 1\}\}$.

5. Recall that a topology $\mathcal{T}$ is *nested* if for every $U, V \in \mathcal{T}$, either $U \subseteq V$ or $V \subseteq U$. Give a proof or counterexample: If $\mathcal{T}_1$ is a nested topology on $X_1$ and $\mathcal{T}_2$ is a nested topology on $X_2$, then the product topology on $X_1 \times X_2$ is a nested topology.

6. Let $X = \{0, 1\}$ with the nested topology $\{\emptyset, \{0\}, \{0, 1\}\}$ and $Y = \{0, 1, 2\}$ with the nested topology $\{\emptyset, \{0\}, \{0, 1\}, \{0, 1, 2\}\}$. Sketch the basic open sets in the product $X \times Y$ and find the total number of open sets in the product.

7. Describe the subspace topology inherited by the following linear subsets of the digital plane, when viewed as a copy of $\mathbb{Z}$ in the digital plane.
   (a) The diagonal $D = \{(n, n) : n \in \mathbb{Z}\}$.
   (b) $L = \{(n, 2n - 1) : n \in \mathbb{Z}\}$.

8. If $X$ and $Y$ are topological spaces, prove that the product $X \times Y$ is Hausdorff if and only if $X$ and $Y$ are both Hausdorff.

9. Suppose $X$ and $Y$ are topological spaces with $A \subseteq X$ and $B \subseteq Y$. There are two natural ways to obtain a topology on $A \times B$: either as a subspace of the product $X \times Y$, or as the product of the subspaces $A$ of $X$ and $B$ of $Y$. Show that the two ways yield the same topology.

10. Let $\mathbb{R}_l$ denote the real line with the lower limit topology. Which of the sets below are open in $\mathbb{R}_l \times \mathbb{R}_l$ with the product topology?
    $A = [0, 1) \times (0, 1)$
    $B = (0, 1] \times [0, 1)$
    $C = \{(x, y) : x \le y < x + 1\}$

11. Consider the octagonal region $D$ in the plane, as shown below. The Euclidean boundary of this set consists of eight line segments N, S, E, W, NE, NW, SE, and SW, labeled according to the compass directions.

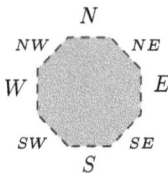

Which edges (including or excluding which endpoints) may be added to $D$ to get an open set in (a) $\mathbb{R}_l \times \mathbb{R}_l$, (b) $\mathbb{R}_l \times \mathbb{R}$, and (c) $\mathbb{R} \times \mathbb{R}_l$, where $\mathbb{R}_l$ is the real line with the lower limit topology, $\mathbb{R}$ is the real line with the Euclidean topology, and each product has the product topology?

12. The particular point topology on $\mathbb{R}$ determined by the particular point $p = 1$ is the collection $\mathcal{P}_1 = \{U \subseteq \mathbb{R} : 1 \in U\} \cup \{\emptyset\}$. Thus, a nonempty set is open in $(\mathbb{R}, \mathcal{P}_1)$ if

and only if it contains the particular point $p = 1$. Let $\mathcal{T}_{\mathcal{E}}$ be the Euclidean topology on $\mathbb{R}$. Which of the sets shown below are open in $(\mathbb{R}, \mathcal{T}_{\mathcal{E}}) \times (\mathbb{R}, \mathcal{P}_1)$ with the product topology? Justify your answers.

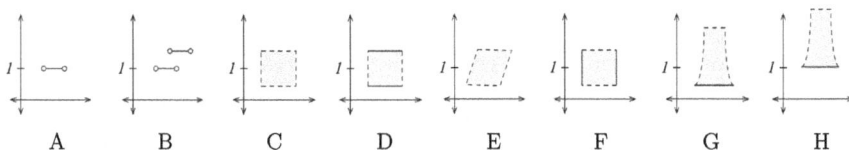

A      B      C      D      E      F      G      H

13. Find the boundary of each of the subsets of $(\mathbb{R}, \mathcal{T}_{\mathcal{E}}) \times (\mathbb{R}, \mathcal{P}_1)$ given in Exercise 12.

14. Consider $\mathbb{R}_l \times \mathbb{R}_l$ with the product topology, where $\mathbb{R}_l$ is the real line with the lower limit topology.
    (a) Describe the subspace topology inherited by the line $y = x$.
    (b) Describe the subspace topology inherited by the line $y = -x$.
    (c) Describe the subspace topology inherited by an arbitrary line $y = mx + b$ or $x = k$.

15. Consider the two topologies on $\mathbb{R} \times \{0, 1\}$ given in Example 2.2.3. Is either topology finer than the other?

16. Suppose $\mathcal{T}_Y$ is a topology on $Y$ and $\mathcal{T}_F$, $\mathcal{T}_C$ are topologies on $X$ with $\mathcal{T}_F$ finer than $\mathcal{T}_C$. Prove that the product topology $(X, \mathcal{T}_F) \times (Y, \mathcal{T}_Y)$ is finer than the product topology $(X, \mathcal{T}_C) \times (Y, \mathcal{T}_Y)$.

17. If $X$ and $Y$ are Hausdorff topological spaces, is their disjoint union $X \mathbin{\mathring{\cup}} Y$ Hausdorff? Give a proof or counterexample.

18. Let $X = (-\infty, 0]$, $Y = (0, \infty)$, and $Z = [0, \infty)$, each with the Euclidean topology.
    (a) Is $(-1, 0]$ open in $X \mathbin{\mathring{\cup}} Y$?
    (b) Is $(0, 1]$ closed in $X \mathbin{\mathring{\cup}} Y$?
    (c) Is $(0, 1]$ closed in $X \mathbin{\mathring{\cup}} Z$?
    (d) Is $(-1, 0] \subseteq X$ open in $X \mathbin{\mathring{\cup}} Z$?
    (e) Is $\{x \in X : x \in (-1, 0]\} \cup \{z \in Z : z = 0\}$ open in $X \mathbin{\mathring{\cup}} Z$?

19. (a) Suppose $(A, \mathcal{T}_A)$ and $(B, \mathcal{T}_B)$ are disjoint topological spaces, and $(x_n)_{n \in \mathbb{N}}$ is a sequence in $A \cup B$ with the union topology. Show that if $(x_n)_{n \in \mathbb{N}}$ converges to $a \in A$, then $(x_n)_{n \in \mathbb{N}}$ is eventually in $A$ and cannot converge to any $b \in B$.
    (b) If $(A, \mathcal{T}_A)$ and $(B, \mathcal{T}_B)$ are not disjoint, carefully state the property analogous to (a) for sequences in the disjoint union $A \mathbin{\mathring{\cup}} B = (A \times \{1\}) \cup (B \times \{2\})$.

## 2.3 Quotient spaces

If we view the interval $[0, 1]$ with the Euclidean topology as a piece of a rubber band, we may bend or stretch the rubber band, and the open sets of $[0, 1]$ bend and stretch along with the points in a natural way. Furthermore, we may bend the interval into a circle and glue the endpoints together, with a natural topology defined on the circle in terms of the open sets from $[0,1]$. The bending and twisting of the line is accomplished by a *homeomorphism*, and the gluing is accomplished by a *quotient space*, as we will see in this section.

The function $h : [0,1] \to [0,\pi]$ defined by $h(x) = \pi x$ is a bijection, and thus gives a one-to-one correspondence between the points of $[0,1]$ and of $[0,\pi]$. If these intervals have the Euclidean topology, $h$ also gives a one-to-one correspondence between the basic open sets $(a,b) \cap [0,1]$ of $[0,1]$ and the basic open sets $(\pi a, \pi b) \cap [0,\pi]$ of $[0,\pi]$, and thus between the open sets of $[0,1]$ and $[0,\pi]$. As a one-to-one correspondence between the points of the spaces which also gives a one-to-one correspondence between their open sets, the function $h$ shows that the spaces $[0,1]$ and $[0,\pi]$ are *topologically equivalent*.

**Definition 2.3.1.** Topological spaces $X$ and $Y$ are *homeomorphic* or *topologically equivalent* if there exists a one-to-one onto function $h : X \to Y$ which also provides a one-to-one correspondence between the open sets of $X$ and $Y$, in the sense that $U$ is open in $X$ if and only if $h(U)$ is open in $Y$. Such a function $h$ is a *homeomorphism*.

The identity function on $X$ shows $X$ is homeomorphic to $X$. If $h : X \to Y$ and $g : Y \to Z$ are homeomorphisms, then $g \circ h : X \to Z$ is a homeomorphism. If $h : X \to Y$ is a homeomorphism, then $h^{-1} : Y \to X$ is a bijection and $V = h(U)$ is open in $Y$ if and only if $U = h^{-1}(V)$ is open in $X$, so $h^{-1}$ is a homeomorphism. Thus, "is homeomorphic to" is an equivalence relation on the set of all topological spaces.

The function $g : [0,\pi] \to \{(r,\theta) \in \mathbb{R}^2 : r = 1, \theta \in [0,\pi]\}$ from the interval to the upper half of the unit circle in $\mathbb{R}^2$ in polar coordinates defined by $g(\theta) = (1,\theta)$ is easily seen to be a homeomorphism, providing a one-to-one correspondence between the points and open sets of the interval and those of the semicircle. If $h : [0,1] \to [0,\pi]$ is the homeomorphism $h(x) = \pi x$, then the composition $g \circ h$ is a homeomorphism showing that $[0,1]$ is homeomorphic to the semicircle.

Figure 2.7 shows several subspaces of the Euclidean plane which are homeomorphic to the interval $[0,1]$. While all of these are topologically equivalent, they do not have the same length, shape, or curvature. Length, shape, and curvature are geometric properties, but are not topological properties. A property is a *topological property* if whenever a topological space $X$ has the property, every topological space $Y$ homeomorphic to $X$ also has the property. The Hausdorff property is a topological property. We will study other topological properties such as connectedness and compactness later.

**Figure 2.7:** Some subsets of $\mathbb{R}^2$ homeomorphic to $[0,1]$.

Homeomorphisms allow us to bend or twist topological spaces into topologically equivalent spaces. Now let us turn to gluing points together.

Suppose we wish to glue together the endpoints 0 and 1 of the Euclidean interval $[0, 1]$. That is, we wish to glue together distinct points 0 and 1 to make them into one identical point $p$. This process is called *identifying* the points 0 and 1, or forming an *identification space* or a *quotient space*. It is easy to see that we would want a neighborhood of $p$ to contain $p$ and a neighborhood of each point 0 and 1 which was identified to form $p$. This is suggested in Figure 2.8. This is the simplest possible example of a quotient space, identifying only one pair of distinct points. We turn to a more general discussion.

**Figure 2.8:** Identification of the endpoints of $[0, 1]$ to form a circle.

To form a quotient space from $(X, \mathcal{T})$, we must specify which points are to be identified. This is done by giving an equivalence relation on $X$, with $x \approx y$ if and only if $x$ and $y$ are to be identified (along with all other elements of the equivalence class $[x]$). Now the collection of identified points will be the set $Y = \{[x] : x \in X\}$ of all equivalence classes. There is a natural function $f : X \to Y$ defined by $f(x) = [x]$. To define the topology on $Y$ in terms of the topology on $X$, we want a set $V$ to be open in $Y$ if and only if it was made up of an open neighborhood of each point of $x$ that was collapsed to a point $[z] \in V$. That is, $V$ should be open in $Y$ if and only if $f^{-1}(V)$ is an open neighborhood of each of its points, which occurs if and only if $f^{-1}(V)$ is open in $X$. Not every open set $U$ in $X$ collapses to an open set $f(U)$ in $Y$; only those open sets of form $f^{-1}(V)$ for some $V \subseteq Y$ do. Open sets of form $f^{-1}(V)$ cannot split an equivalence class: If $[z] \in V \subseteq Y = \{[x] : x \in X\}$, then $f^{-1}([z]) \subseteq f^{-1}(V)$. But $f^{-1}([z]) = \{x \in X : f(x) = [z]\} = \{x \in X : [x] = [z]\} = [z] \subseteq X$. Thus, every set $W$ in $X$ of form $f^{-1}(V)$ has the property that $x \in W$ if and only if $[x] \subseteq W$. Recall that such sets are called saturated sets with respect to the equivalence relation. Conversely, any saturated set $W \subseteq X$ has the form $W = \bigcup_{x \in W}[x] = f^{-1}(V)$ where $V \subseteq Y$ has the form $V = \{[x] : x \in W\}$. Thus, $f^{-1}(V)$ is open in $X$ if and only if it is a saturated open set in $X$. We also note that, since $f : X \to Y$ is onto, $f(f^{-1}(V)) = V$, so $f$ collapses the saturated open sets $W \subseteq X$ to open sets $V$ in $Y$.

In summary, to form a quotient space, we start with a topological space $(X, \mathcal{T})$, define an equivalence relation on $X$ telling which (equivalent) points will become equal in the quotient space $Y$, collapse each equivalence class of $X$ to a single point of $Y$, and collapse each saturated open set of $X$ to an open set of $Y$.

This discussion provides the motivation for the formal definition of a quotient space.

**Definition 2.3.2.** If $(X, \mathcal{T})$ is a topological space, $Y$ is any set, and $f : X \to Y$ is an onto function, then the *quotient topology on Y induced by f* is $\mathcal{T}_f = \{V \subseteq Y : f^{-1}(V) \in \mathcal{T}\}$. The space $(Y, \mathcal{T}_f)$ is a *quotient space* of $X$, and $f$ is the associated *quotient map*.

This definition did not specifically start with an equivalence relation on $X$. The equivalence relation is implicitly defined by $x \approx y$ if and only if $f(x) = f(y)$, so that the equivalence classes in $X$ are of form $f^{-1}(\{y\})$ for $y \in Y$. In most elementary applications, a set $X$ and an equivalence relation will be given, $f$ will be defined $f(x) = [x]$, and $Y = \{[x] : x \in X\}$ will be the set of equivalence classes.

It is easy to verify that $\mathcal{T}_f$ is indeed a topology on $Y$ using the facts that $f^{-1}(\bigcup_{i \in I} A_i) = \bigcup_{i \in I} f^{-1}(A_i)$ and $f^{-1}(\bigcap_{i \in I} A_i) = \bigcap_{i \in I} f^{-1}(A_i)$.

By the definition, $V$ is open in the quotient topology on $Y$ induced by $f : X \to Y$ if and only if $f^{-1}(V)$ is open in $X$. Furthermore, since $f$ is onto and $f^{-1}(Y - A) = f^{-1}(Y) - f^{-1}(A) = X - f^{-1}(A)$, it follows that $A$ is closed in the quotient space $Y$ if and only if $f^{-1}(A)$ is closed in $X$.

**Example 2.3.3.** Let $X = \{a, b, c, d, e\}$ have the topology $\mathcal{T}$ with basis $\mathcal{B} = \{\{a, b\}, \{b\}, \{c, d, e\}, \{d\}\}$. To form the quotient space identifying $b$ and $c$, we consider the equivalence relation defined by $x \approx x$ for all $x \in X$ and $b \approx c$. The associated quotient space is the set $Y = \{\{a\}, \{b, c\}, \{d\}, \{e\}\}$ of all equivalence classes, and the associated quotient map $f : X \to Y$ mapping each point of $X$ to its equivalence class is $f(a) = \{a\}$, $f(b) = \{b, c\} = f(c)$, $f(d) = \{d\}$, and $f(e) = \{e\}$, as suggested in Figure 2.9. Since $f(x) = [x]$, the equivalence classes in $X$ are the subsets of form $f^{-1}(y)$ for $y \in Y$.

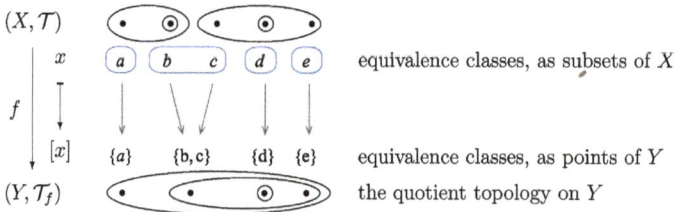

**Figure 2.9:** The quotient space identifying $b$ and $c$.

Since $X$ is a finite topological space, each point has a smallest neighborhood, so it suffices to describe the minimal neighborhoods in $X$ and in $Y$. In the quotient space $Y$, if $V$ is the minimal neighborhood of $[b] = [c] = \{b, c\}$, then $f^{-1}(V)$ is the minimal saturated open set in $X$ and containing $\{b, c\}$, namely $\{b, c, d, e\}$. Now $f$ collapses this saturated open set to $f(\{b, c, d, e\}) = \{\{b, c\}, \{d\}, \{e\}\}$, and this is the minimal neighborhood of $\{b, c\}$ in the quotient.

The minimal neighborhood of $\{a\}$ in the quotient space $Y$ corresponds to the minimal saturated open set in $X$ containing $\{a\}$. Every open set containing $a$ contains $b$, every saturated set containing $b$ contains $c$, and every open set containing $c$ contains

$\{c, d, e\}$. Thus, the minimal saturated open set containing $a$ is $X$, so the minimal neighborhood of $\{a\}$ in $Y$ is $f(X) = Y$.

Similarly, the minimal neighborhood of $[e] = \{e\} \in Y$ is the minimal saturated open set in $X$ containing $e$. Every open set containing $e$ contains $\{c, d, e\}$ and every saturated set containing $c$ contains $[c] = \{b, c\}$, so the minimal saturated open set containing $e$ contains $\{b, c, d, e\}$, and indeed $\{b, c, d, e\}$ is a saturated open set, so the minimal neighborhood of $\{e\}$ in $Y$ is $f(\{b, c, d, e\}) = \{\{b, c\}, \{d\}, \{e\}\}$.

It is easy to see that $\{d\}$ is the minimal open neighborhood of $\{d\}$ in $Y$.

**Example 2.3.4.** Let $X = [0, 2\pi] \times [0, 1]$ with the Euclidean topology, and let $\sim$ be the equivalence relation on $X$ defined by $(x, y) \sim (x, y)$ for all $(x, y) \in X$ and $(0, y) \sim (2\pi, y)$ for all $y \in [0, 1]$. The quotient space $Y$ determined by this equivalence relation (that is, by the quotient map $f(x) = [x]$) identifies points $(0, y)$ on the left edge of the rectangle with the corresponding point $(2\pi, y)$ on the right edge. We may view the quotient space, formed by gluing the left and right edges of the rectangle together, as a cylinder in $\mathbb{R}^3$ with the subspace topology, as suggested by Figure 2.10.

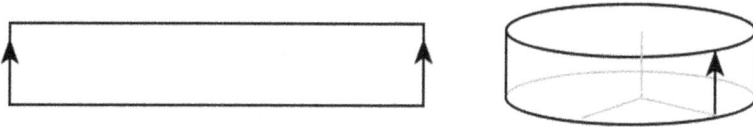

**Figure 2.10:** A quotient space producing a cylinder.

To be precise, the quotient space $Y$ of the last example is not equal to a cylinder $Z$ in $\mathbb{R}^3$, but is homeomorphic to $Z$. The points of the quotient space are equivalence classes like $\{(2, \frac{1}{2})\}$ or $\{(0, \frac{1}{3}), (2\pi, \frac{1}{3})\}$, while the points of the cylinder in $\mathbb{R}^3$ have the form $(x, y, z)$. However, the function $h$ from the quotient space $Y$ to the cylinder $Z = \{(\cos x, \sin x, y) : x \in [0, 2\pi], y \in [0, 1]\}$ in $\mathbb{R}^3$ defined by $h([(x, y)]) = (\cos x, \sin x, y)$ is a homeomorphism. Note that this is a well-defined function: If $[(x, y)] = [(z, w)]$, then $x = z$ or $x = z \pm 2\pi$ and $y = w$, so $h([(x, y)]) = h([(z, w)])$.

**Example 2.3.5.** Let $X = [0, 5] \times [0, 1]$ with the Euclidean topology.
(a) Identify the points $(0, y)$ with $(5, 1 - y)$ for all $y \in [0, 1]$. This can be visualized by putting a half twist in the band $X$ and gluing the ends together. The quotient space is homeomorphic to a Möbius strip (see Figure 2.11).

**Figure 2.11:** The Möbius strip.

(b) Identify the points $(x, 0)$ with $(x, 1)$ for all $x \in [0, 5]$, and identify the points $(0, y)$ with $(5, y)$ for all $y \in [0, 1]$. Note then, that $[(0, 0)] = \{(0, 0), (0, 1), (5, 0), (5, 1)\}$. This glues the top and bottom edges of the rectangle $X$ to get a cylinder, and by making two u-turn bends in the cylinder, we may then glue the ends of the cylinder to obtain a torus. The quotient space is homeomorphic to a torus (see Figure 2.12).

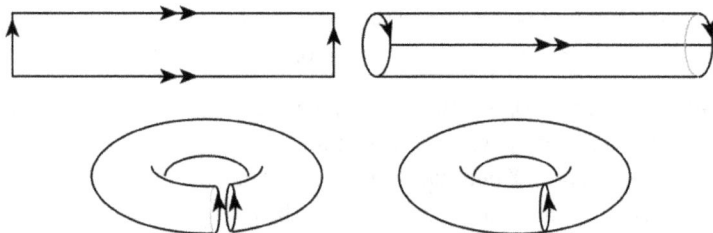

**Figure 2.12:** The torus.

(c) Identify the points $(x, 0)$ with $(x, 1)$ for all $x \in [0, 5]$, and identify the points $(0, y)$ with $(5, 1 - y)$ for all $y \in [0, 1]$. Note then that $[(0, 0)] = \{(0, 0), (0, 1), (5, 0), (5, 1)\}$. This glues the top and bottom edges of the rectangle $X$ to get a cylinder, but after two u-turn bends to bring the ends of the cylinder close together, the orientation on the ends do not match. We must glue the ends of the cylinder after only one u-turn in the cylinder. This quotient space is called the *Klein bottle*. It cannot be realized in 3-dimensional real space, but a suggestive model is shown in Figure 2.13. In the figure, the neck of the bottle appears to pass through the side of the bottle. A point at this apparent intersection in $\mathbb{R}^3$ would have a neighborhood which is topologically equivalent to two open disks glued along a diameter. However, no such point exists in the quotient topology where every point has a neighborhood topologically equivalent to an open disk.

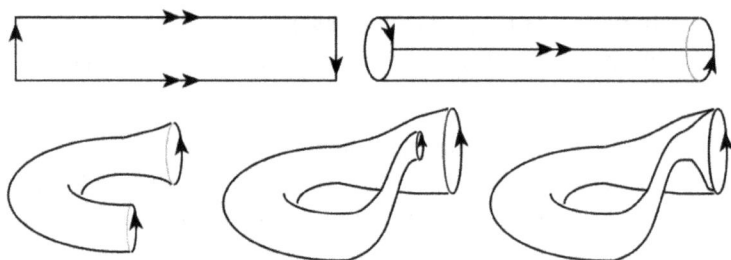

**Figure 2.13:** The Klein bottle.

In effect, we want to pass the neck of the bottle through the 2-dimensional wall of the bottle without intersecting that 2-dimensional barrier. Considering similar scenarios in lower dimensions is instructive. If you live in a 1-dimensional world

$\mathbb{R}$ and wish to get from one side of a 0-dimensional barrier (a point) to the other side—say you wish to get from 1 to –1 and avoid the barrier 0—it is impossible in $\mathbb{R}$, but by adding a second dimension, it is easily accomplished, as seen in Figure 2.14.

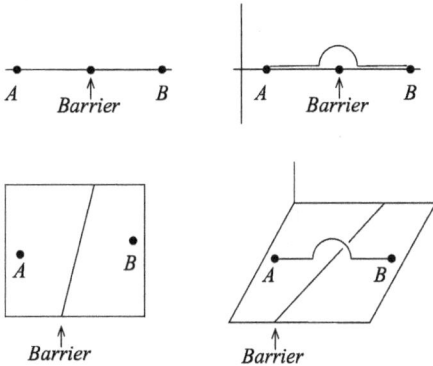

**Figure 2.14:** A 0-dimensional point creates a barrier in 1-dimensional line. A 1-dimensional line creates a barrier in a 2-dimensional plane. Adding a dimension allows us to cross the barrier.

Similarly, in $\mathbb{R}^2$, we may not cross from one side of a 1-dimensional barrier (a line) to the other side without crossing the line, but by adding a third dimension, we can hop over the line. The suggestive figure for the Klein bottle involves the next higher dimensional analog. We wish to get from one side of a 2-dimensional barrier to the other in 3-dimensional space. This is impossible in $\mathbb{R}^3$, but by adding a dimension, which cannot be visualized, we can easily hop over the barrier in the new dimension.

In the example of the Klein bottle, the quotient topology allows us to easily describe a space which cannot be accurately visualized in $\mathbb{R}^3$. For another such quotient space, start with a sphere with an open circular disk removed. The boundary of this set is a circle. Glue onto this circle the circular boundary of a Möbius strip. The resulting space is called the *projective plane*.

## Exercises

1. Consider $\mathbb{Z}$ with the topology $\mathcal{T}$ having basis $\{\{2n, 2n+1\} : n \in \mathbb{Z}\}$. Find three homeomorphisms from $(\mathbb{Z}, \mathcal{T})$ to $(\mathbb{Z}, \mathcal{T})$ other than the identity function id. Include at least one homeomorphism $h_1$ with $h_1 \circ h_1 = $ id and at least one homeomorphism $h_2$ with $h_2 \circ h_2 \neq $ id.
2. On the set $X = [0, 1]$, consider the Euclidean topology $\mathcal{T}_{\mathcal{E}}$ and the topology $\mathcal{T}_B$ described in Exercise 21 of Section 1.4. Find a homeomorphism $h : (X, \mathcal{T}_{\mathcal{E}}) \to (X, \mathcal{T}_B)$.
3. Let $X = \{a, b, c, d, e\}$ have the topology described in Example 2.3.3.

(a) Give a basis for the quotient space formed by identifying $b$ and $d$ to a single point.

(b) Give a basis for the quotient space formed by identifying $a$ and $e$ to a single point.

4. Consider the topology on $X = \{a, b, c, d, e, f\}$ having basis $\mathcal{B} = \{\{a, b, c, d, e\}, \{b\}, \{b, c\}, \{d, e\}, \{e\}, \{e, f\}\}$ as shown below.

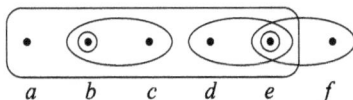

Give a similar sketch showing a basis for the quotient spaces determined by the identifications described.

(a) $d$ and $f$ are identified.

(b) $b$ and $c$ are identified, and $e$ and $f$ are identified.

(c) $a$ and $b$ are identified, and $d, e$, and $f$ are identified.

(d) $b$ and $e$ are identified.

5. Describe partitions or equivalence relations on the square $X = [-1, 1]^2$ in the Euclidean plane which produce quotient spaces homeomorphic to (a) a cylinder, (b) a cone, (c) a sphere, (d) two parallel cylinders intersecting along a line segment, and (e) a cylinder with a slit half way, as shown.

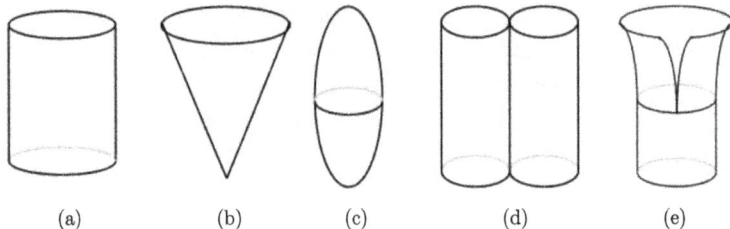

(a)　　　　(b)　　　　(c)　　　　(d)　　　　(e)

6. Let $X = \{1, 2, 3, 4, 5, 6, 7, 8\}$ with the subspace topology from the digital line. Sketch the quotient topology obtained by:

(a) Identifying all the odd points in $X$ to a point.

(b) Identifying all the even points in $X$ to a point.

(c) Identifying all the even points in $X$ to a point and all the odd points in $X$ to a point.

(d) Identifying 1 with 7, and 2 with 8.

7. Consider the cylinder $X = \{(x, y, z) \in \mathbb{R}^3 : x^2 + y^2 = 1, 0 \le z \le 2\}$ in Euclidean space. Describe the quotient space obtained by:

(a) Identifying the set $\{(x, y, z) \in X : z = 0\}$ to a point.

(b) Identifying the sets $\{(x, y, z) \in X : z = 0\}$ and $\{(x, y, z) \in X : z = 2\}$ to points.

(c) Identifying the sets $\{(x, y, z) \in X : z = 0\}$, $\{(x, y, z) \in X : z = 1\}$, and $\{(x, y, z) \in X : z = 2\}$ to points.

(d) Identifying the sets $\{(1, 0, z), (-1, 0, z)\}$ to a point for every $z \in [0, 2]$.

8. Consider the circle $S^1 = \{(x,y) \in \mathbb{R}^2 : x^2 + y^2 = 1\}$ in the Euclidean plane. Describe the quotient space resulting from the following equivalence relations or partitions given.

   (a) $\{\{(x,y),(x,-y)\} : (x,y) \in S^1\}$

   (b) $\{\{(x,y)\} : (x,y) \in S^1, x \neq 0\} \cup \{\{(0,1),(0,-1)\}\}$

   (c) $\{\{(x,y)\} : (x,y) \in S^1, xy \neq 0\} \cup \{\{(0,1),(0,-1)\}, \{(1,0),(-1,0)\}\}$

   (d) For $x \leq 0$, $(x,y) \approx (z,w)$ if and only if $x = z$, and for $x > 0$, $(x,y) \approx (z,w)$ if and only if $(x,y) = (z,w)$.

   (e) $(x,y) \approx (z,w)$ if and only if $\{x,z\} \subseteq \{\frac{1}{2}, \frac{-1}{2}\}$ or $(x,y) = (z,w)$.

9. Consider the topological space $X = \{a,b,c,d,e,f\}$ having basis $\mathcal{B} = \{\{a,b,c\}, \{b,c\}, \{b,c,d,e,f\}, \{e,f\}\}$ as shown below. Give a similar sketch showing the quotient space resulting from the equivalence relation on $X$ defined by $x \approx y$ if and only if cl$\{x\}$ = cl$\{y\}$.

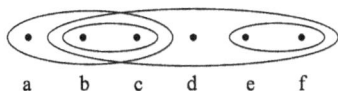

   a    b    c    d    e    f

10. Describe all equivalence relations on $X$ whose associated quotient maps are homeomorphisms.

11. Below is a rectangle with identification of edges indicated to produce the Klein bottle, as in Figure 2.13. Split the rectangle horizontally into two rectangles, as shown below, and (a) on each rectangle, perform the identification indicated by the vertical ends, and (b) perform the remaining identification indicated by the horizontal lines. The result will be the Klein bottle. Explicitly describe the result of step (a) and the subsequent identification occurring in step (b).

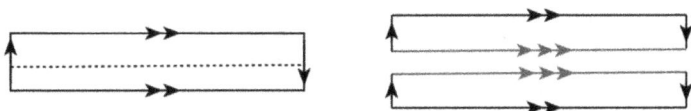

12. *Quotients of Hausdorff spaces need not be Hausdorff.* Let $X = \{(\frac{1}{n}, 1) \in \mathbb{R}^2 : n \in \mathbb{N}\} \cup \{(\frac{1}{n}, -1) \in \mathbb{R}^2 : n \in \mathbb{N}\} \cup \{(0,1), (0,-1)\}$ with the subspace topology from the Euclidean plane. Let $Y$ be the quotient space resulting from the equivalence relation $(x,y) \approx (x,y)$ for every $(x,y) \in X$ and $(\frac{1}{n}, 1) \approx (\frac{1}{n}, -1)$ for every $n \in \mathbb{N}$.

   (a) Sketch $X$ and give a suggestive sketch for $Y$.

   (b) Explain why $X$ is Hausdorff.

   (c) Show that $Y$ is not Hausdorff by exhibiting distinct points which cannot be separated by disjoint open sets.

   (d) Show that $Y$ is not Hausdorff by exhibiting a sequence in $Y$ which does not have a unique limit.

# 3 Continuity

## 3.1 Continuity

You may recall a definition of continuity of a function $f : \mathbb{R} \to \mathbb{R}$ from calculus. In calculus, a pointwise definition is given, first defining continuity at a point, then saying $f$ is continuous if it is continuous at every point of its domain. Loosely speaking, $f : \mathbb{R} \to \mathbb{R}$ is continuous at a point $x = a$ of its domain if, as input values get closer and closer to $a$, the corresponding output values get closer and closer to $f(a)$. Specifically, $f$ is continuous at $a$ if, for any $\varepsilon > 0$ to specify closeness of $f(x)$ to $f(a)$, there exists a $\delta > 0$, specifying closeness of $x$ to $a$ such that $f(x)$ is within $\varepsilon$ of $f(a)$ whenever $x$ is within $\delta$ of $x$. In symbols, $f$ is continuous at $a$ if

$$\forall \varepsilon > 0 \; \exists \delta > 0 \quad \text{such that } |x - a| < \delta \Rightarrow |f(x) - f(a)| < \varepsilon.$$

The statement $|x-a| < \delta$ simply says $x$ is within $\delta$ units of $a$, or $x \in B(a, \delta)$. With a similar interpretation of the second absolute value inequality, the displayed line above can be phrased as

$$\forall \varepsilon > 0 \; \exists \delta > 0 \quad \text{such that } x \in B(a, \delta) \Rightarrow f(x) \in B(f(a), \varepsilon),$$

or

$$\forall \varepsilon > 0 \; \exists \delta > 0 \quad \text{such that } f(B(a, \delta)) \subseteq B(f(a), \varepsilon).$$

The same phrasing would define continuity of $f$ at $a$ for a function $f : (X, d_X) \to (Y, d_Y)$ from any metric space $X$ to any metric space $Y$. To define continuity at a point for a function between two topological spaces, we note that the balls in the displayed line above represent neighborhoods of the points $a$ and $f(a)$. The last statement of continuity above says: for every $\varepsilon$-ball neighborhood $V$ of $f(a)$, there exists a $\delta$-ball neighborhood $U$ around $a$ with $f(U) \subseteq V$. In an arbitrary topological space (that is, ones which may not be metrizable), we simply drop the reference to the balls and use neighborhoods.

**Definition 3.1.1.** If $(X, \mathcal{T}_X)$ and $(Y, \mathcal{T}_Y)$ are topological spaces and $f : X \to Y$ is a function, then $f$ is *continuous at* $a \in X$ if and only if for every $\mathcal{T}_Y$-neighborhood $V$ of $f(a)$, there exists a $\mathcal{T}_X$-neighborhood $U$ of $a$ with $f(U) \subseteq V$.

Figure 3.1(a) suggests the $\varepsilon$-$\delta$ definition of pointwise continuity. Figure 3.1(b) suggests the topological definition of pointwise continuity.

Note that it is equivalent to say $f$ is continuous at $a \in X$ if and only if for every *basic* $\mathcal{T}_Y$-neighborhood $V$ of $f(a)$, there exists a $\mathcal{T}_X$-neighborhood $U$ of $a$ with $f(U) \subseteq V$.

Our goal will be to give a global definition of continuity of a function $f : (X, \mathcal{T}_X) \to (Y, \mathcal{T}_Y)$ between topological spaces in terms of the open sets in $X$ and $Y$, rather than

https://doi.org/10.1515/9783110686579-004

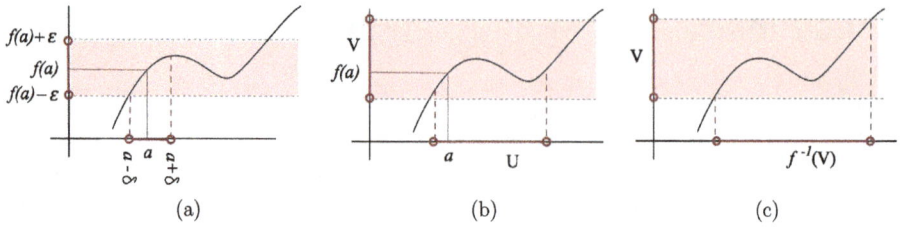

**Figure 3.1:** (a) Pointwise continuity from $\mathbb{R}$ to $\mathbb{R}$. (b) Pointwise continuity between two topological spaces. (c) Global continuity.

a pointwise definition. Of course, we would anticipate that $f$ should be continuous if and only if it is continuous at every point of its domain.

Suppose $f : (X, \mathcal{T}_X) \rightarrow (Y, \mathcal{T}_Y)$ is continuous at every point of its domain and $V$ is open in $Y$. For any $x \in f^{-1}(V)$, $V$ is a $\mathcal{T}_Y$-neighborhood of $f(x)$. Since $f$ is continuous at $x$, there exists an open $\mathcal{T}_X$-neighborhood $U_x$ of $x$ with $f(U_x) \subseteq V$, so

$$\{x\} \subseteq U_x \subseteq f^{-1}(V).$$

Taking the union over all $x \in f^{-1}(V)$ shows that $f^{-1}(V) = \bigcup\{U_x : x \in f^{-1}(V)\}$ and, as a union of $\mathcal{T}_X$-open sets, $f^{-1}(V)$ is $\mathcal{T}_X$-open. This shows that if $f$ is continuous at every point of its domain, then the inverse image of every open set in $Y$ is open in $X$.

Conversely, if the inverse image of every open set in $Y$ is open in $X$, then, for any $x \in X$ and any open neighborhood $V$ of $f(x)$, $U = f^{-1}(V)$ is an open neighborhood of $x$ with $f(U) \subseteq V$, so $f$ is continuous at $x$.

This gives the global characterization of continuity in terms of open sets, which we will take as the definition of continuity of a function between two topological spaces.

**Definition 3.1.2.** If $(X, \mathcal{T}_X)$ and $(Y, \mathcal{T}_Y)$ are topological spaces and $f : X \rightarrow Y$ is a function, then $f$ is *continuous* if and only if

$$V \in \mathcal{T}_Y \Rightarrow f^{-1}(V) \in \mathcal{T}_X.$$

That is, $f$ is continuous if and only if the inverse image of every open set in $Y$ is open in $X$.

Figure 3.1(c) suggests this global definition of continuity. The discussion above proves that our concepts of continuity defined globally and defined pointwise coincide, and we state this as our next theorem.

**Theorem 3.1.3.** *If $(X, \mathcal{T}_X)$ and $(Y, \mathcal{T}_Y)$ are topological spaces and $f : X \rightarrow Y$ is a function, then $f$ is continuous if and only if $f$ is continuous at every point of $X$.*

Note that the topological definition of continuity in terms of inverse images of open sets is a rather more elegant statement than the pointwise definition of continuity of a function between topological spaces, and, in particular, is more elegant than the $\varepsilon$–$\delta$ definition of continuity from calculus.

Since continuity was defined in terms of open sets, it should not be surprising that it may be characterized in terms of basic open sets. Also, taking complements, we should be able to characterize continuity in terms of closed sets. We give these characterizations in the following theorem.

**Theorem 3.1.4.** *Suppose X and Y are topological spaces, and $f : X \to Y$ is a function. Then the following are equivalent.*

(a) *$f$ is continuous.*

(b) *The inverse image of every open set in Y is open in X.*

(c) *The inverse image of every basic open set in Y (relative to any basis) is open in X.*

(d) *The inverse image of every closed set in Y is closed in X.*

*Proof.* The equivalence of (a) and (b) is the definition of continuity. (b) clearly implies (c) since every basic open set is open. If the inverse image of every basic open set is open and $V$ is an arbitrary open set in $Y$, then we may write $V$ as a union $\bigcup_{i \in I} B_i$ of basic open sets, and $f^{-1}(V) = f^{-1}(\bigcup_{i \in I} B_i) = \bigcup_{i \in I} f^{-1}(B_i)$, which is open in $X$ since each $f^{-1}(B_i)$ is open. Thus, (c) implies (b). The equivalence of (b) and (d) is left as an exercise. □

In part (c) of the theorem above, we must be careful to observe what the theorem does not say. It does not say that, for continuity, the inverse image of every basic open set in $Y$ has to be a *basic* open set in $X$. Indeed, given the standard basis of open intervals in the Euclidean line $\mathbb{R}$, the function $f(x) = \sin x$ is continuous, but the inverse image of the basic open set $(0, 2)$ is $\bigcup_{n \in \mathbb{Z}} (2n\pi, (2n+1)\pi)$, which is open, but not a basic open set.

We list some immediate facts about continuous functions, whose proofs are left as exercises.

**Theorem 3.1.5.** *Suppose X, Y, and Z are topological spaces.*

(a) *Any constant function $f : X \to Y$ defined by $f(x) = c$, where c is a fixed element of Y, is continuous.*

(b) *The identity function $f : X \to X$ is continuous.*

(c) *If $f : X \to Y$ and $g : Y \to Z$ are continuous, the composition $g \circ f : X \to Z$ is continuous.*

(d) *Any quotient map $q : X \to Y$ where Y has the quotient topology is continuous.*

(e) *If X has the discrete topology and Y has any topology, then every function $f : X \to Y$ is continuous.*

(f) *If X has any topology and Y has the indiscrete topology, then every function $f : X \to Y$ is continuous.*

(g) *If $f : X \to Y$ is continuous and $A \subseteq X$ has the subspace topology, then the restriction $f|_A$ of $f$ to A (defined by $f|_A(x) = f(x)$ for all $x \in A$) is continuous.*

Generally, your familiarity with Euclidean space will provide good motivation and intuition for the proper definitions in terms of open sets. However, once the definitions are clarified, your intuition from Euclidean space regarding which sets are open—and therefore which functions are continuous and which sequences converge—cannot be extended to general topological spaces. The point of defining other topologies is to characterize other sorts of continuity and convergence. The example below shows one continuous function and one discontinuous function, but determining which is which will depend on the given topologies.

**Example 3.1.6.** Let $\mathbb{R}$ denote the real line with the Euclidean topology and $\mathbb{R}_l$ denote the real line with the lower limit topology. Consider the map $f : \mathbb{R} \to \mathbb{R}_l$ defined by $f(x) = x$ and the function $g : \mathbb{R}_l \to \mathbb{R}$ defined by $g(x) = 1$ if $x \geq 0$ and $g(x) = -1$ if $x < 0$, shown in Figure 3.2. Note that the domains and codomains of $f$ and $g$ are interchanged.

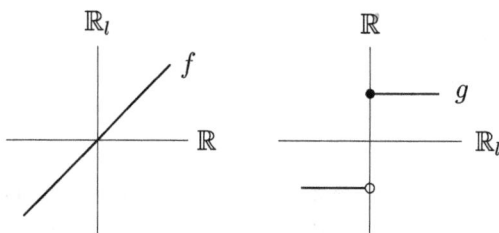

**Figure 3.2:** Two functions. Which is continuous?

As a function between distinct topological spaces $\mathbb{R}$ and $\mathbb{R}_l$, $f$ is not the identity function since the domain is a different topological space from the codomain. As a function on the underlying sets, $f : \mathbb{R} \to \mathbb{R}$ would be the identity function, but without specifying the topologies, we cannot ask whether $f$ is continuous since continuity depends on the open sets. The function $f : \mathbb{R} \to \mathbb{R}_l$ is not continuous: $[1, 2)$ is open in the lower limit topology but $f^{-1}([1, 2)) = [1, 2)$ is not open in the Euclidean topology.

The function $g : \mathbb{R}_l \to \mathbb{R}$ is continuous. Suppose $V$ is an open set in the codomain $\mathbb{R}$. If $V \cap \{1, -1\} = \{1, -1\}$, then $g^{-1}(V) = \mathbb{R}$, which is open in the domain $\mathbb{R}_l$. If $V \cap \{1, -1\} = \{-1\}$, then $g^{-1}(V) = (-\infty, 0)$, which is open. If $V \cap \{1, -1\} = \{1\}$, then $g^{-1}(V) = [0, \infty)$, which is open. If $V \cap \{1, -1\} = \emptyset$, then $g^{-1}(V) = \emptyset$, which is open. Considering all cases, the inverse image of an arbitrary open set is open, so $g$ is continuous.

You may recall from calculus that a function $f$ is continuous at $x = a$ if $\lim_{x \to a} f(x)$ exists, $f(a)$ exists, and $\lim_{x \to a} f(x) = f(a)$; in other words, if $\lim_{x \to a} f(x) = f(\lim_{x \to a} x)$. That is, the limit can be brought inside a continuous function. In terms of sequences, $f$ is continuous if and only if $x_n \to a$ implies $f(x_n) \to f(a)$, so $f$ "preserves limits." This result remains true for functions $f$ between any two metric spaces. Indeed, the result that continuous functions are those that preserve limits remains true for arbitrary topological spaces, if we can define convergence without reference to sequences.

The inadequacy of sequences in arbitrary topological spaces hinges on the fact that sequences are indexed by a countable set, and there are topological spaces which require an uncountable indexing set to achieve convergence to some points. See Section 6.3 for further discussion of this topic.

**Theorem 3.1.7.** *Suppose $X$ and $Y$ are metric spaces. A function $f : X \to Y$ is continuous if and only if whenever a sequence $(x_n)_{n=1}^{\infty}$ converges to $x$ in $X$, the sequence $(f(x_n))_{n=1}^{\infty}$ converges to $f(x)$ in $Y$. That is, $f$ is continuous if and only if it preserves limits.*

*Proof.* Suppose $f : X \to Y$ is a continuous function between metric spaces $X$ and $Y$, and $x_n \to x$ in $X$. To see $f(x_n) \to f(x)$, let $V$ be any open neighborhood of $f(x)$. By continuity of $f$, $f^{-1}(V)$ is an open neighborhood of $x$. Since $x_n \to x$, $(x_n)$ is eventually in $f^{-1}(V)$, so $(f(x_n))$ is eventually in $V$. This proves that $f(x_n) \to f(x)$.

  Conversely, suppose $x_n \to x$ implies $f(x_n) \to f(x)$. To see $f$ is continuous, suppose $V$ is open in $Y$. We will show $f^{-1}(V)$ is open. Suppose to the contrary that $f^{-1}(V)$ is not open. Then $f^{-1}(V)$ contains one of its boundary points $x$. For every $n \in \mathbb{N}$, $B(x, \frac{1}{n})$ intersects both $f^{-1}(V)$ and $X - f^{-1}(V)$. Pick $x_n \in B(x, \frac{1}{n}) \cap (X - f^{-1}(V))$. Now $(x_n)_{n=1}^{\infty}$ is a sequence in $X - f^{-1}(V)$ converging to $x \in f^{-1}(V)$. Applying the function $f$, we see that $f(x_n) \notin V$ for every $n \in \mathbb{N}$, and $f(x) \in V$. But by the hypothesis, $x_n \to x$ should imply $f(x_n) \to f(x)$. However, the sequence $(f(x_n))_{n=1}^{\infty}$ is never in the neighborhood $V$ of $f(x)$, so $f(x_n)$ does not converge to $f(x)$. This contradiction completes the proof.  □

  This theorem shows that continuity between metric spaces is equivalent to the implication $x_n \to x$ implies $f(x_n) \to f(x)$. Continuity is not equivalent to the converse of the implication. That is, if $f$ is continuous, it need not be true that $f(x_n) \to f(x)$ implies $x_n \to x$. For example, consider $f : \mathbb{R} \to \mathbb{R}$ defined by $f(x) = \sin x$, and $x_n = (-1)^n \pi$. Now $f(x_n) \to f(0)$ but $x_n \nrightarrow 0$. Or, with $y_n = \frac{1}{n}$, $f(y_n) \to f(3\pi)$ but $y_n \nrightarrow 3\pi$.

  Suppose $X$ and $Y$ are topological spaces, $X = A \cup B$, and functions $f : A \to Y$ and $g : B \to Y$ are continuous. May we use $f$ and $g$ to define a continuous function on $X = A \cup B$? That is, is

$$h(x) = \begin{cases} f(x) & \text{if } x \in A, \\ g(x) & \text{if } x \in B, \end{cases}$$

continuous? It is immediately obvious that, for $h(x)$ to be a well-defined function, we must have $f(x) = g(x)$ for all $x \in A \cap B$. This necessary condition is not sufficient. If $X = Y = \mathbb{R}$ with the Euclidean topology, $A = (-\infty, 0)$, $B = [0, \infty)$, $f(x) = -1$, and $g(x) = 1$, then $h(x)$ is not continuous. Further assumptions on $A$ and $B$ are needed.

**Theorem 3.1.8** (The pasting lemma). *Suppose $X$ and $Y$ are topological spaces, $X = A \cup B$, functions $f : A \to Y$ and $g : B \to Y$ are continuous, $f(x) = g(x)$ for all $x \in A \cap B$, and $h : X \to Y$ is defined by*

$$h(x) = \begin{cases} f(x) & \text{if } x \in A, \\ g(x) & \text{if } x \in B. \end{cases}$$

*Then $h$ is continuous if $A$ and $B$ are both open or are both closed in $X$.*

*Proof.* Suppose $A$ and $B$ are both open. To see $h$ is continuous, suppose $V$ is open in $Y$. We wish to show $h^{-1}(V)$ is open in $X$. But $h^{-1}(V) = f^{-1}(V) \cup g^{-1}(V)$. Since $f : A \rightarrow Y$ is continuous, $f^{-1}(V)$ is open in $A$. Since $A$ is open in $X$, we have $f^{-1}(V)$ is open in $X$ (by Theorem 2.1.5). Similarly, $g^{-1}(V)$ is open in $B$ and $B$ is open in $X$, so $g^{-1}(V)$ is open in $X$. Thus, $h^{-1}(V) = f^{-1}(V) \cup g^{-1}(V)$ is open in $X$, so $h$ is continuous. If $A$ and $B$ are both closed, then a similar argument using the characterization of continuity in terms of inverse images of closed sets and the second part of Theorem 2.1.5 applies. □

The theorem below characterizes continuity in terms of closures.

**Theorem 3.1.9.** *A function $f : X \rightarrow Y$ is continuous if and only if $f(\text{cl } A) \subseteq \text{cl}(f(A))$ for every $A \subseteq X$.*

*Proof.* Suppose $f : X \rightarrow Y$ satisfies $f(\text{cl } A) \subseteq \text{cl}(f(A))$ for every $A \subseteq X$. To show $f$ is continuous, we will show that, for any closed set $F$ in $Y$, the set $D = f^{-1}(F)$ is closed in $X$. Recalling that $f(f^{-1}(F)) \subseteq F$, we have $f(\text{cl } D) \subseteq \text{cl}(f(D)) = \text{cl}(f(f^{-1}(F))) \subseteq \text{cl } F = F$, and thus $\text{cl } D \subseteq f^{-1}(F) = D$. This shows $\text{cl } D = D$, so $D$ is closed.

To see the converse, suppose to the contrary that it fails. That is, suppose $f : X \rightarrow Y$ is continuous, but there exists a set $A \subseteq X$ with $f(\text{cl } A) \nsubseteq \text{cl}(f(A))$. Then there exists $x \in \text{cl } A$ with $f(x) \notin \text{cl}(f(A))$, and $V = Y - \text{cl}(f(A))$ is an open neighborhood of $f(x)$. Note in particular that $V$ is disjoint from $f(A)$. By continuity of $f$, $f^{-1}(V)$ is an open set, and it contains $x \in \text{cl } A$. Thus, the neighborhood $f^{-1}(V)$ of $x$ intersects $A$. If $a \in A \cap f^{-1}(V)$, then $f(a) \in f(A) \cap V$, contrary to $V$ being disjoint from $f(A)$. Thus, the converse cannot fail. □

## Exercises

1. Show that $f : X \rightarrow Y$ is continuous at $a \in X$ if and only if for every *basic* $\mathcal{T}_Y$-neighborhood $V$ of $f(a)$, there exists a $\mathcal{T}_X$-neighborhood $U$ of $a$ with $f(U) \subseteq V$. That is, show that it suffices to consider basic neighborhoods of $f(a)$ in Definition 3.1.1.

2. Show by example that if $f : X \rightarrow Y$ is a continuous function and $U$ is open in $X$, then $f(U)$ need not be open in $Y$.

3. The functions $f_1, f_2$, and $f_3$ from the Euclidean line to the Euclidean line given below are not continuous.

$$f_1(x) = \begin{cases} x & \text{if } x \neq 0, \\ 1 & \text{if } x = 0, \end{cases} \qquad f_2(x) = \begin{cases} x^2 & \text{if } x \leq 2, \\ x^2 + 1 & \text{if } x > 2, \end{cases}$$

$$f_3(x) = \begin{cases} x^2 & \text{if } x^2 < 4, \\ 1 & \text{if } x^2 \geq 4. \end{cases}$$

For each, find:
(a) an open set $V$ whose inverse image is not open;
(b) a closed set $C$ whose inverse image is not closed;

(c) a sequence $(x_n)_{n \in \mathbb{N}}$ converging to $x$ for which $(f_i(x_n))_{n \in \mathbb{N}}$ does not converge to $f_i(x)$.

4. Shown below are some bases for topologies on $\{1, 2, 3, 4\}$ and $\{1, 2, 3\}$ and some functions $f, g, h, j$. Determine which of the functions shown are continuous.

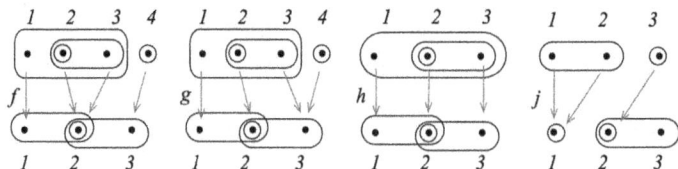

5. Complete the proof of Theorem 3.1.4 by showing that $f : X \to Y$ is continuous if and only if the inverse image of every closed set in $Y$ is closed in $X$.

6. If $\mathbb{R}$ has the Euclidean topology, consider the function $f : \mathbb{R} \to \{1, 2, 3\}$ defined by

$$f(x) = \begin{cases} 1 & \text{if } x \in (-\infty, 1] \\ 2 & \text{if } x \in (1, 2) \\ 3 & \text{if } x \in [2, \infty). \end{cases}$$

(a) Find the finest topology on $\{1, 2, 3\}$ which makes $f$ continuous.
(b) Find all topologies on $\{1, 2, 3\}$ which make $f$ continuous.

7. Give $X = \{1, 2, 3, 4\}$ the topology having basis $\mathcal{B} = \{\{1, 2, 3\}, \{2, 3\}, \{3\}, \{3, 4\}\}$. Show that there is no continuous function $f$ from $X$ to the Euclidean line which maps the closed set $A = \{1, 2\}$ to 0 and maps the closed set $B = \{4\}$ to 1.

8. Suppose $f : X \to Y$ is continuous and $B \subseteq Y$. Show
(a) $\text{cl}(f^{-1}(B)) \subseteq f^{-1}(\text{cl}\, B)$
(b) $f^{-1}(\text{int}\, B) \subseteq \text{int}(f^{-1}(B))$

9. Suppose $f : X \to Y$ is a continuous function, $A \subseteq X$, and $B \subseteq Y$.
(a) Show that if $a$ is a boundary point of $A$, then $f(a)$ need not be a boundary point of $f(A)$.
(b) Show that if $a$ is a boundary point of $f^{-1}(B)$, then $f(a)$ is a boundary point of $B$. That is, show that $\partial f^{-1}(B) \subseteq f^{-1}(\partial B)$. Hint: You may use the result of Exercise 8.

10. Let $X = Y = \mathbb{R}$ with the cofinite topology. Plot the functions from $X$ to $Y$ given below and determine which are continuous.
(a) $f(x) = x^2$
(b) $g(x) = \sin x$
(c) $h(x) = 4 \sin x + x$
(d) The *floor function* $j(x) = \lfloor x \rfloor = $ the greatest integer less than or equal to $x$
(e) $k(x) = \begin{cases} x & \text{if } \lfloor x \rfloor \text{ is even} \\ -x & \text{if } \lfloor x \rfloor \text{ is odd} \end{cases}$

11. Characterize the continuous functions $f : \mathbb{R} \to \mathbb{R}$ where each copy of $\mathbb{R}$ has the cofinite topology.

12. Recall that, for any $a \in \mathbb{R}$, the particular point topology on $\mathbb{R}$ determined by $a$ is $P_a = \{U \subseteq \mathbb{R} : a \in U\} \cup \{\emptyset\}$, and the excluded point topology on $\mathbb{R}$ determined by $a$ is $E_a = \{U \subseteq \mathbb{R} : a \notin U\} \cup \{\mathbb{R}\}$. Describe
    (a) all continuous non-constant functions $f : (\mathbb{R}, P_1) \to (\mathbb{R}, P_3)$;
    (b) all continuous non-constant functions $f : (\mathbb{R}, E_1) \to (\mathbb{R}, E_3)$;
    (c) all continuous non-constant functions $f : (\mathbb{R}, P_1) \to (\mathbb{R}, E_3)$;
    (d) all continuous non-constant functions $f : (\mathbb{R}, E_1) \to (\mathbb{R}, P_3)$.

13. Let $X = \mathbb{R}$ with the right ray topology $\mathcal{T} = \{\emptyset, \mathbb{R}\} \cup \{(a, \infty) : a \in \mathbb{R}\}$. Plot the functions from $X$ to $X$ given below and determine which are continuous.
    (a) $f(x) = x$
    (b) $g(x) = -x$
    (c) $h(x) = -\lfloor -x \rfloor$
    (d) $s(x) = \sin x$

14. Let $X = \mathbb{R}$ with the right ray topology $\mathcal{T} = \{\emptyset, \mathbb{R}\} \cup \{(a, \infty) : a \in \mathbb{R}\}$. Show that if $f : X \to X$ is continuous then it is increasing. Is the converse true?

15. (a) Prove part (a) of Theorem 3.1.5.
    (b) Prove part (b) of Theorem 3.1.5.
    (c) Prove part (c) of Theorem 3.1.5.
    (d) Prove part (d) of Theorem 3.1.5.
    (e) Prove part (e) of Theorem 3.1.5.
    (f) Prove part (f) of Theorem 3.1.5.
    (g) Prove part (g) of Theorem 3.1.5.

16. Give $\mathbb{R}$ the Euclidean topology generated by the standard basis $\mathcal{B} = \{(a, b) \subseteq \mathbb{R} : a < b\}$. Suppose $f : \mathbb{R} \to \mathbb{R}$ is continuous and there exist $a < b < c$ with $f(a) = f(c) \neq f(b)$. Show that there exists a basic open set $B \in \mathcal{B}$ whose inverse image is not a basic open set in $\mathcal{B}$.

17. Suppose $\mathcal{T}_F$ and $\mathcal{T}_C$ are two topologies on the set $X$. Show that $\mathcal{T}_F$ is finer than $\mathcal{T}_C$ if and only if the function $f : (X, \mathcal{T}_F) \to (X, \mathcal{T}_C)$ defined by $f(x) = x$ is continuous.

18. If $X = \mathbb{R}$ with the Euclidean topology, the function $f : X \to X$ defined by

$$f(x) = \begin{cases} 1 & \text{if } x \geq 0 \\ -1 & \text{if } x < 0 \end{cases}$$

is not continuous, yet for the sequence $(x_n) = (1 + \frac{1}{n})$, it is true that $x_n \to 1$ implies $f(x_n) \to f(1)$. Why does this not contradict Theorem 3.1.7?

19. Show that if $D$ is dense in $X$ and $f : X \to Y$ is continuous and onto, then $f(D)$ is dense in $Y$.

20. Suppose $f$ and $g$ are continuous functions from a topological space $X$ to a Hausdorff space $Y$.
    (a) Show that the set $\{x \in X : f(x) = g(x)\}$ is closed.
    (b) Show that if $f(x) = g(x)$ for all $x$ in a dense subset $D$ of $X$, then $f(x) = g(x)$ for all $x \in X$.

21. Suppose $\mathbb{R}$ has the Euclidean topology, $f : \mathbb{R} \to \mathbb{R}$ is continuous, and $f(x + y) = f(x) + f(y)$ for all $x, y \in \mathbb{R}$. Use Exercise 20(b) to show $f$ is a linear function of form $f(x) = mx$.

22. Suppose $\mathbb{R}$ has the Euclidean topology.
    (a) Show that if $f : \mathbb{R} \to \mathbb{R}$ is a continuous function with $f(x) = f(\frac{x}{2})$ for all $x \in \mathbb{R}$, then $f(x)$ is constant.
    (b) Show that continuity is necessary in part (a) by finding a non-constant, discontinuous function $f : \mathbb{R} \to \mathbb{R}$ with $f(x) = f(\frac{x}{2})$ for all $x \in \mathbb{R}$.
    (c) If $f : \mathbb{R} \to \mathbb{R}$ is a continuous function with $2 \int_0^x f(t)\,dt = \int_0^{2x} f(t)\,dt$ for every $x \in \mathbb{R}$, show that $f$ is constant.

23. Show that with the usual topologies, addition is continuous. That is, show that the function $\text{Add} : \mathbb{R}^2 \to \mathbb{R}$ defined by $\text{Add}(x, y) = x + y$ is continuous, where $\mathbb{R}^2$ and $\mathbb{R}$ carry the Euclidean topologies.

24. Show that the addition function $\text{Add} : \mathbb{R}^2 \to \mathbb{R}_l$ defined by $\text{Add}(x, y) = x + y$ is not continuous, where $\mathbb{R}^2$ has the Euclidean topology and $\mathbb{R}_l$ is the real line with the lower limit topology.

25. Suppose $f : X \to Y$ is a function, $I$ is an arbitrary index set, $X = \bigcup_{i \in I} A_i$, and $f$ is continuous on each $A_i$.
    (a) If each $A_i$ is open, must $f$ be continuous on $X$? What if $I$ is finite?
    (b) If each $A_i$ is closed, must $f$ be continuous on $X$? What if $I$ is finite?

## 3.2 Special continuous functions

### 3.2.1 Homeomorphisms

Recall that a homeomorphism is a function $h : X \to Y$ between two topological spaces which is a bijection between the points and provides a bijection between the open sets of $X$ and the open sets of $Y$ in the sense that $U$ is open in $X$ if and only if $h(U)$ is open in $Y$. Topological spaces $X$ and $Y$ are topologically equivalent, or homeomorphic (denoted $X \approx Y$), if there exists a homeomorphism $h$ from $X$ to $Y$. Such a homeomorphism essentially relabels each point $x$ of $X$ as $h(x)$ in $Y$, and relabels each open set $U$ of $X$ as the open set $h(U)$ in $Y$. If $h : X \to Y$ is a homeomorphism, then $h^{-1} : Y \to X$ gives a bijection between the points of $Y$ and $X$ and between the open sets of $Y$ and those of $X$, so $h^{-1}$ is also a homeomorphism.

The following important observation is often given as the definition of a homeomorphism.

**Theorem 3.2.1.** *A function $h : X \to Y$ is a homeomorphism if and only if $h$ is one-to-one, onto, continuous, and has a continuous inverse $h^{-1}$.*

*Proof.* Suppose $h : X \to Y$ is a bijection. Now $h$ is a homeomorphism if and only if (a) if $U$ is open in $X$ then $h(U) = (h^{-1})^{-1}(U)$ is open in $Y$, and (b) if $h(U) = V$ is open in $Y$

then $U = h^{-1}(V)$ is open in $X$. But (a) is the definition of the continuity of $h^{-1}$ and (b) is the definition of the continuity of $h$. Thus, a bijection $h$ is a homeomorphism if and only if $h$ and $h^{-1}$ are continuous. □

**Example 3.2.2.** Consider the topological spaces $X = \{a, b, c, d\}$ and $Y = \{1, 2, 3, 4\}$ whose topologies (excluding the empty set) are shown in Figure 3.3. The representation for $Y$ could be rearranged, repositioning the points in the plane but keeping their neighborhoods the same, to obtain the representation for $X$. Thus, $X$ and $Y$ are homeomorphic. Note that $X$ and $Y$ each have exactly one open set of cardinality $n$ for each $n = 1, 2, 3, 4$. Since a homeomorphism from $X$ to $Y$ must map an $n$-element open set to an $n$-element open set, there is a unique homeomorphism $h : X \to Y$ defined by $h(a) = 2$, $h(b) = 1$, $h(c) = 4$, and $h(d) = 3$.

**Figure 3.3:** Two homeomorphic topological spaces.

We note that since a homeomorphism $h : X \to Y$ gives a bijection between the open sets of $X$ and $Y$, it also gives a bijection between the closed sets of $X$ and $Y$. That is, if $h : X \to Y$ is a homeomorphism, $F$ is closed in $X$ if and only if $h(F)$ is closed in $Y$. This observation also follows from the characterization of the continuity of $h$ and $h^{-1}$ in terms of inverse images of closed sets being closed.

If $h : X \to Y$ is a homeomorphism and $A \subseteq X$ has the subspace topology, then the restriction $h|_A$ of $h$ to $A$ is one-to-one, onto $h(A)$, and continuous. Now $(h|_A)^{-1} = (h^{-1})|_{f(A)}$ is also continuous (as the restriction of the continuous function $h^{-1}$), so $h|_A : A \to h(A)$ is a homeomorphism.

Since a homeomorphism $h : X \to Y$ is a bijection, just from the set-theoretic properties of bijections (without using the topologies), we have $h(\bigcap_{i \in I} A_i) = \bigcap_{i \in I} h(A_i)$, $h(\bigcup_{i \in I} A_i) = \bigcup_{i \in I} h(A_i)$, and $h(A - B) = h(A) - h(B)$ for any $A, B, A_i \subseteq X$. Since $h$ is essentially a relabeling of the points and open sets, the following topological results should be anticipated.

**Theorem 3.2.3.** *Suppose $h : X \to Y$ is a homeomorphism and $A \subseteq X$.*
(a) $h(\operatorname{cl} A) = \operatorname{cl}(h(A))$.
(b) $h(\operatorname{int} A) = \operatorname{int}(h(A))$.
(c) $h(\partial A) = \partial(h(A))$.

*Proof.* Since $h$ is a homeomorphism, $F$ is a closed set if and only if $h(F)$ is a closed set. Since $h$ is a bijection, $A \subseteq F$ if and only if $h(A) \subseteq h(F)$, and $h(\bigcap_{i \in I} F_i) = \bigcap_{i \in I} h(F_i)$ for any index set $I$. Now since $\operatorname{cl} A = \bigcap \{F : F$ is closed and $A \subseteq F\}$, we have $h(\operatorname{cl} A) = \bigcap \{h(F) : h(F)$ is closed and $h(A) \subseteq h(F)\}$, and the latter set is $\operatorname{cl}(h(A))$. Thus, $h(\operatorname{cl} A) = \operatorname{cl}(h(A))$.

The corresponding result for interiors can be shown similarly, or by using the result for closures after noting that $\operatorname{int} A = X - \operatorname{cl}(X - A)$ (Corollary 1.5.6), and recalling that $h(C{-}D) = h(C){-}h(D)$ for bijections. Then $h(\operatorname{int} A) = h(X){-}\operatorname{cl}(h(X{-}A)) = Y{-}\operatorname{cl}(Y{-}h(A)) = \operatorname{int}(h(A))$. The result for boundaries follows from the characterization of boundaries in terms of closures: $\partial A = \operatorname{cl}(A) \cap \operatorname{cl}(X - A)$. Applying $h$ and using properties of bijections gives $h(\partial A) = h(\operatorname{cl}(A) \cap \operatorname{cl}(X - A)) = h(\operatorname{cl}(A)) \cap h(\operatorname{cl}(X - A)) = \operatorname{cl}(h(A)) \cap \operatorname{cl}(h(X - A)) = \operatorname{cl}(h(A)) \cap \operatorname{cl}(Y - h(A)) = \partial(h(A))$. □

If topological spaces $X$ and $Y$ are homeomorphic, we write $X \approx Y$. In the Euclidean line $\mathbb{R}$, the open interval $(0, 1)$ is homeomorphic to any other nonempty open interval $(a, b)$. Indeed, if $h(x)$ is the linear function with $h(0) = a$ and $h(1) = b$, then $h|_{(0,1)} : (0, 1) \to (a, b)$ is a homeomorphism. Using the properties of the equivalence relation $\approx$, we see that any two nonempty open intervals $(a, b)$ and $(c, d)$ in the Euclidean line are homeomorphic. Furthermore, since $h : (-\pi/2, \pi/2) \to \mathbb{R}$ defined by $h(x) = \tan x$ is a homeomorphism, $(-\pi/2, \pi/2)$ and every other nonempty open interval $(a, b)$ are homeomorphic to the real line $\mathbb{R}$. We may visualize the homeomorphism $\tan x$ from $(-\pi/2, \pi/2)$ to $\mathbb{R}$ by bending the interval into the right half of a unit circle, as shown in Figure 3.4, and mapping a point on this arc to its shadow from a light source placed at the origin. This is an example of a *stereographic projection*.

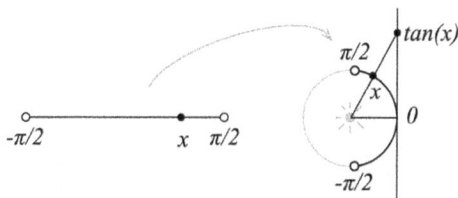

**Figure 3.4:** The interval $(-\pi/2, \pi/2)$ is homeomorphic to $\mathbb{R}$.

Technically, the function $f : (-\pi/2, \pi/2) \to \{(x, y) \in \mathbb{R}^2 : x > 0, x^2 + y^2 = 1\}$ defined by $f(x) = (\cos x, \sin x)$ is a homeomorphism, which we follow by the homeomorphism mapping $(\cos x, \sin x)$ on the unit circle to $\tan x$ on $\mathbb{R}$. The function $f$ gives us a homeomorphic copy of $(-\pi/2, \pi/2)$ appearing as a subspace of the Euclidean plane. That is, $f$ *embeds* the interval $(-\pi/2, \pi/2)$ into the plane.

**Definition 3.2.4.** If $X$ and $Y$ are topological spaces, a function $e : X \to Y$ is an *embedding of $X$ into $Y$* if $e : X \to e(X)$ is a homeomorphism, where $e(X) \subseteq Y$ has the subspace topology. In this case, we say $e(X)$ is a *homeomorphic copy* of $X$ *embedded* in $Y$.

An embedding of $X$ into $Y$ is sometimes called a *homeomorphism into $Y$*.

Other examples of stereographic projection showing $(a, b) \approx \mathbb{R}$ involve embedding $(a, b)$ into $\mathbb{R}^2$ as a circle with the north pole missing, placing a light source at the north pole, and projecting onto a line in $\mathbb{R}^2$ parallel but not equal to the tangent at the north pole on the same side of the tangent as the circle. The ray from the light source to a point $x$ on the circle crosses the line at $f(x)$. Two common illustrations of this (where the line is tangent at the south pole or includes a diameter), and one uncommon one, are shown in Figure 3.5.

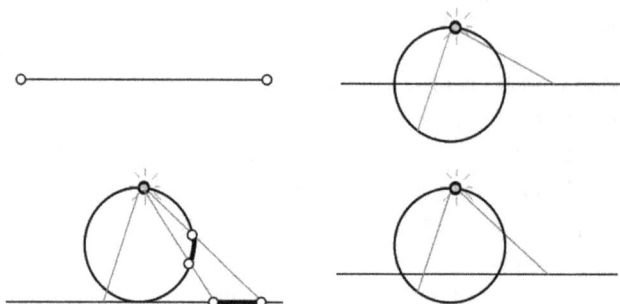

**Figure 3.5:** Stereographic projection showing that an interval $(a, b)$ is homeomorphic to the real line.

A similar stereographic projection in a higher dimension shows that an open disk in $\mathbb{R}^2$, which, by "drawstringing" the boundary, can be bent into the surface of a sphere with the north pole missing, is homeomorphic to the plane $\mathbb{R}^2$ (see Figure 3.6).

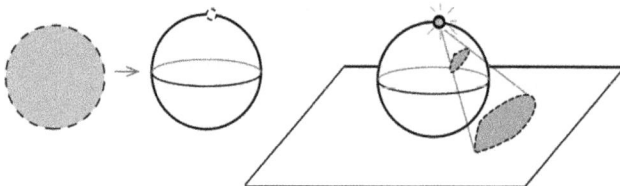

**Figure 3.6:** Stereographic projection showing that an open disc is homeomorphic to the plane.

Often in practice, we will simply need to recognize when bending, stretching, embedding, or projecting gives a homeomorphism, without actually exhibiting the functions doing these contortions. We should recognize, for example, that the interval $[0, 2\pi]$ in the Euclidean line is homeomorphic to the set of points $\{(x, \sin x) \in \mathbb{R}^2 : x \in [0, 2\pi]\}$, which is the graph of one period of the sine function. Embedding an interval in $\mathbb{R}$ into a higher dimensional Euclidean space by viewing it at the graph of a continuous function is an important and useful technique.

**Theorem 3.2.5.** *Suppose $X$ and $Y$ are topological spaces, and $f : X \to Y$ is a continuous function with domain $X$. Then $X$ is homeomorphic to $G = \{(x, f(x)) : x \in X\}$ as a subspace of $X \times Y$. The set $G$ is called the* graph *of $f$.*

*Proof.* We will show that $h : X \to G$ defined by $h(x) = (x, f(x))$ is a homeomorphism. Clearly $h$ is one-to-one and onto. $h^{-1}$ is continuous since, if $U$ is open in $X$, $(h^{-1})^{-1}(U) = h(U) = \{(x, f(x)) : x \in U\} = (U \times Y) \cap G$, which is open in $G$ with the subspace topology. To see that $h$ is continuous, we will show that it is continuous at every point $x \in X$ by showing that, for any neighborhood $W$ of $h(x)$ in $G$, there exists a neighborhood $U'$ of $x$ in $X$ with $h(U') \subseteq W$. Suppose $x \in X$ is given and $W$ is a neighborhood of $h(x)$ in the subspace $G$ of $X \times Y$. Then $W$ contains a neighborhood of form $(U \times V) \cap G$ where $U$ is a neighborhood of $x$ in $X$ and $V$ is a neighborhood of $f(x)$ in $Y$. By the continuity of $f$ at $x$, there exists a neighborhood $U'$ of $x$ in $X$ with $f(U') \subseteq V$. Without loss of generality, we may assume $U' \subseteq U$, for otherwise, replace $U'$ by $U' \cap U$. Now $h(U') = \{(y, f(y)) : y \in U'\} \subseteq U' \times f(U') \subseteq U \times V$, and $h(U') \subseteq G$, so $h(U') \subseteq (U \times V) \cap G \subseteq W$, as needed. $\square$

If a property is defined in terms of the topology, relabeling the points and open sets (by running the space through a homeomorphism) should not destroy the property. This prompts the following definition.

**Definition 3.2.6.** A property $P$ is a *topological property* if whenever a topological space $X$ has the property, $h(X)$ also has the property for every homeomorphisms $h$. That is, a property $P$ is a topological property if it is *preserved by homeomorphisms*.

Topological properties can be used to show that two topological spaces are not homeomorphic. If $X$ has the property and $Y$ does not, then $X$ and $Y$ cannot be topologically equivalent. Of course, if two topological spaces both satisfy some topological property, they need not be homeomorphic. The property of having exactly seven elements is a topological property: If $X$ has exactly seven elements, anything homeomorphic to $X$ also has exactly seven elements. However, it is not the case that any two topological spaces with exactly seven elements are homeomorphic.

In the Euclidean plane, we know many metric or geometric properties which are not topological properties. The homeomorphic image of a right triangle need not be a right triangle, so the property of being a right triangle is not a topological property. Indeed, the geometric property of being a triangle is not a topological property. We may stretch a triangular rubber band into a circle or square through a homeomorphism. Boundedness is not a topological property. The bounded space $(0, 1)$ with the Euclidean topology is homeomorphic to the unbounded space $\mathbb{R}$. Boundedness is defined in terms a metric, and not a topology. Properties which depend on measurements of lengths or measurements of angles will not be topological properties.

Generally, properties defined in terms of open sets, closed sets, and the cardinality of certain subsets of a topological space will be topological properties since homeomorphisms provide bijections between the open sets, the closed sets, and the points of two spaces. The next example combines closed sets and cardinality considerations.

**Example 3.2.7.** A topological space is said to be *separable* if it has a countable dense subset. For example, the Euclidean line $\mathbb{R}$ is separable since $\mathbb{Q}$ is a countable dense subset. We will show that separability is a topological property. If $X$ has a countable dense subset $D$ and $h : X \to Y$ is a homeomorphism, then $h(D)$ is countable since $h$ is a bijection. Since $X = \operatorname{cl} D$, by Theorem 3.2.3, $Y = h(X) = h(\operatorname{cl}(D)) = \operatorname{cl}(h(D))$, so $h(D)$ is dense in $Y$. Thus, $h(D)$ is a countable dense subset of $Y$, so $Y$ is separable.

### 3.2.2 Projection functions

An element of a finite product $\prod_{i=1}^{n} X_i$ is a vector $(x_1, x_2, \ldots, x_n)$ whose $i$th coordinate $x_i$ is an element of the $i$th factor $X_i$ of the product. Given a vector $(x_1, x_2, \ldots, x_n)$ in $\prod_{i=1}^{n} X_i$, we may wish to only focus on the $j$th coordinate, discarding the other coordinates. This process is called projecting the vector onto the $j$th coordinate. The function $\pi_j : \prod_{i \in I} X_i \to X_j$ defined by $\pi_j((x_i)_{i \in I}) = x_j$ is the *projection function onto the $j$th coordinate*. For example, in $\mathbb{R}^3 = \prod_{i \in \{1,2,3\}} \mathbb{R}$, we have $\pi_1((3, -1, 5)) = 3$, $\pi_2((3, -1, 5)) = -1$, and $\pi_3((3, -1, 5)) = 5$.

If each $X_i$ ($i = 1, \ldots, n$) is a topological space and $Z = \prod_{i=1}^{n} X_i$ has the product topology, then projection functions are continuous. For example, to see that $\pi_2 : Z \to X_2$ is continuous, suppose $U_2$ is open in $X_2$. The inverse image $\pi_2^{-1}(U_2)$ consists of all vectors whose projection onto the second coordinate falls in $U_2$. Since there is no restriction on the other coordinates, $\pi_2^{-1}(U_2) = \{(x_1, x_2, \ldots x_n) \in \prod_{i=1}^{n} X_i : x_2 \in U_2\} = X_1 \times U_2 \times X_3 \times \cdots \times X_n$, which, as a product of open sets, is open in the product topology on $Z$. Thus, $\pi_2$ is continuous.

Continuous functions are those for which inverse images of open sets are open. Functions for which the forward images of open sets are open also have a name.

**Definition 3.2.8.** If $X$ and $Y$ are topological spaces, a function $f : X \to Y$ is an *open mapping* or *open function* if and only if $f(U)$ is open in $Y$ for every open set $U$ in $X$. Similarly, $f : X \to Y$ is a *closed mapping* or *closed function* if and only if $f(F)$ is closed in $Y$ for every closed set $F$ in $X$.

Note that any homeomorphism is both an open mapping and a closed mapping.

If an open mapping $f : X \to Y$ is bijective, then it has an inverse function $f^{-1} : Y \to X$, and saying $f(U)$ is open for every open $U$ simply says $(f^{-1})^{-1}(U)$ is open for every open $U$, which is the definition of $f^{-1}$ being continuous. That is, if an open mapping has an inverse function, then the inverse is continuous. The result below provides an example of open mappings which are not invertible.

**Theorem 3.2.9.** *If $X_i$ ($i = 1, \ldots, n$) are topological spaces and $Z = \prod_{i=1}^{n} X_i$ has the product topology, then the projection functions are continuous open mappings.*

*Proof.* The continuity of the projection functions was justified above Definition 3.2.8. To show $\pi_i : Z \to X_i$ is open, suppose $W$ is open in $Z$ and $x \in \pi_i(W)$. Now $x = \pi_i(w_1, \ldots, w_n)$ for some $(w_1, \ldots, w_n) \in W$. Since $W$ is open, it contains a basic neighborhood $B = U_1 \times \cdots \times U_n$ of $(w_1, \ldots, w_n)$, where each $U_i$ is open in $X_i$. Now $U_i = \pi_i(B) \subseteq \pi_i(W)$

is a neighborhood of $x$ contained in $\pi_i(W)$. Since $x$ was an arbitrary element of $\pi_i(W)$, this shows that $\pi_i(W)$ is open, as needed. $\qquad\square$

We note that projections need not be closed maps: The set $F = \{(x, \frac{1}{x}) \in \mathbb{R}^2 : x > 0\}$ is a closed set in $\mathbb{R}^2$, but $\pi_1(F) = (0, \infty)$ is not closed.

### 3.2.3 Coordinate functions

In multivariable calculus, you study vector valued functions of a real variable such as $f : \mathbb{R} \rightarrow \mathbb{R}^3$ defined by $f(t) = (t, \cos t, \sin t)$. For this function, we may write $f(t) = (f_1(t), f_2(t), f_3(t))$ where $f_1(t) = t$, $f_2(t) = \cos t$, and $f_3(t) = \sin t$ are the *coordinate functions* of $f$. You may recall that $f : \mathbb{R} \rightarrow \mathbb{R}^n$ is continuous if and only if each of its coordinate functions is a continuous function from $\mathbb{R}$ to $\mathbb{R}$. This result holds in the more general setting of a function $f$ from any topological space to a product of topological spaces.

**Theorem 3.2.10.** *If $X, Y_1, \ldots, Y_n$ are topological spaces, $Y = Y_1 \times \cdots \times Y_n$ has the product topology, and $f : X \rightarrow Y$ is a function, then $f(x) = (f_1(x), \ldots, f_n(x))$ is continuous from $X$ to $Y$ if and only if each of the coordinate functions $f_i : X \rightarrow Y_i$ is continuous $(i = 1, \ldots, n)$.*

*Proof.* Suppose $f : X \rightarrow Y$ is continuous. Since each projection function $\pi_i : Y \rightarrow Y_i$ is continuous, the composition $\pi_i \circ f = f_i$ is continuous. Thus, each coordinate function is continuous.

Conversely, suppose each coordinate function is continuous and $V = V_1 \times \cdots \times V_n$ is a basic open set in $Y$. Now $f^{-1}(V) = \{x \in X : f(x) \in V\} = \{x \in X : f_i(x) \in V_i \text{ for each } i = 1, \ldots, n\} = \bigcap_{i=1}^n f_i^{-1}(V_i)$, which, as a finite intersection of open sets, is open in $X$. Thus, $f$ is continuous. $\qquad\square$

Thus, the function $f : \mathbb{R} \rightarrow \mathbb{R}^3$ defined by $f(t) = (t, \cos t, \sin t)$ is continuous since each of its coordinate functions is. Recalling that the domain of a function is homeomorphic to the graph of the function (Theorem 3.2.5), we can now conclude that the real line $\mathbb{R}$ is homeomorphic to the helix $H = \{(t, \cos t, \sin t) : t \in \mathbb{R}\}$ in $\mathbb{R}^3$, since $H$ is the graph of $f$.

### Exercises

1. Give $\mathbb{R}$ the Euclidean topology. Which of the following maps from $\mathbb{R}$ to $\mathbb{R}$ are homeomorphisms? Among those that are not homeomorphism, which are open mappings?
   
   (a) $i(x) = x$
   
   (b) $f(x) = x^2$
   
   (c) $g(x) = x^3$

(d) $h(x) = x^3 - 4x$

(e) $j(x) = x + 3$

2. Let $X = \{a, b, c, d, e\}$, $Y = \{1, 2, 3, 4, 5\}$ and $Z = \{r, s, t, u, v\}$ have the topologies generated by the bases shown. Which spaces are homeomorphic? Exhibit a homeomorphism for each homeomorphic pair.

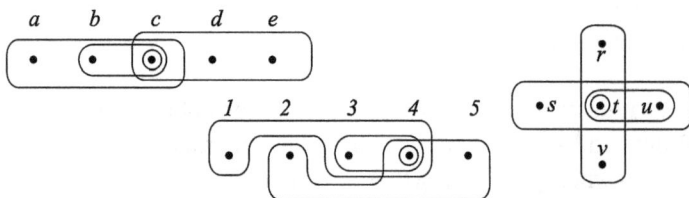

3. For the functions between the finite topological spaces given in Exercise 4 of Section 3.1, determine which are open mappings, which are closed mappings, and which are homeomorphisms.

4. Prove:
   (a) If $X = \{1, 2, 3\}$ and $Y = \{a, b, c, d\}$ have any topologies, $X$ and $Y$ are not homeomorphic.
   (b) The subspaces $\mathbb{Q}$ and $\mathbb{R} - \mathbb{Q}$ of the Euclidean line are not homeomorphic.
   (c) If $X = \{a, b, c\}$ has the discrete topology and $Y = \{a, b, c\}$ has the topology $\{\emptyset, \{a\}, \{a, b\}, \{a, b, c\}, \{c\}, \{a, c\}\}$, then $X$ and $Y$ are not homeomorphic.

5. Suppose that $A, B$, and $C$ are topological spaces.
   (a) Show that if $B \approx C$, then $A \times B \approx A \times C$.
   (b) If $A \times B \approx A \times C$, then it does not follow that $B \approx C$. (Cancellation does not hold for products of topological spaces.) Prove this using $A = \mathbb{N}$, $B = \{0\}$, and $C = \{0, 1\}$ with the usual Euclidean topologies.

6. Recall that $x \in X$ is an isolated point if and only if $\{x\}$ is an open set. Show that if $X$ has exactly $k$ isolated points and $X \approx Y$, then $Y$ has exactly $k$ isolated points. That is, show that "has exactly $k$ isolated points" is a topological property.

7. Prove that the following are topological properties.
   (a) Has a set $A$ whose boundary has exactly two points.
   (b) Has a set $A$ with $\mathrm{cl}\, A = \mathrm{int}\, A$.
   (c) Has an open set which intersects exactly three other open sets.

8. Prove that the following are topological properties.
   (a) Has a nested topology. (Recall that a topology $\mathcal{T}$ is *nested* if $U, V \in \mathcal{T}$ implies $U \subseteq V$ or $V \subseteq U$.)
   (b) Has an open set $U$ with exactly two open sets properly contained in $U$.

9. A topological space $X$ has the *fixed point property* if every continuous function $f : X \to X$ has a fixed point (that is, a point $x \in X$ such that $f(x) = x$.) Show that the fixed point property is a topological property.

10. For which of the spaces $X$ given below is $X \approx X \times X$? Justify your answers.

(a) $X = \{1, 2, 3\}$ with the discrete topology.

(b) $X = \mathbb{N}$ with the discrete topology.

(c) $X = \mathbb{R}$ with the right ray topology $\{(a, \infty) : a \in \mathbb{R}\} \cup \{\emptyset, \mathbb{R}\}$.

11. Find a necessary and sufficient condition on $Y$ to guarantee that $X \times Y \approx X$ for every topological space $X$. Justify your answer.

12. Provide figures to illustrate the steps of proof of Theorem 3.2.5 that $h$ and $h^{-1}$ are continuous, in the case $X = Y = \mathbb{R}$ with the Euclidean topology.

13. Suppose $X$ and $Y$ are topological spaces, $X \times Y$ has the product topology, and $y \in Y$. Show that $X \times \{y\}$ is a homeomorphic copy of $X$ embedded in $X \times Y$.

14. If the north pole of a circle is removed and a light source is placed there, explain why stereographic projection does not provide a homeomorphism to a line $\ell$ not parallel to the tangent at the north pole.

15. Use stereographic projection to explain why the unit sphere in $\mathbb{R}^3$ is homeomorphic to the surface of a cube and to the surface of a pyramid.

16. Suppose $f : X \to Y$ is an open mapping and $A \subseteq X$ has the subspace topology. Is the restriction of $f$ to $A$ an open mapping? What if $A$ is open in $X$?

17. Let $X$ be the real line with the cofinite topology. Which of the functions from $X$ to $X$ given below are open? Which are closed? Justify your answers.

(a) $f(x) = 7$

(b) $g(x) = \arctan x$

18. Let $X$ be the real line with the cofinite topology.

(a) Show that every surjection $f : X \to X$ is a closed mapping.

(b) Must every surjection $f : X \to X$ be an open mapping? Give a proof or counterexample.

19. If $P_a = \{U \subseteq \mathbb{R} : a \in U\} \cup \{\emptyset\}$ is the particular point topology on $\mathbb{R}$ determined by $a$ and $E_b = \{U \subseteq \mathbb{R} : b \notin U\} \cup \{\mathbb{R}\}$ is the excluded point topology on $\mathbb{R}$ determined by $b$, describe:

(a) all open functions $f : (\mathbb{R}, E_a) \to (\mathbb{R}, E_b)$;

(b) all closed functions $f : (\mathbb{R}, E_a) \to (\mathbb{R}, E_b)$;

(c) all open functions $f : (\mathbb{R}, E_a) \to (\mathbb{R}, P_b)$;

(d) all closed functions $f : (\mathbb{R}, E_a) \to (\mathbb{R}, P_b)$.

20. Show that any bijective open mapping is a closed mapping.

21. Suppose $f : X \to Y$ is continuous and $Y$ is Hausdorff. Show that the graph $G = \{(x, f(x)) : x \in X\}$ of $f$ is closed in the product $X \times Y$.

22. The remark before Theorem 3.2.9 suggests that projection functions $\pi_i : \prod_{i=1}^{n} X_i \to X_i$ need not be invertible. Under what circumstances would such a projection function be one-to-one? onto?

23. Sketch three different sets $A$ in $\mathbb{R}^2$ with $\pi_1(A) = \pi_2(A) = (-1, 1)$.

24. Let $h : \mathbb{R} \to \mathbb{R}$ be defined by $h(x) = \frac{x}{2}$, and let $f : \mathbb{R} \to \mathbb{R}^3$ be the function $f(x) = (h(x), (h \circ h)(x), (h \circ h \circ h)(x))$. Describe the set $A = f([0, 4])$ and give $\pi_1(A), \pi_2(A)$, and $\pi_3(A)$.

25. The ellipse $x^2 + \frac{y^2}{4} = 1$ is revolved around the line $x = 2$ to get a surface of revolution $S$ in $\mathbb{R}^3$. Find $\pi_1(S), \pi_2(S)$, and $\pi_3(S)$.

26. In $\mathbb{R}^3$, consider the sets $A = \bigcup_{x \in [0,2)} B((x,0,0),1)$ and $B = B((0,0,0),3) \cup ([0,1] \times (0,5] \times \{3\})$, where $B((x,y,z),\varepsilon)$ represents a ball in the Euclidean metric. Find the projections $\pi_i(A)$ and $\pi_i(B)$ for $i = 1, 2, 3$.

27. Suppose $X, Y$, and $Z$ are sets and $f : X \to Y \times Z$ has coordinate functions $f_1$ and $f_2$, so $f(x) = (f_1(x), f_2(x))$. Consider the statements (a) $f : X \to Y \times Z$ is onto, and (b) $f_1 : X \to Y$ and $f_2 : X \to Z$ are both onto. Does either statement imply the other? Give proofs or counterexamples.

28. Suppose $X, Y$, and $Z$ are sets and $f : X \to Y \times Z$ has coordinate functions $f_1$ and $f_2$, so $f(x) = (f_1(x), f_2(x))$. Consider the statements (a) $f : X \to Y \times Z$ is one-to-one, and (b) $f_1 : X \to Y$ and $f_2 : X \to Z$ are both one-to-one. Does either statement imply the other? Give proofs or counterexamples.

29. *Separate vs. joint continuity.* A function $f : X \times Y \to Z$ is *separately continuous* if for every $a \in X$ and every $b \in Y$, the functions $f(a,y) : Y \to Z$ and $f(x,b) : X \to Z$ are continuous. Separate continuity is not equivalent to $f$ being continuous. To emphasize the difference, if $f$ continuous we may say it is *jointly continuous*.
    (a) Show that (joint) continuity implies separate continuity.
    (b) Show that the real-valued function of two variables $f : \mathbb{R}^2 \to \mathbb{R}$ defined by

    $$f(x,y) = \begin{cases} \frac{2xy}{x^2+y^2} & \text{if } (x,y) \neq (0,0) \\ 0 & \text{if } (x,y) = (0,0) \end{cases}$$

    is separately continuous but not (jointly) continuous. (This is a standard example to illustrate that both partial derivatives $f_x$ and $f_y$ of a function $f$ may exist at $(0,0)$ even though $f$ is not differentiable at $(0,0)$.)

# 4 Connectedness

## 4.1 Connectedness

Here is a brief True/False quiz to check your knowledge of functions from calculus.

___ True ___ False 1. If $f'(x) = 0$ on its domain, then $f(x)$ is constant on its domain.

___ True ___ False 2. If $f(x)$ is continuous on its domain, $f(-2) < 0$, and $f(2) > 0$, then $f(x) = 0$ for some $x \in (-2, 2)$.

On first inspection, these seem to be believable. If a function has zero rate of change, then one may expect it to be constant. The second statement should bring to mind the intermediate value theorem. However, both statements are false, as seen by the function $f : (-\infty, -1) \cup (1, \infty) \to \mathbb{R}$ defined by $f(x) = -1$ if $x < -1$ and $f(x) = 1$ if $x > 1$. The problem with these statements is that they have no restrictions on the domain. The statements do not hold if the domain is not a solid interval. The correct statements are:

1. If $f'(x) = 0$ on an interval, then $f(x)$ is constant on that interval.
2. If $f(x)$ is continuous on the interval $[-2, 2]$, $f(-2) < 0$, and $f(2) > 0$, then $f(x) = 0$ for some $x \in (-2, 2)$.

The important thing about an interval which makes these statements true is that an interval is *connected*—it is all in one piece. The domain $(-\infty, -1) \cup (1, \infty)$ of our counterexample above is not connected: It separates into two disjoint open sets $(-\infty, -1)$ and $(1, \infty)$.

**Definition 4.1.1.** A *separation* of a topological space $X$ is a pair $(U, V)$ of disjoint nonempty open subsets $U$ and $V$ of $X$ such that $X = U \cup V$. A topological space $X$ is *connected* if it has no separation. A subset $A$ of $X$ is a *connected subset* if it is a connected topological space when given the subspace topology.

Thus, $X$ is connected if and only if it cannot be partitioned into two open sets. Since the complement of an open set is closed, we have the following immediate result.

**Theorem 4.1.2.** *For a topological space $X$, the following are equivalent.*
(a) *$X$ is connected.*
(b) *$X$ cannot be written as the union of two disjoint nonempty open sets $U$ and $V$.*
(c) *$X$ cannot be written as the union of two disjoint nonempty closed sets $V$ and $U$.*
(d) *$X$ has no nonempty proper subset $U$ which is both open and closed.*

**Example 4.1.3.** Consider the following subsets of $\mathbb{R}$ with the usual topology.
(a) $X = (0, 1) \cup [2, \infty)$ is not connected since $U = (0, 1)$ and $V = [2, \infty)$ are open sets in $X$ which partition $X$.

https://doi.org/10.1515/9783110686579-005

(b) $\mathbb{Z}$ is not connected since, for example, $U = \{1\}$ and $V = \mathbb{Z} - \{1\}$ form a separation of $\mathbb{Z}$. Indeed, since $\mathbb{Z}$ has the discrete topology, if $U$ is any nonempty, proper subset, then $(U, \mathbb{Z} - U)$ is a separation of $\mathbb{Z}$.

(c) $\mathbb{Q}$ is not connected since $U = (-\infty, \pi) \cap \mathbb{Q}$ and $V = (\pi, \infty) \cap \mathbb{Q}$ form a separation of $\mathbb{Q}$.

**Example 4.1.4.** The real line with the lower limit topology is not connected since, for example, $U = [0, 1)$ is nonempty, proper, closed, and open.

Our next example illustrates a subtlety in dealing with connected subsets of a topological space $X$.

**Example 4.1.5.** Let $X = \{1, 2, 3, 4, 5\}$ with the topology $\mathcal{T} = \{\emptyset, \{1, 2, 3\}, \{3\}, \{3, 4, 5\}, X\}$, and let $A = \{1, 2, 4, 5\} \subseteq X$ as shown in Figure 4.1.

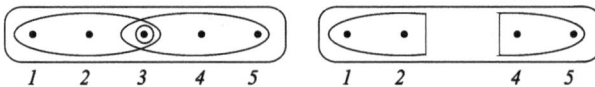

**Figure 4.1:** There is no 'separation' $(U, V)$ of $A = \{1, 2, 4, 5\}$ where $U$ and $V$ are disjoint open sets in $X = \{1, 2, 3, 4, 5\}$, but there is a separation of $A$ in $A$.

Clearly $A$ is not connected, and $(U, V) = (\{1, 2\}, \{4, 5\})$ is the only separation of $A$. Now $U$ and $V$ are open in $A$ and thus they were the intersection of open sets $U', V'$ in $X$ with $A$. The only choices for $U'$ and $V'$ in this example are $U' = \{1, 2, 3\}$ and $V' = \{3, 4, 5\}$, and $U'$ and $V'$ are not disjoint.

This example emphasizes the point that in testing a proper subset $A$ of $X$ for connectedness, any separation must be by disjoint open sets *in the subspace*. We could provide an alternate definition of connectedness of a subspace $A \subseteq X$ by saying $(U, V)$ is a separation of $A$ if $U$ and $V$ are open in $X$, $U \cap A$ and $V \cap A$ are nonempty and disjoint, and $A \subseteq U \cup V$.

The result below will confirm that connectedness is preserved by homeomorphisms and thus is a topological property. In fact, we prove the much stronger result that connectedness is preserved by any continuous function.

**Theorem 4.1.6.** *If* $f : X \to Y$ *is continuous and* $X$ *is connected, then* $f(X)$ *is connected. That is, the continuous image of a connected set is connected.*

*Proof.* Suppose $f : X \to Y$ is continuous and $X$ is connected. We wish to show that $f(X)$ is connected. Suppose to the contrary that $(U, V)$ is a separation of $f(X)$. Now $f^{-1}(U)$ and $f^{-1}(V)$ are disjoint nonempty open subsets of $X$ and $f^{-1}(U) \cup f^{-1}(V) = f^{-1}(U \cup V) = f^{-1}(Y) = X$, so $(f^{-1}(U), f^{-1}(V))$ is a separation of $X$, contrary to $X$ being connected. Thus, $f(X)$ must be connected. $\square$

To prove that a set $A$ is connected in a topological space, we must show that it has no separation. It is common to argue by contradiction, assuming to the contrary that there is a separation of $A$. Proof by contradiction is very useful for proving something is infinite (not finite), uncountable (not countable), connected (no separation), or prime (no nontrivial factorization), since each of these concepts is defined in terms of a negation, that is, in terms of something failing.

The theorem below confirms our intuition about which sets are connected in the real line.

**Theorem 4.1.7.** *A is connected in $\mathbb{R}$ with the usual topology if and only if $A$ is an interval (including rays and $\mathbb{R} = (-\infty, \infty)$).*

*Proof.* If $A$ is not an interval, then there exist points $a, c \in A$ and $b \notin A$ with $a < b < c$. Now $U = (-\infty, b) \cap A$ and $V = (b, \infty) \cap A$ provide a separation of $A$, so $A$ is not connected.

If $A$ is an interval, to see that $A$ is connected, suppose to the contrary that $(U, V)$ is a separation of $A$. Pick $u \in U$ and $v \in V$. Either $u < v$ or $v < u$. Without loss of generality, assume $u < v$. (If $v < u$, we may interchange the labels on $U$ and $V$, making $u < v$.) Consider the set $S = \{y \in \mathbb{R} : [u, y] \subseteq U\}$. Now $S$ is nonempty since $u \in S$, and $S$ is bounded above by $v$. Thus, by the completeness of the reals, $b = \sup S$ exists. Now $u \le b \le v$ and $A$ is an interval containing $u$ and $v$, so $b \in A = U \cup V$. If $b$ is in the open set $U$, then there exists a basic neighborhood $B(b, \varepsilon) = (b - \varepsilon, b + \varepsilon)$ of $b$ contained in $U$. Then $[u, b + \varepsilon/2] \subseteq U$, so $b + \varepsilon/2 \in S$, contrary to $b = \sup S$. If $b$ is in the open set $V$, then there exists a basic neighborhood $B(b, \varepsilon) = (b - \varepsilon, b + \varepsilon)$ of $b$ contained in $V$. Since $b - \varepsilon/2 \in V$, we have $[u, b - \varepsilon/2] \not\subseteq U$, so $b - \varepsilon/2$ is an upper bound of $S$, contrary to $b = \sup S$. $\qquad \square$

The previous two theorems have a significant interpretation for calculus-type functions.

**Corollary 4.1.8** (The intermediate value theorem). *Suppose $[a, b] \subseteq \mathbb{R}$, $[a, b]$ and $\mathbb{R}$ have the Euclidean topology, $f : [a, b] \to \mathbb{R}$ is continuous, and $z$ is any value between $f(a)$ and $f(b)$. Then there exists a point $c \in [a, b]$ with $f(c) = z$.*

*Proof.* The interval $[a, b]$ is connected, so $f([a, b])$ is connected in $\mathbb{R}$. That is, $f([a, b])$ is an interval. Thus, since $f(a)$ and $f(b)$ are in $f([a, b])$, any point $z$ between $f(a)$ and $f(b)$ is in $f([a, b])$, so $z = f(c)$ for some $c \in [a, b]$. $\qquad \square$

As a topological property, connectedness (and variations of it) may be used to tell when two topological spaces are not homeomorphic. The intervals $[0, 1]$ and $(0, 1)$ are both connected, but we would not expect them to be homeomorphic. We should expect that there is no point in $(0, 1)$ which is topologically like the point 1 in $[0, 1]$. To see this, consider what happens when we remove 1 from $[0, 1]$. Now if $h : [0, 1] \to (0, 1)$ were a homeomorphism, then removing $h(1)$ from $h([0, 1]) = (0, 1)$ should give a space homeomorphic to $[0, 1] - \{1\} = [0, 1)$, which is connected. However, removing any point $h(1)$ from $(0, 1)$ leaves a disconnected space. This shows that $[0, 1] \not\approx (0, 1)$. Technically, we used the fact that if $h : X \to Y$ is a homeomorphism, then $h : X - \{a\} \to Y - \{h(a)\}$ is a homeomorphism.

This discussion motivates the following definition.

**Definition 4.1.9.** Suppose $X$ is a connected topological space. An element $a \in X$ is a *cut point* if $X - \{a\}$ is not connected. Otherwise, $x$ is a *non-cut point*. A subset $A \subseteq X$ such that $X - A$ is not connected is called a *cut set* of $X$.

It is easy to see that a homeomorphism carries cut points to cut points and non-cut points to non-cut points, so the cardinality of the set of cut-points (or non-cut points) in a connected topological space is a topological property. Similar remarks apply to cut sets. Thus, since $[0, 1]$ has exactly two non-cut points 0 and 1, and $(0, 1)$ has no non-cut points, these spaces are not homeomorphic. Viewed as subspaces of the Euclidean plane, the letter H is not homeomorphic to the letter T since H has a subset of cardinality 4 which is not a cut set, but every subset of T with cardinality 4 is a cut set.

Our next result shows that if $A$ is a connected subset of a topological space, then adding any boundary points of $A$ results in a connected set.

**Theorem 4.1.10.** *If $A$ is a connected set in a topological space $X$ and $A \subseteq B \subseteq \operatorname{cl} A$, then $B$ is connected.*

*Proof.* Suppose $A$ is connected in $X$ and $A \subseteq B \subseteq \operatorname{cl} A$. To see $B$ is connected, suppose to the contrary that there exists a separation $(U, V)$ of $B$. Now $U \cap A$ and $V \cap A$ are disjoint open sets in $A$ whose union is $A$, so if they are both nonempty, then they would provide a separation of $A$, contradicting the connectedness of $A$. Since $U \neq \emptyset$, there exists $x \in U \subseteq \operatorname{cl} A = A \cup \partial A$, so either $x \in A$ or $x \in \partial A$. In either case, the neighborhood $U$ of $x$ must intersect $A$. Similarly, $V \cap A \neq \emptyset$. □

Note in particular that this theorem says that the closure of a connected set is connected.

Next we will show that the product of two connected spaces is connected. We will need the two nice results below.

**Lemma 4.1.11.** *If $A$ is a connected subset of $B$ and $(U, V)$ is a separation of $B$, then either $A \subseteq U$ or $A \subseteq V$.*

*Proof.* If $U \cap A$ and $V \cap A$ are both nonempty, then they form a separation of $A$, contrary to the connectedness of $A$. Since $A \subseteq B = U \cup V$, if $U \cap A = \emptyset$, then $A \subseteq V$, and if $V \cap A = \emptyset$ then $A \subseteq U$. □

**Theorem 4.1.12.** *Suppose $I$ is an arbitrary index set and for each $i \in I$, $A_i$ is a connected subset of a topological space $X$. If $\bigcap_{i \in I} A_i \neq \emptyset$, then $\bigcup_{i \in I} A_i$ is connected.*

*Proof.* Under the hypotheses, suppose $a \in \bigcap_{i \in I} A_i$ and $\bigcup_{i \in I} A_i$ has a separation $(U, V)$. Without loss of generality, say $a \in U$. By Lemma 4.1.11, each $A_i$ is either contained in

$U$ or in $V$. Since $a \in U$, we have $\bigcup_{i \in I} A_i \subseteq U$ and thus $V \cap \bigcup_{i \in I} A_i = \emptyset$, contrary to $(U, V)$ being a separation of $\bigcup_{i \in I} A_i$. Thus, $\bigcup_{i \in I} A_i$ can have no separation. $\qquad\square$

**Theorem 4.1.13.** *A finite product* $X_1 \times X_2 \times \cdots \times X_n$ *of topological spaces is connected if and only if each factor* $X_i$ *$(i = 1, \ldots, n)$ is connected.*

*Proof.* If $X_1 \times \cdots \times X_n$ is connected, then each factor $X_i = \pi_i(X_1 \times \cdots \times X_n)$ is the image of a connected set under the continuous projection function and thus is connected.

To show the converse, we will show that $X_1 \times X_2$ is connected if $X_1$ and $X_2$ are. The result for any finite number of factors would then follow by iteratively applying this result. Suppose $X_1$ and $X_2$ are connected. Select a point $(a, b) \in X_1 \times X_2$. Now by the previous theorem, the union of all connected subsets of $X_1 \times X_2$ containing $(a, b)$ is connected. Thus, it suffices to show that, for any $(x, y) \in X_1 \times X_2$, there exists a connected set containing $(a, b)$ and $(x, y)$. Now $(a, b)$ is in the set $S_{(a,y)} = (\{a\} \times X_2) \cup (X_1 \times \{y\})$. Since $\{a\} \times X_2$ is homeomorphic to $X_2$, it is connected. Similarly, $X_1 \times \{y\}$ is connected. These two connected sets intersect at the point $(a, y)$, and thus $S_{(a,y)}$ is connected by Theorem 4.1.12. Now $X_1 \times X_2 = \bigcup_{y \in X_2} S_{(a,y)}$ is connected. $\qquad\square$

An important consequence of Theorem 4.1.12 used in the proof above is that, for any point $a$ in a topological space $X$, there exists a largest connected set containing $a$, namely the union of all connected sets in $X$ which contain $a$.

**Definition 4.1.14.** If $a$ is an element of a topological space $X$, the largest connected set containing $a$ is called the *connected component of* $a$.

Clearly every $x \in X$ is in the connected component determined by $x$, so each connected component is nonempty and the union of all connected components is $X$. If $C$ and $D$ are connected components and $c \in C \cap D$, then both $C$ and $D$ are the largest connected set containing $c$, so $C = D$. Thus, the connected components of $X$ are mutually disjoint. This proves the following result.

**Theorem 4.1.15.** *The connected components of a topological space $X$ form a partition of $X$.*

Every partition generates and is generated by an equivalence relation. The equivalence relation generating the connected components of $X$ is $a \approx b$ if and only if there exists a connected set $C$ in $X$ which contains both $a$ and $b$.

It should also be clear that a homeomorphism carries connected components to connected components, so the number of connected components is a topological property. We may use this to show, for example, that the polar graphs of $r = \sin(3\theta)$ and $r = \sin(2\theta)$, depicted in Figure 4.2, are not homeomorphic subsets of the Euclidean plane. Each has exactly one cut point, but removing the cut point leaves a space with three connected components in one case and a space with four connected components in the other case.

**Figure 4.2:** Two roses.

## Exercises

1. Is the empty set connected? Justify your answer.
2. Show that if $A$ is a connected subset of $X$ and $X$ is a subspace of $Y$, then $A$ is a connected subset of $Y$.
3. Show that $X$ is not connected if and only if there exists a continuous onto function $f : X \to \{0, 1\}$ where $\{0, 1\}$ has the discrete topology.
4. Show that $(X, \mathcal{T})$ is not connected if there exists a nonempty proper subset $A$ with $\partial A = \emptyset$.
5. Find all the nonempty connected subsets in each of the topological spaces given below. Justify your answers.
    (a) $\mathbb{R}$ with the discrete topology.
    (b) $\mathbb{R}$ with the indiscrete topology.
    (c) $\mathbb{R}$ with the right ray topology $\{(a, \infty) : a \in \mathbb{R}\} \cup \{\emptyset, \mathbb{R}\}$.
    (d) $\mathbb{R}$ with the cofinite topology.
6. Describe all connected subsets of $\mathbb{R}_l$. If $X$ is a connected topological space, characterize all continuous functions $f : X \to \mathbb{R}_l$.
7. Prove or give a counterexample: If $f : X \to Y$ is continuous and $B$ is a connected subset of $Y$, then $f^{-1}(B)$ is connected.
8. Prove or give a counterexample: If $f : \mathbb{R} \to \mathbb{R}$ is increasing, then the inverse image of every connected set in $f(\mathbb{R})$ is connected.
9. **(One-point connectification)** Suppose $(X, \mathcal{T})$ is a topological space and $Y = X \cup \{\omega\}$ for some point $\omega \notin X$. Give $Y$ the topology $\mathcal{T}_Y = \mathcal{T} \cup \{Y\}$. Show that $(Y, \mathcal{T}_Y)$ is connected. $(Y, \mathcal{T}_Y)$ is called the *one-point connectification* of $X$.
10. Prove that a homeomorphism carries cut points to cut points and non-cut points to non-cut points.
11. Viewing the letters A, E, O, P and W as subspaces of the Euclidean plane, how many cut points and non-cut points does each have?
12. Identify the cut points and non-cut points in the connected subspaces of the Euclidean plane given below. Are any of these topological spaces homeomorphic?
    (a) $\{(x, y) : x^2 + y^2 \le 1\}$
    (b) $\{(x, y) : x^2 + y^2 \le 1\} \cup \{(x, y) : (x - 2)^2 + y^2 \le 1\}$
    (c) $\{(x, y) : x^2 + y^2 \le 1\} \cup \{(x, y) : (x - 3)^2 + y^2 \le 1\} \cup \{(x, 0) : x \in [0, 3]\}$
    (d) $\{(x, y) : -|\sin x| \le y \le |\sin x|\}$

13. Provide a proof or counterexample: The interior of a connected set is connected.
14. Give examples to show that the neither the intersection of connected sets nor the union of connected sets must be connected.
15. Suppose $X$ is a topological space with the property that, for every $a, b \in X$ with $a \neq b$, cl$\{a\} \cap$ cl$\{b\} \neq \emptyset$. Show that $X$ is connected. Does the converse hold? Justify your answer.
16. Suppose that $A_i$ is a connected subset of topological space $X$ for $i = 1, \ldots, n$, and $A_i \cap A_{i+1} \neq \emptyset$ for $i = 1, \ldots, n-1$. Show that $\bigcup_{i=1}^{n} A_i$ is connected.
17. Show that every connected component of a topological space is a closed set.
18. Show that the homeomorphic image of a connected component is a connected component.
19. Use the connected components to show that the following topological spaces are not homeomorphic.
    (a) $A = \{(x, y) \in \mathbb{R}^2 : 0 \leq x < 4\pi, 0 < y \leq \sin x\}$
    (b) $B = \{(x, y) \in \mathbb{R}^2 : 0 \leq x < 4\pi, -2 < y \leq \sin x\}$
    (c) $C = \{(x, y) \in \mathbb{R}^2 : 0 \leq x < 4\pi, 0 < y \leq \cos x\}$
20. With $A = B = [0, 1)$ and $C = (0, 1)$, show that $A \times B \approx A \times C$ does not imply $B \approx C$.
21. Viewing the letters T, H, i, and X as subspaces of the Euclidean plane, use the concept of connected components to show that no two are homeomorphic.
22. For a topological space $(X, \mathcal{T})$, show that the following are equivalent:
    (a) Every nonempty open set in $X$ is dense in $X$.
    (b) Every pair of nonempty open sets in $X$ has a nonempty intersection.
    (c) Every open set in $X$ is connected.
    A topological space satisfying these conditions is called *hyperconnected*.
23. If $X$ is a topological space, show that the relation $a \approx b$ if and only if there exists a connected set $C$ in $X$ which contains both $a$ and $b$ is an equivalence relation and the equivalence classes are the connected components.
24. If $A$ is a closed, connected cut set of a connected space $X$ and $(U, V)$ is a separation of $X - A$, show that $U \cup A$ is connected.
25. Show that every open set in the Euclidean line $\mathbb{R}$ is the union of countably many mutually disjoint open intervals.
26. Prove that if $f : \mathbb{R} \to \mathbb{R}$ is continuous and one-to-one, then $f$ is strictly increasing or strictly decreasing. (Note that $f$ is strictly increasing if and only if $x < y < z$ implies $f(x) < f(y) < f(z)$.)

## 4.2 Path connectedness

Consider the subsets $A = B((0, 0), 2) \cup ([1, 4] \times [-3, 0])$ and $C = B((0, 0), 2) \cup ([3, 4] \times [-3, 0])$ of $\mathbb{R}^2$, as shown in Figure 4.3. Given any two points $a$ and $b$ in the subset $A$, we can connect the points with a continuous path which stays entirely in the set $A$.

However, the set $C$ does not have this property. We will say $A$ is path connected and $C$ is not. The following definition makes this precise.

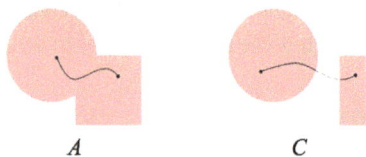

**Figure 4.3:** A is path connected. C is not.

**Definition 4.2.1.** Suppose $X$ is a topological space and $[0, 1]$ has the Euclidean topology. For points $a, b \in X$, a *path* from $a$ to $b$ in $X$ is a continuous function $f : [0, 1] \to X$ with $f(0) = a$ and $f(1) = b$. The topological space $X$ is *path connected* if, for every pair of points $a, b \in X$, there is a path in $X$ from $a$ to $b$. A subset $A \subseteq X$ is said to be path connected if it is a path connected topological space when given the subspace topology.

If $f : [0, 1] \to X$ is a path from $a$ to $b$ in $X$, then $g : [0, 1] \to X$ defined by $g(t) = f(1-t)$ is a path from $b$ to $a$ obtained by tracing the path of $f$ backwards. Thus, if there is a path from $a$ to $b$, we may simply say there is a path connecting $a$ and $b$, without specifying which is the initial point and which is the final point.

The Euclidean space $\mathbb{R}^n$ is path connected. For points $a, b \in \mathbb{R}^n$, the *straight-line path* $f : [0, 1] \to \mathbb{R}^n$ defined by $f(t) = tb + (1 - t)a$ is a path in $\mathbb{R}^n$ from $a$ to $b$. If $A \subseteq \mathbb{R}^n$ and $a, b \in A$, the straight-line path from $a$ to $b$ may or may not lie in $A$. A set $A \subseteq \mathbb{R}^n$ is *(vector-space) convex* if, for every $a, b \in A$, all points $tb + (1 - t)a$ ($t \in [0, 1]$) of the straight-line path from $a$ to $b$ are contained in $A$. Clearly any vector-space convex subset of $\mathbb{R}^n$ is path connected, but not every path connected subset of $\mathbb{R}^n$ is vector-space convex.

The subset $A = B((0, 0), 2) \cup ([1, 4] \times [-3, 0])$ of $\mathbb{R}^2$ shown in Figure 4.3 is easily seen to be not vector-space convex. We formally prove that it is path connected. Given any two points $(a, b)$ and $(c, d)$ in $A$, we need a path in $A$ from $(a, b)$ to $(c, d)$. If $(a, b)$ and $(c, d)$ are both in the ball $B = B((0, 0), 2)$ then, by the convexity of the ball, the straight-line path $f : [0, 1] \to B$ defined by $f(t) = t(c, d) + (1 - t)(a, b)$ connecting $(a, b)$ and $(c, d)$ stays within $B$, and thus within $A$. A similar argument applies if both points were in the rectangle $R = [1, 4] \times [-3, 0]$. Now suppose $(a, b)$ and $(c, d)$ are not both in $B$ and not both in $R$. Without loss of generality, suppose $(a, b) \in B$ and $(c, d) \in R$.

Suppose $(y, z)$ is any point in $B \cap R$. Now the straight-line paths $f$ connecting $(a, b)$ to $(y, z)$ and $g$ connecting $(y, z)$ to $(c, d)$ lie within $B$ and $R$, respectively. Tracing these paths consecutively will give a path from $(a, b)$ to $(c, d)$ inside $A$, but we must adjust the parameters to make this fit our definition of a path. The paths $f$ and $g$ are parameterized by

$f : [0,1] \to B \subseteq A$ defined by $f(t) = t(y,z) + (1-t)(a,b)$

$g : [0,1] \to R \subseteq A$ defined by $g(t) = t(c,d) + (1-t)(y,z)$.

Now to get a path $p : [0,1] \to A$ from $(a,b)$ to $(c,d)$, we would like to trace $f$ during the time interval $[0, \frac{1}{2}]$ and trace $g$ during the time interval $[\frac{1}{2}, 1]$. The function $p : [0,1] \to A$ defined by

$$p(t) = \begin{cases} f(2t) & \text{if } t \in [0, \frac{1}{2}], \\ g(2t - 1) & \text{if } t \in [\frac{1}{2}, 1], \end{cases}$$

does this. Furthermore, by the pasting lemma, $p$ is continuous.

We will next consider how path connectedness is related to connectedness.

**Theorem 4.2.2.** *If a topological space $X$ is path connected, then it is connected.*

*Proof.* Suppose to the contrary that the implication fails. Then there exists a topological space $X$ which is path connected but not connected. Let $(U, V)$ be a separation of $X$. Since $U$ and $V$ are nonempty, there exist $u \in U$ and $v \in V$. Let $f : [0,1] \to X$ be a path in $X$ connecting $u$ and $v$. Now $f([0,1])$ is the continuous image of a connected set and is thus connected. But $(U, V)$ is a separation of $f([0,1])$. □

Connectedness and path connectedness are not equivalent, however, as seen by the following example.

**Example 4.2.3.** In the Euclidean plane, let $C = \{(x,y) : x^2 + y^2 = 4\}$ be the circle of radius 2 and let $S$ be the set described in polar coordinates by $\{(r,\theta) : r = 2 - \frac{1}{\theta}, \theta \geq 1\}$. The graph of $S$ spirals out, approaching $C$ in the limit. Let $A = S \cup C$.

Now $A$ is connected since $A = \text{cl } S$, and, as the continuous image of the connected set $[1, \infty)$ under the polar map $r = 2 - \frac{1}{\theta}$, $S$ is connected.

The set $A$ is not path connected. Intuitively, a path that starts on the circle $C$ cannot jump to the spiral $S$.

Suppose that $A$ is path connected. We will first show that there is a path $f : [0,1] \to A$ from some point $c$ on the circle $C$ to some point $s$ on the spiral $S$ with $f(0) = c$ and $f((0,1]) \subseteq S$. Indeed, suppose $c' \in C$, $s \in S$, and $g$ is a path from $c'$ to $s$ in $A$. Since $C$ is closed and $g$ continuous, $g^{-1}(C)$ is closed in $[0,1]$. Let $t_0 = \sup g^{-1}(C)$. Any closed set in $[0,1]$ contains its supremum, so $t_0 \in g^{-1}(C)$, and thus $g(t_0) \in C$ and $g((t_0, 1]) \subseteq S$. Now $c = g(t_0)$ is the "last point" on the path $g$ which is in $C$, and we may rescale the domain of $g|_{[t_0,1]}$ to make it a path. Specifically, the function $f : [0,1] \to A$ defined by $f = g \circ h$ where $h$ is the linear function mapping $[0,1]$ to $[t_0, 1]$ has the desired properties.

Now $f$ is continuous at the point $t = 0$, so for the neighborhood $V = B(f(0), 1) \cap A$ of $f(0) = c$, there exists a basic neighborhood $U = [0, \varepsilon)$ of 0 with $f(U) \subseteq V$. Now the neighborhood $V$ of $c \in C$ has infinitely many connected components: one on $C$ and infinitely many arcs of the spiral. But $f(U)$ is a connected subset of $V$ and thus must be contained in a single component of $V$. But $f(U)$ contains $c \in C$ and $f(\frac{\varepsilon}{2}) \in S$, which are not in the same component of $V$. This contradiction shows that $A = C \cup S$ is not path connected.

The closure of a connected set is connected, but the example above illustrates that the closure of a path connected set need not be path connected.

Path components will be defined similarly to connected components.

**Definition 4.2.4.** If $X$ is a topological space, the equivalence classes under the equivalence relation $x \sim y$ if and only if there is a path in $X$ from $x$ to $y$ are the *path components* of $X$.

The path component of $x$ is the largest path connected set containing $x$.

## Exercises

1.  Show that the continuous image of a path connected space is path connected (and thus, path connectedness is a topological property).
2.  Let $X$ be the subset of the Euclidean plane described in polar coordinates by $\{(r, \theta) : r \in [1, 2], \theta \in [-3\pi/4, 3\pi/4]\}$. Give explicit equations for paths to prove that $X$ is path connected.
3.  Let $Y$ be the subset of the Euclidean plane $([-1, 0] \times [0, 1]) \cup ([1, 2] \times [0, 2]) \cup \{(x, x) : x \in [0, 1]\}$. Give explicit equations for paths to prove that $Y$ is path connected.
4.  The *comb space* is the subspace of the Euclidean plane

$$C = \left\{ \left\{ \frac{1}{n} \right\} \times [0, 1] : n \in \mathbb{N} \right\} \cup ([0, 1] \times \{0\}) \cup (\{0\} \times [0, 1]\}).$$

The *deleted comb space* is the subspace of the Euclidean plane

$$C^* = \left\{ \left\{ \frac{1}{n} \right\} \times [0, 1] : n \in \mathbb{N} \right\} \cup ([0, 1] \times \{0\}) \cup \{(0, 1)\}$$

obtained by removing $\{0\} \times [0, 1)$ from the comb space.
    (a) Draw the comb space and determine whether it is connected or not and whether it is path connected or not.
    (b) Draw the deleted comb space and show that it is connected but not path connected.
5.  Give examples of the following, or show that no such example can exist.
    (a) Two path connected spaces $X$ and $Y$ such that $X \cap Y$ has three path components.
    (b) Two path connected spaces $X$ and $Y$ such that $X \cup Y$ has three path components.
6.  Consider the following subspaces of the Euclidean plane.
    $S = \{(x, \sin(\frac{1}{x})) \in \mathbb{R}^2 : 0 < x \le 1\} \cup \{(0, 0)\}$

$T = S \cup \{(0,1)\}$

$U = S \cup \{(0,y) \in \mathbb{R}^2 : \frac{1}{2} < y \le 1\}$

$V = S \cup \{(0,y) \in \mathbb{R}^2 : y \in [0,1] \cap \mathbb{Q}\}$

$W = S \cup \{(0,y) \in \mathbb{R}^2 : y \in [0,1] - \mathbb{Q}\}$

The set $S$ is the *topologist's sine curve*. Sketch the sets $S, T, U, V$, and $W$, determine the connected components and the path components of each, and show that no two of these are homeomorphic.

7. Show that if $A$ and $B$ are two path connected sets in a topological space $X$ and $A \cap B \ne \emptyset$, then $A \cup B$ is path connected.

8. Suppose $X$ is a topological space and $\sim$ is the relation on a topological space $X$ defined by $x \sim y$ if and only if there is a path in $X$ from $x$ to $y$. Prove that $\sim$ is an equivalence relation.

9. Prove that the path component $P$ of $x$ is the largest path connected set containing $x$ by showing (a) $P$ is path connected, and (b) if $Q$ is path connected and contains $x$, then $Q \subseteq P$.

10. If $A$ is an open set in the Euclidean space $\mathbb{R}^n$, show that $A$ is connected if and only if it is path connected.

11. If $[0,1]$ has the Euclidean topology and $X$ is any topological space, an *arc* from $a$ to $b$ in $X$ is a continuous one-to-one function $f : [0,1] \to X$ with $f(0) = a$ and $f(1) = b$. Thus, an arc is a one-to-one path. $X$ is *arc connected* if, for any pair of distinct points $a, b \in X$, there exists an arc from $a$ to $b$. Consider the space $X = \{1,2,3\}$ with the topology generated by the basis $\{\{1\}, \{3\}, X\}$. Show that $X$ is path connected but not arc connected.

# 5 Compactness

## 5.1 Compactness

The spaces $X = (0,1)$ and $Y = [0,1]$, each with the Euclidean topology, are not topologically equivalent: every point of $(0,1)$ is a cut point, while $[0,1]$ has two non-cut points. However, cut points and other elementary connectedness arguments are not adequate to distinguish between the open ball $B = \{(x,y) \in \mathbb{R}^2 : x^2 + y^2 < 1\}$ and the closed ball $C = \{(x,y) \in \mathbb{R}^2 : x^2 + y^2 \leq 1\}$ as subspaces of the Euclidean plane, since both sets have no cut points. So, let us consider other ways to show that the topological space $X = (0,1)$ is not topologically equivalent to the space $Y = [0,1]$, which might generalize to show $B$ and $C$ are not homeomorphic.

In terms of the order, $[0,1]$ contains a largest element, while $(0,1)$ does not, but this only shows that these sets are not "order equivalent". From our intuition with the real line, we are tempted to say $[0,1]$ contains its boundary points, but $(0,1)$ does not. This argument requires external knowledge (outside of $X$ or $Y$) of how these intervals are embedded in a larger space $\mathbb{R}$, which was not originally mentioned. Indeed, staying within the sets given, neither topological space $X = (0,1)$ or $Y = [0,1]$ has any boundary points.

The concept of compactness provides a purely internal way to capture the idea that a space like $[0,1]$ is not "missing any limits". A space like $(0,1)$ is "missing some limits", and will not be compact. The idea of compactness is suggested by the following example.

**Example 5.1.1.** Let $X = (0,1)$ and $Y = [0,1]$, each with the topology generated by the Euclidean metric. We will show that $X$ and $Y$ are not homeomorphic. Consider the open subsets $U_n = (\frac{1}{n},1)$ of $X$. We say that the collection $\{U_n : n = 2,3,\ldots\}$ *covers* $X = (0,1)$ since $\bigcup_{n=2}^{\infty} U_n = X$. Furthermore, there is no finite subcollection of the open sets $U_n$ whose union gives $(0,1)$. This is an internal way to capture the idea that the sequence $(\frac{1}{n})_{n=1}^{\infty}$ in $(0,1)$ "converges to the missing boundary point 0" (and thus, $(0,1)$ is not compact). Now suppose $h : X \to Y$ is a homeomorphism. Since $\{U_n : n = 2,3,\ldots\}$ is a collection of nested connected open subsets of $X$ whose union is $X$, it follows that $\{h(U_n) : n = 2,3,\ldots\}$ is a collection of nested connected open subsets of $Y$ whose union is $h(X) = Y$. Furthermore, since there is no finite subcollection of $\{U_n : n = 2,3,\ldots\}$ which covers $X$, there is no finite subcollection of $\{h(U_n) : n = 2,3,\ldots\}$ which covers $Y$. Since $0 \in Y = \bigcup_{n=2}^{\infty} h(U_n)$, there exists $j \geq 2$ with $0 \in h(U_j)$, and similarly, since $1 \in Y = \bigcup_{n=2}^{\infty} h(U_n)$, there exists $k \geq 2$ with $1 \in h(U_k)$. Because $h(U_j)$ and $h(U_k)$ are nested, $\{0,1\} \subseteq h(U_m)$ where $m$ is either $j$ or $k$. Since $h(U_m)$ is a connected set containing $\{0,1\}$, it contains $[0,1]$. Thus, $\{h(U_m)\}$ is a finite subcollection of $\{h(U_n) : n = 2,3,\ldots\}$ which covers $Y = [0,1]$, a contradiction.

While this example includes the crux of the concept of compactness, it also used some unnecessary restrictions which allowed us to easily illustrate the idea. Indeed,

https://doi.org/10.1515/9783110686579-006

the open sets $\{U_n : n = 2, 3, \ldots\}$ covering $X$ need not be nested: given any collection of open subsets, by taking the union of successively more and more of them, we get a nested collection. Also, the use of countable indexing on our collection of open sets was an unnecessary convenience. Furthermore, connectedness plays no role in the definition of compactness. In this specific example, we simply used to our advantage the fact that the topologies in question have bases of connected sets. We are now ready for the definition of compactness.

**Definition 5.1.2.** Suppose $X$ is a topological space and $A \subseteq X$. An *open cover* of $A$ is a collection $\mathcal{C}$ of open sets in $X$ whose union contains $A$. If $\mathcal{C}$ is an open cover of $A$, a *finite subcover* of $A$ from $\mathcal{C}$ is a finite subcollection of $\mathcal{C}$ which is an open cover of $A$. The set $A \subseteq X$ is *compact* if and only if every open cover of $A$ has a finite subcover.

Not only is compactness a topological property (preserved by homeomorphisms), it is preserved by any continuous function.

**Theorem 5.1.3.** *The continuous image of a compact set is compact. That is, if $f : X \to Y$ is a continuous function and $A \subseteq X$ is compact, then $f(A)$ is compact in $Y$.*

*Proof.* Suppose $f : X \to Y$ is a continuous function, $A \subseteq X$ is compact, and $\mathcal{C}$ is an open cover of $f(A)$. Then $f^{-1}(\mathcal{C}) = \{f^{-1}(V) : V \in \mathcal{C}\}$ is an open cover of $A$, so there exists a finite subcover $\{f^{-1}(V_1), \ldots, f^{-1}(V_n)\}$ covering $A$. Now $\{V_1, \ldots, V_n\}$ is a finite subcover from $\mathcal{C}$ covering $f(A)$. □

In Example 5.1.1, we saw that $\mathcal{C} = \{U_n : n = 2, 3, \ldots\} = \{(\frac{1}{n}, 1)\}_{i=2}^{\infty}$ is an open cover of $(0, 1)$ which has no finite subcover, so $(0, 1)$ with the Euclidean topology is not compact. That example did not show that $[0, 1]$ is compact. We only showed that one particular open cover of $[0, 1]$, namely $\{h(U_n) : U_n \in \mathcal{C}\}$, had a finite subcover. The compactness of $[0, 1]$ requires showing that every open cover has a finite subcover, and will be addressed in the next theorem. The collection $\mathcal{C}_1 = \{(x - \frac{1}{4}, x + \frac{1}{4}) : \frac{1}{4} < x < \frac{3}{4}\}$ is another open cover of $(0, 1)$ which has no finite subcover, again showing that $(0, 1)$ is not compact. Of course, some open covers of $(0, 1)$, such as $\mathcal{C}_2 = \{(0, 1)\}$, $\mathcal{C}_3 = \{(0, \frac{3}{4}), (\frac{2}{3}, 1)\}$, or $\mathcal{C}_4 = \{(0, b) : \frac{1}{2} < b < 1\} \cup \{(a, 1) : \frac{1}{2} < a < 1\}$ will have finite subcovers. Note that $\mathcal{C}_3$ is a finite subcover of $\mathcal{C}_3$ and of $\mathcal{C}_4$. Typically, a topological space will have so many open sets that it will be impossible to list every open cover, so compactness cannot be verified by an exhaustive check for finite subcovers. Techniques such as those in the result below will be helpful.

**Theorem 5.1.4.** *If $a < b$ in $\mathbb{R}$, the closed interval $[a, b]$ is a compact subset of the Euclidean line.*

*Proof.* Since every closed interval $[a, b]$ is homeomorphic to $[0, 1]$, it suffices to show that $[0, 1]$ is compact. Suppose $\mathcal{C}$ is an arbitrary open cover of $[0, 1]$. Our goal is to show that we can cover the entire interval $[0, 1]$ with a finite subcollection from $\mathcal{C}$.

Now there exists at least one open set $C_1 \in C$ which contains 0. This set will contain a basic neighborhood of 0, so the finite subcollection $\{C_1\}$ of $C$ covers $[0, \varepsilon)$ for some $\varepsilon > 0$. Our strategy will be to see how far from 0 we can extend to the right and still be covered by a finite subcollection. Let

$$y = \sup\{x \in [0,1] : [0,x) \text{ can be covered by a finite subcollection of } C\}.$$

As the supremum of a nonempty set bounded above by 1, $y$ exists by the completeness property of $\mathbb{R}$. Choose $C_0 \in C$ with $y \in C_0$. If $y < 1$, since $C_0$ is open, there exists $\varepsilon > 0$ such that $(y - \varepsilon, y + \varepsilon) \subseteq C_0 \subseteq (0,1)$. By the definition of $y$, there is a finite subcollection $\{C_1, C_2, \ldots, C_n\}$ from $C$ which covers $[0, y - \varepsilon/2)$. Now $\{C_1, \ldots, C_n, C_0\}$ covers $[0, y + \varepsilon)$, contrary to the definition of $y$. Thus, $y = 1$. Now since $C_0$ is an open neighborhood of $y = 1$, there exists $\varepsilon > 0$ with $(1 - \varepsilon, 1] \subseteq C_0$. By the definition of $y$, there is a finite subcollection $\{C_1, C_2, \ldots, C_n\}$ from $C$ which covers $[0, y - \varepsilon/2) = [0, 1 - \varepsilon/2)$, and thus $\{C_1, \ldots, C_n, C_0\}$ is a finite subcollection of $C$ which covers $[0,1]$. □

Thus, since $(0,1)$ is not compact and $[0,1]$ is compact, these spaces are not homeomorphic.

Note that compactness of $A \subseteq X$ was defined in terms of covers of $A$ by sets open in $X$ having finite subcovers. If $C_X$ is a cover of $A$ by sets open in $X$, then $C_A = \{C \cap A : C \in C_X\}$ is a cover of $A$ by sets open in the subspace $A$ of $X$. Conversely, if $C_A$ is a cover of $A$ by sets open in the subspace $A$, then each $C \in C_A$ is of form $C \cap U_C$ for some set $U_C$ open in $X$, and $C_X = \{U_C : C \in C_A\}$ is a cover of $A$ by sets open in $X$. Furthermore, in either case, $C_X$ has a finite subcover if and only if $C_A$ does.

This proves the following important result.

**Theorem 5.1.5.** *A is a compact subset of a topological space X if and only if A with the subspace topology is a compact topological space.*

This shows that *compactness of A is an absolute property, not dependent on whether we view A as a subspace of a larger space or not.* Openness and closedness are not absolute properties: $A = [0,1]$ is open in $A$ with the subspace topology from the Euclidean line $\mathbb{R}$, but $A = [0,1]$ is not open in $\mathbb{R}$. $B = (0,1)$ is closed in the subspace $B$ of $\mathbb{R}$, but not closed in $\mathbb{R}$. However, if $A \subseteq X$, the subspace $A$ is compact if and only if $A$ is a compact subset of $X$. We may iterate this to subspaces of subspaces. $[3,4)$ is not closed in $\mathbb{R}$, but is closed in the subspace $(0,4)$ of $\mathbb{R}$. However, $[3,4)$ is not compact in $\mathbb{R}$, in $(0,4)$, nor in $[3,4)$.

Below is another important result about compact sets.

**Theorem 5.1.6.** *If A is a compact set in a Hausdorff topological space X, then A is closed.*

*Proof.* Suppose $A$ is a compact set in a Hausdorff topological space $X$. We will show that $X - A$ is open. If $x \in X - A$, then, for every $a \in A$, there exist disjoint open neighborhoods $U_a$ of $a$ and $V_a$ of $x$, as suggested in Figure 5.1. Now $\{U_a : a \in A\}$ is an open cover of $A$, so there exists a finite subcover $\{U_a : a \in F\}$ where $F$ is some finite subset of $A$.

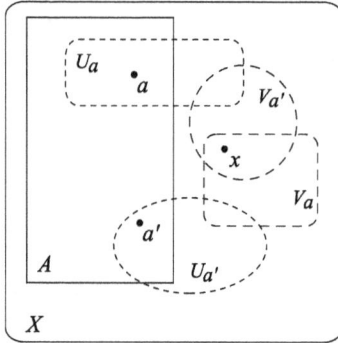

**Figure 5.1:** $\{U_a : a \in A\}$ is an open cover of $A$.

Now $U = \bigcup\{U_a : a \in F\}$ is an open set containing $A$, and $V = \bigcap\{V_a : a \in F\}$ is an open set containing $x$. Furthermore, $U \cap V = \emptyset$, for $y \in U \cap V$ would imply $y \in U_i \cap V \subseteq U_i \cap V_i$ for some $i \in F$, contrary to $U_i \cap V_i = \emptyset$. In particular, $V$ is an open neighborhood of $x$ contained in $X - U \subseteq X - A$. Since $x$ was an arbitrary point of $X - A$, this shows $X - A$ is open. □

In general, compactness is not a hereditary property. That is, if $X$ is compact and $Y \subseteq X$, then the subspace $Y$ need not be compact. The example of $(0, 1) \subseteq [0, 1]$ in the Euclidean line shows this. However, compactness is hereditary to closed subsets.

**Theorem 5.1.7.** *If $A$ is a closed set in a compact topological space $X$, then $A$ is compact.*

*Proof.* Suppose $A$ is a closed set in a compact topological space $X$ and $\mathcal{C}$ is an open cover of $A$. The collection $\mathcal{C} \cup \{X - A\}$, suggested in Figure 5.2, is an open cover of $X$ which has, by the compactness of $X$, a finite subcover $\mathcal{F}$. Now $\mathcal{F} - \{X - A\}$ is a finite subcollection of $\mathcal{C}$ which covers $A$. Thus, $A$ is compact. □

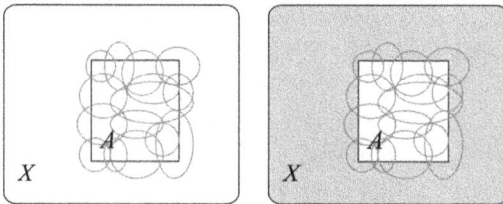

**Figure 5.2:** If $A$ is closed, an open cover $\mathcal{C}$ of $A$ gives an open cover $\mathcal{C} \cup \{X - A\}$ of $X$.

The previous two results are often stated as "compact in Hausdorff is closed" and "closed in compact is compact". While compactness is not generally hereditary, it is productive. That is, a product of compact spaces is compact. This result holds for arbitrary products, but the proof for infinite products is much more complicated and will be presented in Section 7.4. We present the proof for finite products. To simplify the proof, we would like to work with covers by basic open sets, and will need the following results.

**Theorem 5.1.8.** *A topological space X is compact if and only if every open cover of X by basic open sets has a finite subcover.*

*Proof.* If every open cover of $X$ has a finite subcover, then every open cover consisting of basic open sets has a finite subcover. Conversely, suppose every cover of $X$ by basic open sets has a finite subcover. If $C$ is an open cover of $X$, then consider the collection $\mathcal{B} = \{B : B$ is a basic open set and $B \subseteq C$ for some $C \in C\}$. Now since each $C \in C$ is the union of the basic open sets it contains, we see that $\bigcup C = \bigcup \mathcal{B}$, so $\mathcal{B}$ is a cover of $X$ by basic open sets. Let $\{B_1, \ldots, B_n\}$ be a finite subcover from $\mathcal{B}$, and pick $C_i \in C$ such that $B_i \subseteq C_i$. Now $\{C_1, \ldots, C_n\}$ is a finite subcollection of $C$ and $X \subseteq B_1 \cup \cdots \cup B_n \subseteq C_1 \cup \cdots \cup C_n$, so $\{C_1, \ldots, C_n\}$ is a finite subcover of $X$ from $C$. Thus, every open cover of $X$ has a finite subcover. $\square$

**Theorem 5.1.9.** *If $X_1, X_2, \ldots, X_n$ are compact topological spaces, then the product $X_1 \times \cdots \times X_n$ is compact.*

*Proof.* We will prove that if $X$ and $Y$ are compact, then $X \times Y$ is compact. The result for finite products will then follow from iterative applications of this result.

Suppose $X$ and $Y$ are compact and $C$ is an open cover of $X \times Y$ by basic open sets of form $U \times V$ where $U$ and $V$ are open in $X$ and $Y$, respectively. For any given $x \in X$, the set $\{x\} \times Y$ is a homeomorphic copy of $Y$ embedded in $X \times Y$. Thus, $C$ is an open cover of the compact set $\{x\} \times Y$. If $C_x$ is a finite subcover of $\{x\} \times Y$ from $C$, then $C_x$ actually covers not just the "line" $\{x\} \times Y$ but an open band $U_x \times Y$ for some open neighborhood $U_x$ of $x$. (This is often called the *tube lemma*.) To see this, if $C_x = \{U_1 \times V_1, \ldots, U_n \times V_n\} \subseteq C$ covers $\{x\} \times Y$, then without loss of generality, we may assume each $U_i \times V_i$ intersects $\{x\} \times Y$; those that do not intersect $\{x\} \times Y$ are not needed in a cover of $\{x\} \times Y$ and may be discarded. Now if $U_x = U_1 \cap \cdots \cap U_n$, then $U_x \times V_i \subseteq U_i \times V_i$ for each $i = 1, \ldots, n$, so $C_x = \{U_1 \times V_1, \ldots, U_n \times V_n\}$ covers the open band $U_x \times Y$, as depicted in Figure 5.3. So far, we have used only the compactness of $Y$. Now we will use the compactness of $X$.

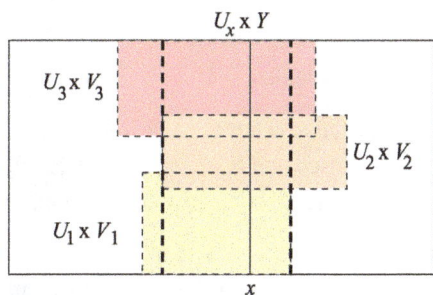

**Figure 5.3:** A covering of the line $\{x\} \times Y$ by a finite number of basic open sets $\{U_1 \times V_1, \ldots, U_n \times V_n\}$ actually covers a band $U_x \times Y$.

Consider the open cover $\{U_x : x \in X\}$ of $X$, where $U_x$ is as constructed in the previous paragraph. Since $X$ is compact, there exists a finite subcover $\{U_x : x \in F\}$ where $F$ is a finite subset of $X$. But each $U_x$ corresponds to an open band which is covered by finitely many members of $\mathcal{C}$, so this will result in a finite subcollection of $\mathcal{C}$ covering $X \times Y$. Specifically, $\bigcup\{\mathcal{C}_x : x \in F\}$ is a finite subcover of $X \times Y$ from $\mathcal{C}$ since it is a finite subcollection of $\mathcal{C}$, $\mathcal{C}_x$ covers $U_x \times Y$, and $\{U_x : x \in F\}$ covers $X$. $\qquad\square$

Before presenting a criterion to easily recognize compact sets in Euclidean space, we give a definition.

**Definition 5.1.10.** A subset $A$ of a metric space $(X, d)$ is *bounded* if there exists a real number $M$ such that $d(a, b) < M$ for all $a, b \in A$.

It is easy to see that a subset $A$ of a metric space $X$ is bounded if and only if for any $x \in X$, there exists a finite radius $N$ with $A \subseteq B(x, N)$. Indeed, if $A \subseteq B(x, N)$, then, for any $a, b \in A$, the triangle inequality gives $d(a, b) \leq d(a, x) + d(x, b) < N + N = 2N$, so $A$ is bounded. Conversely, if $A$ is a nonempty bounded set with $d(a, b) < M$ for all $a, b \in A$, fix a point $b_0 \in A$. Then $d(a, b_0) < M$ for all $a \in A$, so $A \subseteq B(b_0, M)$. Furthermore, given any $x \in X$, $A \subseteq B(b_0, M) \subseteq B(x, N)$ where $N = d(x, b_0) + M$. If $A$ is empty, then it is trivially bounded and contained in any ball $B(x, N)$ with positive radius.

**Theorem 5.1.11.** *In $\mathbb{R}^n$ with the Euclidean topology, $A$ is compact if and only if it is closed and bounded.*

*Proof.* Suppose $A$ is compact in Euclidean space $\mathbb{R}^n$. Let $\mathbf{0} \in \mathbb{R}^n$ be the zero vector. Since $\mathbb{R}^n$ is Hausdorff, $A$ is closed by Theorem 5.1.6. Furthermore, since the nested open cover $\{B(\mathbf{0}, n) : n \in \mathbb{N}\}$ of $A$ has a finite subcover, $A \subseteq B(\mathbf{0}, N)$ for some $N > 0$, so $A$ is bounded. Conversely, suppose $A \subseteq \mathbb{R}^n$ is closed and bounded. Then there exists $N > 0$ such that $A \subseteq B(\mathbf{0}, N)$. Furthermore, $B(\mathbf{0}, N) \subseteq [-N, N]^n$. (Since $|x_i| > N$ implies $\sqrt{x_1^2 + \cdots + x_n^2} > N$, $x \notin [-N, N]^n$ implies $x \notin B(\mathbf{0}, N)$.) Thus, $A$ is a closed subset of $[-N, N]^n$. Since $[-N, N]$ is homeomorphic to $[0, 1]$, it is compact, and $[-N, N]^n$ is compact by Theorem 5.1.9. Now $A$ is a closed subset of a compact space, so by Theorem 5.1.7, $A$ is compact. $\qquad\square$

This immediately gives the following result from calculus.

**Corollary 5.1.12** (The extreme value theorem). *If $[a, b]$ and $\mathbb{R}$ have the Euclidean topology and $f : [a, b] \to \mathbb{R}$ is a continuous function on a closed and bounded interval $[a, b]$, then $f$ has a maximum value and a minimum value over $[a, b]$. That is, there exist points $c, d \in [a, b]$ with $f(c) \leq f(x) \leq f(d)$ for all $x \in [a, b]$.*

*Proof.* The domain $[a, b]$ is a closed and bounded interval in $\mathbb{R}$, and thus compact and connected. Since $f$ is continuous, $f([a, b])$ is compact and connected. That is, $f([a, b])$ is a closed and bounded interval $[m, M]$. Now the minimum value $m \in f([a, b])$ must equal $f(c)$ for some $c \in [a, b]$, and similarly, the maximum value $M = f(d)$ for some $d \in [a, b]$. $\qquad\qquad\qquad\qquad\qquad\qquad\qquad\qquad\qquad\qquad\qquad\qquad\qquad$ □

**Example 5.1.13.** As subspaces of the Euclidean plane, the open ball $B = \{(x, y) \in \mathbb{R}^2 : x^2 + y^2 < 1\}$ and the closed ball $C = \{(x, y) \in \mathbb{R}^2 : x^2 + y^2 \leq 1\}$ are not homeomorphic. $C$ is closed and bounded in the Euclidean plane, so it is compact. But $B$ is not compact, since the open cover $\{B((0, 0), 1 - \frac{1}{n}) : n = 2, 3, \ldots\}$ of $B$ has no finite subcover. Or, $B$ is not a compact set in the Hausdorff space $\mathbb{R}^2$ since it is not closed.

Compactness is defined in terms of collections of open sets. Taking complements, we should be able to characterize compactness in terms of closed sets. A (finite or arbitrary) collection $\mathcal{C}$ of open sets covers $X$ if and only if $\bigcup \mathcal{C} = X$, which occurs if and only if $X - \bigcup\{C : C \in \mathcal{C}\} = \bigcap\{X - C : C \in \mathcal{C}\} = \emptyset$. Thus, a collection $\mathcal{C}$ of open sets covers $X$ if and only if the associated collection $\mathcal{F} = \{X - C : C \in \mathcal{C}\}$ of closed sets has empty intersection.

Now the following are equivalent:
(a) $X$ is compact.
(b) If $\mathcal{C}$ is a collection of open sets and every finite subcollection fails to covers $X$, then $\mathcal{C}$ fails to cover $X$.
(c) If $\mathcal{F}$ is a collection of closed sets such that every finite subcollection $\{F_1, \ldots, F_n\}$ has nonempty intersection, then $\bigcap \mathcal{F} \neq \emptyset$.

After introducing some terminology, the equivalence of (a) and (c) above is stated in the next theorem.

**Definition 5.1.14.** A collection $\mathcal{F}$ of closed subsets of a topological space $X$ has the *finite intersection property* if every finite subcollection of $\mathcal{F}$ has nonempty intersection.

**Theorem 5.1.15.** *A topological space $X$ is compact if and only if every collection $\mathcal{F}$ of closed sets with the finite intersection property has $\bigcap \mathcal{F} \neq \emptyset$.*

## Exercises

1. Each set below is noncompact. Exhibit an open cover which has no finite subcover.
   (a) $A = [0, 1]$ in $\mathbb{R}$ with the discrete topology.
   (b) $B = [1, \infty)$ in the Euclidean line.
   (c) $C = \{(x, y) \in \mathbb{R}^2 : x > 0\}$ in the Euclidean plane.
   (d) $D = ([0, 2] \times [-1, 1]) - \{(0, 0)\}$ in the Euclidean plane.
2. For $X = (0, 3)$ in the Euclidean line, find open covers with the indicated property.
   (a) $\mathcal{C}$ has exactly one finite subcover.
   (b) $\mathcal{D}$ is an infinite collection, and has exactly one finite subcover of cardinality 3.
   (c) $\mathcal{E}$ has no finite subcover.

3. Suppose $X$ is a set. Describe the compact subsets of $X$ if
   (a) $X$ has the discrete topology.
   (b) $X$ has the indiscrete topology.
   (c) $X$ has the cofinite topology.
   (d) $X$ is finite and has any topology.
   (e) $X = \mathbb{R}$ with the right ray topology.

4. Give $(0,1)$ and $[0,1]$ the Euclidean topologies. Find the functions described or prove that no such function can exist.
   (a) A continuous surjection $f : (0,1) \to [0,1]$.
   (b) A continuous surjection $g : [0,1] \to (0,1)$.

5. Show that no two of subsets of the Euclidean plane shown below are homeomorphic.

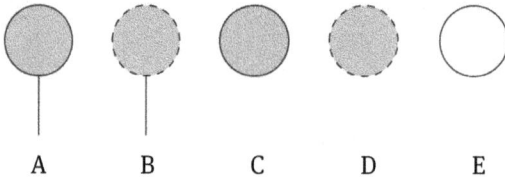

A     B     C     D     E

6. Give an example of a compact set which is not closed.

7. Give an example of a Hausdorff topological space $X$ and a nonempty proper subset $A$ which is open and compact. Can such an example be connected? Explain.

8. Let $d$ be the Euclidean metric on $\mathbb{R}^2$ and define the *post office metric* $p$ on $\mathbb{R}^2$ by

$$p(a, b) = \begin{cases} 0 & \text{if } a = b, \\ d(a, 0) + d(0, b) & \text{if } a \neq b. \end{cases}$$

   (a) Which subspaces of $(\mathbb{R}^2, p)$ inherit the discrete topology?
   (b) Discuss the limits of the sequences $((\frac{1}{n}, 1))_{n=1}^{\infty}$ and $((\frac{1}{n}, 0))_{n=1}^{\infty}$.
   (c) Show that the subset $[-1, 1]^2$ is closed and bounded, but not compact in $(\mathbb{R}^2, p)$.

9. (a) Prove that the union of finitely many compact sets is compact.
   (b) Show that the union of arbitrarily many compact sets need not be compact.

10. Prove that the intersection of arbitrarily many compact subsets of a Hausdorff space is compact.

11. Let $X = [0,1]$ be the unit interval in $\mathbb{R}$ with the topology having basis $\{\{x\} : 0 < x < 1\} \cup \{[0,1]\}$. Find two compact subsets of $X$ whose intersection is not compact.

12. Suppose $\mathcal{T}_C$ and $\mathcal{T}_F$ are topologies on $X$ with $\mathcal{T}_C$ coarser than $\mathcal{T}_F$. Show that if $(X, \mathcal{T}_F)$ is compact, then $\mathcal{T}_C$ is compact.

13. Does the converse of Theorem 5.1.9 hold? That is, if $X_1 \times \cdots \times X_n$ is compact, must each $X_i$ ($i = 1, \ldots, n$) be compact? Provide a proof or counterexample.

14. Prove that any continuous bijection $f$ from a compact space $X$ to a Hausdorff space $Y$ is a homeomorphism.

15. Suppose $A$ is a compact subset of the Euclidean line $\mathbb{R}$. Show that every sequence $(x_n)$ in $A$ has a subsequence converging to a point of $A$.

16. Suppose $A$ is a compact subset of $X$, $B$ is a compact subset of $Y$, and $W$ is an open set in the product $X \times Y$ which contains $A \times B$. Show that there exist open sets $U \subseteq X$ and $V \subseteq Y$ with $A \times B \subseteq U \times V \subseteq W$.

17. Show that every compact metric space is separable. That is, show that every compact metric space has a countable dense subset.

18. Suppose the sets $X_1$ and $X_2$ are countably infinite and $(X_1, \mathcal{T}_1)$ and $(X_2, \mathcal{T}_2)$ are compact Hausdorff topological spaces with exactly one point which is not isolated. Show that $X_1$ and $X_2$ are homeomorphic.

19. (a) Show that the assumption that $(X_1, \mathcal{T}_1)$ and $(X_2, \mathcal{T}_2)$ are Hausdorff is necessary for the result of Exercise 18.

    (b) Show that the assumption that $(X_1, \mathcal{T}_1)$ and $(X_2, \mathcal{T}_2)$ are compact is necessary for the result of Exercise 18.

20. The extreme value theorem was stated for real valued functions whose domain was a compact and connected set in $\mathbb{R}$. Prove the following more general version: $A$ is a compact subset of the Euclidean space $\mathbb{R}^n$ if and only if every continuous function $f : A \to \mathbb{R}$ has a maximum value and a minimum value over $A$.

21. If $X$ is any topological space and $Y$ is compact, show that the projection function $\pi_1 : X \times Y \to X$ is a closed mapping.

22. Use the result of Exercise 21 to show that if $Y$ is compact and $f : X \to Y$ is a function, then $f$ is continuous if its graph $G = \{(x, f(x)) : x \in X\}$ is closed in $X \times Y$. (Compare to Exercise 21 of Section 3.2.)

23. Prove that the converse of Exercise 22 holds if $Y$ is also assumed to be Hausdorff. That is, prove the *closed graph theorem:* If $Y$ is compact and Hausdorff and $X$ is any topological space, $f : X \to Y$ is continuous if and only if its graph $G = \{(x, f(x)) : x \in X\}$ is closed in $X \times Y$.

24. Suppose $X$ is a nonempty compact Hausdorff space and $f : X \to X$ is continuous. If $f^{(n)}$ represents the composition of $f$ with itself $n$ times, show that $\mathcal{F} = \{f(X), f^{(2)}(X), f^{(3)}(X), \ldots\}$ has the finite intersection property, and show that there exists a nonempty closed set $A$ with $f(A) = A$.

25. (**The Cantor set**) The Cantor set is the subspace of the Euclidean line defined as follows. Let $C_1 = [0, 1]$. Remove the open middle third to get $C_2 = [0, \frac{1}{3}] \cup [\frac{2}{3}, 1]$. Remove the open middle thirds of the two intervals in $C_2$ to get $C_3 = [0, \frac{1}{9}] \cup [\frac{2}{9}, \frac{1}{3}] \cup [\frac{2}{3}, \frac{7}{9}] \cup [\frac{8}{9}, 1]$. Having defined $C_n$, we remove the open middle thirds of each interval in $C_n$ to form $C_{n+1}$. Now the Cantor set (or Cantor middle-thirds set) is $C = \bigcap_{n \in \mathbb{N}} C_n$. Note that in ternary (base 3) notation, a number in $[0, 1]$ has representation $0.d_1 d_2 d_3 \cdots = d_1 \cdot 3^{-1} + d_2 \cdot 3^{-2} + d_3 \cdot 3^{-3} + \cdots$. Removing the middle third $(\frac{1}{3}, \frac{2}{3})$ of $C_1$ removes the numbers with $d_1 = 1$, leaving those with $d_1 = 0$ or 2. Removing the middle thirds of $C_2$ removes those numbers with $d_2 = 1$, and so on. Thus, the Cantor set (or Cantor ternary set) consists of the real num-

bers in $[0,1]$ which can be represented in base 3 as $\sum_{n \in \mathbb{N}} d_n 3^{-n}$ where $d_n \in \{0,2\}$ for all $n$. (Note that non-uniqueness of representation allows that some of these numbers might be represented using 1's as well. For example, in base 3, $0.0\bar{2} = 0.1\bar{0}$.)

(a) Show that the Cantor set $C$ is compact.

(b) Show that $C$ is uncountable.

(c) If $L_n$ is the combined length of the intervals in $C_n$, show that $\lim_{n \to \infty} L_n = 0$. (This shows that the Lebesgue measure of the Cantor set is zero.)

(d) Show that the connected components of $C$ are the singletons.

(e) Show that $C$ is homeomorphic to the product $\prod_{n \in \mathbb{N}} \{0,2\}$ with the product topology, where $\{0,2\}$ has the discrete topology.

26. Let $\mathcal{B} = \{(a,b) \cup (\frac{1}{b}, \frac{1}{a}) : 0 < a < b\} \cup \{[0,b) \cup (\frac{1}{b}, \infty) : b > 0\}$.

(a) Show that $\mathcal{B}$ is a basis for a topology $\mathcal{T}$ on $X = [0, \infty)$.

(b) Show that $(X, \mathcal{T})$ is not Hausdorff by exhibiting distinct points which cannot be separated by disjoint neighborhoods and by exhibiting a sequence which has more than one limit.

(c) Which convergent sequences have unique limits?

(d) Determine whether $(X, \mathcal{T})$ is compact or not. Prove your answer.

## 5.2 Compactness in metric spaces

In the previous section, compactness was motivated as one way to quantify that a topological space has no "missing points". The open cover definition of compactness is a powerful tool developed over time. It is perhaps no surprise that historically, the open cover definition was not the first attempt to quantify that a space has no missing points. Many of the first instances motivating the need to quantify compactness occurred in metric spaces, and particularly in Euclidean spaces. In this section, we will investigate some other useful characterizations of compactness in metric spaces.

In Section 1.6, we defined a *limit point* of a subset $A$ of a topological space $X$ to be a point $a \in X$ such that every neighborhood of $a$ intersects $A$ in a point other than $a$. Thus, $a$ is a limit point of $A$ if and only if $a \in \text{cl}(A - \{a\})$.

**Definition 5.2.1.** A subset $Y$ of a topological space $X$ is *limit point compact* if every infinite subset of $Y$ has a limit point in $Y$. A subset $Y$ of a topological space $X$ is *sequentially compact* if every sequence in $Y$ has subsequence converging to a point of $Y$.

Both of these new versions of compactness may be interpreted as conditions to eliminate missing "limits". For example, let $Y = (0,1)$ with the Euclidean topology. The infinite subset $A = \{1/n : n \in \mathbb{N}, n \geq 2\}$ is an infinite set in $Y$ with no limit point in $Y$, so $(0,1)$ is not limit point compact. Similarly, the sequence $(1/n)_{n \in \mathbb{N}}$ in $Y$ has no subsequence converging to a point of $Y$, so $Y$ is not sequentially compact. We will see that in a metric space, limit point compactness, sequential compactness, and compactness are equivalent. Before turning to metric spaces, we note some implications which hold in arbitrary topological spaces.

**Theorem 5.2.2.**
(a) *If a topological space X is compact, then it is limit point compact.*
(b) *If a topological space X is sequentially compact, then it is limit point compact.*

*Proof.* (a) We will show the contrapositive. If $X$ is not limit point compact, then there exists an infinite subset $A$ of $X$ which has no limit points. Then, for every $x \in X$, there exists an open neighborhood $U_x$ of $x$ with $U_x \cap A \subseteq \{x\}$. Now $C = \{U_x : x \in X\}$ is an open cover of $X$. Since each $U_x$ covers at most one element of $A$, any finite subcollection of $C$ cannot cover the infinite set $A$, so $C$ has no finite subcover. Thus, $X$ is not compact.

(b) Suppose $X$ is sequentially compact and $A$ is an infinite subset of $X$. Now $A$ must contain a countably infinite sequence $(a_n)$ of distinct terms. If $a$ is a limit of a subsequence of $(a_n)$, then every neighborhood of $a$ contains a tail of the sequence $(a_n)$ and thus intersects $A$ in a point other than $a$. Thus, $a$ is a limit point of $A$. $\qquad\square$

**Example 5.2.3** (Limit point compactness does not imply compactness). Give $\mathbb{Z}$ the topology having basis $\mathcal{B} = \{\{2n, 2n+1\} : n \in \mathbb{Z}\}$. The basis $\mathcal{B}$ is an open cover of $\mathbb{Z}$ with no finite subcover, so this space is not compact. However, given any infinite subset $A$ of $\mathbb{Z}$, if $2n \in A$, then $2n+1$ is a limit point of $A$ and if $2n+1 \in A$, then $2n$ is a limit point of $A$. Thus, the space is limit point compact.

**Theorem 5.2.4.** *A metric space X is limit point compact if and only if it is sequentially compact.*

*Proof.* By Theorem 5.2.2, we need only show that limit point compactness in a metric space implies sequential compactness. Suppose the metric space $X$ is limit point compact and $(x_n)$ is a sequence in $X$. If the set of terms $\{x_n\}$ is a finite set, then by the pigeonhole principle, one term must be repeated infinitely often, and this gives a constant subsequence, which must converge. So, suppose the set of terms $A = \{x_n\}$ is infinite. Then $A$ has a limit point $a$. Not only does every neighborhood $U$ of $a$ intersect $A - \{a\}$, but this intersection must be infinite, for otherwise, if $\varepsilon$ is the distance from $a$ to the closest point of $U \cap (A - \{a\})$, then we have the contradiction that $U \cap B(a, \varepsilon) \cap (A - \{a\}) = \emptyset$. Pick $a_1 \in B(a, 1) \cap (A - \{a\})$. Since $a_1 \in A$, we have $a_1 = x_{m_1}$ for some $m_1 \in \mathbb{N}$. Among the infinitely many elements of $B(a, 1/2) \cap (A - \{a\})$, pick one $a_2$ such that $a_2 = x_{m_2}$ for some $m_2 \geq m_1$. Continuing inductively, picking $a_j \in (B(a, 1/j) \cap (A - \{a\})) - \{x_1, x_2, \ldots, a_1, \ldots, a_2, \ldots, a_{j-1}\}$, we define a subsequence $(a_n)$ of $(x_n)$ which clearly converges to $a$. $\qquad\square$

By Theorems 5.2.2 and 5.2.4, to show that limit point compactness, sequential compactness, and compactness agree in a metric space, it only remains to show that sequential compactness implies compactness. For this step, we present a definition and some lemmas.

**Definition 5.2.5.** A metric space $X$ is *totally bounded* if for every $\varepsilon > 0$, the open cover $\{B(x,\varepsilon) : x \in X\}$ of $X$ has a finite subcover.

**Lemma 5.2.6.** *Every sequentially compact metric space is totally bounded.*

*Proof.* Suppose to the contrary that $X$ is a sequentially compact metric space and there exists $\varepsilon > 0$ such that the open cover $\{B(x,\varepsilon) : x \in X\}$ of $X$ has no finite subcover. Select a sequence $(x_n)$ in $X$ by choosing $x_1 \in X, x_2 \in X-B(x_1,\varepsilon), x_3 \in X-(B(x_1,\varepsilon)\cup B(x_2,\varepsilon))$, and in general, $x_n \in X-\bigcup\{B(x_i,\varepsilon) : i = 1,\ldots,n-1\}$. Note that, since the cover $\{B(x,\varepsilon) : x \in X\}$ of $X$ has no finite subcover, we are guaranteed that $X - \bigcup\{B(x_i,\varepsilon) : i = 1,\ldots,n - 1\} \neq \emptyset$. Furthermore, by the selection, $x_n$ is not within $\varepsilon$ of any of the points $x_1,\ldots,x_{n-1}$, and thus $d(x_j,x_k) < \varepsilon$ if and only if $j = k$. Now we claim that the sequence $(x_n)$ has no convergent subsequence. If $(x_n)$ has a subsequence converging to $a \in X$, then the ball $B(a,\varepsilon/2)$ should contain distinct terms $x_j, x_k$ of the sequence $(x_n)$. Then $d(x_j,x_k) \leq d(x_j,a) + d(a,x_k) < \varepsilon/2 + \varepsilon/2 = \varepsilon$, contrary to $j \neq k$. □

**Lemma 5.2.7.** *If $X$ is a sequentially compact metric space and $C$ is an open cover of $X$, then there exists a number $\delta > 0$ such that, for any $x \in X$, there exists $U \in C$ with $B(x,\delta) \subseteq U$. The number $\delta$ is called the* Lebesgue number *of $C$.*

*Proof.* Assume to the contrary that $X$ is sequentially compact, $C$ is an open cover of $X$, but for any $\delta > 0$ there exists $x \in X$ with $B(x,\delta) \not\subseteq U$ for any $U \in C$. In particular, for any $n \in \mathbb{N}$, we may select $x_n \in X$ with $B(x_n,1/n) \not\subseteq U$ for any $U \in C$. Let $a$ be a limit of a subsequence of $(x_n)$, and pick $U_a \in C$ with $a \in U_a$. Since $U_a$ is open, it contains $B(a,1/k)$ for some $k \in \mathbb{N}$. Now $B(a,1/(2k))$ must contain a tail of the subsequence of $(x_n)$ converging to $a$. Pick $n > 2k$ with $x_n \in B(a,1/(2k))$. Now $y \in B(x_n,1/n)$ implies $d(y,x_n) < 1/n < 1/(2k)$ and $x_n \in B(a,1/(2k))$ implies $d(x_n,a) < 1/(2k)$. It follows that $d(y,a) \leq d(y,x_n) + d(x_n,a) < 1/k$, so $y \in B(a,1/k) \subseteq U_a$. Thus $B(x_n,1/n) \subseteq U_a$, contrary to the choice of $x_n$. □

**Theorem 5.2.8.** *In a metric space $X$, compactness, limit point compactness, and sequential compactness are equivalent.*

*Proof.* With Theorems 5.2.2 and 5.2.4, it remains to show that sequential compactness implies compactness. Suppose the metric space $X$ is sequentially compact. To show $X$ is compact, suppose $C$ is an open cover of $X$. By Lemma 5.2.7, there exists $\delta > 0$ such that, for any $x \in X$, $B(x,\delta) \subseteq U$ for some $U \in C$. Since $X$ is totally bounded (Lemma 5.2.6), the open cover $\{B(x,\delta) : x \in X\}$ has a finite subcover $\{B(x_i,\delta)\}_{i=1}^{n}$. For each $i = 1,\ldots,n$, pick $U_i \in C$ with $B(x_i,\delta) \subseteq U_i$. Now $\{U_i\}_{i=1}^{n}$ is a finite subcover of $C$. Thus, $X$ is compact. □

The three notions of compactness discussed all say that, in some sense, the space has nothing missing. The three notions of compactness were topological properties, and they were equivalent in metric spaces. We now turn to the notion of completeness, which is also way to say that, in another sense, no points are missing from a metric space. Completeness is only defined for metric spaces, and requires the following definition.

**Definition 5.2.9.** A sequence $(x_n)$ in a metric space $X$ is a *Cauchy sequence* if for every $\varepsilon > 0$, there exists $n \in \mathbb{N}$ such that $d(x_j, x_k) < \varepsilon$ for all $j, k > n$.

For comparison, recall that a sequence $(x_n)$ in a metric space converges to a limit $L$ if for every $\varepsilon > 0$, there exists $n \in \mathbb{N}$ such that $d(x_j, L) < \varepsilon$ for $j > n$. Thus, the terms of a convergent sequence are eventually arbitrarily close to the limit $L$, while the terms of a Cauchy sequence are eventually arbitrarily close to each other.

**Theorem 5.2.10.** *Every convergent sequence in a metric space is a Cauchy sequence.*

*Proof.* Suppose the sequence $(x_n)$ converges to $L$ in a metric space. Given $\varepsilon > 0$, there exists $n \in \mathbb{N}$ such that $j > n$ implies $d(x_j, L) < \varepsilon/2$. Now for $j, k > n$, we have $d(x_j, x_k) \le d(x_j, L) + d(L, x_k) < \varepsilon/2 + \varepsilon/2 = \varepsilon$. Thus, $(x_n)$ is a Cauchy sequence. $\square$

The converse of Theorem 5.2.10 fails: A Cauchy sequence may not converge. For example, in the set $\mathbb{Q}$ of rational numbers with the Euclidean metric, consider the sequence $(p_n)_{n=1}^{\infty} = (3.1, 3.14, 3.141, 3.1415, 3.14159, \dots)$ whose $n$th term $p_n$ is the value of $\pi$ truncated $n$ places to the right of the decimal. Now as a sequence in the Euclidean line $\mathbb{R}$, $(p_n)$ converges to $\pi$, and thus by Theorem 5.2.10, $(p_n)$ is a Cauchy sequence in $\mathbb{R}$. Since $\mathbb{Q}$ carries the same metric as $\mathbb{R}$, $(p_n)$ is a Cauchy sequence in $\mathbb{Q}$. However, $(p_n)$ does not converge in $\mathbb{Q}$; its "limit" $\pi$ is not an element of $\mathbb{Q}$.

This example is typical. If a Cauchy sequence does not converge in a metric space $X$, then in some sense the "intended limit" is missing from the metric space. This prompts the following definition.

**Definition 5.2.11.** A metric space $X$ is *complete* if every Cauchy sequence in $X$ converges to a point of $X$.

The discussion above shows that $\mathbb{Q}$ with the Euclidean metric is not complete.

We have already discussed completeness of the real line in order-theoretic terms. Axiom 0.7.4 stated that every nonempty set of real numbers bounded above has a least upper bound, and dually. This result was called completeness of the real line, and was given without proof. The theorem below shows that the order-theoretic concept of completeness of the real line is equivalent to the Euclidean metric concept of completeness of the real line.

**Theorem 5.2.12.** *The following are equivalent characterizations of the completeness of the real line with the Euclidean topology.*

(a) *Every nonempty subset of* $\mathbb{R}$ *bounded above has a least upper bound, and dually, every nonempty subset of* $\mathbb{R}$ *bounded below has a greatest lower bound.*

(b) *Every Cauchy sequence in* $\mathbb{R}$ *converges.*

*Proof.* Suppose (a). If $(x_n)$ is a Cauchy sequence, the set $\{x_n : n \in \mathbb{N}\}$ of terms is bounded above and below (see Exercise 8). For $n \in \mathbb{N}$, let $b_n = \text{lub}\{x_n, x_{n+1}, x_{n+2}, \ldots\}$. The sequence $(b_n)$ is a decreasing sequence, since each successive term is the supremum of a smaller set. The set of terms $\{b_n : n \in \mathbb{N}\}$ are bounded below, so $L = \text{glb}\{b_n : n \in \mathbb{N}\}$ exists. We will show that $(x_n)$ converges to $L$. Given $\varepsilon > 0$, there exists $m \in \mathbb{N}$ such that $j, k \geq m$ implies $|x_j - x_k| < \varepsilon/4$, so $\{x_m, x_{m+1}, \ldots\} \subseteq [x_m - \varepsilon/4, x_m + \varepsilon/4]$. Now for $j \geq m$, we have $x_m - \varepsilon/4 \leq x_j \leq b_j \leq x_m + \varepsilon/4$, so $|x_j - b_j| < \varepsilon/2$. Furthermore, since $L = \text{glb}\{b_n : n \in \mathbb{N}\}$, and $(b_n)$ is a decreasing sequence, there exists $s \in \mathbb{N}$ such that $j \geq s$ implies $b_j \in [L, L + \varepsilon/2)$, so $|b_j - L| < \varepsilon/2$. Now for $j \geq \max\{m, s\}$, we have $|x_j - L| \leq |x_j - b_j| + |b_j - L| < \varepsilon/2 + \varepsilon/2 = \varepsilon$. Thus, $(x_n)$ converges to $L$.

Suppose (b). Let $A$ be nonempty set bounded above. We wish to show that $A$ has a least upper bound. Pick $a_1 \in A$ and $b_1 \in \text{ub}\,A$. (Note that if $a_1 = b_1$, then $b_1 = \text{lub}\,A$.) Having defined an interval $[a_n, b_n]$ with endpoints $a_n \in A$ and $b_n \in \text{ub}\,A$, if $[\frac{a_n + b_n}{2}, b_n] \cap A \neq \emptyset$, chose $a \in [\frac{a_n + b_n}{2}, b_n] \cap A$ and set $[a_{n+1}, b_{n+1}] = [a, b_n]$. Otherwise, take $[a_{n+1}, b_{n+1}] = [a_n, \frac{a_n + b_n}{2}]$. Either way, we have $a_{n+1} \in A$, $b_{n+1} \in \text{ub}\,A$, $a_n \leq a_{n+1} \leq b_{n+1} \leq b_n$, and $|b_{n+1} - a_{n+1}| \leq \frac{|b_1 - a_1|}{2^n}$, and consequently, $|b_k - a_j| \leq \frac{|b_1 - a_1|}{2^n}$ for all $j, k > n$. We will show $(a_n)$ is a Cauchy sequence. Given $\varepsilon > 0$, pick $m \in \mathbb{N}$ such that $\frac{|b_1 - a_1|}{2^m} < \varepsilon/2$. Now for $j, k > m$, we have $|a_j - a_k| \leq |a_j - b_{m+1}| + |b_{m+1} - a_k| \leq \varepsilon/2 + \varepsilon/2 = \varepsilon$. Thus, $(a_n)$ is a Cauchy sequence, and similarly, $(b_n)$ is Cauchy. Thus, both sequences $(a_n)$ and $(b_n)$ converge, and $|b_{n+1} - a_{n+1}| \leq \frac{|b_1 - a_1|}{2^n}$ implies $\lim_{n \to \infty} |b_n - a_n| = 0$, so $(a_n)$ and $(b_n)$ converge to the same limit, which we will call $L$. Now if $x < L < y$, then there are points $a_n \in A$, $b_n \in \text{ub}\,A$ with $x < a_n < L < b_n < y$, so $x \notin \text{ub}\,A$ and $y \neq \text{lub}\,A$. Thus, $L = \text{lub}\,A$. Dually, every nonempty set of real numbers bounded below has a greatest lower bound. $\square$

Note that we have not proved that the real line is complete; we have merely shown that the two descriptions of this property are equivalent. The proof that $\mathbb{R}$ is complete would require a formal construction of the real numbers. The rational numbers are easily constructed from $\mathbb{N}$ by introducing an additive identity, additive inverses, and multiplicative inverses of non-zero elements. In one formal construction, the real numbers are defined to be equivalence classes of Cauchy sequences of rational numbers under the equivalence $(x_n) \approx (y_n)$ if and only if $\lim_{n \to \infty} |y_n - x_n| = 0$ in $\mathbb{Q}$. Thus, the real number $7/8$ is defined to be the collection of equivalent Cauchy sequence of rationals, including $(7/8)_{n=1}^{\infty}$ and $(7/8 - 1/n)_{n=1}^{\infty}$, which converge to $7/8$, and $\pi$ is defined to be the collection of equivalent Cauchy sequence of rationals, including $(3.1, 3.14, 3.141, 3.1415, \ldots)$, which after having defined $\mathbb{R}$, would be described as "converging to $\pi$". With much effort, it can be shown that this construction of the real numbers satisfies condition (a) of Theorem 5.2.12, and the topology arises from the expected metric.

The connection between completeness and compactness in a metric space is given by the following result.

**Theorem 5.2.13.** *A metric space is compact if and only if it is complete and totally bounded.*

*Proof.* Suppose the metric space $X$ is compact. Then it is sequentially compact (Theorem 5.2.8) and thus is totally bounded (Lemma 5.2.6). To see that the sequentially compact metric space $X$ is complete, suppose $(x_n)$ is a Cauchy sequence in $X$. This sequence must have a subsequence $(x_{\sigma(n)})$ converging to a limit $L$ (where $\sigma : \mathbb{N} \to \mathbb{N}$ is a strictly increasing function). We will show that the sequence $(x_n)$ itself must converge to $L$. Given $\varepsilon > 0$, there exists $n \in \mathbb{N}$ such that $k > n$ implies $d(x_{\sigma(k)}, L) < \varepsilon/2$. Also, there exists $m \in \mathbb{N}$ such that $j, k > m$ implies $d(x_j, x_k) < \varepsilon/2$. Now for $j, k > \max\{m, n\}$, $d(x_j, L) \le d(x_j, x_{\sigma(k)}) + d(x_{\sigma(k)}, L) < \varepsilon/2 + \varepsilon/2 = \varepsilon$. Thus, $(x_n)$ converges to $L$ and $X$ is complete.

Conversely, suppose $X$ is complete and totally bounded. To see that $X$ is compact, we will show that it is sequentially compact. Suppose $(x_n)$ is a sequence in $X$. By total boundedness, the open cover $\{B(x, 1) : x \in X\}$ by balls of radius 1 has a finite subcover $\{B(x_i, 1)\}_{i=1}^n$. By the pigeonhole principle, at least one element of this finite subcover, say $B(x_1, 1)$, contains $x_n$ for infinitely many values of $n$. Thus, $B(x_1, 1)$ contains an infinite subsequence $(x_{\sigma_1(n)})$ of $(x_n)$. Now the open cover $\{B(x, 1/2) : x \in X\}$ by balls of radius $1/2$ has a finite subcover $\{B(x_i, 1/2)\}_{i=1}^m$, and one element of this subcover must contain an infinite subsequence $(x_{\sigma_2(n)})$ of $(x_{\sigma_1(n)})$. Iterating, the open cover $\{B(x, 1/3) : x \in X\}$ by balls of radius $1/3$ has a finite subcover $\{B(x_i, 1/3)\}_{i=1}^k$, one element of which contains an infinite subsequence $(x_{\sigma_3(n)})$ of $(x_{\sigma_2(n)})$. Continuing in this manner, consider the subsequence $(x_{\sigma_1(1)}, x_{\sigma_2(2)}, x_{\sigma_3(3)}, \ldots)$. By the construction, for any $j, k \ge 2n$, both $x_{\sigma_j(j)}$ and $x_{\sigma_k(k)}$ lie in a ball $B(x_i, 1/(2n))$ so $d(x_{\sigma_j(j)}, x_{\sigma_k(k)}) < 1/n$. Thus, the subsequence $(x_{\sigma_j(j)})$ of $(x_n)$ is a Cauchy sequence, and by the completeness of $X$, it converges. Thus, $X$ is sequentially compact. $\square$

## Exercises

1. Let $X$ be an infinite set with the cofinite topology. Is $X$ limit point compact? Is $X$ sequentially compact? Prove your answers directly from the definitions.

2. Compactness does not imply sequential compactness. Find the flaw in this "proof" that compactness implies sequential compactness:

   > Suppose $X$ is not sequentially compact. Then there exists a sequence $(x_n)$ in $X$ which has no convergent subsequence. In particular, the sequence $(x_n)$ itself does not converge to any $x \in X$, so for any $x \in X$ there exists an open neighborhood $U_x$ of $x$ which contains only finitely many terms of $(x_n)$. Now $\{U_x : x \in X\}$ is an open cover of $X$, but any finite subcollection only

covers finitely many terms of the sequence $(x_n)$ and thus is not a cover of $X$. Thus, $X$ is not compact.

3. Show that every closed subset of a sequentially compact topological space is sequentially compact.

4. Suppose $X$ is a sequentially compact metric space and $A \subseteq X$. Show that $A$ is sequentially compact if and only if $A$ is closed.

5. Show that every closed subset of a limit point compact topological space is limit point compact.

6. Suppose $X$ is a limit point compact metric space and $A \subseteq X$. Show that $A$ is limit point compact if and only if $A$ is closed.

7. Show that the continuous image of a limit point compact topological space is limit point compact.

8. Suppose $(x_n)$ is a Cauchy sequence in a metric space $X$. Show that the set $\{x_n : n \in \mathbb{N}\}$ of terms is bounded.

9. (a) Show that the Euclidean line $\mathbb{R}$ is complete if and only if $[a, b]$ is complete for every compact interval $[a, b] \subseteq \mathbb{R}$.
   (b) Together with (a), Theorem 5.1.4 and Theorem 5.2.13 may seem to prove that the Euclidean line is complete. Looking over the proofs of those theorems, explain why this does not give a valid logical proof that the Euclidean line is complete.

10. Suppose $X$ is a metric space in which every closed and bounded subset is compact. Show that $X$ is complete.

11. (a) Show that every totally bounded metric space is bounded.
    (b) Find an example of a metric space which is bounded but not totally bounded.
    (c) Show that a subset of Euclidean space $\mathbb{R}^n$ is bounded if and only if it is totally bounded.

12. Prove that $(0, 1)$ with the Euclidean metric is totally bounded but not compact.

13. (a) Prove that $(0, 1)$ with the Euclidean metric is not sequentially compact directly from the definition of sequentially compact.
    (b) Prove that $(0, 1)$ with the Euclidean metric is not sequentially compact by showing that the conclusion of Lemma 5.2.7 fails.

14. Let $X = [0, 1]$ with the Euclidean metric. For the each open cover of $X$ below, find a Lebesgue number.
    (a) $\mathcal{C} = \{(1/n, 1]\}_{n=1}^{\infty} \cup \{[0, 1/4)\}$
    (b) $\mathcal{D} = \{[0, 1/2 + 1/n)\}_{n=1}^{\infty} \cup \{(1/2 - 1/n, 1]\}_{n=1}^{\infty}$

15. Give an example of a complete metric space which is not totally bounded.

16. Let $X = (0, 1)$ with the discrete metric, $Y = (0, 1)$ with the Euclidean metric, and $f : X \to Y$ be the function $f(x) = x$. Complete the details that this shows the continuous image of a complete metric space need not be complete.

17. Suppose $X$ is a complete metric space and $A \subseteq X$. Must $A$ be complete? If $A$ is closed, must $A$ be complete?

18. Suppose $(x_n)$ and $(y_n)$ are Cauchy sequences in $\mathbb{Q}$. Without appealing to the fact that $(x_n)$ and $(y_n)$ must converge to real numbers which may not be in $\mathbb{Q}$, show
    (a) $(x_n - y_n)$ is a Cauchy sequence in $\mathbb{Q}$.
    (b) $(|x_n|)$ is a Cauchy sequence in $\mathbb{Q}$.
19. Repeat Exercise 18, appealing to the fact that $(x_n)$ and $(y_n)$ must converge to real numbers which may not be in $\mathbb{Q}$.

# 6 Metric spaces and real analysis

## 6.1 Metric spaces

The Euclidean metric topology generated by a basis of balls is the most basic motivating example for the study of topology. We start with a brief review of some results on metric spaces. Recall that a *metric* on a set $X$ is a function $d : X \times X \to [0, \infty)$ such that
(a)  $d(x, y) \geq 0$ for all $x, y \in X$ (nonnegativity);
(b)  $d(x, y) = 0$ if and only if $x = y$;
(c)  $d(x, y) = d(y, x)$ for all $x, y \in X$ (symmetry);
(d)  $d(x, y) \leq d(x, z) + d(z, y)$ for all $x, y, z \in X$ (triangle inequality).

The pair $(X, d)$ is a metric space. The $\varepsilon$-ball $B(x, \varepsilon)$ consists of all points $y \in X$ which are less than $\varepsilon$ away from $x$. If $(X, d)$ is a metric space, the collection $\mathcal{B} = \{B(x, \varepsilon) : x \in X, \varepsilon > 0\}$ of all balls is a basis for the *metric topology* on $X$.

Metric spaces were introduced by Maurice Fréchet in his Ph. D. thesis in 1906.

In the discussion and exercises of Section 1.1, we defined three metrics on the plane:

$$\text{the Euclidean metric} \quad d((x, y), (a, b)) = \sqrt{(x - a)^2 + (y - b)^2},$$
$$\text{the sup metric} \quad m((x, y), (a, b)) = \sup\{|x - a|, |y - b|\}, \quad \text{and}$$
$$\text{the taxicab metric} \quad t((x, y), (a, b)) = |x - a| + |y - b|.$$

The $\varepsilon$-balls in these three metrics on $\mathbb{R}^2$ are shown in Figure 6.1.

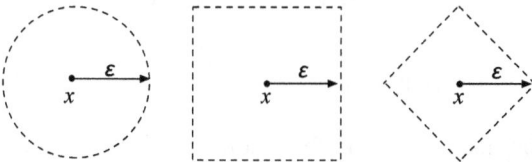

**Figure 6.1:** $B(x, \varepsilon)$ in the Euclidean, sup, and taxicab metrics.

Recall (Theorem 1.4.10) that one topology $\mathcal{T}_F$ on $X$ is finer than another $\mathcal{T}_C$ if for each basic $\mathcal{T}_C$ neighborhood $V$ of $x \in X$, there exists a basic $\mathcal{T}_F$ neighborhood of $x$ contained in $V$. Since each round (Euclidean) ball around $x$ contains a square (sup-metric) ball, which contains a diamond (taxicab-metric) ball, which contains a round ball, we see that these three metrics on $\mathbb{R}^2$ generate the same topology.

In Section 5.1, we defined a bounded subset of a metric space $X$ to be a set $A$ such that $A$ is contained in some ball $B(x, N)$. It is easy to show that $A$ is bounded if and only if the set $S = \{d(x, y) : x, y \in A\}$ is a bounded subset of $\mathbb{R}$, that is, if there exists $M \in \mathbb{R}$

https://doi.org/10.1515/9783110686579-007

with $S \subseteq [-M, M]$. If $(X, d)$ is a metric and the set $X$ itself is bounded, then we say $d$ is a *bounded metric* on $X$. The Euclidean metric $d$ on $\mathbb{R}$ is not a bounded metric. Given any potential bound $M > 0$, we can find points such as $x = 0$ and $y = 2M$ which are farther than $M$ units apart. However, if we truncate all distances greater than 1 by taking

$$\bar{d}(x, y) = \min\{d(x, y), 1\},$$

then one can verify that $\bar{d}$ is a bounded metric. (Only the triangle inequality is non-trivial to verify.) This metric $\bar{d}$ is called the *standard bounded metric* associated with $d$. Furthermore, if $\varepsilon < 1$, then the $\varepsilon$-balls in $d$ and $\bar{d}$ agree. Since it is the balls with small radii which are most important in defining the basis for a topology, $d$ and $\bar{d}$ generate the same topology. (See Exercise 2.)

Thus, metrics with very different properties may generate the same topology.

A significant question in topology has been the question of which topological spaces are generated by some metric; that is, which topological spaces are *metrizable*. Since every metric space is Hausdorff, any topological space which is not Hausdorff will not be metrizable. The idea of the taxicab metric provides the basis for the proof of the next result.

**Theorem 6.1.1.** *If $(X, d_X)$ and $(Y, d_Y)$ are metric spaces, then $X \times Y$ with the product topology is metrizable.*

*Proof.* Consider the function $d : (X \times Y)^2 \rightarrow [0, \infty)$ defined by $d((x, y), (a, b)) = d_X(x, a) + d_Y(y, b)$. It is easy to see that $d$ is nonnegative, symmetric, and $d((x, y), (a, b)) = 0$ if and only if $(x, y) = (a, b)$. The triangle inequality follows by applying the triangle inequality in both coordinates: for any $(x, y), (a, b), (z, w) \in X \times Y$,

$$d((x, y), (a, b)) + d((a, b), (z, w)) = d_X(x, a) + d_Y(y, b) + d_X(a, z) + d_Y(b, w)$$
$$\geq d_X(x, z) + d_Y(y, w)$$
$$= d((x, y), (z, w)).$$

For $(x, y) \in X \times Y$, a basic neighborhood of $(x, y)$ in the product topology has form $B_X(x, \varepsilon) \times B_Y(y, \varepsilon)$. Now $(z, w) \in B((x, y), \varepsilon)$ implies $z \in B_X(x, \varepsilon)$ and $w \in B_Y(y, \varepsilon)$, so $B((x, y), \varepsilon) \subseteq B_X(x, \varepsilon) \times B_Y(y, \varepsilon)$, so the metric topology on $X \times Y$ is finer than the product topology. However, $(z, w) \in B_X(x, \varepsilon/2) \times B_Y(y, \varepsilon/2)$ implies $d((x, y), (z, w)) < \varepsilon$, so $B_X(x, \varepsilon/2) \times B_Y(y, \varepsilon/2) \subseteq B((x, y), \varepsilon)$, showing that the product topology is finer than the metric topology. Thus, the topologies agree, and this shows that the product topology is metrizable. □

In any interpretation of sequential convergence, $x_n \rightarrow x$ should occur if and only if the terms $x_n$ are getting close to $x$. Thus, the following result is expected.

**Theorem 6.1.2.** *If $(x_n)_{n \in \mathbb{N}}$ is a sequence in a metric space $(X, d)$, then $x_n \rightarrow x$ if and only if $d(x_n, x) \rightarrow 0$.*

*Proof.* The following statements are equivalent:

$x_n \to x$.

For any $\varepsilon > 0$, there exists $N \in \mathbb{N}$ with $x_n \in B(x, \varepsilon)$ for all $n \geq N$.

For any $\varepsilon > 0$, there exists $N \in \mathbb{N}$ with $d(x_n, x) < \varepsilon$ for all $n \geq N$.

For any $\varepsilon > 0$, there exists $N \in \mathbb{N}$ with $d(x_n, x) \in B(0, \varepsilon)$ for all $n \geq N$.

$d(x_n, x) \to 0$. □

Thus, if $(x_n, x) \to (x, x)$, then $d(x_n, x) \to d(x, x)$. In the proof of the next result, we show that a metric $d$ preserves limits of all sequences and thus (by Theorem 3.1.7) is a continuous function.

**Theorem 6.1.3.** *Suppose $(X, d)$ is a metric space and $(x_n, y_n) \to (x, y)$ in $X \times X$. Then $d(x_n, y_n) \to d(x, y)$, and thus the metric $d : X \times X \to [0, \infty)$ is continuous, where $[0, \infty)$ carries the Euclidean metric.*

*Proof.* Suppose $(x_n, y_n) \to (x, y)$ in $X \times X$. Since the projection functions are continuous, $x_n \to x$ and $y_n \to y$ in $X$. Thus, given $\varepsilon > 0$, there exist $N_1, N_2 \in \mathbb{N}$ such that $n \geq N_1$ implies $d(x_n, x) < \varepsilon/2$ and $n \geq N_2$ implies $d(y_n, y) < \varepsilon/2$. Now for $n \geq N = \max\{N_1, N_2\}$,

$$d(x_n, y_n) \leq d(x_n, x) + d(x, y) + d(y, y_n) < \varepsilon/2 + d(x, y) + \varepsilon/2.$$

Subtracting $d(x, y)$ from both sides of this inequality shows $d(x_n, y_n) - d(x, y) < \varepsilon$. Similarly, for $n \geq N$,

$$d(x, y) \leq d(x, x_n) + d(x_n, y_n) + d(y_n, y) < \varepsilon/2 + d(x_n, y_n) + \varepsilon/2,$$

which gives $d(x_n, y_n) - d(x, y) > -\varepsilon$. Together, this shows that, for every $\varepsilon > 0$, there exists $N \in \mathbb{N}$ such that $|d(x_n, y_n) - d(x, y)| < \varepsilon$ for all $n \geq N$. Thus, $d(x_n, y_n) \to d(x, y)$, and since $d$ preserves limits, $d$ is continuous. □

Most of the metrics we have seen have described distances between points of $\mathbb{R}^n$. We now look at metrics on sets of other kinds of objects.

**Example 6.1.4** (The Hamming metric). Let $X = \{0, 1\}^n$ be the set of all binary $n$-tuples. An element of $X$ has form $(x_1, x_2, \ldots, x_n)$ where each $x_i$ is either 0 or 1, and is called a *binary word* of length $n$. The Hamming distance between two binary words of length $n$ is the number of places in which the words differ. That is, if $x, y \in X$ with $x = (x_1, \ldots, x_n)$ and $y = (y_1, \ldots, y_n)$, then $d_H(x, y) = |\{i \in \{1, \ldots, n\} : x_i \neq y_i\}|$. For example, on the set of binary words of length 11, comparing

$$\begin{array}{rccccccccccc}
x & = & (1, & 1, & 1, & 1, & 0, & 0, & 0, & 1, & 1, & 1, & 1) \quad \text{and} \\
y & = & (0, & 1, & 0, & 1, & 0, & 1, & 0, & 1, & 0, & 1, & 1), \\
& & \times & & \times & & & \times & & & \times & &
\end{array}$$

we see that $x$ and $y$ differ in four places, so $d_H(x, y) = 4$.

The Hamming metric is the basis for *error-correcting codes.* Suppose we wish to transmit combinations of the letters $A, B, C, D$, and $E$. We may agree beforehand to represent these letters by the binary words

$$A = (1,1,1,1,1,1,1,1,1,1,1)$$
$$B = (1,1,1,1,1,0,0,0,0,0,0)$$
$$C = (0,0,0,0,0,1,1,1,1,1,1)$$
$$D = (1,1,1,0,0,0,0,0,1,1,1)$$
$$E = (0,0,0,1,1,1,1,1,0,0,0).$$

Since we will only send these letters, we are restricting our attention to the subspace $Y = \{A, B, C, D, E\}$ of $X = \{0, 1\}^{11}$. Now it is easy to check that the smallest distance between any distinct pair of these binary words is 5, so for distinct words $y, z$ in our subspace $Y$, we have $B(y, 3) \cap B(z, 3) = \emptyset$. That is, for any $y \in X$, we have $B(y, 3) \cap Y = \{y\}$. If a received message includes the binary word $w = (1, 0, 0, 0, 0, 1, 1, 1, 1, 0, 1) \notin Y = \{A, B, C, D, E\}$, we would observe that $w$ differs from $C$ in two positions, so $w \in B(C, 3)$. Since $B(C, 3) \cap Y = \{C\}$, if we assume that the received word has fewer than three incorrect binary digits, we must conclude that the incorrect $w$ should have been $C$.

We present an application of the Hamming metric to the mathematical study of social choice. Suppose several individuals each rank $k$ options. For example, fellow travelers may rank points of interest to visit on a 3-day trip to Paris, employees may rank items for budgetary preference, or voters may rank candidates. For our purposes, a ranking will be a strict total order $<$, so each individual specifies the first choice, second choice, and so on until the last choice, with no ties. Such a ranking $P$ on $k$ options $\{a_1, a_2, \ldots, a_k\}$ can be specified by the binary vector $\langle P \rangle$ formed by going through the $\binom{k}{2}$ distinct pairs $(a_i, a_j)$ in a specified order and recording 1 if $a_i < a_j$ and 0 otherwise. Then, we may use the Hamming metric to measure the distance between two rankings. Formally, if $P$ is a ranking of $\{a_1, a_2, \ldots, a_k\}$, we take

$$\langle P \rangle = \langle P(a_1, a_2), \quad P(a_1, a_3), \quad P(a_1, a_4), \qquad \cdots \qquad , P(a_1, a_k),$$
$$P(a_2, a_3), \quad P(a_2, a_4), \qquad \cdots \qquad , P(a_2, a_k),$$
$$\ddots \qquad \qquad \vdots$$
$$P(a_{k-2}, a_{k-1}), \quad P(a_{k-2}, a_k),$$
$$P(a_{k-1}, a_k) \rangle,$$

where $P$ is viewed as the strict total order $<$ and

$$P(a_i, a_j) = \begin{cases} 1 & \text{if } a_i < a_j \\ 0 & \text{if } a_i \not< a_j. \end{cases}$$

Given two rankings $P, Q$ of the $k$ options, the *Kemeny distance* (or *Kendall tau distance*) between them is $K(P, Q) = d_H(\langle P \rangle, \langle Q \rangle)$. Table 6.1 shows the six strict total orders $P_1, \ldots, P_6$ on $\{a_1, a_2, a_3\}$ and their associated binary vectors.

**Table 6.1:** Binary vectors associated with the six rankings of $\{a_1, a_2, a_3\}$.

| Label | Order | $\langle P_i(a_1, a_2),$ | $P_i(a_1, a_3),$ | $P_i(a_2, a_3) \rangle$ |
|-------|-------|------|------|------|
| $P_1$ | $a_1 < a_2 < a_3$ | $\langle$ 1, | 1, | 1 $\rangle$ |
| $P_2$ | $a_1 < a_3 < a_2$ | $\langle$ 1, | 1, | 0 $\rangle$ |
| $P_3$ | $a_2 < a_1 < a_3$ | $\langle$ 0, | 1, | 1 $\rangle$ |
| $P_4$ | $a_2 < a_3 < a_1$ | $\langle$ 0, | 0, | 1 $\rangle$ |
| $P_5$ | $a_3 < a_1 < a_2$ | $\langle$ 1, | 0, | 0 $\rangle$ |
| $P_6$ | $a_3 < a_2 < a_1$ | $\langle$ 0, | 0, | 0 $\rangle$ |

Now we see, for example, that

$$K(P_2, P_5) = d_H(\langle P_2 \rangle, \langle P_5 \rangle) = d_H(\langle 1, 1, 0 \rangle, \langle 1, 0, 0 \rangle) = 1.$$

If eight individuals place their ranking of three candidates into a ballot box, the contents of the ballot box might be: $P_1, P_1, P_3, P_3, P_4, P_6, P_6, P_6$. This collection of rankings, with appropriate multiplicities, is called a *profile*, and is denoted $\pi = \{\{P_1, P_1, P_3, P_3, P_4, P_6, P_6, P_6\}\}$ or $\pi = \{P_i\}_{i \in \{\{I\}\}}$, where $\{\{I\}\} = \{\{1, 1, 3, 3, 4, 6, 6, 6\}\}$. The double set-brackets indicate that we are not merely denoting a set of elements, but a set of elements listed with the appropriate multiplicities; that is, a *multiset*. A standard goal would be to take the rankings of all individuals, that is, a profile $\pi$, and aggregate them into a single societal ranking which is "most acceptable". Since the distance $K(R, P_i)$ gives a measure of how unhappy someone who prefers the ranking $P_i$ would be with the ranking $R$, we may say that a ranking $R$ is the most acceptable societal ranking from a profile $\pi = \{P_i\}_{i \in \{\{I\}\}}$ if the sum $u(R, \pi) = \sum_{i \in \{\{I\}\}} K(R, P_i)$ of the unhappiness is minimum. Such a ranking $R$ is called a *Kemeny ranking* for the profile.

**Example 6.1.5.** Given the rankings of $\{a_1, a_2, a_3\}$ listed in Table 6.1 and the profile $\pi = \{\{P_1, P_1, P_3, P_3, P_4, P_6, P_6, P_6\}\}$, we note that, for $i = 1, \ldots, 6$, if $m_i$ is the multiplicity of $P_i$,

$$u(P_i, \pi) = m_1 K(P_i, P_1) + m_2 K(P_i, P_2) + \cdots + m_6 K(P_i, P_6)$$
$$= 2K(P_i, P_1) + 2K(P_i, P_3) + K(P_i, P_4) + 3K(P_i, P_6).$$

Calculating the distances $K(P_i, P_j)$ from Table 6.1, we find

$$u(P_1, \pi) = 2(0) + 2(1) + 2 + 3(3) = 13,$$
$$u(P_2, \pi) = 2(1) + 2(2) + 3 + 3(2) = 15,$$
$$u(P_3, \pi) = 2(1) + 2(0) + 1 + 3(2) = 9,$$

$$u(P_4, \pi) = 2(2) + 2(1) + 0 + 3(1) = 9,$$
$$u(P_5, \pi) = 2(2) + 2(3) + 2 + 3(1) = 15,$$
$$u(P_6, \pi) = 2(3) + 2(2) + 1 + 3(0) = 11.$$

Since $P_3$ and $P_4$ produce the minimum value of the unhappiness sum $u(P_i, \pi)$, each of them is a Kemeny ranking for the profile $\pi$. In particular, we note that a Kemeny ranking need not be unique.

It is known that Kemeny rankings are NP-hard to compute, even for profiles of four rankings.

The Hamming metric measured the distance between words of the same length formed from a binary alphabet $\{0, 1\}$. We turn to a metric which measures distances between words of arbitrary finite length formed from larger alphabets.

**Example 6.1.6** (The Levenshtein metric). If you misspell a word in a word processor, the spell-checking software may suggest words in its dictionary which it thinks are close to what you typed. One way to measure closeness of two words is to find the minimal number of insertions, deletions, and replacements to get from one word to the other. For example, if you intended to transmit "CAT" but "CAGE" was received, then the error introduced one replacement ("G" for "T") and one insertion ("E"). There are other sequences of errors which could have resulted in the transformation of CAT to CAGE, such as changing T to N, changing N to G, inserting Z and changing Z to E. The minimal number of insertions, deletions, and replacements to transform CAT to CAGE is 2, and this will be the Levenshtein distance between the words. The Levenshtein distance between finite words $x$ and $y$ is

$$d_L(x, y) = \text{the minimum number of (deletions, insertions, and replacements)}$$
$$\text{required to change } x \text{ to } y.$$

For example, $d_L(\text{CAT}, \text{CAGE}) = 2$, $d_L(\text{CAT}, \text{HAT}) = 1$, $d_L(\text{CAT}, \text{AT}) = 1$, and $d_L(\text{CAT}, \text{IT}) = 2$.

Vladimir Levenshtein introduced this metric in a 1966 paper in Russian [33]. In 1964, Fred Damerau [12] introduced a very similar metric in which the distance between two words is the minimum number of deletions, insertions, replacements, or transpositions of adjacent letters. In a trial for his paper, Damerau found that 80 % of misspelled words had distance 1 from the intended word using his metric.

A more classical usage of metrics is to measure the distance between two functions. The study of functions from subsets of $\mathbb{R}$ to $\mathbb{R}$, both with the Euclidean topology, is an important part of *real analysis*. Let $C[a, b]$ be the collection of continuous functions $f : [a, b] \to \mathbb{R}$, where the compact interval $[a, b]$ and $\mathbb{R}$ carry the Euclidean topology. For the three functions $f, g, h \in C[a, b]$ shown in Figure 6.2, is $g$ or $h$ closer to $f$?

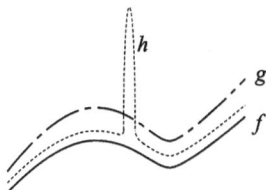

**Figure 6.2:** Which function, $g$ or $h$, is closer to $f$?.

Suppose $f$ represents a heating element in a windshield. Since heat disperses uniformly from $f$, the function $g$ will be completely defrosted before $h$, so $g$ would be considered closer to $f$. However, if the functions represent boundary fences of a lawn, the region between $f$ and $h$ can be mowed faster than the region between $f$ and $g$, so in this interpretation, $h$ would be considered closer. In the latter interpretation, we are simply taking the distance between two curves to be the area between them. We are using the *areawise metric* (or the $L^1$ *metric*) metric $d$ on $C[a, b]$ defined by

$$d(f, g) = \int_a^b |f(x) - g(x)|\, dx.$$

In the former interpretation, the distance between two functions is the greatest vertical distance between them, which is the *sup metric* $\rho$ on $C[a, b]$ defined by

$$\rho(f, g) = \sup_{x \in [a,b]} \{|f(x) - g(x)|\}.$$

Note that by considering continuous functions over the compact interval $[a, b]$, the extreme value theorem implies that the supremum indicated exists and is actually realized at a point in $[a, b]$, so in this case, the supremum is in fact a maximum. In the sup metric, $g$ is close to $f$ if it is uniformly close; indeed, the $\varepsilon$-ball $B_\rho(f, \varepsilon) = \{g \in C[a, b] : f(x) - \varepsilon < g(x) < f(x) + \varepsilon$ for all $x \in [a, b]\}$ consists of all functions $g \in C[a, b]$ whose graph lies in the band between $f(x) - \varepsilon$ and $f(x) + \varepsilon$, as suggested by Figure 6.3.

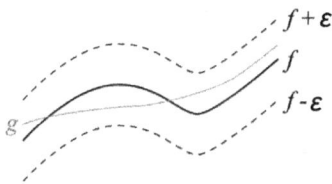

**Figure 6.3:** $B_\rho(f, \varepsilon)$ consists of all functions like $g$, whose graph falls within the $\varepsilon$-band above and below $f$.

The sup metric is well-defined on the collection of bounded real-valued functions on $\mathbb{R}$. For classes of functions which contain unbounded functions, $\sup\{|f(x) - g(x)| :$

$x \in \mathbb{R}\}$ may not exist (that is, may "$= \infty$"). Still, we should be able to recognize when two functions $f, g : \mathbb{R} \to \mathbb{R}$ are uniformly close. We can do this with the *bounded sup metric* $\bar{\rho}$ on the set of all functions from $\mathbb{R}$ to $\mathbb{R}$ defined by

$$\bar{\rho}(f,g) = \min\Big\{1, \sup_{x \in \mathbb{R}}\{|f(x) - g(x)|\}\Big\}$$

$$= \sup_{x \in \mathbb{R}}\{\min\{1, |f(x) - g(x)|\}\} = \sup_{x \in \mathbb{R}}\{\bar{d}(f(x), g(x))\},$$

where $\bar{d}$ is the standard bounded metric associated with the Euclidean metric on $\mathbb{R}$, and where we take $\min\{1, \infty\} = 1$. For $\varepsilon < 1$, the $\varepsilon$-balls in $\rho$ and $\bar{\rho}$ agree, so these metrics generate the same topology.

A sequence of functions $f_n : \mathbb{R} \to \mathbb{R}$ *converges uniformly* to a function $f$ if it converges to $f$ in the metric space of all functions from $\mathbb{R}$ to $\mathbb{R}$ with the bounded sup metric. The topology generated by the bounded sup metric is called the *uniform topology*. For this reason, the bounded sup metric is also called the *uniform metric*.

**Example 6.1.7.** The sequence $(f_n)_{n \in \mathbb{N}}$ where $f_n(x) = x^2 + \cos(nx)/n$ converges uniformly to $f(x) = x^2$. Given any $\varepsilon > 0$, pick $N \in \mathbb{N}$ such that $\frac{1}{N} < \varepsilon$. Then, for $n > N$, we have

$$\sup\{|f_n(x) - f(x)| : x \in \mathbb{R}\} = \sup\{|\cos(nx)/n| : x \in \mathbb{R}\} = \frac{1}{n} < \frac{1}{N} < \varepsilon.$$

That is, for $n > N$ we have $f_n \in B(f, \varepsilon)$. Thus, $(f_n)_{n \in \mathbb{N}}$ converges uniformly to $f(x) = x^2$. Some terms of the sequence are shown in Figure 6.4.

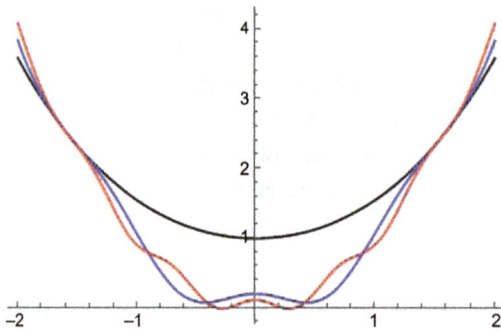

**Figure 6.4:** $f_n(x) = x^2 + \cos(nx)/n$ for $n = 1, 5, 9$.

## Exercises

1. Show that if $A$ is a bounded subset of a metric space $(X, d)$, then $\mathrm{cl}\, A$ is bounded.
2. Suppose $(X, d)$ is a metric space and $\delta > 0$. Show that $\mathcal{B}_\delta = \{B(x, \varepsilon) : x \in X, 0 < \varepsilon \le \delta\}$ is a basis for the metric topology on $X$.

3. If $d$ is a metric on $X$, prove that the associated standard bounded metric $\bar{d}$ defined by $\bar{d}(x, y) = \min\{1, d(x, y)\}$ is a metric on $X$.

4. Suppose $(X, d)$ is a metric space, $x \in X$, and $\varepsilon \geq 0$. The *closed ball* centered at $x$ with radius $\varepsilon$ is $\overline{B}(x, \varepsilon) = \{y \in X : d(x, y) \leq \varepsilon\}$.
   (a) Show that the closed ball $\overline{B}(x, \varepsilon)$ is a closed set in the metric topology on $X$.
   (b) Show that $\text{cl}(B(x, \varepsilon)) \subseteq \overline{B}(x, \varepsilon)$, and give an example of a metric space $(X, d)$ in which $\text{cl}(B(x, \varepsilon)) \neq \overline{B}(x, \varepsilon)$ for some $x \in X$ and some $\varepsilon > 0$.

5. Suppose $d_1$ and $d_2$ are two metrics on a set $X$.
   (a) Give a proof or counterexample: The sum $d_1 + d_2$ is always a metric.
   (b) Give a proof or counterexample: The product $d_1 d_2$ is always a metric.

6. Consider the function $p : \mathbb{R}^2 \to [0, \infty)$ defined by $p(x, y) = \max\{|x - y|, |x + y|\}$. Which properties of the definition of a metric are satisfied by $p$?

7. Show that an infinite connected metric space must be uncountably infinite.

8. Using the error-correcting code given in Example 6.1.4, decode the following erroneous messages.
   (a) $(1, 1, 1, 1, 0, 1, 1, 0, 1, 1, 1)$ $(0, 0, 0, 1, 1, 1, 1, 1, 1, 1, 1)$ $(0, 1, 0, 1, 0, 1, 1, 1, 0, 0, 0)$
   (b) $(1, 1, 1, 1, 1, 0, 1, 0, 0, 1, 0)$ $(0, 0, 0, 0, 0, 1, 1, 1, 0, 0, 0)$ $(1, 1, 0, 1, 1, 1, 1, 1, 0, 0, 0)$

9. Label the binary 5-tuple $(d_4, d_3, d_2, d_1, d_0) \in X = \{0, 1\}^5$ by the base 10 integer $n = d_4 2^4 + d_3 2^3 + d_2 d^2 + d_1 2^1 + d_0 2^0$ having binary digits $d_4, d_3, d_2, d_1, d_0$. With these labels, $X = \{0, 1, \ldots, 31\}$. In the Hamming metric, find $B(31, 2)$, $B(20, 2)$, and $B(0, 3)$.

10. With the orders on $\{a_1, a_2, a_3\}$ as in Table 6.1, find all Kemeny rankings for the profile $\pi = \{\{P_1, P_3, P_3, P_5, P_6, P_6\}\}$. Is each Kemeny ranking an element of the profile?

11. With the orders on $\{a_1, a_2, a_3\}$ as in Table 6.1, find all Kemeny rankings for the profile $\pi = \{\{P_1, P_1, P_2, P_2, P_2, P_3, P_4, P_4, P_4, P_5, P_6, P_6\}\}$.

12. If $P$ is a ranking of $\{a_1, \ldots, a_k\}$, let $P^\leftarrow$ be the ranking obtained by reversing $P$. That is, $P$'s ranking of the elements from best to worst is $P^\leftarrow$'s ranking of the elements from worst to best; $a_i < a_j$ in $P$ if and only if $a_j < a_i$ in $P^\leftarrow$. Show that if $\pi$ is a profile on $\{a_1, \ldots, a_k\}$ in which the multiplicity of $P_i$ equals the multiplicity of $P_i^\leftarrow$ for every $P_i \in \pi$, then every ranking of $\{a_1, \ldots, a_k\}$ is a Kemeny ranking.

13. Suppose $\pi$ is a profile of rankings of $\{a_1, a_2, a_3\}$ and $i, j$ are distinct elements of $\{1, 2, 3\}$. If $\pi$ only includes rankings which place $a_i < a_j$, show that any Kemeny ranking for $\pi$ also has $a_i < a_j$. (That is, the Kemeny ranking of a profile of $\{a_1, a_2, a_3\}$ satisfies the *Pareto condition*.)

14. In the Levenshtein metric, find a word whose distance from COUNC is one, and find five words in $B(\text{COUNC}, 3)$.

15. In the Levenshtein metric,
   (a) Is SIXTY closer to SEVENTY or EIGHTY?
   (b) Is TOPOLOGY closer to TOPOGRAPHY or BIOLOGY?
   (c) Is CONNECTED closer to COMPACT or COUNTABLE?

16. **(The prefix metric)** Let $S = \{(x_i)_{i=1}^{\infty} : x_i \in \{0, 1\}$ for all $i \in \mathbb{N}\}$ be the set of all binary sequences. Two sequences $x = (x_i)_{i=1}^{\infty}$ and $y = (y_i)_{i=1}^{\infty}$ are said to have a *common prefix of length $n$*, denoted $\text{pre}(x, y) = n$, if the two sequences agree in the

first $n$ terms and differ in the $(n + 1)$st term. The prefix metric $p$ on $X$ is defined by $p(x, y) = 0$ if $x = y$, and for distinct sequences $x$ and $y$ with a common prefix of length $n \in \mathbb{N} \cup \{0\}$, $p(x, y) = \frac{1}{2^n} = \frac{1}{2^{\text{pre}(x,y)}}$.

Suppose $a = (1, 0, 1, 1, 1, 1, 1, 1, 1, \ldots)$, $b = (1, 0, 1, 1, 1, 0, 0, 0, 0, \ldots)$, $c = (1, 1, 0, 1, 0, 1, 0, 1, 0, \ldots)$, and $d = (0, 0, 1, 1, 1, 1, 1, 1, 1, \ldots)$.

(a) Find $p(a, b)$, $p(a, c)$, and $p(a, d)$.

(b) If $x = (1)_{n=1}^{\infty}$, describe the points $y$ with $p(x, y) = \frac{1}{16}$.

(c) If $z$ is an arbitrary element of $S$, and $n \in \mathbb{N} \cup \{0\}$, give a verbal description of $B(z, \frac{1}{2^n})$.

(d) Show that, for every $x, y, z \in S$, $\text{pre}(x, z) \geq \min\{\text{pre}(x, y), \text{pre}(y, z)\}$.

(e) Show that the prefix metric is indeed a metric on $S$.

17. An *isometry* is a distance-preserving function from one metric space to another. That is, if $(X, d_X)$ and $(Y, d_Y)$ are metric spaces, an isometry from $X$ to $Y$ is a function $f : X \to Y$ such that, for every $a, b \in X$, $d_X(a, b) = d_Y(f(a), f(b))$. Show that if $f$ is an isometry from $(X, d)$ onto $(Y, m)$, then $f$ is a homeomorphism and $f^{-1} : Y \to X$ is an isometry.

18. Let $f : \mathbb{R}^2 \to \mathbb{R}^2$ be the function which rotates a vector $(x, y) \in \mathbb{R}^2$ around the origin through a fixed angle $\alpha$. That is, in polar coordinates, $f((r, \theta)) = (r, \theta + \alpha)$. Is $f$ an isometry (see Exercise 17) if:

(a) The domain and codomain $\mathbb{R}^2$ carry the Euclidean metric?

(b) The domain and codomain $\mathbb{R}^2$ carry the taxicab metric?

19. For $r > 0$, let $S_r = \{(x, y) \in \mathbb{R}^2 : x^2 + y^2 = r^2\} = \{(r, \theta) : \theta \in [0, 2\pi)\}$ be the circle of radius $r$. Let $d : S_r \times S_r \to \mathbb{R}$ be the function that gives the length of the shortest arc of $S_r$ between two points. Show that $d$ is a metric on $S_r$.

20. If $(X, d)$ is a metric space, we say that $\{x, y, z\} \subseteq X$ is a *degenerate triangle* if $d(x, z) = d(x, y) + d(y, z)$. If every triangle in a metric space is degenerate, one might expect that the metric space is isometric to a subspace of the Euclidean line (see Exercise 17). This is true except for one class of 4-point spaces. (For more details, see [40].)

Let $d$ be the arc length metric on the unit circle $S_1$ defined in Exercise 19. Consider the 4-point subspace $X = \{(r, \theta) : r = 1, \theta \in \{0, \pi/4, \pi, 5\pi/4\}\}$ of $S_1$.

(a) Show that every triangle in $(X, d)$ is degenerate.

(b) Show that $(X, d)$ cannot be realized as a subspace of the Euclidean line.

21. Confirm that $d(f, g) = \int_a^b |f(x) - g(x)| \, dx$ defines a metric on $C[a, b]$.

22. Confirm that $\rho(f, g) = \sup\{|f(x) - g(x)| : x \in [a, b]\}$ defines a metric on $C[a, b]$.

23. The bounded sup metric $\bar{\rho}$ on the set of functions from $\mathbb{R}$ to $\mathbb{R}$ was defined by

$$\bar{\rho}(f, g) = \min\left\{1, \sup_{x \in \mathbb{R}}\{|f(x) - g(x)|\}\right\} = \sup_{x \in \mathbb{R}}\{\min\{1, |f(x) - g(x)|\}\}.$$

Confirm that these two expressions for $\bar{\rho}(f, g)$ are equal.

24. Consider the function $h : \mathbb{R} \to \mathbb{R}$ defined by

$$h(x) = \begin{cases} 1 & \text{if } x \in \mathbb{Q}, \\ -1 & \text{if } x \notin \mathbb{Q}. \end{cases}$$

   (a) Give two continuous functions in $B_\rho(h, 1.25)$.
   (b) Prove that there is no continuous function in $B_\rho(h, 1)$.

25. Consider the functions $f(x) = x^2$, $g(x) = x^3$, and $h(x) = x^5$ in $C[0,1]$.
   (a) Determine whether $f$ or $h$ is closer to $g$ in the sup metric $\rho$.
   (b) Determine whether $f$ or $h$ is closer to $g$ in the areawise metric $d$ defined by $d(f,g) = \int_0^1 |f(x) - g(x)|\, dx$.
   (c) Is $f$ closer to $g$ in the sup metric or in the areawise metric?

26. (Metric-preserving functions) A function $f : [0, \infty) \to [0, \infty)$ is *metric-preserving* if for any metric $d$ on $X$, $f \circ d$ is also a metric on $X$.
   (a) Show that $f : [0, \infty) \to [0, \infty)$ is metric preserving if $f$ is increasing, $f(x) = 0$ if and only if $x = 0$, and $f(x + y) \le f(x) + f(y)$ for any $x, y \in X$.
   (b) If $f$ meets the hypotheses of (a) and is continuous, show that the metrics $d$ and $f \circ d$ generate the same topology on $X$.
   (c) Show that $f(x) = \frac{x}{1+x}$ meets the hypotheses of (a) and is continuous, so $f \circ d$ is a metric for any metric $d$ on $X$. Note that $(f \circ d)(a, b) = \frac{d(a,b)}{1+d(a,b)}$ defines a bounded metric generating the same topology as the metric $d$.

## 6.2 Infinite products and functional analysis

In the Cartesian coordinate system, to visualize a point of $\mathbb{Z}^3$, we draw three perpendicular copies of $\mathbb{Z}$ and find $(a, b, c) \in \mathbb{Z}^3$ by finding the coordinates $a$ in the first copy, $b$ in the second copy, and $c$ in the third copy. This technique will not work for products of more than three factors since we have only 3 spatial dimensions in which to draw perpendicular axes. For a visual interpretation which can generalize to products of more than three factors, we may draw three *parallel* copies of $\mathbb{Z}$ and find $(a, b, c) \in \mathbb{Z}^3$ by finding the coordinates $a$ in the first copy, $b$ in the second copy, and $c$ in the third copy, as suggested in Figure 6.5.

We may write the product $\mathbb{Z}^3$ in various forms, including $\prod_{i \in \{1,2,3\}} \mathbb{Z} = \mathbb{Z}^{\{1,2,3\}}$. Now a point of $\mathbb{Z}^3$ has form $(x_1, x_2, x_3) = (f(1), f(2), f(3))$ where $x_i = f(i) \in \mathbb{Z}$ for each $i$ in the index set $\{1, 2, 3\}$. Thus, we may view $\mathbb{Z}^3 = \mathbb{Z}^{\{1,2,3\}}$ as the collection of all functions $f : \{1, 2, 3\} \to \mathbb{Z}$.

For another example, the product $\prod_{n \in \mathbb{N}} \mathbb{R} = \mathbb{R}^{\mathbb{N}}$ consists of a copy of $\mathbb{R}$ for each natural number $n \in \mathbb{N}$. An element of $\prod_{n \in \mathbb{N}} \mathbb{R}$ may be viewed as a vector $\langle a_n \rangle_{n \in \mathbb{N}}$, a sequence $(a_n)_{n \in \mathbb{N}}$, or a function $f : \mathbb{N} \to \mathbb{R}$ defined by $f(n) = a_n$, where $a_n \in \mathbb{R}$ for every $n \in \mathbb{N}$. This is the usual interpretation of a sequence $(a_n)_{n \in \mathbb{N}}$ of real numbers as a function $f : \mathbb{N} \to \mathbb{R}$, and we graph such functions by finding the $n$th coordinate $a_n$ in

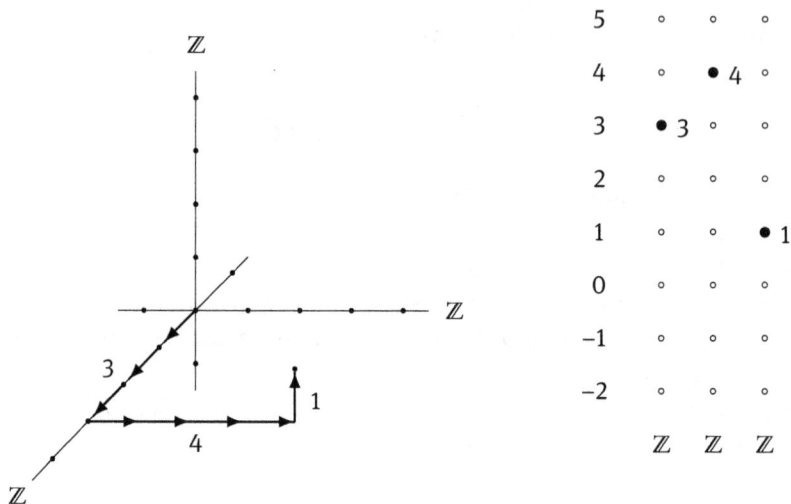

**Figure 6.5:** The point $(3, 4, 1)$ in $\mathbb{Z} \times \mathbb{Z} \times \mathbb{Z}$ visualized with perpendicular coordinate axes and parallel coordinate axes.

the $n$th parallel copy of $\mathbb{R}$. That is, given $n \in \mathbb{N}$ to specify the coordinate in question, we plot $(n, f(n)) = (n, a_n)$.

Similarly, $\prod_{x \in \mathbb{R}}[0, 1] = [0, 1]^{\mathbb{R}}$ provides a copy of $[0, 1]$ for each $x \in \mathbb{R}$, and a point of this product may be viewed as a vector $\langle a_x \rangle_{x \in \mathbb{R}} = \langle f(x) \rangle_{x \in \mathbb{R}}$ or a function $f : \mathbb{R} \to [0, 1]$, where for $x \in \mathbb{R}$, $f(x)$ represents the $x$th coordinate $a_x$. Notice that, since the indexing set $\mathbb{R}$ is uncountable, we may not think of this vector as a sequence. Sequences are indexed by countable sets.

We formally define our notation.

**Definition 6.2.1.** If $X$ is any set and $I$ is any index set, then the *Cartesian product* $\prod_{i \in I} X = X^I$ of $I$ copies of $X$ is the collection $\{f : I \to X\}$ of all functions from $I$ to $X$. A function $f \in X^I = \prod_{i \in I} X$ may be thought of as a vector $\prod_{i \in I} f(i) = \langle f(i) \rangle_{i \in I}$ whose $i$th coordinate is $f(i) = \pi_i(f)$.

Our discussion of infinite products so far has been purely set-theoretic. We now turn to topology and introduce the product topology on an infinite product. While the notation of the previous definition considered products with all factors equal, we now consider arbitrary products which may have different factors.

The fundamental topological property that we wish to preserve for arbitrary products is that the projection maps should be continuous.

**Definition 6.2.2.** The *product topology* on an arbitrary product $\prod_{i \in I} X_i$ of topological spaces $(X_i, \mathcal{T}_i)$ is the smallest topology on $\prod_{i \in I} X_i$ which makes the projection maps $\pi_j : \prod_{i \in I} X_i \to X_j$ continuous for every $j \in I$. That is, the product topology on $\prod_{i \in I} X_i$ is the coarsest topology which includes $\mathcal{S} = \{\pi_j^{-1}(U_j) : j \in I, U_j \in \mathcal{T}_j\}$.

Now $S = \{\pi_j^{-1}(U_j) : j \in I, U_j \in T_j\}$ need not be a basis for a topology on $\prod_{i \in I} X_i$, but it is a subbasis since, for any given $j \in I$, $\bigcup S$ includes $\pi_j^{-1}(X_j) = \prod_{i \in I} X_i$. Any topology which makes each projection function continuous must contain $S$. If $U_j$ is open in $X_j$, then $\pi_j^{-1}(U_j)$ has form $\prod_{i \in I} W_i$ where $W_i = X_i$ if $i \neq j$, and $W_j = U_j$. To project into $U_j \subseteq X_j$, the other coordinates are completely unrestricted; only the $j$th coordinate must fall in $U_j$. In $\prod_{i \in \mathbb{N}} \mathbb{R}$, Figure 6.6 shows $\pi_4^{-1}((1, 2))$.

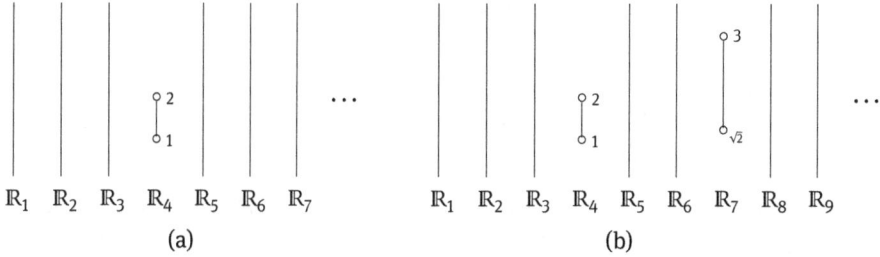

**Figure 6.6:** (a) Subbasic element $\pi_4^{-1}((1, 2))$ in $\prod_{i \in \mathbb{N}} \mathbb{R}$. (b) Basic element $\pi_4^{-1}((1, 2)) \cap \pi_7^{-1}((\sqrt{2}, 3))$ in $\prod_{i \in \mathbb{N}} \mathbb{R}$.

Now a basis is obtained by taking finite intersections of the subbasis elements from $S$. This gives the following result.

**Theorem 6.2.3.** *The product topology on $\prod_{i \in I} X_i$ has a basis consisting of sets of form $\prod_{i \in I} W_i$ where $W_i$ is open in $X_i$ for every $i \in I$ and $W_i = X_i$ for all but finitely many indices $i \in I$.*

Many familiar properties for finite products remain true for infinite products. We list some of them here.

**Theorem 6.2.4.** *Suppose $(X_i, T_i)$ is a topological space for each $i$ in an arbitrary index set $I$, and $Y = \prod_{i \in I} X_i$ with the product topology.*
(a) *Note that a function $f : Z \to Y = \prod_{i \in I} X_i$ has form $f(z) = \prod_{i \in I} f_i(z)$ where, for $i \in I$, the function $f_i : Z \to X_i$ is the $i$th coordinate function. A function $f : Z \to Y$ is continuous if and only if each of its coordinate functions $f_i = \pi_i \circ f$ is continuous.*
(b) *$Y$ is Hausdorff if and only if $X_i$ is Hausdorff for every $i \in I$.*
(c) *$Y$ is connected if and only if $X_i$ is connected for every $i \in I$.*
(d) *$Y$ is path connected if and only if $X_i$ is path connected for every $i \in I$.*

*Proof.* Most of these results are straightforward generalizations of the case for finite products and are left as exercises. We will only give one of the more involved proofs, to show that a product of connected spaces is connected.

Suppose $X_i$ is connected for every $i \in I$. Fix a vector $\langle a_i \rangle_{i \in I} \in X = \prod_{i \in I} X_i$. For each finite set $F \subseteq I$, let $Z_F = \{\langle x_i \rangle_{i \in I} \in X : x_i = a_i \ \forall i \in I - F\}$ be the set of vectors in $x$ which agree with $\langle a_i \rangle_{i \in I}$ except possibly in the finite number of coordinates determined by $F$. Now each $Z_F$ is connected since $Z_F \approx \prod_{i \in F} X_i \times \langle a_i \rangle_{i \in I - F} \approx \prod_{i \in F} X_i$, which is a finite product of connected spaces (see Theorem 4.1.13). Let $Z = \bigcup \{ Z_F : F \text{ is a finite subset of } I \}$. As a union of connected spaces with a point $\langle a_i \rangle_{i \in I}$ in common, $Z$ is connected (Theorem 4.1.12). Since the closure of a connected set is connected (Theorem 4.1.10), $\mathrm{cl}\, Z$ is connected. We will complete the proof by showing $\mathrm{cl}\, Z = X$, that is, by showing $Z$ is dense in $X$. For this, it suffices to show that every basic open set in $X$ intersects $Z$. Suppose $B = \prod_{i \in I} B_i$ is a basic open set in $X$ with $B_i \neq X_i$ only for indices $i$ in a finite set $F \subseteq I$. Pick $z_i \in B_i$ for $i \in F$ and set $z_i = a_i$ for $i \in I - F$. Now $\langle z_i \rangle_{i \in I} \in B \cap Z_F \subseteq B \cap Z$, as needed. $\qquad \square$

We mention one other result in this vein separately. In Theorem 5.1.9 we saw that a finite product of compact sets is compact. A. N. Tychonoff (1906–1993) proved that this result holds for arbitrary products.

**Theorem 6.2.5** (The Tychonoff theorem). *A product $\prod_{i \in I} X_i$ of topological spaces is compact if and only if every factor $X_i$ is compact.*

Tychonoff proved a special case of the theorem in 1930 [48], and stated the general case in a 1935 paper. Indeed, it was his 1935 paper that introduced the definition of the product topology on infinite products. Besides being a deep theorem, the Tychonoff theorem has many ramifications and applications. We will prove this theorem in Section 7.4.

An easier way to define a topology on a product $\prod_{i \in I} X_i$ of topological spaces $X_i$ is to take the collection $\{ \prod U_i : U_i \text{ is open in } X_i \}$ as a basis, without requiring any $U_i$ to be $X_i$. This gives the *box topology* on $\prod_{i \in I} X_i$. Now every basic open set in the product topology is a product of open sets (with all but finitely many of them being the entire coordinate), so the product topology is contained in the box topology. Unfortunately, many of the expected results of Theorem 6.2.4 do not hold if the product is given the box topology, as we see in Exercises 11–12 and the next example.

**Example 6.2.6.** Consider $Y = \prod_{n \in \mathbb{N}} \mathbb{R}$, and $f : \mathbb{R} \to Y$ defined by $f(x) = (x, x, x, x, \ldots)$, where $\mathbb{R}$ has the Euclidean topology and $Y$ has the box topology. The $n$th coordinate function of $f$ is $f_n(x) = x$, which is continuous. However, $f$ is not continuous. The set $W = \prod_{n \in \mathbb{N}} (-\frac{1}{n}, \frac{1}{n})$ is an open neighborhood of $(0, 0, 0, 0, \ldots)$ in the box topology. Now $f^{-1}(W) = \{ x \in \mathbb{R} : f(x) \in W \} = \{ x \in \mathbb{R} : x \in (-\frac{1}{n}, \frac{1}{n}) \text{ for every } n \in \mathbb{N} \} = \{0\}$, which is not open in $\mathbb{R}$.

We will now discuss a basic topic from functional analysis: convergence of sequences of functions from $D$ to $\mathbb{R}$, where the domain $D$ is a subset of the Euclidean line, and the codomain $\mathbb{R}$ also has the Euclidean topology. A sequence $(f_n)_{n \in \mathbb{N}}$ of functions $f_n : D \to \mathbb{R}$ is just a sequence of elements $f_n \in \mathbb{R}^D = \prod_{d \in D} \mathbb{R}$. Now any topology on the product $\mathbb{R}^D$ will allow us to define convergence of such sequences $(f_n)_{n \in \mathbb{N}}$. Before pursuing this, we introduce a more elementary approach to defining convergence of sequences of functions $f_n : D \to \mathbb{R}$.

For the remainder of this section, $\mathbb{R}$ and all subsets of $\mathbb{R}$ will carry the Euclidean topology.

**Definition 6.2.7.** Suppose $D \subseteq \mathbb{R}$. A sequence $(f_n)_{n \in \mathbb{N}}$ of functions $f_n : D \to \mathbb{R}$ *converges pointwise* to a function $f : D \to \mathbb{R}$ if for each point $x_0 \in D$, the sequence $(f_n(x_0))_{n \in \mathbb{N}}$ of real numbers converges to $f(x_0)$.

We illustrate this definition with an example.

**Example 6.2.8.** Consider the functions $f_n : [0, \pi] \to \mathbb{R}$ defined by $f_n(x) = \sin^n x$. The first five terms of this sequence are shown in Figure 6.7.

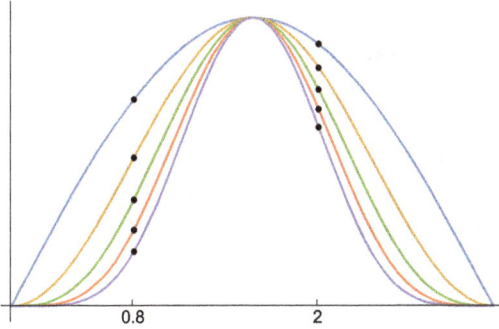

**Figure 6.7:** The first five terms of $(f_n)_{n \in \mathbb{N}}$, $(f_n(0.8))_{n \in \mathbb{N}}$, and $(f_n(2))_{n \in \mathbb{N}}$.

For the point $x_0 = \pi/2 \in D$, the sequence $(f_n(x_0))_{n \in \mathbb{N}} = (\sin^n(\pi/2))_{n \in \mathbb{N}} = (1^n)_{n \in \mathbb{N}}$ converges to the real number 1. For any point $x_0 \in D - \{\pi/2\}$, we have $0 \leq \sin x_0 < 1$, so the sequence $(f_n(x_0))_{n \in \mathbb{N}} = (\sin^n(x_0))_{n \in \mathbb{N}}$ is a geometric sequence with ratio $r = \sin x_0 \in [0, 1)$, and converges to 0. Thus, for any point $x_0$ in the domain $D$ of the functions $f_n$, the sequence of real numbers $(f_n(x_0))_{n \in \mathbb{N}}$ converges, so the sequence of functions $(f_n)_{n \in \mathbb{N}}$ converges pointwise. The pointwise limit of $(f_n)_{n \in \mathbb{N}}$ is the function $f : [0, \pi] \to \mathbb{R}$ defined by $f(x) = 0$ for $x \in D - \{\pi/2\}$ and $f(\pi/2) = 1$.

The theorem below provides another important property of the product topology. It says that pointwise convergence of a sequence $(f_n)_{n \in \mathbb{N}}$ of functions $f_n : \mathbb{R} \to D$ is just convergence in $\mathbb{R}^D$ with the product topology.

**Theorem 6.2.9.** *If $D \subseteq \mathbb{R}$, a sequence $(f_n)_{n \in \mathbb{N}}$ of functions $f_n : D \to \mathbb{R}$ converges pointwise to a function $f$ if and only if $(f_n)_{n \in \mathbb{N}}$ converges to $f$ in $\mathbb{R}^D$ with the product topology.*

*Proof.* Suppose the sequence $(f_n)_{n \in \mathbb{N}}$ converges pointwise to $f$. To see that the sequence converges to $f$ in the product topology, let $V = \prod_{x \in D} V_x$ be a basic neighborhood of $f = \langle f(x) \rangle_{x \in D} \in \mathbb{R}^D$. Then the set $F = \{x \in D : V_x \neq \mathbb{R}\}$ is finite. For each $z \in F$, pointwise convergence implies $(f_n(z))_{n \in \mathbb{N}}$ converges to $f(z)$, so there exists $N_z \in \mathbb{N}$ such that $f_n(z) \in V_z$ for $n \geq N_z$. If $N = \max\{N_z : z \in F\}$, then $n \geq N$ implies $f_n(x) \in V_x$ for all $x \in D$, so the sequence $(f_n)_{n \in \mathbb{N}}$ is eventually in the neighborhood $V$ of $f$.

Conversely, suppose $f_n \to f$ in $\mathbb{R}^D$ with the product topology. Given an arbitrary $z \in D$, we want to show that $f_n(z) \to f(z)$ in $\mathbb{R}$. Suppose $U_z$ is an arbitrary neighborhood of $f(z)$. Now in $\mathbb{R}^D$, the sequence $(f_n)_{n\in\mathbb{N}}$ is eventually in the neighborhood $V = \prod_{x\in D} V_x$ of $f$, where $V_x = \mathbb{R}$ if $x \neq z$ and $V_z = U_z$. Projecting onto the $z$th coordinate, we see that $(f_n(z))_{n\in\mathbb{N}}$ is eventually in $U_z$. Thus, $f_n(z) \to f(z)$, as needed. □

Pointwise convergence of functions from $\mathbb{R}$ to $\mathbb{R}$ can be compared with uniform convergence of functions, that is convergence of functions in the *uniform topology*, generated by the bounded sup metric and having a basis of balls $B(f, \varepsilon) = \{g : \mathbb{R} \to \mathbb{R} : f(x) - \varepsilon < g(x) < f(x) + \varepsilon\}$. Thus, the basic neighborhoods $B(f, \varepsilon)$ in the uniform topology consist of those functions $g$ which, for every $x \in \mathbb{R}$, pass through the hoop $\{x\} \times (f(x) - \varepsilon, f(x) + \varepsilon)$. In the product topology on $\mathbb{R}^\mathbb{R}$, a basic neighborhood $V$ of $f$ is determined by a finite set $F \subseteq \mathbb{R}$ of indices and a neighborhood $U_{f(x)}$ of $f(x)$ for each $x \in F$. The neighborhood $V$ consists of all the functions $g$ which pass through the hoops $\{x\} \times U_{f(x)}$ for each $x$ in the finite set $F$. Figure 6.8 illustrates this. Indeed, for a basis, we may take every neighborhood $U_{f(x)}$ to be a basic neighborhood $(f(x) - \varepsilon_x, f(x) + \varepsilon_x)$ centered at $f(x)$ for each $x \in F$. Now for $\varepsilon = \min\{\varepsilon_x : x \in F\}$, the basic neighborhood $B(f, \varepsilon)$ in the uniform topology is clearly contained in the basic neighborhood $V$ of $f$ in the product topology. This proves the following result.

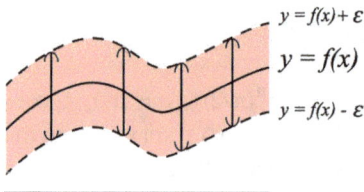

Basic neighborhood $B(f, \varepsilon)$ of $f$ in the uniform topology on $\mathbb{R}^\mathbb{R}$: All functions passing through $\{x\} \times (f(x) - \varepsilon, f(x) + \varepsilon)$ for every $x \in \mathbb{R}$.

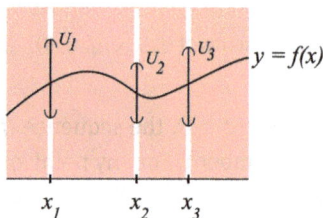

Basic neighborhood $V$ of $f$ in the product topology on $\mathbb{R}^\mathbb{R}$: All functions passing through $\{x_i\} \times U_i$ for $i = 1, \ldots, n$ where $U_i$ is a neighborhood of $f(x_i)$.

**Figure 6.8:** Uniform vs. product neighborhoods of $f$.

**Theorem 6.2.10.** *The uniform topology is finer than the product topology on $\mathbb{R}^D$. That is, the topology on $\mathbb{R}^D$ generated by the bounded sup metric is finer than the product topology.*

It is easy to see that if a sequence converges in one topology, then it converges in any coarser topology. (See Exercise 22 of Section 1.4.) This provides the following corollary.

**Corollary 6.2.11.** *If a sequence of functions $(f_n)_{n\in\mathbb{N}}$ in $\mathbb{R}^D$ converges uniformly to $f$ in $\mathbb{R}^D$, then it converges pointwise to $f$.*

In Example 6.2.8, we saw that the sequence of functions $f_n : [0, \pi] \to \mathbb{R}$ defined by $f_n(x) = \sin^n x$ converged pointwise to the function $f$ defined by $f(x) = 0$ for $x \in D-\{\pi/2\}$ and $f(\pi/2) = 1$. If the sequence of functions $f_n$ converges uniformly, then the "uniform limit" must equal the "pointwise limit" $f$. But, the sequence does not converge uniformly to $f$ since, for example, the uniform ball $B(f, \frac{1}{4})$ contains no functions with $y$-values in $[\frac{1}{4}, \frac{3}{4}]$ and thus contains no $f_n$.

**Example 6.2.12.** Consider the sequence of functions $g_n : \mathbb{R} \to \mathbb{R}$ where $g_n(x)$ is zero for $x$ outside $[n-1, n+1]$ and on $[n-1, n+1]$, $g_n(x)$ is the linear function from $(n-1, 0)$ to $(n, 1)$ followed by the linear function from $(n, 1)$ to $(n+1, 0)$, as shown below.

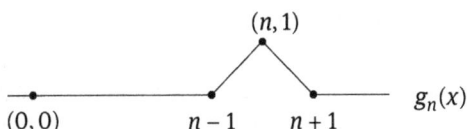

Now $(g_n)$ converges pointwise to $g(x) = 0$, since given any $x \in \mathbb{R}$ and any neighborhood $U$ of $0 = g(x)$, $g_n(x) = 0 \in U$ for all $n > x+1$. However, this sequence does not converge uniformly. If it converges uniformly, it would have to converge to the pointwise limit $g(x) = 0$. But, for $\varepsilon < 1$, the functions $g_n(x)$ never fall in the $\varepsilon$-ball $(-\varepsilon, \varepsilon) \times \mathbb{R}$ around $g(x) = 0$.

The product topology is the topology of pointwise convergence. The bounded sup metric topology is the topology of uniform convergence (and thus, is called the uniform topology). Uniform convergence implies pointwise convergence, but not conversely.

We close this section with two very important theorems.

**Theorem 6.2.13.** *If a sequence of continuous functions $f_n \in \mathbb{R}^D$ converges uniformly to $f$, then $f$ is continuous.*

*Proof.* Suppose $f_n$ is a sequence of continuous functions in $\mathbb{R}^D$ converging uniformly to $f$. To show $f$ is continuous at each $x_0$ in its domain $D$, suppose a basic neighborhood $B(f(x_0), \varepsilon)$ of $f(x_0) \in \mathbb{R}$ is given. We need to find a neighborhood $U$ of $x_0 \in D$ such that $f(U) \subseteq B(f(x_0), \varepsilon)$. That is, we need a neighborhood $U$ of $x_0$ such that $|f(x) - f(x_0)| < \varepsilon$ for all $x \in U$. Since $f_n$ converges uniformly to $f$, there exists $k \in \mathbb{N}$ such that $f_k \in B(f, \frac{\varepsilon}{4})$ in the uniform metric on $\mathbb{R}^D$, and thus $|f_k(x) - f(x)| < \frac{\varepsilon}{4}$ for all $x \in \mathbb{R}$. Since $f_k$ is continuous, for the neighborhood $B(f_k(x_0), \frac{\varepsilon}{2})$, there exists a neighborhood $U$ of $x_0$ with $f_k(U) \subseteq B(f_k(x_0), \frac{\varepsilon}{2})$, so $|f_k(x) - f_k(x_0)| < \frac{\varepsilon}{2}$ for all $x \in U$. Now for $x \in U$, the triangle inequality gives

$$|f(x) - f(x_0)| \leq |f(x) - f_k(x)| + |f_k(x) - f_k(x_0)| + |f_k(x_0) - f(x_0)|$$
$$< \frac{\varepsilon}{4} + \frac{\varepsilon}{2} + \frac{\varepsilon}{4}$$
$$= \varepsilon.$$

Thus, $f(U) \subseteq B(f(x_0), \varepsilon)$, as needed. $\qquad\square$

The next theorem was published in 1885 by the prolific German mathematician Karl Weierstrass (1815–1897).

**Theorem 6.2.14** (Weierstrass approximation theorem). *If $[a, b]$ is a compact interval in $\mathbb{R}$, then the collection $P[a, b]$ of polynomials over $[a, b]$ is dense in the set $C[a, b]$ of continuous functions over $[a, b]$ with the uniform topology.*

The Weierstrass approximation theorem says that any continuous function $f : [a, b] \rightarrow \mathbb{R}$ is in the closure of $P[a, b]$, so every neighborhood $B(f, \varepsilon)$ contains a point of $P[a, b]$. That is, given any continuous function $f : [a, b] \rightarrow \mathbb{R}$ and any $\varepsilon > 0$, there exists a polynomial $p$ with $f(x) - \varepsilon < p(x) < f(x) + \varepsilon$ for all $x \in [a, b]$. Thus, any continuous function over a compact interval is uniformly approximated to any desired degree of accuracy by a polynomial. For a continuous function $f : [a, b] \rightarrow \mathbb{R}$ which has derivatives of all orders, if the Taylor series for $f$ converges to $f$ on $[a, b]$, then it converges uniformly and a Taylor polynomial may be used as the uniform approximation. However, the Weierstrass approximation theorem applies to any continuous function, including those which are not differentiable. Indeed, in 1872, Weierstrass exhibited a class of functions, including

$$W(x) = \sum_{n=0}^{\infty} \frac{\cos(5^n \pi x)}{2^n},$$

which are continuous but nowhere differentiable. Weierstrass's example was perhaps the first such example to be published, but the Czech mathematician Bernard Bolzano knew of such an example around 1830. The proof that this function $W$ is continuous is based on the fact that $W$ is the uniform limit of a sequence of continuous functions, namely, the sequence of partial sums of the infinite series. Indeed, most examples of continuous nowhere differentiable functions are defined as a limit of a sequences of functions.

The proof of the Weierstrass approximation theorem is lengthy and may be found in [3].

## Exercises

1. Is $X \times Y - (A \times B) = (X - A) \times (Y - B)$? Provide a proof or counterexample.
2. (a) If $Y = \prod_{i \in I} X_i$ has the product topology and $F_i$ is closed in $X_i$ for every $i \in I$, show that $F = \prod_{i \in I} F_i$ is closed.
   (b) More generally, show that if $Y = \prod_{i \in I} X_i$ has the product topology and $A_i \subseteq X_i$ for every $i \in I$, then cl $\prod_{i \in I} A_i = \prod_{i \in I}$ cl $A_i$.
3. If $\prod_{i \in I} X_i$ has the product topology, prove that a function $f : Z \rightarrow \prod_{i \in I} X_i$ is continuous if and only if each of its coordinate functions $f_i = \pi_i \circ f$ is continuous.
4. If $J \subseteq I$, then we may consider the projection function $\pi_J : \prod_{i \in I} X_i \rightarrow \prod_{i \in J} X_i$ defined by $\pi_J(\langle x_i \rangle_{i \in I}) = \langle x_i \rangle_{i \in J}$. If both products have the product topology, show that every projection function $\pi_J$ is continuous.

5. For $J \subseteq I$, consider the projection function $\pi_J$ as in Exercise 4. Must $\pi_J$ be an open mapping? Give a proof or counterexample.

6. Suppose $I$ is infinite and $\langle a_i \rangle_{i \in I}$ is an isolated point in $\prod_{i \in I} X_i$ with the product topology. What can be said about the topological spaces $X_i$?

7. If $I$ is an infinite index set, prove that $\prod_{i \in I} X_i$ with the product topology is Hausdorff if and only if $X_i$ is Hausdorff for every $i \in I$.

8. If $I$ is an infinite index set, prove that if $\prod_{i \in I} X_i$ with the product topology is connected, then $X_i$ is connected for every $i \in I$.

9. If $I$ is an infinite index set, prove that $\prod_{i \in I} X_i$ with the product topology is path connected if and only if $X_i$ is path connected for every $i \in I$.

10. Give $\{0, 1\}$ the discrete topology and let $\mathcal{T}$ be the product topology on $S = \prod_{n \in \mathbb{N}} \{0, 1\} = \{0, 1\}^{\mathbb{N}}$. The prefix metric topology $\mathcal{T}_{\text{pre}}$ (see Exercise 16 of Section 6.1) gives another topology on the set $S$.
    (a) Is one of these topologies finer than the other, are they equal, or are they non-comparable?
    (b) Is $S$ compact in none, one, or both of these topologies? Justify your answer.

11. If $\{0, 1\}$ has the discrete topology, and $Y = \{0, 1\}^{\mathbb{N}}$ has the box topology, show that $Y$ is not compact. Does this violate the Tychonoff theorem?

12. A product of connected spaces need not be connected if the product has the box topology. Let $\mathbb{R}$ have the Euclidean topology and consider the product $\mathbb{R}^{\mathbb{N}}$ with the box topology. Each factor $\mathbb{R}$ is connected, but the (box) product is not. Prove this by showing that the set $B$ of bounded real valued sequences is a nonempty, proper, closed and open subset of $\mathbb{R}^{\mathbb{N}}$ with the box topology.

13. Suppose $D_i$ is dense in $X_i$ for $i \in I$. Is $\prod_{i \in I} D_i$ dense in $\prod_{i \in I} X_i$ with the product topology? Is $\prod_{i \in I} D_i$ dense in $\prod_{i \in I} X_i$ with the box topology?

14. Suppose $D$ is dense in $\prod_{i \in I} X_i$ with the product topology. Must $\pi_i(D)$ be dense in $X_i$ for all $i \in I$? What if $\prod_{i \in I} X_i$ has the box topology?

15. If $X$ is a topological space and $I$ an index set, consider the product topology $\mathcal{T}$, the uniform topology $\mathcal{T}_p$, and the box topology $\mathcal{T}_b$ on $X^I$. Which of these topologies are finer than which, and which are not finer than which? Justify your answers.

16. For $n \in \mathbb{N}$, define $f_n : \mathbb{R} \to \mathbb{R}$ by $f_n(x) = \frac{1}{n}$. Determine whether the sequence $(f_n)_{n \in \mathbb{N}}$ converges:
    (a) in $\mathbb{R}^{\mathbb{R}}$ with the product topology.
    (b) In $\mathbb{R}^{\mathbb{R}}$ with the box topology.
    (c) In $\mathbb{R}^{\mathbb{R}}$ with the uniform topology.

17. Let $C[a, b]$ be the set of continuous functions $f : [a, b] \to \mathbb{R}$. Show that $C[a, b]$ is not compact in the uniform topology by finding an open cover with no finite subcover.

18. Let $C$ be the set of continuous bounded functions $f : [0, 1] \to [0, 1]$. That is, $C = \{f \in [0, 1]^{[0,1]} : f \text{ is continuous}\}$.
    (a) Show that, as a subspace of $[0, 1]^{[0,1]} = \prod_{x \in [0,1]} [0, 1]$ with the product topology, $C$ is not compact.

(b) Use part (a) to show that, as a subspace of $[0,1]^{[0,1]} = \prod_{x\in[0,1]}[0,1]$ with the uniform topology, $C$ is not compact.

(c) Show directly that $C$ is not compact in the uniform topology by finding an open cover with no finite subcover.

19. Consider the sequence $(f_n)_{n\in\mathbb{N}}$ of functions $f_n : [0,2\pi] \to \mathbb{R}$ defined by $f_n(x) = \cos^n x$. Find the largest subset $D$ of $[0,2\pi]$ on which the sequence converges pointwise and determine whether the sequence converges uniformly on $D$. Give the pointwise and uniform limits, if they exist.

20. Consider the sequence $(g_n)_{n\in\mathbb{N}}$ of functions $g_n : (0,1) \to \mathbb{R}$ defined by $g_n(x) = x^n$. Determine whether the sequence converges pointwise and whether the sequence converges uniformly. Give the pointwise and uniform limits, if they exist.

21. In the areawise metric (or $L^1$ metric) $d(f,g) = \int_{-\infty}^{\infty} |f(x) - g(x)|\, dx$ on the set of all continuous functions $h : \mathbb{R} \to \mathbb{R}$ with $\int_{-\infty}^{\infty} h(x)\, dx < \infty$, consider the functions $g_n$ and $g$ of Example 6.2.12. Does the sequence $(g_n)$ converge to $g$ areawise?

22. Find a sequence of functions in $C[-1,1]$ which converges pointwise to $f(x) = 0$, but does not converge in the areawise metric $d(f,g) = \int_{-1}^{1} |f(x) - g(x)|\, dx$.

23. Prove or give a counterexample: If $f_n$ converges uniformly to $f$ in $C[a,b]$, then $f_n$ converges to $f$ in the areawise metric $d(f,g) = \int_a^b |f(x) - g(x)|\, dx$ on $C[a,b]$.

24. Prove or give a counterexample: If $f_n$ converges to $f$ in $C[a,b]$ with the areawise metric $d(f,g) = \int_a^b |f(x) - g(x)|\, dx$ on $C[a,b]$, then $f_n$ converges uniformly to $f$.

25. Define $m : \mathbb{R}^{\mathbb{N}} \times \mathbb{R}^{\mathbb{N}} \to \mathbb{R}$ by

$$m((x_n)_{n=1}^{\infty}, (y_n)_{n=1}^{\infty}) = \sup_{k\in\mathbb{N}} \frac{\min\{1, |x_k - y_k|\}}{k}.$$

(a) Show that $m$ is a metric.

(b) Show that the metric topology $\mathcal{T}_m$ is the product topology on $\mathbb{R}^{\mathbb{N}}$

26. Exercise 25 showed that $\mathbb{R}^{\mathbb{N}}$ with the product topology is metrizable. Complete the steps below to show that $\mathbb{R}^{\mathbb{N}}$ with the box topology is not metrizable.

(a) Show that $\mathbf{0} = \prod_{n\in\mathbb{N}} 0$ is in the closure of $A = (\mathbb{R} - \{0\})^{\mathbb{N}}$.

(b) Show that there is no sequence in $A$ converging to $\mathbf{0}$, and invoke Theorem 1.5.11.

27. The Lagrange remainder theorem states that if a function $f(x)$ has derivatives of all orders and is approximated by its $n$th-degree Taylor polynomial,

$$T_n(x) = f(c) + f'(c)(x - c) + \frac{f''(c)(x - c)^2}{2!} + \cdots + \frac{f^{(n)}(c)(x - c)^n}{n!},$$

then the error $f(x) - T_n(x)$ is equal to

$$R_n(x) = \frac{f^{(n+1)}(z)(x - c)^{n+1}}{(n + 1)!} \quad \text{for some } z \text{ between } x \text{ and } c.$$

This error function $R_n(x)$ is called the *Lagrange remainder*.

(a) Use the Lagrange remainder theorem to show that the series $\sum_{n=0}^{\infty} \frac{x^n}{n!}$ (that is, the sequence of partial sums of the series) converges uniformly to $e^x$ for all $x \in [-1, 1]$.

(b) Show that if the Lagrange remainders $R_n(x)$ for $f$ converge uniformly to $z(x) = 0$ on an interval $I$, then the sequence $(T_n(x))_{n=1}^{\infty}$ of partial sums of the Taylor series for $f$ converges uniformly to $f$ on the interval $I$.

## 6.3 The cocountable topology

Since this section is neither about metric spaces nor real analysis (that is, analysis on $\mathbb{R}^n$ with the Euclidean metric), its placement in this chapter may need an explanation. To appreciate the good things that happen in metric spaces, this section provides an example which serves as a reminder that many of those good things need not happen in arbitrary topological spaces.

Recall that in a metric space $(X, d)$, $x \in \operatorname{cl} A$ if and only if there exists a sequence in $A$ converging to $x$ (Theorem 1.5.11), and a function $f$ between two metric spaces is continuous if and only if it preserves limits of sequences, that is, if and only if $(x_n)_{n=1}^{\infty}$ converges to $x$ in the domain of $f$ implies $(f(x_n))_{n=1}^{\infty}$ converges to $f(x)$ in the codomain (Theorem 3.1.7).

These facts do not hold in arbitrary topological spaces. That is, in arbitrary topological spaces, "sequences do not suffice". These sequence-based facts work in a metric space $(X, d)$ since every point $x \in X$ has a countable neighborhood base $\{B(x, \frac{1}{n}) : n \in \mathbb{N}\}$, that is, since $X$ is *first countable*.

In Example 1.4.16, we have seen that $\mathbb{R}$ with the cofinite topology is not first countable. We introduce a similar topology with is also not first countable.

**Definition 6.3.1.** The *cocountable topology* (or *countable complement topology*) on a set $X$ is $\mathcal{T}_{cc} = \{U \subseteq X : X - U \text{ is countable}\} \cup \{\emptyset\}$.

If $X$ is countable, then every set has a countable complement and the cocountable topology is the discrete topology. Thus, in our further consideration of the cocountable topology, we will generally assume the underlying set is uncountable. The verification that $\mathcal{T}_{cc}$ is a topology and the proof of the following result are straightforward and are left to the exercises.

**Theorem 6.3.2.** *If $X$ is uncountable, $(X, \mathcal{T}_{cc})$ is not Hausdorff, and thus not metrizable.*

Recall that if $(X, \mathcal{T})$ is Hausdorff, then every convergent sequence in $X$ has a unique limit. We next show that the Hausdorff condition is not necessary for unique limits.

Suppose $X$ is uncountable and carries the cocountable topology $\mathcal{T}_{cc}$, and $(x_n)_{n \in \mathbb{N}} \to a$. Then $(x_n)_{n \in \mathbb{N}}$ is eventually in every neighborhood of $a$. Now $X - \{x_n : n \in \mathbb{N}\}$ is an open neighborhood of any point $a$ which is not a term $x_j$ of the sequence, and the sequence is never in this neighborhood. Thus, if $(x_n)_{n \in \mathbb{N}}$ converges, it must converge to a term of the sequence. Furthermore, if $(x_n)_{n\in\mathbb{N}}$ converges to $x_j$, then the sequence is eventually in the neighborhood $X - \{x_n : n \in \mathbb{N}, x_n \neq x_j\}$ of $x_j$, and thus the sequence is eventually constantly $x_j$. In particular, every convergent sequence in the cocountable topology on an uncountable set has a unique limit, even though the topology is not Hausdorff.

The theorem below shows that in a first countable space, the Hausdorff condition is not only sufficient for uniqueness of limits of sequences, but also necessary.

**Theorem 6.3.3.** *If $X$ is a first countable space and every convergent sequence has a unique limit, then $X$ is Hausdorff.*

*Proof.* Suppose $X$ is first countable. We will show the contrapositive: if $X$ is not Hausdorff, then there exists a convergent sequence with distinct limits $a \neq b$. Suppose $X$ is not Hausdorff. Then there exist distinct points $a \neq b$ in $X$ such that every neighborhood of $a$ intersects every neighborhood of $b$. Suppose $\{N_1, N_2, \ldots\}$ is a countable neighborhood base at $a$ and $\{M_1, M_2, \ldots\}$ is a countable neighborhood base at $b$. Pick $x_1 \in N_1 \cap M_1$, $x_2 \in N_1 \cap M_1 \cap N_2 \cap M_2$, and in general, $x_n \in \bigcap_{i=1}^{n}(N_i \cap M_i)$. Now for any neighborhood $U$ of $a$, there exists $N_j \subseteq U$, and for $k \geq j$, we have $x_k \in N_j \subseteq U$, so $(x_n)_{n\in\mathbb{N}}$ is eventually in $U$. Since $U$ was arbitrary, $(x_n)_{n\in\mathbb{N}}$ converges to $a$. Similarly, $(x_n)_{n\in\mathbb{N}}$ converges to $b \neq a$. □

Having seen that an uncountable set with the cocountable topology has unique limits, Theorems 6.3.2 and 6.3.3 give the following result.

**Corollary 6.3.4.** *The cocountable topology on an uncountable set is not first countable.*

Now we return to showing that sequences do not suffice in spaces which are not first countable.

In a metric space, $x \in \mathrm{cl}(B)$ if and only if there exists a sequence $(b_n)_{n\in\mathbb{N}}$ of points in $B$ which converges to $x$. We show that this fails in $(X, \mathcal{T}_{cc})$ if $X$ is uncountable.

Suppose $X$ is uncountable and $A$ is a countable subset of $(X, \mathcal{T}_{cc})$. Then $A$ is closed. Any nonempty open set must be uncountable, so the largest open set contained in the countable set $A$ is $\emptyset$. Thus, $\partial A = \mathrm{cl}\,A - \mathrm{int}\,A = A - \emptyset = A$. Since $A = \partial A = \mathrm{cl}\,A \cap \mathrm{cl}(X - A)$, every element $a \in A$ is an element of $\mathrm{cl}(X - A)$. However, there is no sequence in $X - A$ converging to $a \in A$, since every convergent sequence of terms in $X - A$ is eventually constant and has a unique limit, and thus converges to a point in $X - A$ and cannot converge to $a \in A$.

Finally, we show that, for arbitrary topological spaces $X$ and $Y$, the continuity of $f : X \to Y$ is not equivalent to $f$ preserving limits of sequences. If $(x_n)_{n\in\mathbb{N}}$ converges to $a$ in $(\mathbb{R}, \mathcal{T}_{cc})$, then $(x_n)_{n\in\mathbb{N}}$ is eventually constantly $a$, so for any function $f$ from $(\mathbb{R}, \mathcal{T}_{cc})$ to any topological space $(Y, \mathcal{T})$, $(f(x_n))_{n\in\mathbb{N}}$ is eventually constantly $f(a)$ and thus converges to $f(a)$. Thus, every function $f$ with domain $(\mathbb{R}, \mathcal{T}_{cc})$ preserves limits of sequences. However, not every such function is continuous. For example, if $\mathcal{T}_D$ is the discrete topology, $f : (\mathbb{R}, \mathcal{T}_{cc}) \to (\mathbb{R}, \mathcal{T}_D)$ defined by $f(x) = x$ is not continuous, since $\{0\} \in \mathcal{T}_D$ but $f^{-1}(\{0\}) \notin \mathcal{T}_{cc}$. Specifically, this shows that if a function $f$ between two arbitrary topological spaces preserves limits of sequences, then $f$ need not be continuous.

The following theorem gives a more precise link between continuity and preservation of limits of sequences.

**Theorem 6.3.5.**
(a) *Continuous functions always preserve limits of sequences.*
(b) *If $X$ and $Y$ are topological spaces, $X$ is first countable, and $f : X \to Y$ preserves limits of sequences, then $f$ is continuous.*

The proof is almost identical to the proof of Theorem 3.1.7.

## Exercises

1. Prove that the cocountable topology $\mathcal{T}_{cc} = \{U \subseteq X : X - U \text{ is countable}\} \cup \{\emptyset\}$ is indeed a topology on $X$.

2. Given a set $X$, are either of the collections $\mathcal{T}_1 = \{U \subseteq X : X - U \text{ is infinite}\} \cup \{\emptyset\}$ or $\mathcal{T}_2 = \{U \subseteq X : X - U \text{ is finite or infinite}\}$ a topology on $X$?

3. Among the functions from $\mathbb{R}$ to $\mathbb{R}$ given below, which are continuous if the domain and codomain have the cocountable topology? Which are continuous if the domain and codomain have the cofinite topology?
   (a) $f(x) = \arctan x$
   (b) $g(x) = \begin{cases} \lfloor x \rfloor & \text{if } x \in \mathbb{Q} \\ x & \text{if } x \notin \mathbb{Q} \end{cases}$
   (c) $h(x) = \begin{cases} x & \text{if } x \in \mathbb{Q} \\ \lfloor x \rfloor & \text{if } x \notin \mathbb{Q} \end{cases}$
   (d) $s(x) = \begin{cases} \sin(1/x) & \text{if } x \neq 0 \\ 0 & \text{if } x = 0 \end{cases}$

4. Suppose $X$ is uncountable.
   (a) Show that if $U, V$ are nonempty sets in $\mathcal{T}_{cc}$, then $U \cap V \neq \emptyset$. (That is, $(X, \mathcal{T}_{cc})$ is *hyperconnected*.)
   (b) Prove Theorem 6.3.2: If $X$ is uncountable, $(X, \mathcal{T}_{cc})$ is not Hausdorff.
   (c) If $(Y, \mathcal{T})$ is a Hausdorff space and $f : (X, \mathcal{T}_{cc}) \to (Y, \mathcal{T})$ is continuous, show that $f$ is constant. (In particular, every continuous real-valued function $f : (X, \mathcal{T}_{cc}) \to (\mathbb{R}, \mathcal{T}_{\mathcal{E}})$ is bounded, which is the definition of $(X, \mathcal{T}_{cc})$ being *pseudocompact*.)

5. If $X$ has the cocountable topology $\mathcal{T}_{cc}$, show that $A \subseteq X$ is compact if and only if $A$ if finite. (In particular, every compact subset of $(X, \mathcal{T}_{cc})$ is closed, even though $(X, \mathcal{T}_{cc})$ is not Hausdorff.)

6. Provide a proof of Corollary 6.3.4 without appealing to Theorems 6.3.2 and 6.3.3 by showing that if $\{B_i\}_{i \in \mathbb{N}}$ is a countable $\mathcal{T}_{cc}$-neighborhood base at $x$ in an uncountable set $X$, then $X - \{x\} = X - \bigcap_{i \in \mathbb{N}} B_i = \bigcup_{i \in \mathbb{N}} (X - B_i)$. Why is this a contradiction?

7. Describe the modifications of the proof of Theorem 3.1.7 required to give a proof of Theorem 6.3.5.

# 7 Separation axioms and compactifications

## 7.1 Separation axioms

We have seen the Hausdorff separation axiom: $X$ is Hausdorff if and only if every pair of distinct points in $X$ can be separated by disjoint neighborhoods. In 1914, before the definitions of topology had become standardized, Felix Hausdorff required what is now known as the Hausdorff separation axiom in his definition of a topology. In their 1935 book *Topologie*, written in German, Pavel Alexandroff and Heinz Hopf listed the *Trennungsaxioms* (that is, the separation axioms) which became known as $T_0, T_1, T_2, T_3$, and $T_4$. They associated the names Kolmogorov, Fréchet, Hausdorff, Vietoris, and Tietze, respectively, with these axioms.

**Definition 7.1.1.** Suppose $X$ is a topological space.
(a) $X$ is $T_0$ if for every pair of distinct points $x, y$ in $X$, there exists a neighborhood $U$ of $x$ which does not contain $y$ OR there exists a neighborhood $V$ of $y$ which does not contain $x$.
(b) $X$ is $T_1$ if for every pair of distinct points $x, y$ in $X$, there exists a neighborhood $U$ of $x$ which does not contain $y$ AND there exists a neighborhood $V$ of $y$ which does not contain $x$.
(c) $X$ is $T_2$ or *Hausdorff* if for every pair of distinct points $x, y$ in $X$, there exist disjoint open sets $U$ and $V$ with $x \in U$ and $y \in V$.
(d) $X$ is $T_3$ if for every closed set $A$ and every $x \notin A$, there exist disjoint open sets $U$ and $V$ with $x \in U$ and $A \subseteq V$.
(e) $X$ is *regular* if $X$ is $T_3$ and $T_1$.
(f) $X$ is $T_4$ if for every pair of disjoint closed sets $A, B \subseteq X$, there exist disjoint open sets $U$ and $V$ with $A \subseteq U$ and $B \subseteq V$.
(g) $X$ is *normal* if $X$ is $T_4$ and $T_1$.

Figure 7.1 suggests these separation axioms.

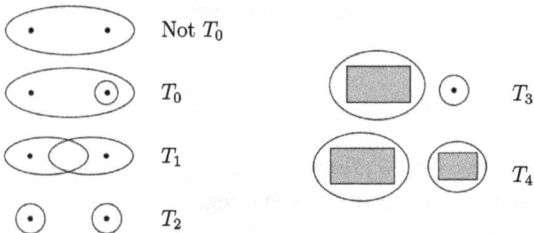

**Figure 7.1:** The separation axioms.

**Caution:** Topologists universally agree that the definitions of regular and $T_3$ and the definitions of normal and $T_4$ are identical except that one includes the hypothesis of

https://doi.org/10.1515/9783110686579-008

$T_1$ and the other does not. Unfortunately, there is not universal agreement on which is which. We define regular and normal to include $T_1$. Some authors define $T_3$ and $T_4$ to include $T_1$. Thus, whenever you encounter these separation axioms, carefully check the author's definitions to see whether the $T_1$ condition is included or not.

Also, these separation axioms are unrelated to the use of the word separation for a pair of sets $U$ and $V$ which shows that a topological space is not connected.

In Definition 7.1.1(b) for the $T_1$ property, the "AND" condition is unnecessary: If $x$ and $y$ are distinct points, so are $y$ and $x$, and applying the condition before the "AND" to the pair $y, x$ gives the second condition. The redundant second condition was included for emphasis to contrast it with the definition of $T_0$.

The separation axioms involving only points (namely, $T_0$, $T_1$, and $T_2$) are called the *lower separation axioms* and those involving closed sets are called *higher separation axioms*. For the $T_3$ condition, it is sufficient to check that, for every *nonempty* closed set $A$ and every point $x \notin A$, there exist disjoint open sets $U$ and $V$ separating them, for if $A = \emptyset$, the sets $U = X$ and $V = \emptyset$ are open sets separating $x$ and $A$. Similarly, in checking the $T_4$ condition it suffices to consider disjoint closed *nonempty* sets $A$ and $B$.

**Example 7.1.2.** The real line $\mathbb{R}$ with the right ray topology is $T_0$ but not $T_1$. Given $x < y$ in $\mathbb{R}$, there is a neighborhood of $y$ excluding $x$, but there is no neighborhood of $x$ excluding $y$.

The real line $\mathbb{R}$ with the cofinite topology is $T_1$ but not $T_2$. Given $x \neq y$ in $\mathbb{R}$, $\mathbb{R} - \{y\}$ is a neighborhood of $x$ excluding $y$ and $\mathbb{R} - \{x\}$ is a neighborhood of $y$ excluding $x$. The cofinite topology on $\mathbb{R}$ has no disjoint nonempty open sets, so it is not $T_2$.

If $X$ is $T_1$ and $y \neq x$ in $X$, then there exists a neighborhood of $y$ which does not intersect $x$, so $y \notin \text{cl}\{x\}$. Thus, if $X$ is $T_1$, then singleton sets $\{x\}$ are closed. Conversely, if every singleton set is closed, then, for $x \neq y$, $y \notin \text{cl}\{x\}$ and $x \notin \text{cl}\{y\}$, so there exists a neighborhood of $y$ not containing $x$ and there exists a neighborhood of $x$ not containing $y$, and thus $X$ is $T_1$. This proves the following result.

**Theorem 7.1.3.** *A topological space $X$ is $T_1$ if and only if every singleton set $\{x\}$ is closed.*

As an immediate consequence, we note the following implications between the separation axioms:

$$\text{Normal} \Rightarrow \text{Regular} \Rightarrow T_2 \Rightarrow T_1 \Rightarrow T_0.$$

There are fairly simple spaces to show that $T_2$ does not imply regular (see Exercise 13), and there are fairly complicated examples to show that regular does not imply normal. Every separation axiom is a topological property. We introduce some terminology.

**Definition 7.1.4.** A topological property is *hereditary* if whenever $X$ has the property, then every subspace of $X$ has the property. A topological property is *productive* if arbitrary products of spaces with the property have the property.

For example, the Tychonoff theorem says that compactness is a productive property. Theorem 6.2.4 shows that Hausdorff, connected, and path connected are productive topological properties.

**Theorem 7.1.5.** *The properties $T_0, T_1, T_2, T_3$, and regularity are hereditary and productive.*

*Proof.* Suppose $X$ is $T_i$ for $i = 0, 1$, or 2, $Y$ is a subspace of $X$, and $x \neq y$ in $Y$. Then $x \neq y$ in $X$. If $U$ and $V$ are neighborhoods of $x$ and $y$ in $X$ guaranteed by the definition of $X$ being $T_i$, (only $U$ is needed for $T_0$) then $U \cap Y$ and $V \cap Y$ are neighborhoods of $x$ and $y$ in $Y$ showing that $Y$ is $T_i$. Suppose $X$ is $T_3$, $Y$ is a subspace of $X$, $A$ is closed in $Y$, and $x \in Y - A$. Now $A = A' \cap Y$ for some closed set $A'$ in $X$. If $x$ were in $A'$, it would be in $A' \cap Y = A$, so $x \notin A'$. Now if $U$ and $V$ are the disjoint open sets in $X$ separating $x$ and $A'$, then $U \cap Y$ and $V \cap Y$ are disjoint open sets in $Y$ separating $x$ and $A$. Regularity is hereditary since $T_1$ and $T_3$ both are.

The proof that these properties are productive is left to the exercises. □

Normality is neither productive nor hereditary, but we have the following result regarding subspaces.

**Theorem 7.1.6.** *If $X$ is $T_4$ and $Y$ is a closed subset of $X$, then $Y$ is $T_4$. Thus, $T_4$ and normality are hereditary to closed subspaces.*

The proof is left as an exercise.
The next theorems tell us some spaces which are normal.

**Theorem 7.1.7.** *Every metric space $(X, d)$ is normal.*

*Proof.* Suppose $(X, d)$ is a metric space and $A, B$ are disjoint closed sets in $X$. For each $a \in A$, since $a \notin \operatorname{cl} B = B$, there exists a basic neighborhood $B(a, \varepsilon_a)$ of $a$ which does not intersect $B$. Similarly, for each $b \in B$, there exists a neighborhood $B(b, \varepsilon_b)$ of $b$ which does not intersect $A$. Let $U = \bigcup \{B(a, \frac{\varepsilon_a}{2}) : a \in A\}$ and $V = \bigcup \{B(b, \frac{\varepsilon_b}{2}) : b \in B\}$. Now $U$ and $V$ are open sets containing $A$ and $B$, respectively. It only remains to show that $U$ and $V$ are disjoint. Suppose $x \in U \cap V$. Then $x \in B(a, \frac{\varepsilon_a}{2}) \cap B(b, \frac{\varepsilon_b}{2})$ for some $a \in A$ and some $b \in B$. Either $\varepsilon_a \leq \varepsilon_b$ or $\varepsilon_b \leq \varepsilon_a$. Without loss of generality, assume $\varepsilon_a \leq \varepsilon_b$. Now $d(a, b) \leq d(a, x) + d(x, b) < \frac{\varepsilon_a}{2} + \frac{\varepsilon_b}{2} \leq \varepsilon_b$, so $a \in B(b, \varepsilon_b)$, contrary to $B(b, \varepsilon_b) \cap A = \emptyset$. Thus, $U \cap V$ must be empty. Since $A$ and $B$ were arbitrary disjoint closed subsets of $X$, it follows that $X$ is $T_4$. Since metric spaces are Hausdorff, they are also $T_1$, and thus are normal. □

**Theorem 7.1.8.** *Every compact Hausdorff space is normal.*

*Proof.* Suppose $X$ is compact and Hausdorff. Since Hausdorff implies $T_1$, to show $X$ is normal we need only check that $X$ is $T_4$. We will first show that $X$ is $T_3$. Recall that a closed subset of a compact space is compact. If $A \subseteq X$ is closed and $b \notin A$, then, for each $a \in A$, by the Hausdorff property, there exist disjoint open neighborhoods $U_a$ of

$a$ and $V_a$ of $b$. Now $\{U_a : a \in A\}$ is an open cover of the compact set $A$. If $\{U_a : a \in F\}$ is a finite subcover, then $U = \bigcup\{U_a : a \in F\}$ is an open set containing $A$ and $V = \bigcap\{V_a : a \in F\}$ is an open set containing $b$. Furthermore, $U \cap V = \emptyset$, for $x \in U \cap V$ would give the contradiction that $x \in U_{a_0} \cap V_{a_0}$ for some $a_0 \in F$. Thus, $X$ is $T_3$. Now suppose $A$ and $B$ are disjoint closed sets in $X$. For every $b \in B$, we apply the $T_3$ property to find disjoint open sets $U_b$ and $V_b$ with $A \subseteq U_b$ and $b \in V_b$. Now $\{V_b : b \in B\}$ is an open cover of the compact set $B$, so there exists a finite subcover $\{V_b : b \in F\}$. With $U = \bigcap\{U_b : b \in F\}$ and $V = \bigcup\{V_b : b \in F\}$, we find that $U$ and $V$ are disjoint open sets containing $A$ and $B$, respectively. Thus, $X$ is $T_4$, as needed. □

The previous results support the argument that metric spaces and compact Hausdorff spaces are some of the nicest, most well-behaved topological spaces you will ever meet.

We now give characterizations of $T_3$ and $T_4$ spaces.

**Theorem 7.1.9.**
(a) *A topological space $X$ is $T_3$ if and only if for every $x \in X$ and every neighborhood $U$ of $x$, there exists a neighborhood $V$ of $x$ with cl $V \subseteq U$.*
(b) *A topological space $X$ is $T_4$ if and only if for every closed set $A$ in $X$ and every open set $U$ containing $A$, there exists an open set $V$ with $A \subseteq V \subseteq$ cl $V \subseteq U$.*

*Proof.* We will prove the first part. The second part is similar. Suppose $X$ is a $T_3$ space, $x \in X$, and $U$ is a neighborhood of $x$. Let $U'$ be an open neighborhood of $x$ contained in $U$. To find the desired neighborhood $V$ of $x$, note that $X - U'$ is a closed set not containing $x$, so there exist disjoint open sets $V$ and $W$ with $x \in V$ and $X - U' \subseteq W$. Since $V$ is disjoint from $W$, we have $V$ is contained in the closed set $X - W$, and thus cl $V \subseteq X - W$. Now $X - U' \subseteq W$ implies $X - W \subseteq U' \subseteq U$, and thus $V$ is a neighborhood of $x$ whose closure is contained in $U$.

Conversely, suppose every neighborhood $U$ of an arbitrary point $x \in X$ contains a neighborhood $V$ of $x$ with cl $V \subseteq U$. To see that $X$ is $T_3$, suppose $A$ is a closed set in $X$ and $x \notin A$. The neighborhood $U = X - A$ of $x$ contains a neighborhood $V$ of $x$ with $V \subseteq$ cl $V \subseteq U$. Taking complements, we get $A = X - U \subseteq X -$ cl $V \subseteq X - V$. Thus, $X -$ cl $V$ is an open set containing $A$ which is disjoint from the neighborhood $V$ of $x$. This shows that $X$ is $T_3$. □

A collection $\mathcal{N}(x)$ of neighborhoods of $x$ is a *neighborhood base at $x$* if every neighborhood of $x$ contains a neighborhood of $x$ from $\mathcal{N}(x)$. By definition, the open neighborhoods of $x$ form a neighborhood base at $x$ for any point $x$. The condition in Theorem 7.1.9 characterizing the $T_3$ property says that, for every $x \in X$, every neighborhood $U$ of $x$ contains a closed neighborhood cl $V$. That is, Theorem 7.1.9(a) states that a topological space $X$ is $T_3$ if and only if every point $x \in X$ has a neighborhood base of closed sets.

It is of general interest when every point of a topological space has a neighborhood base of special kinds of sets.

**Definition 7.1.10.** A topological space $X$ is *locally compact (respectively, locally connected, locally path connected)* if every $x \in X$ has a neighborhood base of compact (respectively, connected, path connected) neighborhoods.

For example, even though the Euclidean line $\mathbb{R}$ is not compact, it is locally compact since any neighborhood of an arbitrary $x \in \mathbb{R}$ contains an open neighborhood $(x - \varepsilon, x + \varepsilon)$ which contains a compact neighborhood $[x - \varepsilon/2, x + \varepsilon/2]$. The same argument shows that the Euclidean line is locally connected and locally path connected. The set $\mathbb{Q}$ of rationals with the Euclidean topology is neither compact nor locally compact: any neighborhood $U$ of $0$ must contain an interval $(-\varepsilon, \varepsilon) \cap \mathbb{Q}$, and if $\alpha$ is an irrational in $(-\varepsilon, \varepsilon)$, the open cover $\{(-\infty, \alpha) \cap \mathbb{Q}\} \cup \{(\beta, \infty) \cap \mathbb{Q} : \beta > \alpha\}$ is an open cover of $U$ which has no finite subcover, so $U$ is not compact. Thus, in $\mathbb{Q}$, no neighborhood of $0$ is compact.

## Exercises

1. For the topological spaces below, determine which of the separation axioms $T_0$, $T_1$, and $T_2$ are satisfied.
   (a) The digital line topology on $\mathbb{Z}$.
   (b) The particular point topology $P_0 = \{U \subseteq \mathbb{R} : 0 \in U\} \cup \{\emptyset\}$ on $\mathbb{R}$.
   (c) The excluded point topology $E_0 = \{U \subseteq \mathbb{R} : 0 \notin U\} \cup \{\mathbb{R}\}$ on $\mathbb{R}$.
2. Show that if $X$ is not $T_0$, then there exists a convergent sequence in $X$ which does not have a unique limit. Show that the converse fails.
3. Show that a topological space $X$ is $T_0$ if and only if for all $x, y \in X$, $\mathrm{cl}\{x\} = \mathrm{cl}\{y\}$ implies $x = y$.
4. Show that the following are equivalent:
   (a) $Y$ is $T_0$.
   (b) If $f : X \to Y$ is a function with $f^{-1}(V) = \emptyset$ or $X$ for every open set $V$ in $Y$, then $f$ is a constant function.
5. Suppose $(X, \mathcal{T})$ is a $T_0$ topological space and $V \in \mathcal{T}$. Show that $\mathcal{T}' = \{U \in \mathcal{T} : U \subseteq V \text{ or } V \subseteq U\}$ is a topology on $X$ and is $T_0$.
6. Show that a topological space $(X, \mathcal{T})$ is $T_1$ if and only if for every $x \in X$, $\bigcap\{U \in \mathcal{T} : x \in U\} = \{x\}$.
7. Show that the only $T_1$ topology on a finite set is the discrete topology.
8. In the definitions of $T_0$, $T_1$, and $T_2$, replace each reference to a "neighborhood" by a "basic neighborhood", relative to some fixed basis. Are the resulting statements equivalent to the original definitions? What if "neighborhoods" are replaced by "subbasic neighborhoods"?

9.  Show that $X$ is Hausdorff if and only if the diagonal $\Delta = \{(x,x) : x \in X\}$ is a closed subset of the product $X \times X$ with the product topology.

10. Let $X$ be an infinite set with the indiscrete topology. Which separation axioms $(T_0, T_1, T_2, T_3, T_4,$ regular, normal) does $X$ satisfy?

11. Which separation axioms $(T_0, T_1, T_2, T_3, T_4,$ regular, normal) does $\mathbb{R}_l$ satisfy?

12. Give $X = \{1,2,3\}$ the topology having basis $\{\{1,2\},\{3\}\}$. Which of the separation axioms $T_3$, regular, $T_4$, and normal does $X$ satisfy?

13. Give $X = \mathbb{R}^2$ the deleted radius topology defined before Exercise 23 of Section 1.4.
    (a) Show that $F = \{(0,y) : y \in \mathbb{R}, y \neq 0\}$ is a closed set not including the point $(0,0)$.
    (b) Show that $X$ is $T_2$ but not $T_3$.
    (c) Is $\mathbb{R}^2$ with the bow-tie topology $T_3$?

14. If $(X, \mathcal{T}_C)$ is a $T_2$ space and $\mathcal{T}_F$ is a finer topology on $X$, then $(X, \mathcal{T}_F)$ is $T_2$, since $\mathcal{T}_F$ has more than enough open sets to separate points. Show that this result does not hold if $T_2$ is replaced by $T_3$. (Hint: see Exercise 13.) Why would having more open sets prevent the finer topology from being $T_3$?

15. Determine which of the following topological properties are productive and which are hereditary. Justify your answers.
    (a) Is finite.
    (b) Has cardinality 10.
    (c) Is infinite.
    (d) Has exactly two connected components.
    (e) Has an isolated point.

16. Show that $\prod_{i \in I} X_i$ is $T_0$ if and only if $X_i$ is $T_0$ for every $i \in I$. Show that the statement remains true if $T_0$ is replaced by $T_1$.

17. Show that $T_3$ is productive.

18. Consider $Y = \prod_{x \in \mathbb{R}} \mathbb{R}_x$ where $\mathbb{R}_0$ is the real line with the cofinite topology and for $x \neq 0$, $\mathbb{R}_x$ is the real line with the Euclidean topology. By Theorem 6.2.4, $Y$ is not $T_2$ since $\mathbb{R}_0$ is not $T_2$. Find two elements of $Y$ which have no disjoint neighborhoods.

19. From the definition, $X$ is regular if and only if it is $T_3$ and $T_1$. Show that $X$ is regular if and only if it is $T_3$ and $T_0$.

20. Show that if $X$ is $T_4$ and $Y$ is a closed subset of $X$, then $Y$ is $T_4$. Show the same result with $T_4$ replaced by normal.

21. A topological space $X$ satisfies the $R_0$ separation axiom if, for every $x \in X$ and every neighborhood $U$ of $x$, $\mathrm{cl}\{x\} \subseteq U$. Prove that $X$ is $T_1$ if and only if it is $T_0$ and $R_0$.

22. Consider the $R_0$ separation axiom defined in Exercise 21. Show that the following are equivalent.
    (a) $X$ is $R_0$.
    (b) For every $x \in X$, $\mathrm{cl}\{x\} = \bigcap\{U : U$ is a neighborhood of $x\}$.
    (c) $\{\mathrm{cl}\{x\} : x \in X\}$ is a partition of $X$.

23. Show that any compact Hausdorff space is locally compact.

24. In general, a compact space need not be locally compact. Let $\mathcal{T}_Q$ be the Euclidean topology on $Q$. Let $Y = Q \cup \{\pi\}$ and give $Y$ the topology $\{Y\} \cup \mathcal{T}_Q$. Show that $Y$ is compact but not locally compact.

25. Suppose $A$ is a subset of a locally connected topological space, $x \in \text{int} A$, and $A_x$ is the connected component of $x \in A$. Show that $x \in \text{int} A_x$.

26. (a) Let $X = \{(x, \sin(1/x)) \in \mathbb{R}^2 : x \in (0, 4]\} \cup \partial([0, 4] \times [-2, 2])$ with the Euclidean topology. Is $X$ connected? path connected? locally connected? locally path connected?

    (b) Show that neither connectedness nor locally connectedness implies the other.

    (c) Show that neither path connectedness nor locally path connectedness implies the other.

27. If $X$ is $T_1$, connected, and $|X| > 2$, prove that there exists a connected subset $B$ of $X$ with $1 < |B| < |X|$. (Hint: If the result fails, fix $a \in X$, and for a separation $(U, V)$ of $X - \{a\}$, consider closed sets $(E, F)$ which give a separation of cl $U$.)

## 7.2 Separation by continuous functions

The definitions of the separation axioms $T_i$ ($i = 0, 1, 2, 3, 4$) used separation by open sets. We now consider separation by a continuous function.

**Definition 7.2.1.** Let $X$ be a topological space and $\mathbb{R}$ the Euclidean line. Two sets $A$ and $B$ in $X$ are *separated by a continuous function* if there exists a continuous function $f : X \to \mathbb{R}$ with $f(A) = 0$ and $f(B) = 1$.

Now if sets $A$ and $B$ in a topological space $X$ are separated by a continuous function, then they are separated by open sets, for $f^{-1}((-0.25, 0.25))$ and $f^{-1}((0.75, 1.25))$ are disjoint open sets containing $A$ and $B$. However, if $A$ and $B$ are separated by open sets, they need not be separated by a continuous function, as Example 7.2.3 below illustrates. Before that, we make a quick observation about the range of a function which separates two sets.

**Theorem 7.2.2.** *Two sets $A$ and $B$ in a topological space $X$ can be separated by a continuous function $f : X \to \mathbb{R}$ if and only if they can be separated by a continuous function $\bar{f} : X \to [0, 1]$.*

*Proof.* If $A$ and $B$ are separated by $\bar{f} : X \to [0, 1]$, then $f : X \to \mathbb{R}$ defined by $f(x) = \bar{f}(x)$ is a continuous real-valued function separating $A$ and $B$. If $f : X \to \mathbb{R}$ is a continuous function separating $A$ and $B$, define

$$\bar{f} = \begin{cases} 1 & \text{if } f(x) \geq 1, \\ f(x) & \text{if } 0 \leq f(x) \leq 1, \\ 0 & \text{if } f(x) \leq 0. \end{cases}$$

Clearly $\bar{f}$ separates $A$ and $B$, and $\bar{f}$ is continuous by the pasting lemma, since it is obtained by pasting together $f|_{f^{-1}((0,1))}$ and the constant functions 0 and 1 on the closed sets $f^{-1}([0,1])$, $f^{-1}((-\infty,0])$, and $f^{-1}([1,\infty))$, and the functions agree on the intersections. □

**Example 7.2.3** (Simplified Arens square). Let $S = (-1,1) \times (0,1)$, $a = (-1,0)$, $b = (1,0)$, and $X = S \cup \{a,b\}$. For $(x,y) \in S$, take the Euclidean neighborhoods having a base of sets of form $(x-1/n, x+1/n) \times (y-1/n, y+1/n) \cap S$. Take the basic neighborhoods of $a$ and $b$ to be sets of form $\{a\} \cup (-1,0) \times (0,1/n)$ and $\{b\} \cup (0,1) \times (0,1/m)$, respectively. It is easily verified that this is a basis for a topology, and the topology is Hausdorff. However, there is no continuous function $f : X \to \mathbb{R}$ with $f(a) = 0$ and $f(b) = 1$. If there were such a function, then $U = f^{-1}((-.25,.25))$ would be a neighborhood of $a$ containing a set of form $(1,0) \times (0,1/n)$ and $V = f^{-1}((.75,1.25))$ would be a neighborhood of $b$ containing a set of form $(0,1) \times (0,1/m)$. With $k = \max\{m,n\}$, the point $c = (0,1/(2k))$ is in the closure of $U$ and the closure of $V$. Now the sequence $u_j = (-1/j, 1/(2k))_{j \in \mathbb{N}}$ converges to $c$ from within $U$ and the sequence $v_j = (1/j, 1/(2k))_{j \in \mathbb{N}}$ converges to $c$ from within $V$. Since $f$ preserves limits (see Theorem 6.3.5) $f(u_j)$ is a sequence in $f(U) \subseteq (-.25,.25)$ converging to $f(c)$, and thus $f(c) \in [-.25,.25]$. Similarly, $f(v_j)$ is a sequence in $f(V) \subseteq (.75,1.25)$ converging to $f(c)$, so $f(c) \in [.75,1.25]$, contrary to $f(c) \in [-.25,.25]$. Thus, there can be no continuous function $f$ separating $a$ and $b$. See [45] for more details on this example.

Thus, using continuous functions rather than open sets for our separation will give stronger separation axioms. We define these separation axioms below.

**Definition 7.2.4.** A topological space $X$ is *functionally Hausdorff* if for every pair of distinct points $x, y \in X$, there exists a continuous function $f : X \to \mathbb{R}$ with $f(x) = 0$ and $f(y) = 1$.

A topological space $X$ is $T_{3.5}$ if for every closed set $A$ in $X$ and every $x \notin A$, there exists a continuous function $f : X \to \mathbb{R}$ with $f(A) = 0$ and $f(x) = 1$.

A topological space $X$ is *completely regular* if it is $T_1$ and $T_{3.5}$.

Thus, $X$ is functionally Hausdorff if distinct points are separated by continuous functions, and is completely regular if it is $T_1$ and disjoint points and closed sets are separated by continuous functions. The example of the simplified Arens square shows that functionally Hausdorff is a strictly stronger property than the Hausdorff property. Similarly, complete regularity is strictly stronger than regularity. Completely regular spaces are also called *Tychonoff spaces*.

Suppose $X$ is completely regular and $Y \subseteq X$ has the subspace topology. If $A$ is a closed set in $Y$ and $b \in Y - A$, then there exists a closed set $A'$ in $X$ with $A' \cap Y = A$ and $b \notin A'$. If $f$ is a continuous function separating $A'$ and $b$ in $X$, then the restriction $f|_Y$ of $f$ to $Y$ is a continuous function separating $A$ and $b$ in $Y$. Thus, complete regularity is hereditary. Similarly, functionally Hausdorff is a hereditary property.

A version of normality based on separation by continuous functions was conspicuously missing from Definition 7.2.4. It is missing due to the following theorem by

Pavel S. Urysohn, which says closed sets can be separated by open sets if and only if they can be separated by a continuous function. Urysohn was a brilliant Russian mathematician who, with Pavel Alexandroff, first defined compactness in terms of open covers. In 1924, at the age of 26, Urysohn drowned while swimming in rough seas in Brittany, France.

**Theorem 7.2.5** (Urysohn's lemma). *A $T_1$ topological space X is normal if and only if for every pair of disjoint closed sets A, B in X, there exists a continuous function $f : X \to \mathbb{R}$ with $f(A) = 0$ and $f(B) = 1$.*

*Sketch of proof.* If a $T_1$ topological space $X$ has the property that disjoint closed sets $A$ and $B$ can be separated by a continuous function $f$, then $f^{-1}((-0.25, 0.25))$ and $f^{-1}((0.75, 1.25))$ are disjoint open sets separating $A$ and $B$, so $X$ is normal.

Suppose $A$ and $B$ are disjoint closed sets in a normal space $X$. We must create from nothing a continuous function $f : X \to \mathbb{R}$ with $f(A) = 0$ and $f(B) = 1$. How do we define $f(x)$ for $x \notin A \cup B$? We will use Theorem 7.1.9(b) repeatedly. Since $U_1 = X - B$ is an open set containing $A$, by Theorem 7.1.9, there exists an open set $U_0$ with

$$A \subseteq U_0 \subseteq \operatorname{cl} U_0 \subseteq U_1.$$

Now we insert an open set $U_{1/2}$ between $U_0$ and $U_1$. Focusing on the fact that $\operatorname{cl} U_0 \subseteq U_1$, we may apply Theorem 7.1.9 to find an open set $U_{1/2}$ with

$$\operatorname{cl} U_0 \subseteq U_{1/2} \subseteq \operatorname{cl} U_{1/2} \subseteq U_1. \tag{7.1}$$

Now having defined $U_0, U_{1/2}$, and $U_1$, we apply Theorem 7.1.9 to the first and third inclusions in Equation 7.1 to get an open set $U_{1/4}$ between $U_0$ and $U_{1/2}$ and an open set $U_{3/4}$ between $U_{1/2}$ and $U_1$ with

$$\operatorname{cl} U_0 \subseteq U_{1/4} \subseteq \operatorname{cl} U_{1/4} \subseteq U_{1/2} \subseteq \operatorname{cl} U_{1/2} \subseteq U_{3/4} \subseteq \operatorname{cl} U_{3/4} \subseteq U_1.$$

Having defined open sets $U_r$ for all values of $r$ which are multiples of $1/4$ in $[0, 1]$, we repeat the process, inserting open sets $U_r$ for the missing multiples of $1/8$, then for the missing multiples of $1/16$, and so on. Continuing in this manner, we will obtain open sets $U_r$ for every $r \in \{\frac{m}{2^n} \in [0, 1] : m, n \in \mathbb{N} \cup \{0\}\} = D$. The set $D$ is called the set of dyadic fractions in $[0, 1]$ and is dense in $[0, 1]$. Furthermore, for $p < q$ in $D$, we have

$$A \subseteq U_0 \subseteq U_p \subseteq \operatorname{cl} U_p \subseteq U_q \subseteq U_1 = X - B.$$

Now we are ready to define our function $f : X \to [0, 1] \subseteq \mathbb{R}$. For $x \in X$, take

$$f(x) = \begin{cases} \inf\{r \in D : x \in U_r\} & \text{if } x \in X - B = U_1, \\ 1 & \text{if } x \in B. \end{cases}$$

Now $A \subseteq U_0$ implies $f(x) = 0$ for $x \in A$, and clearly $f(B) = 1$.

It remains to show that $f$ is continuous at each $x \in X$. Given $x \in X$, suppose first that $f(x) \in (0, 1)$. Now any neighborhood $U$ of $f(x)$ contains a basic open neighborhood $(p, q)$ of $f(x)$, where $p, q \in D$. Furthermore, there exist points $p', q' \in D$ with $p < p' < f(x) < q' < q$. Now it can be checked that $V = U_{q'} - \text{cl}\, U_{p'}$ is a neighborhood of $x$ with $f(V) \subseteq [p', q'] \subseteq (p, q) \subseteq U$, so $f$ is continuous at $x$. Continuity at points $x$ with $f(x) \in \{0, 1\}$ can be shown similarly. $\qquad\square$

We give two immediate consequences of Urysohn's lemma.

**Corollary 7.2.6.** *If $X$ is normal then it is completely regular.*

*Proof.* Normal spaces are $T_1$, so singletons are closed. If any disjoint closed sets can be separated by a continuous function, then taking one of those sets to be a singleton proves that $X$ is completely regular. $\qquad\square$

Thus, we have

$$\text{normal} \Rightarrow \text{completely regular} \Rightarrow \text{regular} \Rightarrow T_2 \Rightarrow T_1 \Rightarrow T_0.$$

**Corollary 7.2.7.** *Every subspace of a compact $T_2$ space is completely regular.*

*Proof.* Any compact $T_2$ space is normal, and thus completely regular. Complete regularity is hereditary. $\qquad\square$

Besides the Urysohn lemma, another very significant result on separation axioms is that the converse of Corollary 7.2.7 also holds. Every completely regular space is a subspace of a compact $T_2$ space. This is shown in the next section.

## Exercises

1.  A *zero set* in a topological space $X$ is a set of form $f^{-1}(\{0\})$ for some continuous function $f : X \to \mathbb{R}$ (where $\mathbb{R}$ has the Euclidean topology). A *cozero set* in $X$ is the complement of a zero set. Show that a $T_1$ space $X$ is completely regular if and only if the collection $\mathcal{B}$ of all cozero sets of $X$ is a basis for the topology on $X$.
2.  Complete the missing details from Example 7.2.3, showing that the basis for the topology describe there really is a basis, and that the topology is Hausdorff.
3.  Is the simplified Arens square given in Example 7.2.3 completely regular or normal? Justify your answer.
4.  Give a proof or counterexample for this statement: Functionally Hausdorff is a productive property.
5.  Give a proof or counterexample for this statement: Completely regular is a productive property.
6.  Let $\mathcal{T}$ be the topology on $\mathbb{R}$ generated by the basis $\mathcal{B} = \{[x, x + \varepsilon) : x \in \mathbb{Q}, \varepsilon > 0\} \cup \{(x - \varepsilon, x] : x \in \mathbb{R} - \mathbb{Q}, \varepsilon > 0\}$. Which of the following properties does $(\mathbb{R}, \mathcal{T})$ have? $T_1, T_2, T_3, T_{3.5}$, compactness. Justify your answers.

7. Show that $\mathbb{R}$ with the right ray topology is not $T_{3.5}$ by exhibiting a point and a closed set which cannot be separated by a continuous function.
8. Show that the digital line is not $T_{3.5}$ by exhibiting a point and a closed set which cannot be separated by a continuous function.
9. Give a proof or counterexample for this statement: If $X$ is a $T_{3.5}$ space, $U$ is open in $X$, and $x \notin U$, then there exists a continuous function $f : X \to \mathbb{R}$ with $f(U) = 0$ and $f(x) = 1$.
10. Define $(X, \mathcal{T})$ to be $\mathbb{R}_l$-$T_{3.5}$ if for every closed $A \subseteq X$ and every $x \notin A$, there exists a continuous function $f : X \to \mathbb{R}_l$ with $f(A) = 0$ and $f(x) = 1$.
    (a) Show that $\mathbb{Q}$ with the Euclidean topology is $\mathbb{R}_l$-$T_{3.5}$.
    (b) Does $\mathbb{R}_l$-$T_{3.5}$ imply $T_{3.5}$? Justify your answer.
    (c) Does $T_{3.5}$ imply $\mathbb{R}_l$-$T_{3.5}$? Justify your answer.
    (d) Characterize the connected $\mathbb{R}_l$-$T_{3.5}$ spaces.
11. Complete the proof of Urysohn's lemma by showing that the function $f$ defined there is continuous.

## 7.3 Compactifications

A (Hausdorff) compactification of $X$ is a compact $T_2$ space $\alpha X$ which contains a homeomorphic copy of $X$ as a dense subspace. The subspace $X = (0, 1)$ of the Euclidean line is not compact, but we may add the points 0 and 1 to make it compact, forming the compactification $\alpha X = [0, 1]$ of $X = (0, 1)$. Adding points—in some minimal way—to get a compact space is the concept of forming a compactification. Since $(0, 1)$ is a homeomorphic copy of $\mathbb{R}$, the compact $T_2$ space $[0, 1]$ contains a homeomorphic copy of $\mathbb{R}$ as a dense subspace, so $[0, 1]$ may also be viewed as a compactification of $\mathbb{R}$. To help recognize the homeomorphic copy of $X$ embedded in a compactification $\alpha X$ of $X$, we will include the homeomorphism as part of the formal definition of a compactification.

**Definition 7.3.1.** A *compactification* of a topological space $X$ is a pair $(\alpha X, \alpha)$ where $\alpha X$ is a compact $T_2$ space and $\alpha : X \to \alpha X$ is a homeomorphic embedding of $X$ into $\alpha X$, such that the homeomorphic copy $\alpha(X)$ of $X$ is a dense subspace of $\alpha X$. The subspace $\alpha X - \alpha(X)$ is the *remainder* associated with this compactification. If the embedding $\alpha$ is understood, we may simply refer to the compactification $\alpha X$.

Now $\alpha X = [0, 1]$ is a compactification of $X = (0, 1)$, or more formally, $(\alpha X, \alpha)$ is a compactification of $X$ where $\alpha : (0, 1) \to [0, 1]$ is the inclusion map $\alpha(x) = x$. Similarly, $([0, 5], y)$ is a compactification of $X = (0, 1)$, where $y : (0, 1) \to (0, 5)$ is the homeomorphism $y(x) = 5x$. In this case, $yX = [0, 5]$ and $y(X) = (0, 5)$. The space $\delta X = [0, 5] \cup \{6\}$ is also a compact $T_2$ space which contains a homeomorphic copy of $(0, 1)$, but $\delta X$ is not a compactification of $(0, 1)$ since no homeomorphic image of $(0, 1)$ is dense in $\delta X$.

**Example 7.3.2.**

(a) The real line $\mathbb{R}$ is homeomorphic to $(-\frac{\pi}{2}, \frac{\pi}{2})$ by the homeomorphism $\alpha(x) = \arctan(x)$, and $[-\frac{\pi}{2}, \frac{\pi}{2}]$ is a compact $T_2$ space containing $\alpha(\mathbb{R}) = (-\frac{\pi}{2}, \frac{\pi}{2})$ as a dense subspace, so $[-\frac{\pi}{2}, \frac{\pi}{2}]$ is a compactification of $\mathbb{R}$. The remainder consists of the two point set $\{-\frac{\pi}{2}, \frac{\pi}{2}\}$ added to make $(-\frac{\pi}{2}, \frac{\pi}{2}) \approx \mathbb{R}$ compact, so this is called a *two-point compactification* of $\mathbb{R}$. As noted above, $[0, 1]$ is another two-point compactification of $\mathbb{R}$. In Exercise 14 we will see that every two-point compactification of $\mathbb{R}$ is equivalent to these in a natural way. Essentially, one point must compactify the open left end and one point must compactify the open right end of $\mathbb{R}$. The two-point compactification of $\mathbb{R}$ is sometimes denoted $\mathbb{R} \cup \{\pm\infty\}$ and called the *extended real line*.

(b) The real line $\mathbb{R}$ is homeomorphic to $(0, 2\pi)$. The function $f : (0, 2\pi) \to \mathbb{R}^2$ defined by $f(x) = (\cos x, \sin x)$ is an embedding of the interval $(0, 2\pi)$ into the plane $\mathbb{R}^2$, suggested by Figure 7.2.

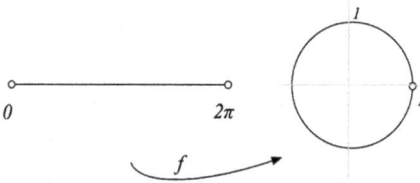

**Figure 7.2:** $f : (0, 2\pi) \to \mathbb{R}^2$ defined by $f(x) = (\cos x, \sin x)$.

This tells us that the unit circle with one point missing is a homeomorphic copy of $\mathbb{R}$. To compactify this copy of $\mathbb{R}$, since it is embedded in the Euclidean space $\mathbb{R}^2$ as a bounded set, we need only take its closure to obtain a closed and bounded set, which will be compact. Taking the closure adds the missing point of the circle, giving the *one-point compactification* of $\mathbb{R}$. In effect, the one-point compactification wraps the two open ends of $\mathbb{R}$ around to the same "limit".

The technique of the last example may be used to find other compactifications of $\mathbb{R}$. We may embed $\mathbb{R}$ into Euclidean space $\mathbb{R}^2$ or $\mathbb{R}^3$ as a bounded set, then simply take the closure. The copy of $\mathbb{R}$ will be dense in its closure, and as a closed and bounded set in Euclidean space, the closure will be compact and $T_2$. The example below illustrates this technique further. To simplify the illustrations, we will consider $[0, 1) \approx [0, \infty)$ so we have only one end of the line to compactify.

**Example 7.3.3.** Consider the interval $[0, 1)$ with the Euclidean topology. Note that $[0, 1) \approx (0, 1] \approx [a, \infty) \approx (-\infty, b]$, so a compactification of any of these homeomorphic spaces will give a compactification of $[0, 1)$.

(a) Embed $(0,1]$ in $\mathbb{R}^2$ as the set $A = \{(x, \sin\frac{1}{x}) : x \in (0,1]\}$, as suggested by Figure 7.3(a). Taking the closure of $A$ gives a compactification of $[0,1)$ obtained by adding an interval $\{0\} \times [-1,1]$. Thus, $(0,1]$ has a compactification whose remainder is an interval.

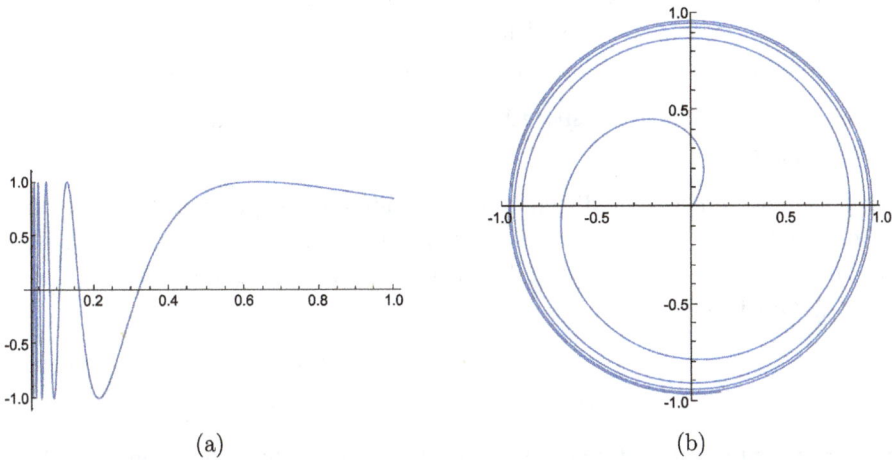

**Figure 7.3:** Compactifications of $(0,1]$ which add (a) an interval and (b) a circle.

(b) Embed $(0,1] \approx [1,\infty)$ into $\mathbb{R}^2$ as the spiral $B$ given in polar coordinates by $r = 1 - \frac{1}{\theta}$ for $\theta \in [1,\infty)$, as suggested by Figure 7.3(b). Taking the closure of $B$ gives a compactification of $[0,1)$ obtained by adding a circle.

(c) Embed $(0,1]$ into $\mathbb{R}^3$ by the cylindrical coordinate function $\boldsymbol{g}(t) = (r(t), \theta(t), z(t)) = (2 + \sin(\frac{1}{t}), \frac{\pi}{t}, t)$. We claim that taking the closure of $\boldsymbol{g}((0,1])$ will give a compactification whose remainder is the annulus $1 \le r \le 3$ in the plane $z = 0$. For this, it suffices to show that given any $\varepsilon > 0$, for $t \in (0, \varepsilon)$, the points $(r(t), \theta(t)) = (2 + \sin(\frac{1}{t}), \frac{\pi}{t})$ are dense in the annulus. Given $r_0 \in [1, 3]$, there is a $t_0 \in [0, \varepsilon)$ such that $2 + \sin(\frac{1}{t_0}) = r_0$, and thus $r_0$ is realized by $r(t)$ for $\frac{1}{t} \in \{\frac{1}{t_0} + 2\pi m : n \in \mathbb{N}\}$, or, since $\theta(t) = \frac{\pi}{t}$, for $\theta \in \{\frac{\pi}{t_0} + 2\pi(\pi n) : n \in \mathbb{N}\}$. It is well known (see Exercise 21) that if $y$ is an irrational number and $k \in \mathbb{N}$, then $\{ny : n \ge k\}$ is dense in $[0,1]$ modulo 1. (That is, the fractional parts $\{ny - \lfloor ny \rfloor : n \ge k\}$ are dense in $[0,1]$.) Thus, if $y$ is irrational, the set $\{2\pi(yn) : n \ge k\}$ will be dense in $[0, 2\pi]$ mod $2\pi$, as will any fixed translation $\{a + 2\pi(yn) : n \ge k\}$ of such a set of angles. Applied in our situation, this tells us that each $r_0 \in [1, 3]$ is realized by $t \in [0, \varepsilon)$ on a dense set of angles $\{\frac{\pi}{t_0} + 2\pi(\pi n) : n \in \mathbb{N}\}$, so $(r(t), \theta(t)) = (2 + \sin(\frac{1}{t}), \frac{\pi}{t})$ is dense in the annulus.

The example above showed that $(0,1]$ has compactifications with remainders of an interval, a circle, and an annulus.

Not every topological space has a compactification. If $X$ has a compactification $\alpha X$, then $\alpha X$ is a compact $T_2$ space, and thus is normal and completely regular. Since completely regular is a hereditary property, $X$ must be completely regular. Thus, completely regular is a necessary condition for a topological space to have a compactification. In fact, completely regular is necessary and sufficient. This significant result was given in different degrees of generality in papers by A. N. Tychonoff in 1930 and 1935. In these same papers, Tychonoff defined the product topology for infinite products and proved that with this topology, products of compact spaces are compact.

**Theorem 7.3.4.** *A topological space $X$ has a compactification if and only if it is completely regular.*

*Proof.* We have proved one direction of implication above. Suppose that $X$ is completely regular. Let $C^*(X)$ be the collection of continuous, bounded real valued functions $f : X \to \mathbb{R}$. For each $f \in C^*(X)$, there exists a closed and bounded interval $I_f$ with $f(X) \subseteq I_f$. Consider the product

$$\prod_{f \in C^*(X)} I_f \subseteq \prod_{f \in C^*(X)} \mathbb{R} = \mathbb{R}^{C^*(X)}.$$

An element of this product is an element of $\mathbb{R}^{C^*(X)}$, and is thus a function from $C^*(X)$ to $\mathbb{R}$. Such a function $e$ from $C^*(X)$ to $\mathbb{R}$ must give a real number for each $f \in C^*(X)$. One obvious way to define such a function is to pick an $x_0 \in X$, and for each $f \in C^*(X)$, take the real number $f(x_0)$. Furthermore, we may do this for each $x_0 \in X$. This gives the *evaluation function*

$$e : X \to \prod_{f \in C^*(X)} I_f$$

defined by

$$e(x_0) = \langle f(x_0) \rangle_{f \in C^*(X)} = \prod_{f \in C^*(X)} f(x_0) \subseteq \prod_{f \in C^*(X)} I_f.$$

For a given $x_0 \in X$, the evaluation function evaluates every $f \in C^*(X)$ at $x_0$ to get an element of $\prod_{f \in C^*(X)} I_f$.

We will need the following technical detail about the evaluation function.

**Lemma 7.3.5.** *If $X$ is completely regular, then the evaluation function is an embedding of $X$ into $\prod_{f \in C^*(X)} I_f$.*

We continue with the main argument before proving this lemma. We now have

$$e : X \to e(X) \subseteq \mathrm{cl}(e(X)) \subseteq \prod_{f \in C^*(X)} I_f.$$

Now $e(X)$ is a homeomorphic copy of $X$ by the lemma, and it is dense in its closure. By the Tychonoff theorem, $\prod_{f \in C^*(X)} I_f$ is compact, and is $T_2$ since products of $T_2$ spaces are $T_2$. As a closed subset of a compact $T_2$ space, $cl(e(X))$ is a compact $T_2$ space. Thus, $cl(e(X))$ is a compactification of the completely regular space $X$. □

The compactification $cl(e(X))$ described above is called the *Stone–Čech compact-ification* and is denoted $\beta X$. Though the construction is due to Tychonoff, the compactification bears the names of Marshall Stone and Eduard Čech, who independently published papers in 1937 exhibiting useful properties of the compactification.

Any completely regular topological space is a dense subspace of its Stone–Čech compactification. Together with Corollary 7.2.7, we have the following result.

**Corollary 7.3.6.** *A topological space is completely regular if and only if it is a subspace of a compact $T_2$ space.*

The proof of Theorem 7.3.4 was incomplete for two reasons. We must prove the lemma used, and we must prove the Tychonoff theorem. The first of these is an easy task and is given here. The proof of the Tychonoff theorem is a deep result and is given in Section 7.4.

*Proof of Lemma 7.3.5.* The $f$th coordinate function of $e(x) = \prod_{f\in C^*(X)} f(x)$ is $f(x)$. Since each coordinate function is continuous, $e$ is continuous. Suppose $x \neq y$ in $X$. Since $X$ is completely regular, there exists a bounded continuous function $g : X \to [0,1]$ which separates the point $x$ from the closed set $\{y\}$. It follows that $e(x) \neq e(y)$ since they differ in the $g$th coordinate. This shows that $e$ is one-to-one. Thus, $e$ is continuous, one-to-one, and onto its range $e(X)$. It only remains to show that $e^{-1}$ is continuous, or equivalently, that $e$ is an open map. Suppose $U$ is open in $X$. Now $e(U)$ will be open in $e(X)$ if and only if for each $e(u) \in e(U)$, there exists an open neighborhood $V$ of $e(u)$ contained in $e(U)$. Suppose $u \in U$. Since $X$ is completely regular, there exists a bounded continuous function $g : X \to [0,1]$ with $g(u) = 0$ and $g(X - U) = 1$. Let $V = \pi_g^{-1}([0,1)) \cap e(X)$. Since $\pi_g^{-1}([0,1))$ is open in $\prod_{f\in C^*(X)} I_f$ and $e(X)$ is a subspace of this product, $V$ is open in $e(X)$. The $g$th coordinate of $e(u) = \prod_{f\in C^*(X)} f(u)$ is $g(u) = 0 \in [0,1)$, so $e(u) \in V$. Finally, we will show that $V \subseteq e(U)$. Suppose $p \in V$. From the definition of $V$, $p = e(z)$ for some $z \in X$ and $g(z) \in [0,1)$. Now we must have $z \in U$, for $z \in X - U$ would imply $g(z) = 1 \notin [0,1)$. Thus, $p = e(z)$ for some $z \in U$, so $p \in e(U)$, as needed. □

Many nice properties of the Stone–Čech compactification have been studied extensively, including the following two nice extension properties.

First, we consider extending a continuous function on $X$ to its compactification. Note that the continuous bounded function $f : (0,1) \to \mathbb{R}$ defined by $f(x) = \sin\frac{1}{x}$ cannot be continuously extended to the compactification $[0,1]$ of $(0,1)$. A continuous bounded function $f : (0,1) \to \mathbb{R}$ can be extended to $[0,1]$ if and only if both limits $\lim_{x\to 0^+} f(x)$ and $\lim_{x\to 1^-} f(x)$ exist. The theorem below shows that the $f(x) = \sin\frac{1}{x}$ on

$(0,1)$ can be extended to the Stone–Čech compactification $\beta(0,1)$, as can any continuous bounded real-valued function on $(0,1)$. First, we must understand what is meant by an extension of a function on $X$ to a compactification $Y$.

Recall that if $A \subseteq B$ and $f : A \to \mathbb{R}$ is a continuous function, then $f_B : B \to \mathbb{R}$ is an extension of $f$ if the restriction of $f_B$ to $A$ equals $f$; that is, if $f_B(x) = f(x)$ for all $x \in A$.

**Definition 7.3.7.** Suppose $X$ and $Z$ are topological spaces, $(\alpha X, \alpha)$ is a compactification of $X$, and $f : X \to Z$ is a continuous function. A *continuous extension* of $f$ to $\alpha X$ is a continuous function $f_\alpha : \alpha X \to Z$ with $f_\alpha(\alpha(x)) = f(x)$ for all $x \in X$.

Thus, a function $f_\alpha$ defined on $\alpha X$ is an extension of $f$ defined on $X$ if, for each $x \in X$, $f_\alpha$ maps the homeomorphic copy $\alpha(x)$ of $x$ to $f(x)$. In the diagram of Figure 7.4, starting from any point in $X$ and following either path along the indicated arrows from $X$ to $Z$ gives the same result. A diagram showing maps between several sets with this property that following any of the arrows from one set to another gives the same result is called a *commutative diagram*, or we may say that the diagram *commutes*.

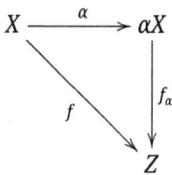

**Figure 7.4:** An extension of $f : X \to Z$ to $\alpha X$ is a function $f_\alpha$ which makes the diagram commute.

**Theorem 7.3.8.** *If $X$ is a completely regular space, then any continuous bounded real-valued function $f \in C^*(X)$ has a continuous extension to $\beta X$.*

**Proof.** Suppose $g \in C^*(X)$. We want to find a continuous function $g_\beta : \beta X \to \mathbb{R}$ with $g_\beta(e(x)) = g(x)$ for all $x \in X$. The proof is based on the fact that $\beta X = \mathrm{cl}\,e(X) \subseteq \prod_{f \in C^*(X)} I_f$ contains a copy of $g$ in the $g$th coordinate of this product. Indeed, the projection function $\pi_g : \prod_{f \in C^*(X)} I_f \to I_g$ is continuous, and the restriction of $\pi_g$ to $\beta X$ is also continuous. Furthermore, for $x_0 \in X$, $e(x_0) = \langle f(x_0) \rangle_{f \in C^*(X)}$, so $\pi_g(e(x_0)) = g(x_0)$, and thus $g_\beta = \pi_g$ is an extension of $g$. $\qquad\square$

This result tells us that if $X$ is completely regular, any continuous function $f$ from $X$ into a compact subspace $\mathrm{cl}\,f(X)$ of $\mathbb{R}$ can be continuously extended to $\beta X$. The next theorem is stronger.

**Theorem 7.3.9.** *If $X$ is a completely regular space, $K$ is any compact Hausdorff space, and $f : X \to K$ is continuous, then $f$ has a continuous extension to $\beta X$.*

**Proof.** Under the hypotheses, consider the evaluation maps $e : X \to e(X) \subseteq \beta X \subseteq \prod_{g \in C^*(X)} I_g$ and $e' : K \to e'(K) \subseteq \beta K \subseteq \prod_{h \in C^*(K)} I_h$. Since $K$ is compact, $e' : K \to \beta K$ is a homeomorphism. Note that, for any $h \in C^*(K)$, $h \circ f \in C^*(X)$. We will define a function $G$ which maps a vector in $\prod_{g \in C^*(X)} I_g$ to the vector in $\prod_{h \in C^*(K)} I_h$ whose $h$th coordinate is the $(h \circ f)$th coordinate of the input vector. Specifically, define $G : \prod_{g \in C^*(X)} I_g \to \prod_{h \in C^*(K)} I_h$ by

$$G(\langle x_g \rangle_{g \in C^*(X)}) = \langle x_{h \circ f} \rangle_{h \in C^*(K)}.$$

Now for any $h \in C^*(K)$, $\pi_h \circ G = \pi_{h \circ f}$, which, as a projection function, is continuous. Since every coordinate function $\pi_h \circ G$ of $G$ is continuous, $G$ is continuous. Given $x \in X$,

$$G(e(x)) = G(\langle g(x) \rangle_{g \in C^*(X)}) = \langle h \circ f(x) \rangle_{h \in C^*(K)} = e'(f(x)) \in e'(K),$$

so $e'(K)$ is a closed set containing $G(e(X))$. Since $e(X)$ is dense in $\beta X$, $G(e(X))$ is dense in $G(\beta X)$ (see Exercise 19 of Section 3.1), and thus $\operatorname{cl} G(e(X)) = G(\beta X) \subseteq e'(K)$. Now consider $F = e'^{-1} \circ G|_{\beta X} : \beta X \to K$. As the composition of continuous functions, $F$ is continuous, and for any $x \in X$,

$$F \circ e(x) = e'^{-1}(G(e(x))) = e'^{-1}(e'(f(x))) = f(x),$$

so $F \circ e = f$, showing that $F : \beta X \to K$ is a continuous extension of $f : X \to K$. $\qquad\square$

Let us consider an arbitrary compactification $\alpha \mathbb{R}$ of the Euclidean line $\mathbb{R}$. Intuitively, the "middle part" $[-M, M]$ of the real line is already compact, so any point $\infty$ in the remainder $\alpha \mathbb{R} - \mathbb{R}$ should have neighborhoods which, when restricted to the homeomorphic copy $\alpha(\mathbb{R})$ of $\mathbb{R}$, should correspond to unbounded sets of $\mathbb{R}$. The following result confirms this.

**Theorem 7.3.10.** *Suppose $\alpha \mathbb{R}$ is a compactification of $\mathbb{R}$ and $\infty \in \alpha \mathbb{R} - \alpha(\mathbb{R})$ is a point in the remainder. If $U$ is a neighborhood of $\infty$ in $\alpha \mathbb{R}$, then $\alpha^{-1}(U)$ is unbounded in $\mathbb{R}$.*

*Proof.* Suppose $\alpha \mathbb{R}$ is a compactification of $\mathbb{R}$, $\infty \in \alpha \mathbb{R} - \alpha(\mathbb{R})$ is a point in the remainder, and $U$ is neighborhood of $\infty$ in $\alpha \mathbb{R}$. First, we will show that $\infty \in \partial(U \cap \alpha(\mathbb{R}))$. Given any neighborhood $V$ of $\infty$, $V \cap U$ is a neighborhood of $\infty \in \operatorname{cl} \alpha(\mathbb{R})$, so $V \cap U \cap \alpha(\mathbb{R}) \neq \emptyset$. Also, $\infty \in V \cap (\alpha \mathbb{R} - (U \cap \alpha(\mathbb{R})))$, and thus $\infty \in \partial(U \cap \alpha(\mathbb{R}))$. Now suppose $\alpha^{-1}(U)$ is bounded. Then there exists a compact set $K$ in $\mathbb{R}$ containing $\alpha^{-1}(U) = \alpha^{-1}(U \cap \alpha(\mathbb{R}))$. Now applying $\alpha$ to these sets gives $\alpha(\alpha^{-1}(U)) = \alpha(\alpha^{-1}(U \cap \alpha(\mathbb{R}))) = U \cap \alpha(\mathbb{R}) \subseteq \alpha(K)$. Now $\alpha(K)$ is compact in the Hausdorff space $\alpha \mathbb{R}$ and thus is closed. Now we have $\infty \in \partial(U \cap \alpha(\mathbb{R})) \subseteq \operatorname{cl}(U \cap \alpha(\mathbb{R})) \subseteq \alpha(K) \subseteq \alpha(\mathbb{R})$, which is a contradiction since $\infty$ was a point of the remainder $\alpha \mathbb{R} - \alpha(\mathbb{R})$. Thus, $\alpha^{-1}(U)$ must be unbounded in $\mathbb{R}$. $\qquad\square$

The description of the one-point compactification of $\mathbb{R}$ in Example 7.3.2 should suggest that the neighborhoods of the remainder point correspond to very particular unbounded sets—those whose complements are compact. Indeed, if any space $X$ has a one-point compactification $X \cup \{\omega\}$, then $\omega$ has a neighborhood base of sets of form

$U \cup \{\omega\}$ where $X - U$ is compact. This is the basis for the Alexandroff one-point compactification explored in the exercises.

Theorem 7.3.10 can be used to prove that $\mathbb{R}$ has no 3-point compactification. This is left to the exercises.

## Exercises

1.  In advanced studies, the requirement that a compactification be $T_2$ may be dropped to get a *non-Hausdorff compactification*. If $(X, \mathcal{T})$ is a topological space, let $\omega X = X \cup \{\omega\}$ where $\omega$ is a point not in $X$, and give $\omega X$ the topology $\mathcal{T}_\omega = \mathcal{T} \cup \{\omega X\}$. Show that $\omega X$ is a non-Hausdorff compactification of $X$. It is called the *trivial one-point compactification* of $X$, or the *one-point connectification* of $X$.

2.  Let $X$ be the subspace $[0, 1]$ of $\mathbb{R}_l$, let $\alpha X = [0, 1] \cup [2, 3]$ with the topology having basis $\mathcal{B} = \{[x, x+\varepsilon) \cap X : x \in [0, 1], 0 < \varepsilon < 1\} \cup \{((x-\varepsilon, x+\varepsilon) \cup (x-2-\varepsilon, x-2) \cup (x-2, x-2+\varepsilon)) \cap X : x \in [2, 3], 0 < \varepsilon < 1\}$, and define $\alpha : X \to \alpha X$ by $\alpha(x) = x$. Show that $(\alpha X, \alpha)$ is a non-Hausdorff compactification of $X$. That is, show that $(\alpha X, \alpha)$ meets all the requirements of a compactification except the $T_2$ condition.

3.  **(Alexandroff one-point compactification)** A topological space $X$ is *locally compact* if for every $x \in X$, every neighborhood $U$ of $x$ contains a compact neighborhood $V$ of $x$, that is, a compact set $V$ with $x \in \text{int } V$. The Alexandroff one-point compactification of a locally compact noncompact space $(X, \mathcal{T})$ is $\omega X = X \cup \{\omega\}$ where $\omega$ is a point not in $X$ and the topology on $\omega X$ is $\mathcal{T} \cup \{U \cup \{\omega\} : X - U$ is compact in $(X, \mathcal{T})\}$. Show that this construction does give a compactification of $X$. Why is local compactness of $X$ needed? Why is the noncompact condition needed?

4.  Suppose $\alpha X = \alpha X \cup \{a\}$ is a one-point compactification of a locally compact space $X$. If $X \cup \{\omega\}$ is the Alexandroff one-point compactification, show that there is a homeomorphism $h : X \cup \{\omega\} \to \alpha X$ with $h(x) = \alpha(x)$ for all $x \in X$. This shows that all one-point compactifications of $X$ are equivalent.

5.  Give a geometric description of the Alexandroff one-point compactification (see Exercise 3) of the spaces $X = [0, 1)$, $Y = (0, 1)$, $Z = (0, 1) \cup (2, 3)$, and $W = (0, 1) \cup (2, 3) \cup (4, 5), \cup (6, 7]$, where each space carries the Euclidean topology.

6.  In Example 7.3.3 we saw that $f : (0, 1] \to \mathbb{R}^2$ defined by $f(x) = (x, \sin \frac{1}{x})$ is an embedding which gives a compactification of $(0, 1]$ whose remainder is an interval. Describe the remainder of the compactifications of $(0, 1]$ obtained by taking the closure of the homeomorphic copy of $(0, 1]$ under the embeddings described below.

    (a) The embedding $g : (0, 1] \to \mathbb{R}^3$ defined by $g(x) = (x, \sin \frac{1}{x}, \sin \frac{1}{x})$.

    (b) The embedding $h : (0, 1] \to \mathbb{R}^3$ defined by $h(x) = (x, \sin \frac{1}{x}, \cos \frac{1}{x})$.

    (c) The embedding $s : (0, 1] \to \mathbb{R}^3$ defined in cylindrical coordinates by $s(t) = (r(t), \theta(t), z(t)) = (1 + t, \frac{\pi}{t}, \sin \frac{1}{t})$.

(d) The embedding $u : (0,1] \to \mathbb{R}^3$ defined in cylindrical coordinates by $u(t) = (r(t), \theta(t), z(t)) = (\sin\frac{1}{t}, \frac{\pi}{t}, t)$.

7. Give an embedding $f : [1,\infty) \to \mathbb{R}^3$ such that $\mathrm{cl}\, f([1,\infty))$ is a compactification of $[1,\infty)$ whose remainder is a sphere.

8. Let $X$ be the subspace $((-1,1)\times\{0\})\cup(\{0\}\times(-1,1))$ of $\mathbb{R}^2$ with the Euclidean topology.
   (a) Describe 1-point, 2-point, 3-point, and 4-point compactifications of $X$.
   (b) Prove or give a counterexample: If $aX$ and $yX$ are 2-point compactifications of $X$, then $aX$ is homeomorphic to $yX$.
   (c) Prove or give a counterexample: If $aX$ and $yX$ are 3-point compactifications of $X$, then $aX$ is homeomorphic to $yX$.

9. Describe a compactification of $[1,\infty)$ obtained by adding a torus.

10. Describe compactifications of $\mathbb{R}$ having the remainder described.
    (a) A closed interval and a point.
    (b) A closed interval.
    (c) A point and a circle.
    (d) Two intervals intersecting at their midpoints.

11. Characterize the functions $f \in C^*(\mathbb{R})$ which have a continuous extension to the one-point compactification of the Euclidean line $\mathbb{R}$.

12. Consider the compactification $Y$ of $X = (0,1]$ formed by taking the closure in $\mathbb{R}^2$ of the homeomorphic copy $a(X) = \{(x, \cos\frac{1}{x}) : x \in (0,1]\}$ of $(0,1]$. The remainder of this compactification is the interval $\{0\} \times [-1,1]$. Since $a(X)$ is dense in $Y$, every point of $Y$ is a limit of a sequence of points in $a(X)$. For example, the remainder point $p = (0,1) \in \{0\} \times [-1,1]$ is the limit of the sequence $(a_n)_{n\in\mathbb{N}} = ((\frac{1}{2n\pi}, 1))_{n\in\mathbb{N}}$ of local maxima of the curve $a(X) = \{(x, \cos\frac{1}{x}) : x \in (0,1]\}$. Find two other sequences $(b_n)_{n\in\mathbb{N}}$ and $(c_n)_{n\in\mathbb{N}}$ in $a(X)$ which also converge to the remainder point $p = (0,1)$, with the terms of $(a_n)_{n\in\mathbb{N}}$, $(b_n)_{n\in\mathbb{N}}$, and $(c_n)_{n\in\mathbb{N}}$ all mutually disjoint.

13. Suppose $aX$ is a compactification of $X$ and $a, b$ are distinct points in the remainder $aX - X$. Show that the quotient space formed by identifying $a$ and $b$ is a compactification of $X$. (This shows that if $X$ has a $n$-point compactification, then $X$ has $k$-point compactifications for $1 \le k \le n$.)

14. Suppose $a\mathbb{R} = a(\mathbb{R}) \cup \{a, b\}$ is a 2-point compactification of the real line $\mathbb{R}$ and $a$ is the limit point of the set $\{a(-n)\}_{n=1}^\infty$. Let $U_a$ and $U_b$ be disjoint open neighborhoods of $a$ and $b$ in $a\mathbb{R}$.
    (a) Show that there exist $m, n \in \mathbb{N}$ with $a((-\infty, -m)) \cup a((n, \infty)) \subseteq U_a \cup U_b$.
    (b) Show that $U_a$ intersects the connected set $a((-\infty, -m))$, so $U_b$ does not intersect $a((-\infty, -m))$, and thus $\{a\} \cup a((-\infty, -m)) \subseteq U_a$ and $\{b\} \cup a((n, \infty)) \subseteq U_b$.
    (c) Show $\{a\} \cup a((-\infty, -m))$ and $\{b\} \cup a((n, \infty))$ are open in $a\mathbb{R}$, so $a$ has a neighborhood base $\{\{a\} \cup a((-\infty, -m)) : m \in \mathbb{N}\}$ and $b$ has a neighborhood base $\{\{b\} \cup a((n, \infty)) : n \in \mathbb{N}\}$.
    (d) Show that for any two 2-point compactifications $a\mathbb{R}$ and $a'\mathbb{R}$ of $\mathbb{R}$ there exists a homeomorphism $h : a\mathbb{R} \to a'\mathbb{R}$ with $h \circ a = a'$.

15. Prove that $\mathbb{R}$ has no 3-point compactification. Does your proof generalize to $n$-point compactifications for $n > 2$ in $\mathbb{N}$? Does your proof generalize to compactifications with infinite remainders?

16. (a) Suppose that, for every $i \in I$, $\alpha_i X_i$ is a compactification of topological space $X_i$. Prove that $\prod_{i \in I} \alpha_i X_i$ is a compactification of $\prod_{i \in I} X_i$.
    (b) Use part (a) to show that the completely regular axiom is productive.

17. If $X$ is a completely regular space, show that $X$ is connected if and only if $\beta X$ is connected.

18. If $X$ and $Y$ are completely regular spaces and $f : X \rightarrow Y$ is continuous, show that there exists a continuous function $\hat{f} : \beta X \rightarrow \beta Y$ with $\hat{f}|_X = f$.

19. Given two compactification $\alpha X$ and $\gamma X$ of a completely regular space $X$, we say $\alpha X \geq \gamma X$ if and only if there exists a continuous function $f : \alpha X \rightarrow \gamma X$ with $\gamma = f \circ \alpha$. That is, $\alpha X \geq \gamma X$ if there is a continuous function from $\alpha X$ to $\gamma X$ which leaves $X$ fixed, if we consider $X = \alpha(X) \subseteq \alpha X$ and $X = \gamma(X) \subseteq \gamma X$.
    (a) Show that if $\alpha X \geq \gamma X$ and $\gamma X \geq \alpha X$, then there is a homeomorphism $h : \alpha X \rightarrow \gamma X$ with $h \circ \alpha = \gamma$. (In this case, we say $\alpha X$ and $\gamma X$ are *equivalent compactifications*.
    (b) Show that the Stone–Čech compactification $\beta X$ is the largest compactification of $X$, in the sense that $\beta X \geq \alpha X$ for every compactification $\alpha X$ of $X$.

20. With $\alpha, \gamma : \mathbb{N} \rightarrow \mathbb{R}^2$ defined by $\alpha(2n) = (1/n, -1)$, $\alpha(2n - 1) = (1/n, 1)$, $\gamma(3n) = (1/n, -1)$, $\gamma(3n - 1) = (1/(2n), 1)$, $\gamma(3n - 2) = (1/(2n - 1), 1)$ for all $n \in \mathbb{N}$, taking the closures in $\mathbb{R}^2$ of $\alpha(\mathbb{N})$ and $\gamma(\mathbb{N})$ gives compactifications $(\alpha\mathbb{N}, \alpha) = \{\{1/n\}_{n=1}^{\infty} \cup \{0\}\} \times \{-1, 1\} = (\gamma\mathbb{N}, \gamma)$. Show that $\alpha X$ and $\gamma X$ are homeomorphic, but not equivalent in the sense of Exercise 19(a).

21. Let $[a]_1 = a - \lfloor a \rfloor$ denote the fractional part of $a \in \mathbb{R}$. Complete the missing details below to prove that, for any irrational number $\gamma$, $\{[n\gamma]_1 : n \in \mathbb{N}\}$ is dense in $[0, 1]$. Given any basic open set $(x, y) \subseteq (0, 1]$, pick $j, k \in \mathbb{N}$ such that $(x, y)$ contains the interval $(\frac{j}{k}, \frac{j+1}{k})$.
    (a) Show that some interval $(\frac{m}{k}, \frac{m+1}{k}]$ (for $m = 0, \ldots, k - 1$) contains two points $[a\gamma]_1, [b\gamma]_1$ for integers $a < b$.
    (b) Show that $[(b - a)\gamma]_1 \in (0, \frac{1}{k}) \cup (\frac{k-1}{k}, 1)$. (That is, modulo 1, $|[(b - a)\gamma]_1| \leq \frac{1}{k}$.)
    (c) Show that, for some natural number $m$, $[m(b-a)\gamma]_1 \in (x, y)$. Thus, every open interval $(x, y)$ contains a point of $\{[n\gamma]_1 : n \in \mathbb{N}\}$.

## 7.4 Proof of the Tychonoff theorem

In mathematics, one must start with certain axioms or postulates. Euclid based his geometry on five postulates. Most mathematics courses start with the assumption that we know what the real numbers are, but some analysis courses may start with a formal construction of the reals as a completion of the rationals. One much-studied axiom in mathematics is the *axiom of choice*, which states that, for any collection of nonempty sets, it is possible to form a new set consisting of one element from each member of the collection. This axiom was proposed by Ernst Zermelo in 1904. While it seems rather innocuous and believable, it produces some rather unbelievable results. One consequence of the axiom of choice appeared in a paper by Stefan Banach and Alfred Tarski in 1924, and is now called the Banach–Tarski paradox [2]. It states that a solid ball in $\mathbb{R}^3$ of radius 1 can be decomposed into a finite number of subsets which can be rearranged through rigid motions (that is, translations, rotations, and reflections, but no compressions or expansions) and reassembled into two disjoint solid balls of radius 1. This is certainly less believable than the statement of the axiom of choice, and has caused some mathematicians to shun the axiom of choice and its applications. We will freely assume the axiom of choice and its consequences.

We will start with an assumption equivalent to the axiom of choice, given by Max Zorn (1906–1993) in 1935, and now known as Zorn's lemma.

**Theorem 7.4.1** (Zorn's lemma). *Suppose $(X, \leq)$ is a nonempty partially ordered set and every totally ordered subset of $X$ has an upper bound in $X$. Then $X$ has a maximal element.*

Zorn's lemma is employed in the proof of the Alexander subbase theorem, proved by James Waddell Alexander II (1888–1971). Alexander is also known for the discovery of the first "knot polynomial", which now bears his name, in 1923.

**Theorem 7.4.2** (Alexander subbase theorem). *Let $X$ be a topological space with subbasis $S$. Then $X$ is compact if and only if every open cover of $X$ by subbasic sets from $S$ has a finite subcover.*

*Proof.* If $X$ is compact, then every cover by arbitrary open sets has a finite subcover, so every cover by subbasic open sets has a finite subcover.

For the converse, we argue by contradiction. Suppose every open cover of $X$ by subbasic sets from $S$ has a finite subcover but $X$ is not compact. Let $\Phi$ be the collection of open covers of $X$ which do not have a finite subcover, partially ordered by set inclusion. If $\{C_i\}_{i \in I}$ is a totally ordered subset of $\Phi$, then, for every $j \in I$, $C_j \subseteq C = \bigcup\{C_i : i \in I\}$, so $C$ is an upper bound of $\Phi$. If $C \in \Phi$, then Zorn's lemma would apply. Now $C$ is clearly an open cover of $X$. If $C$ has a finite subcover $U_1, \ldots, U_n \in C = \bigcup\{C_i : i \in I\}$, then each $U_k$ is contained in some $C_{i_k}$ ($k = 1, \ldots, n$), and because $C_{i_1}, C_{i_2}, \ldots C_{i_n}$ are totally ordered by set inclusion, their union (or maximum) is $C_m$ for some $m \in \{i_1, \ldots, i_n\}$. Now $U_1, \ldots, U_n \in C_m$, contrary to $C_m$ having no finite subcover. Thus, $C \in \Phi$, and Zorn's lemma implies that $\Phi$ has a maximal element $\mathcal{D}$.

Now $\mathcal{D}$ is an open cover with no finite subcover. If $\mathcal{D} \cap S$ covered $X$, as a cover of $X$ by subbasic elements, it would have a finite subcover, contrary to $\mathcal{D}$ having no finite subcover. Thus, $\mathcal{D} \cap S$ does not cover $X$, so there exists $x \in X - \bigcup(\mathcal{D} \cap S)$. Now there exists $U \in \mathcal{D}$ with $x \in U$, and since $S$ is a subbasis, there exist $S_1, \ldots, S_n \in S$ with $x \in S_1 \cap \cdots \cap S_n \subseteq U$. Since $x$ was not covered by $\mathcal{D} \cap S$, $S_i \notin \mathcal{D}$ for any $i = 1, \ldots, n$. For any $j \in \{1, \ldots, n\}$, $\mathcal{D} \cup \{S_j\}$ is an open cover of $X$, and by the maximality of $\mathcal{D}$ in $\Phi$,

$\mathcal{D} \cup \{S_j\}$ has a finite subcover $\mathcal{F}_j \cup \{S_j\}$. Now $\bigcup \{\mathcal{F}_i\}_{i=1}^n \cup \{S_1 \cap \cdots \cap S_n\}$ is a finite open cover of $X$. Since $S_1 \cap \cdots \cap S_n \subseteq U$ and $U \in \mathcal{D}$, $\bigcup \{\mathcal{F}_i\}_{i=1}^n \cup \{U\}$ is a finite subcover of $X$ from $\mathcal{D}$, contrary to $\mathcal{D} \in \Phi$. $\qquad \square$

Now we present the Tychonoff theorem, which was already stated as Theorem 6.2.5.

**Theorem 7.4.3** (The Tychonoff theorem). *A product $\prod_{i \in I} X_i$ of topological spaces is compact if and only if every factor $X_i$ is compact.*

*Proof.* If $X = \prod_{i \in I} X_i$ is compact, since each projection function $\pi_i : X \to X_i$ is continuous and onto, the image $\pi_i(X) = X_i$ is compact for every $i \in I$.

Conversely, suppose $(X_i, \mathcal{T}_i)$ is compact for every $i \in I$, and let $X = \prod_{i \in I} X_i$. Now $\mathcal{S} = \{\pi_i^{-1}(U) : i \in I, U \in \mathcal{T}_i\}$ is a subbasis for $X$. Suppose $\mathcal{C}$ is an open cover of $X$ by subbasic sets from $\mathcal{S}$. For each $i \in I$, let $\mathcal{C}_i = \{U \in \mathcal{T}_i : \pi_i^{-1}(U) \in \mathcal{C}\}$. Next, we show that there must exist $i_0 \in I$ such that $\mathcal{C}_{i_0}$ covers $X_{i_0}$. If not, then, for each $i \in I$, there exists $a_i \in X_i$ not covered by $\mathcal{C}_i$, and $\langle a_i \rangle_{i \in I} \in X$ is not covered by $\mathcal{C} = \bigcup \{\mathcal{C}_i : i \in I\}$, contrary to $\mathcal{C}$ being a cover of $X$. Now if $\mathcal{C}_{i_0}$ covers $X_{i_0}$, then $X_{i_0}$ is covered by a finite subcover $\{U_1, \ldots, U_n\} \subseteq \mathcal{C}_{i_0}$. The collection $\{\pi_{i_0}^{-1}(U_1), \ldots, \pi_{i_0}^{-1}(U_n)\} \subseteq \mathcal{C}_{i_0} \subseteq \mathcal{C}$ is a finite subcover of $X$ from $\mathcal{C}$. Thus, any cover of $X$ by subbasic open sets has a finite subcover, so by the Alexander subbase theorem, $X$ is compact. $\qquad \square$

We have thus proved that the Tychonoff theorem follows from the axiom of choice (in the form of Zorn's Lemma). In 1950, John Kelley [25] proved that the axiom of choice follows from the Tychonoff theorem, so the Tychonoff theorem is equivalent to the axiom of choice.

## Exercises

1. Identify the exact point where the axiom of choice is used in the proof of
   (a) Theorem 0.5.1(c).
   (b) Exercise 19(a) of Chapter 0.
   (c) Theorem 3.1.7.
   (d) Theorem 1.5.11.
   (e) Theorem 5.1.4.
   (f) Theorem 5.1.8.
2. In the Euclidean line, $(0, 1)$ is not compact. Find an open cover by sets from the standard subbasis $\mathcal{S} = \{(-\infty, b) : b \in \mathbb{R}\} \cup \{(a, \infty) : a \in \mathbb{R}\}$ which has no finite subcover.
3. (a) Show that $\mathcal{S} = \{(a, \infty) \times (b, \infty) : a, b \in \mathbb{R}\} \cup \{(-\infty, a) \times (-\infty, b) : a, b \in \mathbb{R}\}$ is a subbasis for the Euclidean topology on $\mathbb{R}^2$.
   (b) Find an open cover of the noncompact set $(0, 1) \times (0, 1)$ by elements of $\mathcal{S}$ which has no finite subcover.

(c) Find an open cover of the noncompact set $\{(x,y) \in \mathbb{R}^2 : x^2+y^2 < 1\}$ by elements of $S$ which has no finite subcover.

4. (a) For $a, b \in \mathbb{R}$, let $S(a, b)$ be the strip $(a, b) \times (-\infty, \infty)$ in $\mathbb{R}^2$ and let $D(b, \varepsilon)$ be the diagonal strip $\{(x,y) \in \mathbb{R}^2 : x + b < y < x + b + \varepsilon\}$. Show that $S = \{S(a, b) : a, b \in \mathbb{R}, a < b\} \cup \{D(b, \varepsilon) : b \in \mathbb{R}, \varepsilon > 0\}$ is a subbasis for the Euclidean topology on $\mathbb{R}^2$.

(b) Find an open cover of the noncompact set $[0, 1] \times (0, 1)$ by elements of $S$ which has no finite subcover.

5. Show that $A = \{f : [-1, 1] \to [-1, 1] : |f(x)| \leq |x|\}$ is a compact subset of $[-1, 1]^{[-1,1]}$ with the product topology.

6. By Exercise 25 of Section 6.2, $[0, 1]^{\mathbb{N}}$ is metrizable, so by Theorem 5.2.8, it is compact if and only if every sequence in $[0, 1]^{\mathbb{N}}$ has a convergent subsequence. Prove that every sequence in $[0, 1]^{\mathbb{N}}$ has a convergent subsequence. Note that this shows $[0, 1]^{\mathbb{N}}$ is compact without using the Tychonoff theorem.

# 8 Alexandroff spaces

Most of the material covered to this point was motivated by classical analysis, and was studied extensively in the first half of the 20th century. The fundamental concepts driving the development of the material were nearness, convergence, and continuity. In Chapters 8–11, we present some more recent material. The two fundamental concepts driving the development of these topics might be summarized as *loss of resolution* arising in computer applications and *topology and order*.

The decimal expansion of $\pi$ has an infinite number of digits, but computer arithmetic with $\pi$ necessarily replaces $\pi$ by a finite (rational) representation. The Euclidean plane contains infinitely many points, but any computer screen representation of the plane must be done with only a finite number of pixels. Thus, convergence, nearness, and continuity on a computer screen must be characterized by topologies on the finite set of pixels. The only $T_1$ topology on a finite set is the discrete topology, in which the only convergent sequences are those which are eventually constant, and since each singleton set $\{x\}$ is open, nothing is "near" $x$ other than $x$ itself. Thus, meaningful convergence on finite topological spaces can only be accomplished by non-$T_1$ (and thus, non-Hausdorff) topologies.

$T_1$ topological spaces have a certain symmetry to them: If there is a neighborhood of $x$ excluding $y$, then there is a neighborhood of $y$ excluding $x$. The study of non-$T_1$ spaces is thus part of *asymmetric topology*. Perhaps the most obvious example of asymmetric topology is the study of quasi-metrics (Section 11.2), which are distance functions $q$ which do not require that $q(x,y) = q(y,x)$. In a more general setting, quasi-uniformities (Section 12.5) continue the study of asymmetric topology.

The study of spaces which are not Hausdorff has its drawbacks, such an non-uniqueness of limits, but such problems naturally arise in the loss of resolution situations where we must represent an infinite set by a finite set. For example, two distinct points in the Euclidean plane determine a unique line, but two distinct pixels on a screen may not determine a unique line, as suggested in Figure 8.1. Section 8.1 investigates a class of topological spaces which includes all topologies on finite sets.

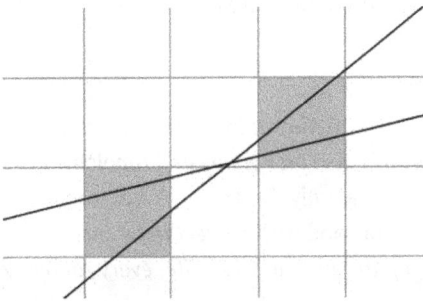

**Figure 8.1:** Two distinct pixels do not determine a unique line.

https://doi.org/10.1515/9783110686579-009

When using a finite set of pixels to represent a rectangular subset $R$ of the Euclidean plane, each pixel corresponds to infinitely many points in the rectangle. This introduces an equivalence relation on the rectangle $R$: two points of $R$ are equivalent if and only if they are represented by the same pixel. Now any work on the set of pixels is actually work on equivalence classes of $R$. In Section 8.2, we investigate partial orders on equivalence classes on a set $X$, or equivalently, on the blocks of a partition of $X$. In Section 11.1, we investigate metrics on blocks of a partition (i. e., equivalence classes) of $X$.

The study of topologies on finite sets can be approached nicely with techniques from discrete mathematics, especially including the theory of ordered sets. The rich interaction between topology and order is investigated in the next few chapters.

## 8.1 Alexandroff topologies

Given a set $X$ and a point $p \in X$, recall that the particular point topology $P_p = \{\emptyset\} \cup \{A \subseteq X : p \in A\}$ on $X$ determined by $p$ consists of the empty set and all subsets of $X$ which contain the particular point $p$. The closed sets in this topology are thus $X$ and the subsets which exclude $p$. That is, the closed sets in $P_p$ are precisely the open sets in the excluded point topology $E_p = \{X\} \cup \{B \subseteq X : p \notin B\}$ determined by $p$. Thus, $P_p$ (and also $E_p$) is a topology whose collection of closed sets forms a topology. Such a topology is called an *Alexandroff topology*.

By De Morgan's laws, the closed sets of any topology on $X$ include $\emptyset$ and $X$ are closed under arbitrary (and thus finite) intersections. The only remaining condition needed to ensure that the closed sets form a topology is that arbitrary unions of closed sets must be closed, or equivalently, arbitrary intersections of open sets must be open.

**Definition 8.1.1.** An *Alexandroff topology* is a topology in which arbitrary intersections of open sets are open.

Alexandroff topologies are sometimes called *principal topologies*. When P. S. Alexandroff introduced these spaces in 1937, he called them "Diskrete Räume", or "discrete spaces".

The following theorem lists some important and immediate properties of Alexandroff spaces.

**Theorem 8.1.2.** *Let $X$ be a set.*
(a) *Alexandroff topologies on $X$ occur in pairs: If $T$ is an Alexandroff topology on $X$, then the collection $\{X - U : U \in T\}$ of $T$-closed sets is also an Alexandroff topology on $X$.*
(b) *A topology $T$ on $X$ is an Alexandroff topology if and only if every $x \in X$ has a smallest neighborhood $N(x)$ (that is, a minimal neighborhood with respect to set inclusion).*
(c) *Every finite topology on $X$ is Alexandroff. In particular, if $X$ is finite, every topology on $X$ is Alexandroff.*
(d) *The only $T_1$ Alexandroff topology on $X$ is the discrete topology.*

*Proof.* (a) follows immediately from De Morgan's laws.

(b) If $(X, \mathcal{T})$ is an Alexandroff topological space and $x \in X$, then the intersection of all neighborhoods of $x$ must be a neighborhood $N(x)$ of $x$, and clearly it is contained in every neighborhood of $x$. Conversely, suppose every point $x$ in $(X, \mathcal{T})$ has a smallest neighborhood $N(x)$. If $U_i \in \mathcal{T}$ for every $i \in I$, we wish to show $B = \bigcap_{i \in I} U_i$ is open. If $B = \emptyset$, then $B$ is open. If $x \in B$, then $x \in U_i$ for each $i \in I$, and since each $U_i$ is open, we have $\{x\} \subseteq N(x) \subseteq \bigcap_{i \in I} U_i = B$. Taking the union over all $x \in B$ gives $B \subseteq \bigcup_{x \in B} N(x) \subseteq B$, so $B$ equals a union of open sets, and thus is open.

(c) If $\mathcal{T}$ is a finite topology, then there are only finitely many open sets, so "arbitrary" intersections of open sets are actually finite intersections, which must be open.

(d) If $\mathcal{T}$ is a $T_1$ Alexandroff topology on $X$, then, for $y \neq x$, there exists a neighborhood of $x$ excluding $y$, so the intersection $N(x)$ of all neighborhoods of $x$ excludes $y$. This holds for every $y \neq x$, so $N(x) = \{x\}$ is open for every $x \in X$ and thus $\mathcal{T}$ is discrete. $\qquad\square$

We consider some examples of Alexandroff topologies.

**Example 8.1.3.** If $\mathcal{T}$ is a partition topology on $X$, that is, if $\mathcal{T}$ has a basis $\mathcal{B} = \{B_i\}_{i \in I}$ which is a partition of $X$, then $\mathcal{T}$ is an Alexandroff topology. The minimal neighborhood of $x \in X$ is the block $B_i$ of the partition which contains $x$. For any $j \in I$, $B_j = X - \bigcup\{B_i \in \mathcal{B} : B_i \neq B_j\}$, so each block $B_j$ is closed. It follows that the topology of $\mathcal{T}$-closed sets is again $\mathcal{T}$. Since the discrete topology is a partition topology, the discrete topology is an Alexandroff topology.

**Example 8.1.4.** The topology $\mathcal{T}$ on $\mathbb{R}$ generated by the basis $\mathcal{B} = \{[-a, a] \subseteq \mathbb{R} : a \geq 0\}$ is an Alexandroff topology. Note that this topology contains the open intervals $(-a, a)$ as well as the closed intervals $[-a, a]$. We can see directly that arbitrary intersections of open or closed intervals centered at the origin are again such intervals, or we can see $\mathcal{T}$ is Alexandroff by noting that every $x \in \mathbb{R}$ has a minimal neighborhood $[-a, a]$ where $a = |x|$. The Alexandroff topology of $\mathcal{T}$-closed sets is $\{\emptyset\} \cup \{(-\infty, -a) \cup (a, \infty) : a \geq 0\} \cup \{(-\infty, -a] \cup [a, \infty) : a \geq 0\}$.

As noted above, the particular point topology on $X$ determined by a point $p$ is an Alexandroff topology. It consists of $\emptyset$ and all the supersets of $\{p\}$. Its associated topology of closed sets, the excluded point topology, consists of $X$ and all sets disjoint from $\{p\}$. Replacing the singleton set $\{p\}$ in the particular point topology by an arbitrary set provides the foundation for other Alexandroff topologies, which we define below.

**Definition 8.1.5.** Suppose $X$ is a set and $S$ is a fixed subset of $X$. The collections

$$\mathrm{Super}(S) = \{U \subseteq X : S \subseteq U\} \cup \{\emptyset\},$$
$$\mathrm{Disjoint}(S) = \{U \subseteq X : U \cap S = \emptyset\} \cup \{X\}, \quad \text{and}$$
$$\mathrm{Sub}(S) = \{U \subseteq X : U \subseteq S\} \cup \{X\}$$

are Alexandroff topologies on $X$, respectively known as the *topology of supersets of S*, the *topology of sets disjoint from S*, and the *topology of subsets of S*.

It is easy to see directly that each of these collections is an Alexandroff topology and that they are closely related.

Since $U$ is a subset of $S$ if and only if $U$ is disjoint from $X - S$, we have $\mathrm{Sub}(S) = \mathrm{Disjoint}(X - S)$.

For any set $S$, the Alexandroff topology of the $\mathrm{Super}(S)$-closed sets is $\mathrm{Disjoint}(S)$, since $X - U$ is a superset of $S$ if and only if $S$ is disjoint from $U$. The nonempty $\mathrm{Sub}(S)$-closed sets are sets $F$ with $X - F \subseteq S$, or equivalently, with $X - S \subseteq F$. Thus, the Alexandroff topology of $\mathrm{Sub}(S)$-closed sets is $\mathrm{Super}(X - S)$.

In any Alexandroff topology on $X$, the collection $\{N(x) : x \in X\}$ of minimal neighborhoods of the points is a basis for the topology. In $\mathrm{Super}(S)$, the minimal neighborhood of $x$ is $N(x) = \{x\} \cup S$. In $\mathrm{Sub}(S)$, if $x \in S$, then $\{x\}$ is a subset of $S$, so $N(x) = \{x\}$; if $x \notin S$, then there is no subset of $S$ containing $x$, so $N(x) = X$. Loosely speaking, this says that $\mathrm{Sub}(S)$ is discrete on $S$ and indiscrete on $X - S$. The minimal neighborhoods in $\mathrm{Disjoint}(S) = \mathrm{Sub}(X - S)$ are determined similarly.

Table 8.1 shows the associated topology of closed sets and the minimal neighborhoods of points in these topologies.

**Table 8.1:** $\mathrm{Super}(S)$, $\mathrm{Disjoint}(S)$, and $\mathrm{Sub}(S)$.

| $\mathcal{T}$ | $\mathcal{T}$-closed sets |
| --- | --- |
| $\mathrm{Super}(S)$ | $\mathrm{Disjoint}(S)$ |
| $\mathrm{Disjoint}(S)$ | $\mathrm{Super}(S)$ |
| $\mathrm{Sub}(S)$ | $\mathrm{Super}(X - S)$ |

$$\mathrm{Disjoint}(S) = \mathrm{Sub}(X - S)$$

| $\mathcal{T}$ | minimal neighborhoods $N(x)$ |
| --- | --- |
| $\mathrm{Super}(S)$ | $N(x) = \{x\} \cup S$ |
| $\mathrm{Disjoint}(S)$ | $N(x) = \begin{cases} X & \text{if } x \in S \\ \{x\} & \text{if } x \notin S \end{cases}$ |
| $\mathrm{Sub}(S)$ | $N(x) = \begin{cases} \{x\} & \text{if } x \in S \\ X & \text{if } x \notin S \end{cases}$ |

**Example 8.1.6.** Consider the Alexandroff topological space $(\mathbb{R}, \mathrm{Super}((-1,1)))$. Since $(-1,1)$ and $[-1,3] \cup \{\pi, 4\}$ are supersets of $(-1,1)$, they are open sets in this space. Since the $\mathrm{Super}((-1,1))$-closed sets are the members of $\mathrm{Disjoint}((-1,1))$, the closure $\mathrm{cl}((0,2))$ is the smallest set containing $(0,2)$ which is either disjoint from $(-1,1)$ or equal to $\mathbb{R}$. No set containing $(0,2)$ is disjoint from $(-1,1)$, so $\mathrm{cl}((0,2)) = \mathbb{R}$. Since this is an Alexandroff topology, every point has a minimal neighborhood $N(x)$. If $x \notin (-1,1)$, the smallest superset of $(-1,1)$ containing $x$ is $N(x) = (-1,1) \cup \{x\}$ and if $x \in (-1,1)$, then $N(x) = (-1,1)$.

The sequence $(2 + \frac{1}{n})_{n=1}^{\infty}$ is not eventually in the minimal neighborhood $N(x) = (-1, 1)$ or $(-1, 1) \cup \{x\}$ of any $x \in \mathbb{R}$, so this sequence does not converge. The sequence $(\frac{1}{n})_{n=1}^{\infty}$ is eventually in $(-1, 1)$, and therefore is eventually in every open set in $\text{Super}((-1, 1))$. Thus, $(\frac{1}{n})_{n=1}^{\infty}$ converges to every point $x \in \mathbb{R}$. With non-unique limits, this topological space is not Hausdorff. Furthermore, for the distinct points $x = 0$ and $y = \frac{1}{2}$, neither has a neighborhood which excludes the other, so this space is not even $T_0$. The space is connected since any two nonempty open sets contain $(-1, 1)$ and cannot be disjoint. The open cover $\mathcal{C} = \{N(x) : x \notin (-1, 1)\}$ of $\mathbb{R}$ has no finite subcover, so the space is not compact.

The next example is a subtle extension of the $\text{Sub}(S)$ topology.

**Example 8.1.7.** Let $Z = \mathbb{R}^{\mathbb{R}}$ be the collection of all functions $\{f : \mathbb{R} \to \mathbb{R}\}$. Define a topology $\mathcal{T}$ on $Z$ by saying a set $U \subseteq Z$ of functions is open if and only if $U = Z$ or for each $f \in U$, the graph of $f$ is contained in the closed upper half-plane. That is, $U \ne Z$ is open if and only if

$$f \in U \Rightarrow G_f = \{(x, f(x)) \in \mathbb{R}^2 : x \in \mathbb{R}\} \subseteq \mathbb{R} \times [0, \infty).$$

Clearly $Z \in \mathcal{T}$, and $\emptyset \in \mathcal{T}$ vacuously. Suppose $U_i \in \mathcal{T}$ for $i \in I$ where $I$ is an arbitrary index set. Now $\bigcup_{i \in I} U_i$ is a union of sets $U_i$ of functions whose graphs are in the upper half-plane, and this union will only contain functions whose graphs are in the upper half-plane, so the union is in $\mathcal{T}$. Similarly, $\bigcap_{i \in I} U_i$ is either empty, $Z$, or a set of functions whose graphs are in the upper half-plane, so $\mathcal{T}$ is closed under arbitrary intersections. Thus, $\mathcal{T}$ is an Alexandroff topology.

The set $U = \{f_n(x) = 3 + n + \sin nx : n \in \mathbb{N}\}$ is an open set in $\mathcal{T}$. The set $A = \{f(x) = -x^2 - 2, g(x) = 4 + x^2\}$ is not open since the graphs of the functions of $A$ are not all contained in the upper half-plane. Also, since $Z - A$ is a proper subset of $Z$ which contains functions like $h(x) = x$ whose graph is not in the upper half-plane, $Z - A$ is not open, so $A$ is not closed.

Suppose $(f_n)_{n=1}^{\infty}$ is a sequence of functions in $Z = \mathbb{R}^{\mathbb{R}}$. If $g$ is a function whose graph is not contained in the closed upper half-plane, then the sequence is eventually in the minimal neighborhood $N(g) = Z$, so every such $g$ is a limit of the sequence. If $g$ is a function whose graph is contained in the closed upper half-plane, then the sequence is eventually in $N(g) = \{g\}$ only if the sequence is eventually constantly $f_n = g$. Thus, the convergence of sequences are completely characterized by saying all eventually constant sequences converge to the constant term and every sequence converges to every function $g$ whose graph is not contained in the closed upper half-plane.

Another nice class of Alexandroff topologies are the *functionally Alexandroff topologies* which were explicitly introduced independently in 2011 as functional Alexandroff spaces [1] and in 2012 as *primal spaces* [14].

If $X$ is a nonempty set and $f : X \to X$ is a function, let

$$\mathcal{T}_f = \{U \subseteq X : f^{-1}(U) \subseteq U\}.$$

Since $f^{-1}(\emptyset) \subseteq \emptyset$ and $f^{-1}(X) \subseteq X$, we have $\emptyset, X \in \mathcal{T}_f$. If $U_i \in \mathcal{T}_f$ for all $i$ in an arbitrary index set $I$, then

$$f^{-1}\left(\bigcup_{i \in I} U_i\right) = \bigcup_{i \in I} f^{-1}(U_i) \subseteq \bigcup_{i \in I} U_i,$$

so $\bigcup_{i \in I} U_i \in \mathcal{T}_f$. Replacing each $\bigcup$ by $\bigcap$, these statements remain valid, so $\bigcap_{i \in I} U_i \in \mathcal{T}_f$. Thus, $\mathcal{T}_f$ is an Alexandroff topology on $X$.

**Definition 8.1.8.** A topology on $X$ which arises as $\mathcal{T}_f = \{U \subseteq X : f^{-1}(U) \subseteq U\}$ for some function $f : X \to X$ is a *functionally Alexandroff topology* (or a *primal topology*).

The following results about functionally Alexandroff spaces are routine to verify, and are left to the exercises. In what follows, $f^2(x) = f(f(x))$ and for $n \geq 2$, $f^{n+1}(x) = f(f^n(x))$.

**Theorem 8.1.9.** *Suppose $\mathcal{T}_f$ is the functionally Alexandroff topology on $X$ arising from $f : X \to X$.*
(a) *$C$ is $\mathcal{T}_f$-closed if and only if $f(C) \subseteq C$. That is, the $\mathcal{T}_f$-closed sets are the $f$-invariant subsets of $X$.*
(b) *For $x \in X$, $\mathrm{cl}\{x\} = \{f^n(x)\}_{n=0}^{\infty}$. That is, $\mathrm{cl}\{x\}$ is the orbit $\mathcal{O}(x) = \{f^n(x)\}_{n=0}^{\infty}$ of $x$ under $f$.*
(c) *For $x \in X$, the smallest neighborhood of $x$ is $N_f(x) = \{y \in X : f^n(y) = x$ for some $n \in \{0, 1, 2, \ldots\}\}$.*

For example, suppose $f : \mathbb{N} \to \mathbb{N}$ is the function defined by taking $f(n)$ to be the sum of the digits of $n$. Now $f(42613) = 4 + 2 + 6 + 1 + 3 = 16$, $f^2(42613) = f(16) = 7$, and for $n \geq 3$, $f^n(42613) = 7$. Thus, $\mathrm{cl}\{42613\} = \{42613, 16, 7\}$. It is well-known that the remainder when $n$ is divided by 9 is the same as the remainder when $f(n)$ is divided by 9. Thus, $N_f(4) = \{m \in \mathbb{N} : m \equiv 4 \bmod 9\}$.

## Exercises

1. Every point $x$ in an Alexandroff topological space $X$ has a minimal neighborhood $N(x) = \bigcap \{U : U$ is a neighborhood of $x\}$. If $(X, \mathcal{T})$ is an Alexandroff space, show that $\mathcal{B} = \{N(x) : x \in X\}$ is a basis for $\mathcal{T}$.
2. If $\mathcal{T}$ is an Alexandroff topology on $X$, let $\mathcal{T}^*$ denote the Alexandroff topology of $\mathcal{T}$-closed sets. For $x \in X$, describe the minimal $\mathcal{T}^*$-neighborhood $N^*(x)$ of $x$ in terms of $\mathcal{T}$.
3. Show that a collection $\mathcal{B}$ of subsets of $X$ is a basis for an Alexandroff topology on $X$ if and only if $\bigcup \mathcal{B} = X$ and whenever $x$ is in the intersection $\bigcap\{B_i : i \in I\}$ of an arbitrary collection of sets $B_i \in \mathcal{B}$, there exists $B \in \mathcal{B}$ with $x \in B \subseteq \bigcap\{B_i : i \in I\}$.
4. Show that the digital line topology $\mathcal{T}_{\mathrm{DL}}$ on $\mathbb{Z}$ is an Alexandroff topology and describe the topology of $\mathcal{T}_{\mathrm{DL}}$-closed sets.

5. Describe the topologies Super($S$), Disjoint($S$), and Sub($S$) on $X$ (a) if $S = \emptyset$ and (b) if $S = X$.

6. Let $X = [0,6] \times [0,2]$ and $S = [0,3] \times [0,2]$. Give $X$ the topology Super($S$). Find the interior, closure, and boundary of the following sets.

$$A = [0,1] \times [0,1] \quad B = [2,4] \times [0,1] \quad C = [5,6] \times [0,1]$$
$$D = [0,5] \times [0,2] \quad E = [1,6] \times [0,2]$$

7. Repeat Exercise 6 using the topology Disjoint($S$) instead of Super($S$).

8. Repeat Exercise 6 using the topology Sub($S$) instead of Super($S$).

9. Suppose $X$ is a set and $S$ is a fixed subset of $X$ which is nonempty and proper. Give $X$ the topology Disjoint($S$). For an arbitrary nonempty subset $A \subseteq X$, describe int $A$, cl $A$, and $\partial A$.

10. Suppose $X$ is a set and $S$ is a fixed subset of $X$ which is nonempty and proper. Give $X$ the topology Sub($S$). For an arbitrary nonempty subset $A \subseteq X$, describe int $A$, cl $A$, and $\partial A$.

11. Suppose $X$ is a set and $S$ is a fixed subset of $X$ which is nonempty and proper. Give $X$ the topology Super($S$). For an arbitrary subset $A \subseteq X$, give rules to determine int $A$ and cl $A$.

12. Suppose $X$ is a set containing distinct points $p$ and $q$. Let $\mathcal{T}_{\{p,q\}} = $ Super($\{p\}$) $\cup$ Disjoint($\{p,q\}$). Thus, $U \in \mathcal{T}_{\{p,q\}}$ if and only if $U$ contains $p$ or excludes both $p$ and $q$. Show that $\mathcal{T}_{\{p,q\}}$ is an Alexandroff topology on $X$ and describe the topology of $\mathcal{T}_{\{p,q\}}$-closed sets. (In Section 9.3, we will see that the only topology on $X$ which is strictly finer than $\mathcal{T}_{\{p,q\}}$ is the discrete topology.)

13. Two topologies $\mathcal{T}_1$ and $\mathcal{T}_2$ on the same set $X$ are said to be *(lattice) complements* if the only topology finer than both $\mathcal{T}_1$ and $\mathcal{T}_2$ is the discrete topology and the only topology coarser than both $\mathcal{T}_1$ and $\mathcal{T}_2$ is the indiscrete topology.
    (a) Show that topologies $\mathcal{T}_1$ and $\mathcal{T}_2$ on $X$ are complements if and only if (i) for every $x \in X$ there exists a $\mathcal{T}_1$-neighborhood $U$ of $x$ and a $\mathcal{T}_2$-neighborhood $V$ of $x$ with $U \cap V = \{x\}$, and (ii) the only sets which are open in both $\mathcal{T}_1$ and $\mathcal{T}_2$ are $\emptyset$ and $X$.
    (b) If $S \subseteq X$, show that Sub($S$) and Disjoint($S$) are complements.

14. Suppose $S$ is a nonempty proper subset of $X$. Discuss the connectedness of $X$ under the topologies Super($S$), Disjoint($S$), and Sub($S$). What if $S$ is allowed to be $\emptyset$ or $X$?

15. Suppose $S$ is a nonempty proper subset of $X$. Discuss the compactness of $X$ under the topologies Super($S$), Disjoint($S$), and Sub($S$). What if $S$ is allowed to be $\emptyset$ or $X$?

16. Let $\mathcal{T} = \{U \subseteq \mathbb{R}^3 : \pi_1(U) \subseteq [0,1]\} \cup \{\mathbb{R}^3\}$, where $\pi_1$ is the projection function onto the first coordinate. Show that $\mathcal{T}$ is an Alexandroff topology on $X = \mathbb{R}^3$.

17. Suppose $(X, \mathcal{T})$ is an Alexandroff topology and $Y \subseteq X$. If $\mathcal{T}_Y$ is the subspace topology on $Y$, must $(Y, \mathcal{T}_Y)$ be an Alexandroff space?

18. Suppose $(X, \mathcal{T})$ is an Alexandroff space, $q : X \to Y$ is a surjection, and $\mathcal{T}' = \{V \subseteq Y : q^{-1}(V) \in \mathcal{T}\}$ is the quotient topology on $Y$. Is the quotient space $(Y, \mathcal{T}')$ necessarily an Alexandroff space?

19. Prove Theorem 8.1.9.

20. If $\mathcal{T}_f$ is the functionally Alexandroff topology arising from $f : X \rightarrow X$, show that $f : (X, \mathcal{T}_f) \rightarrow (X, \mathcal{T}_f)$ is a continuous closed mapping.

21. (a) How are the fixed points of $f : X \rightarrow X$ recognizable in the functionally Alexandroff topology $\mathcal{T}_f$?
    (b) Characterize the functions $f : X \rightarrow X$ for which $\mathcal{T}_f$ is $T_1$.

22. Show that the indiscrete topology on $X$ is functionally Alexandroff if and only if $X$ is finite.

23. Suppose $\mathcal{T}_f$ is the functionally Alexandroff topology arising from $f : X \rightarrow X$. For $x, y \in X$, define $x \approx y$ if and only if there exist $m, n \in \mathbb{N} \cup \{0\}$ with $f^m(x) = f^n(y)$. Show that $\approx$ is an equivalence relation on $X$ and that the $\approx$-equivalence classes are the connected components of $(X, \mathcal{T}_f)$.

24. Show that the Alexandroff topology depicted is not functionally Alexandroff.

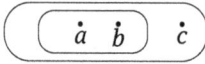

## 8.2 Quasiorders

In this section, we momentarily abandon topology to introduce an elementary topic from discrete mathematics: quasiorders. Recall that a partial order on a set $X$ is a relation $\leq$ on $X$ which is reflexive, transitive, and antisymmetric, and an equivalence relation on $X$ is a relation $\sim$ which is reflexive, transitive, and symmetric. An equivalence relation $\sim$ on $X$ partitions $X$ into equivalence classes $[x] = \{y \in X : x \sim y\}$, and every partition $\mathcal{P} = \{B_i : i \in I\}$ of $X$ into blocks $B_i$ ($i \in I$) gives an equivalence relation defined by $x \sim y$ if and only if $x$ and $y$ fall in the same block $B_i$ of $\mathcal{P}$. Equivalence relations are, in effect, equality of some attribute of the elements of $X$. If a farmer sells pumpkins by the pound, his accountant may say that pumpkin $a$ is equivalent to pumpkin $b$ if and only if their weights are equal.

Every partial order and every equivalence relation is a reflexive, transitive relation. Such relations are called *quasiorders*.

**Definition 8.2.1.** A *quasiorder* on a set $X$ is a reflexive, transitive relation $\leq$ on $X$. That is, a relation $\leq$ on set $X$ is a quasiorder if $x \leq x$ for every $x \in X$, and for every $x, y, z \in X$, if $x \leq y$ and $y \leq z$, then $x \leq z$.

Quasiorders are like partial orders without antisymmetry. If we interpret $a \leq b$ to mean $b$ is as good as or better than $a$, then applications abound where we could have $a \leq b$ and $b \leq a$ with $a \neq b$. To a farmer's accountant, any two different six-pound pumpkins $a$ and $b$ would satisfy $a \leq b$ and $b \leq a$ but $a \neq b$. If a professor is grading essays, two different essays $a$ and $b$ may each be as good as or better than the other, even though they are different essays. Two such essays would receive the same grade, and thus $a \leqslant b$ and $b \leqslant a$ defines an equivalence relation based on equality of the essay grades.

If $a \leqslant b$ and $b \leqslant a$, then our experience with partial orders leads us to expect that $a$ should equal $b$; if $\leqslant$ is a quasiorder which is not a partial order, then they may not be equal, but we may declare them to be equivalent. This introduces antisymmetry, but on the equivalence classes of $X$ instead of the points of $X$. Thus, a quasiorder on $X$ gives a partial order on the equivalence classes of an equivalence relation. Conversely, a partial order on the blocks of a partition of $X$ (that is, on the equivalence classes of some equivalence relation on $X$) produces a quasiorder in a natural way. The following theorem explicitly gives this important connection.

**Theorem 8.2.2.**

(a) *If $\leqslant$ is a quasiorder on $X$, the relation $\approx$ on $X$ defined by $a \approx b$ if and only if $a \leqslant b$ and $b \leqslant a$ is an equivalence relation on $X$.*

(b) *If $[x]$ is the equivalence class of $x \in X$ with respect to the equivalence relation $\approx$ as in (a), then taking $[a] \leqslant [b]$ if and only if $a \leqslant b$ defines a partial order $\leqslant$ on the set $X/\approx$ of equivalence classes.*

(c) *If $\leqslant$ is any partial order on the blocks $\{B_i : i \in I\} = \{[a] : a \in X\}$ of a partition $\mathcal{P}$ of $X$, then the relation $\leqslant$ defined on $X$ by $a \leqslant b$ if and only if $[a] \leqslant [b]$ is a quasiorder on $X$.*

*Proof.* (a) It is easy to see that $\approx$ as defined above is reflexive, transitive, and symmetric, and thus is an equivalence relation.

(b) To see that this gives a partial order on the equivalence classes, we first note that the definition $[a] \leqslant [b]$ if and only if $a \leqslant b$ is well-defined. If $[a] = [a']$ and $[b] = [b']$, then $[a] \leqslant [b]$ if and only if $[a'] \leqslant [b']$, and these two different choices for the equivalence class representatives lead to different—and possibly contradictory— defining conditions, namely $a \leqslant b$ and $a' \leqslant b'$. However, if $[a] = [a']$ and $[b] = [b']$, then $a \leqslant a', a' \leqslant a, b \leqslant b'$, and $b' \leqslant b$, so if $a \leqslant b$, then we have $a' \leqslant a \leqslant b \leqslant b'$, so $a' \leqslant b'$ by transitivity of $\leqslant$. Similarly, if $a' \leqslant b'$ then $a \leqslant b$. Now for any equivalence classes $[a], [b]$, and $[c]$, it is easy to see that reflexivity, transitivity, and symmetry follow for the equivalence classes from the corresponding properties of the representative elements $a, b$, and $c$.

(c) Suppose $\leqslant$ is a partial order on the blocks $\{[a] : a \in X\}$ of a partition $\mathcal{P}$ of $X$, and $\leqslant$ is defined by $a \leqslant b$ if and only if $[a] \leqslant [b]$. For $a \in X$, $[a] \leqslant [a]$ and thus $a \leqslant a$. If $a \leqslant b$ and $b \leqslant c$, then $[a] \leqslant [b]$ and $[b] \leqslant [c]$, so $[a] \leqslant [c]$ and thus $a \leqslant c$. Thus, $\leqslant$ is reflexive and transitive, as needed. $\square$

While an equivalence relation on $X$ partitions $X$ into equivalence classes based on equality of some attribute, a quasiorder on $X$ focuses on some attribute of the elements, partitions $X$ into the corresponding equivalence classes, and provides a partial order on the equivalence classes.

We may iterate the parts of Theorem 8.2.2: a quasiorder $\leq$ on $X$ produces a partial order on the blocks of a partition, which then produces a quasiorder. It is easy to see that the resulting quasiorder is the original. Similarly, starting with a partial order on the blocks of a partition, finding the associated quasiorder, and then the partial order on the blocks of a partition using the process in the theorem, we cycle back to the original partial order on blocks of the original partition. This proves the following corollary.

**Corollary 8.2.3.** *There is a one-to-one correspondence between quasiorders on $X$ and partial orders on blocks of partitions of $X$.*

**Example 8.2.4.** Let $C_r = \{(x, y) \in \mathbb{R}^2 : x^2 + y^2 = r^2\}$ be the circle of radius $r$ centered at the origin. Partition $\mathbb{R}^2$ into $\{C_r : r \geq 0\}$. Partially order the blocks of the partition (that is, the circles) by $C_r \leq C_s$ if and only if $r \leq s$. By Theorem 8.2.2(c), this structure gives a quasiorder on $\mathbb{R}^2$ defined by $(a, b) \leq (x, y)$ if and only if the circle $C_r$ containing $(a, b)$ has radius less than or equal to the circle $C_s$ containing $(x, y)$. That is, $(a, b) \leq (x, y)$ if and only if $\sqrt{a^2 + b^2} \leq \sqrt{x^2 + y^2}$, or equivalently, if and only if $a^2 + b^2 \leq x^2 + y^2$.

**Example 8.2.5.** On $\mathbb{R}^2$, define $(a, b) \leq (x, y)$ if and only if $a \leq x$. It is easily seen that $\leq$ is reflexive and transitive, and thus is a quasiorder on $\mathbb{R}^2$. The equivalence relation $(a, b) \approx (x, y)$ if and only if $(a, b) \leq (x, y)$ and $(x, y) \leq (a, b)$ is $(a, b) \approx (x, y)$ if and only if $a = x$. The $\approx$-equivalence class of $(a, b)$ is $[(a, b)] = \{(x, y) : x = a\}$, which is the vertical line $L_a = \{a\} \times \mathbb{R}$. The partial order on set of equivalence classes $\{L_a : a \in \mathbb{R}\}$ from Theorem 8.2.2(b) is given by $L_a \leq L_b$ if and only if $a \leq b$ in $\mathbb{R}$.

**Example 8.2.6.** Define $\leq$ on $\mathbb{R}$ by $x \leq y$ if and only if $x = y$ or $x \in \mathbb{Z}$. Thus, besides the reflexivity condition that each point is below itself, we have each integer is below everything. This is a quasiorder. Reflexivity follows from the definition. To show transitivity, suppose $x \leq y$ and $y \leq z$. If $x \in \mathbb{Z}$, then $x \leq z$ since integers are below everything. If $x \notin \mathbb{Z}$ and $x \leq y$, then $x = y$, so $y \leq z$ says $x \leq z$.

To find the associated equivalence relation $\approx$, suppose $x \leq y$ and $y \leq x$. If $x \notin \mathbb{Z}$, then $x \leq y$ implies $x = y$, so the $\approx$-equivalence class of $x \notin \mathbb{Z}$ is $\{x\}$. If $x \in \mathbb{Z}$, then $x \leq y$ for every $y \in \mathbb{R}$, and $y \leq x$ if $x = y$ or $y \in \mathbb{Z}$. Thus, for $x \in \mathbb{Z}$, $x \approx y$ if and only if $y \in \mathbb{Z}$.

Now the associated partial order $\leq$ on the equivalence classes $\{\mathbb{Z}\} \cup \{\{x\} : x \in \mathbb{R} - \mathbb{Z}\}$ includes the natural reflexivity $\mathbb{Z} \leq \mathbb{Z}$ and $\{x\} \leq \{x\}$ for all $x \in \mathbb{R} - \mathbb{Z}$, and the condition that $n \leq x$ for any $n \in \mathbb{Z}$ and any $x \in \mathbb{R}$ implies $[n] \leq [x]$, so we have $\mathbb{Z} \leq \{x\}$ for any $x \in \mathbb{R} - \mathbb{Z}$. The Hasse diagram for this partial order is suggested in Figure 8.2.

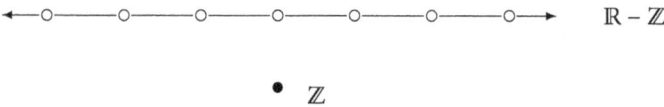

**Figure 8.2:** The partial order on the $\approx$-equivalence classes for Example 8.2.6.

**Example 8.2.7.** On the set of words, define a relation $\leq$ by word1 $\leq$ word2 if and only if the set of consonants used in word1 is a subset of the set of consonants used in word2. Now $\leq$ is a quasiorder since it is reflexive (the set of consonants in word1 is a subset of the set of consonants in word1) and transitive (if the set of consonants in word1 is contained in the set of consonants in word2 and the set of consonants in word2 is contained in the set of consonants in word3, then the set of consonants in word1 is contained in the set of consonants in word3). The associated equivalence relation is word1 $\approx$ word2 if and only if the set of consonants of each word is contained in the set of consonants of the other, which happens if and only if they have the same set of consonants. Then, for example,

$$\text{cool} \leq \text{calculus} \leq \text{class} \leq \text{calculus} \leq \text{school}$$

since

$$\{c,l\} \subseteq \{c,l,s\} \subseteq \{c,l,s\} \subseteq \{c,l,s\} \subseteq \{c,l,s,h\}.$$

Furthermore, calculus $\approx$ class. If we assume the set $C$ of consonants has 21 elements, and if we allow nonsense words like bcdfghyz, then the set of $\approx$-equivalence classes with the partial order induced by $\leq$ is $\mathcal{P}(C)$, the power set of $C$, ordered by set inclusion.

**Example 8.2.8** (Simplex combination locks). A simplex combination lock is a push-button lock with $n$ buttons. To open the lock, the proper sequence of buttons must be pressed. Buttons may be pressed simultaneously, but no button can be pressed more than once. A standard configuration uses $n = 5$ buttons. One possible combination might be $(\{1,4\},\{2\},\{3,5\})$, requiring that buttons 1 and 4 be pressed simultaneously first, then button 2, then buttons 3 and 5 simultaneously. This combination gives a partition of the buttons pressed, and a partial order (in fact, a total order) on the blocks $\{\{1,4\},\{2\},\{3,5\}\}$ of the partition, and thus a quasiorder on $X = \{1,2,3,4,5\}$.

The number of combinations $L_n$ on an $n$-button lock which use all $n$ buttons is called the $n$th *Fubini number*. $L_n$ is the number of total quasiorders on $X$, where a total quasiorder is a quasiorder whose associated partial order on the equivalence classes is a total order. By convention, $L_0 = 1$, and it is easy to see that $L_1 = 1$. There is a recursive formula for the numbers $L_n$. Having defined $L_0, L_1, \ldots, L_{n-1}$ for $n \geq 2$, on the first press, we must press $k$ buttons ($k = 1, \ldots, n$). There are $\binom{n}{k}$ ways to do this. Then, we are left with the $(n-k)$-button problem, which can be solved in $L_{n-k}$ ways. Summing over all $k$ gives

$$L_n = \sum_{k=1}^{n} \binom{n}{k} L_{n-k}.$$

The initial conditions for $L_0$ and $L_1$ give $(L_n)_{n=0}^{5} = (1,1,3,13,75,541)$, and in particular, there are 541 combinations on a 5-button lock which use all five buttons.

Since combinations on a 5-button lock need not use all five of the buttons, to find the total number of combinations, there are more combinations to count. If a combination uses $j$ buttons for $j = 0, \ldots, 5$. There are $\binom{5}{j}$ choices for which $j$ buttons to use, and summing over all values of $j$ gives the total

$$T_5 = \sum_{j=0}^{5} \binom{5}{j} L_j = 1082.$$

Note that the total number $T_5 = 1082 = 2(541) = 2L_5$, and in general $T_n = 2L_n$. Thus, a 5-button Simplex lock has 1082 possible combinations, making it only slightly more secure than a bike lock with three rotors with 10 possible positions for each rotor, which has $10^3 = 1000$ possible combinations. Further analysis of this example is given in [49].

Given a subset $B$ of a partially ordered set or a quasiordered set, it is natural to consider all the points in $B$ or above an element of $B$. The next definition provides our terminology.

**Definition 8.2.9.** Given a subset $B$ of a quasiordered set $(X, \leq)$, the *increasing hull* of $B$ is $i(B) = \{x \in X : b \leq x \text{ for some } b \in B\}$, and $B$ is an *increasing set* if $B = i(B)$. Dually, the *decreasing hull* of $B$ is $d(B) = \{x \in X : x \leq b \text{ for some } b \in B\}$, and $B$ is a *decreasing set* if $B = d(B)$. The increasing and decreasing hulls $i(\{x\})$ and $d(\{x\})$ of a singleton set $\{x\}$ will be denoted $i(x)$ and $d(x)$, respectively. A set is *monotone* if it is either increasing or decreasing.

Thus, the increasing hull of $B$ contains everything in $B$ and everything above any point of $B$. Our notation and terminology follows that of the seminal monograph *Topology and Order* [38] by Leopoldo Nachbin, which appeared in Portuguese in 1950 and in English translation in 1965. Increasing sets appear in the literature under many names, including *upper sets*, *upward closed sets*, or *upsets*, and the increasing hull $i(A)$ may be called the *upward closure* and be denoted $\uparrow A$. In the natural dual terminology, the decreasing hull of $A$ may be called the *downward closure*, denoted $\downarrow A$, and decreasing sets may be called *lower sets*, *downward closed sets*, or *downsets*.

When we are dealing with more than one quasiorder on a set, we may need to include the order $\leq$ in our terminology, saying $B$ is $\leq$-increasing if and only if $B = i_{\leq}(B)$. For example, a quasiorder $\leq$ on $X$ always produces the *dual quasiorder* $\geq$ defined by $x \geq y$ if and only if $y \leq x$. Then, it is clear that $i_{\leq}(B) = d_{\geq}(B)$.

We present some facts about monotone sets.

**Theorem 8.2.10.** *Suppose $B$ is a subset of a quasiordered set $(X, \leq)$.*
(a) *The increasing hull of $B$ is the smallest increasing set containing $B$.*
(b) *$B$ is increasing if and only if $B = \bigcup\{i(x) : x \in B\}$.*
(c) *The complement of an increasing set is a decreasing set.*

*The dual statements about decreasing sets also hold. That is:*

(a′)  *The decreasing hull of B is the smallest decreasing set containing B.*

(b′)  *B is decreasing if and only if $B = \bigcup\{d(x) : x \in B\}$.*

(c′)  *The complement of a decreasing set is an increasing set.*

The straightforward proofs are left as exercises. We note that given a statement about an ordered set $(X, \leq)$, the *dual statement* is obtained by reversing $\leq$ and any terms defined using $\leq$ to $\geq$ and the corresponding terms defined using $\geq$.

**Example 8.2.11.** Consider the quasiordered space $\mathbb{R}^2$ of Example 8.2.4 with $(a, b) \leq (x, y)$ if and only if the distance from $(a, b)$ to the origin is less than or equal to the distance from $(x, y)$ to the origin. Now the decreasing hull of the singleton set $\{(3, 4)\}$ is $d((3, 4)) = \{(x, y) \in \mathbb{R}^2 : (x, y) \leq (3, 4)\}$, which is the closed disk of radius 5 centered at the origin. Similarly, $i((3, 4))$ is the set of all points in the plane 5 units or further from the origin. Note that $i((3, 4)) \cap d((3, 4))$ is the circle $C_5$ of radius 5.

For $B = \{(1, 1), (2, 0)\}$, $d(B)$ consists of $B$ and every point $(x, y) \in \mathbb{R}^2$ with $(x, y) \leq (1, 1)$ or $(x, y) \leq (2, 0)$, so $d(B) = \{(x, y) \in \mathbb{R}^2 : x^2 + y^2 \leq 4\}$ is a closed disk of radius 4. Similarly, $i(B) = \{(x, y) \in \mathbb{R}^2 : (x, y) \geq (a, b)$ for $(a, b) = (1, 1)$ or $(a, b) = (2, 0)\} = \{(x, y) \in \mathbb{R}^2 : x^2 + y^2 \geq 2\}$, which is the complement of an open disk of radius $\sqrt{2}$.

For $C = \{(x, y) \in \mathbb{R}^2 : x^2 + y^2 < 3\}$, $d(C)$ consists of all points in the plane which are in $C$ or are closer to the origin than some point of $C$. Thus, $d(C) = C$, and $C$ is a decreasing set. Every open disk centered at the origin and every closed disk centered at the origin is a decreasing set, and indeed, every nonempty proper decreasing set is of this form.

## Exercises

1. Define a relation $\leq$ on $\mathbb{R}$ by $x \leq y$ if and only if $x^2 \leq y^2$, where $\leq$ is the usual order on $\mathbb{R}$.

   (a) Show that $\leq$ is a quasiorder on $\mathbb{R}$.

   (b) Describe the associated equivalence relation $a \approx b$ if and only if $a \leq b$ and $b \leq a$.

   (c) Describe the partial order on the $\approx$-equivalence classes associated with the quasiorder $\leq$.

   (d) Find $i(-3)$, $i(\{2, 3\})$, $d(\{6, 15\})$, $i(\{3 + \frac{1}{n} : n \in \mathbb{N}\})$, and $d(\{x \in \mathbb{R} : -4 < x < 3\})$.

2. Define a relation $\leq$ on $X = \mathbb{R} - \{n\pi : n \in \mathbb{Z}\}$ by $x \leq y$ if and only if $\cot x \leq \cot y$, where $\leq$ is the usual order on $\mathbb{R}$.

   (a) Show that $\leq$ is a quasiorder on $\mathbb{R}$.

   (b) Describe the associated equivalence relation $a \approx b$ if and only if $a \leq b$ and $b \leq a$.

(c) Describe the partial order on the ≈-equivalence classes associated with the quasiorder ≤.

(d) Find $i(\frac{\pi}{2})$, and $d((\frac{\pi}{4}, \frac{\pi}{3}))$.

3. The "divides" relation | on the set $\mathbb{Z}-\{0\}$ of positive and negative integers is defined by $a|b$ if and only if there exists $n \in \mathbb{Z}$ with $b = na$.

(a) Show that | is a quasiorder on $\mathbb{Z} - \{0\}$.

(b) Describe the associated equivalence relation $a \approx b$ if and only if $a|b$ and $b|a$.

(c) Describe the partial order on the ≈-equivalence classes associated with the quasiorder |.

(d) Find $i(3)$, $i(\{2, 3\})$, $d(12)$, and $d(\{6, 15\})$.

4. If $X$ is a set, $(Y, \leq)$ is a poset, and $f : X \to Y$ is a function, show that defining $x_1 \leq_f x_2$ if and only if $f(x_1) \leq f(x_2)$ gives a quasiorder $\leq_f$ on $X$. Describe the associated equivalence relation $\approx_f$ on $X$. Among the quasiorders of the previous exercises in this section, which are defined in this manner?

5. Let $X = \{( ), (1), (2), \ldots, (15)\}$ represent a standard set of billiard balls, with their standard colors and numbers. Define a quasiorder on $X$ by $(x) \leq (y)$ if and only if the set of colors on ball $(x)$ is a subset of the set of colors on ball $(y)$.

(a) Sketch the Hasse diagram for $(X, \leq)$.

(b) Find $i((8))$, $i(\{(2), (13)\})$, and $d(\{(1), (7), (12)\})$.

(c) How many increasing sets and how many decreasing sets are there in $(X, \leq)$?

6. Repeat Exercise 5 using the quasiorder ≤ on $X$ defined by $(x) \leq (y)$ if and only if the number of colors on ball $(x)$ is less than or equal to the number of colors on ball $(y)$.

7. In a quasiordered set $X$, if $B \subseteq X$, show that $i(B)$ is an increasing set, and is the smallest increasing set containing $B$.

8. In a quasiordered set $X$, show that $\emptyset$ and $X$ are increasing.

9. In a quasiordered set, show that the complement of an increasing set is a decreasing set.

10. In a quasiordered set $X$, show that $B$ is increasing if and only if $B = \bigcup \{i(b) : b \in B\}$.

## 8.3 Alexandroff topologies as quasiorders

After having introduced quasiorders in the last section, the next theorem brings us back to topology.

**Theorem 8.3.1.** *Suppose $(X, \leq)$ is a quasiordered set.*

(a) *$\emptyset$ and $X$ are increasing sets.*

(b) *Arbitrary intersections of increasing sets are increasing sets.*

(c) *Arbitrary unions of increasing sets are increasing sets.*

*That is, the increasing sets in the quasiordered set* $(X, \leq)$ *form an Alexandroff topology* $\mathcal{T}[\leq]$ *on X called the* specialization topology *from* $\leq$. *The smallest neighborhood of x is the smallest increasing set containing x; that is,* $N(x) = i(x)$.

The Alexandroff topology of $\mathcal{T}[\leq]$-closed sets consists of the complements of the increasing sets, that is, the decreasing sets in $(X, \leq)$. Since the smallest closed set containing $A \subseteq X$ is the smallest decreasing set containing $A$, we have $\mathrm{cl}\, A = d(A)$.

The proofs are left to the exercises.

**Example 8.3.2.** In the space $\mathbb{R}$ with $x \leq y$ if and only if $x = y$ or $x \in \mathbb{Z}$ considered in Example 8.2.6, the open sets of the specialization topology $\mathcal{T}[\leq]$ are the increasing sets. Since an integer $n \in \mathbb{Z}$ is below everything in $\mathbb{R}$, $i(n) = \mathbb{R}$ is the minimal neighborhood $N(n)$ of $n \in \mathbb{Z}$. For $x \notin \mathbb{Z}$, $i(x) = \{x\}$, so $N(x) = \{x\}$. The minimal neighborhoods form a basis for the Alexandroff topology, so the $\mathcal{T}[\leq]$-open sets other than $\mathbb{R}$ are unions of singletons $\{x\}$ for $x \notin \mathbb{Z}$. That is, the proper open sets are those that are disjoint from $\mathbb{Z}$. Thus, $\mathcal{T}[\leq] = \mathrm{Disjoint}(\mathbb{Z})$.

Our previous theorem showed that any quasiorder $\leq$ on a set $X$ produces an Alexandroff topology $\mathcal{T}[\leq]$ consisting of the increasing sets. The theorem below shows that the converse holds as well. Every Alexandroff topology on $X$ is the specialization topology generated from some quasiorder on $X$. Indeed, either of the observations in Theorem 8.3.1 that $i(x) = N(x)$ and $d(x) = \mathrm{cl}(\{x\})$ provides a link both ways between quasiordered sets and Alexandroff topologies, shown in Table 8.2.

**Table 8.2:** The link between quasiorders and Alexandroff topologies.

| Order Theory | | | Topology |
|---|---|---|---|
| $x \leq y$ | $\Longleftrightarrow$ | $x \in d(y)$ $\Longleftrightarrow$ | $x \in \mathrm{cl}\{y\}$ |
| $\updownarrow$ | | | $\updownarrow$ |
| $y \geq x$ | $\Longleftrightarrow$ | $y \in i(x)$ $\Longleftrightarrow$ | $y \in N(x)$ |

For completeness, we state both directions in the theorem.

**Theorem 8.3.3.** *Every quasiorder* $\leq$ *on X produces an Alexandroff topology* $\mathcal{T}[\leq]$ *consisting of the* $\leq$*-increasing sets in* $(X, \leq)$.

*Every Alexandroff topology* $\mathcal{T}$ *on X produces a quasiorder* $\leq_\mathcal{T}$ *on X defined by* $x \leq_\mathcal{T} y$ *if and only if* $x \in \mathrm{cl}\{y\}$; *that is, defined by* $x \in d(y)$ *if and only if* $x \in \mathrm{cl}\{y\}$.

*Furthermore, if* $\mathcal{T}$ *is an Alexandroff topology on X, then* $\leq_{\mathcal{T}[\leq]} = \leq$, *and if* $\leq$ *is a quasiorder on X, then* $\mathcal{T}[\leq_\mathcal{T}] = \mathcal{T}$.

*Proof.* The first statement is Theorem 8.3.1.

Suppose $\mathcal{T}$ is a topology on $X$ and $\leq_\mathcal{T}$ is defined by $x \leq_\mathcal{T} y$ if and only if $x \in \mathrm{cl}\{y\}$. For any $x \in X$, $x \in \mathrm{cl}\{x\}$, so $x \leq_\mathcal{T} x$, and thus $\leq_\mathcal{T}$ is reflexive. Suppose for $x, y, z \in X$

we have $x \leq_T y$ and $y \leq_T z$. Then $x \in$ cl$\{y\}$ and $y \in$ cl$\{z\}$. Every open neighborhood $U$ of $x$ must intersect $\{y\}$, and thus is a neighborhood of $y$, and every neighborhood of $y$ must intersect $\{z\}$. It follows that every neighborhood of $x$ intersects $\{z\}$, so $x \in$ cl$\{z\}$ and thus $x \leq_T z$, showing transitivity. Thus, $\leq_T$ is a quasiorder on $X$.

If we start with a quasiorder $\leq$ on $X$, convert to the associated Alexandroff topology $\mathcal{T}[\leq]$, then convert this topology to the associated quasiorder $\leq_{\mathcal{T}[\leq]}$, because the conversions are as described in the circular loop of equivalences in Table 8.2, we return to the original quasiorder $\leq$. Similarly comments apply to the situation of iterating the conversions when starting from an Alexandroff topology. $\qquad\square$

This theorem shows that there is a one-to-one correspondence between quasiorders and Alexandroff topologies on a set $X$. In particular, if $X$ is a finite set, every topology on $X$ is Alexandroff, so there is a one-to-one correspondence between topologies on a finite set $X$ and quasiorders on $X$.

Just as the Alexandroff topology corresponding to a quasiorder is called the specialization topology, the quasiorder $\leq_T$ produced by topology $\mathcal{T}$ is called the *specialization order* produced by $\mathcal{T}$.

In Theorem 8.3.3, the proof that an Alexandroff topology $\mathcal{T}$ on $X$ defines a quasiorder $\leq_T$ on $X$ by $x \leq_T y$ if and only if $x \in$ cl$\{y\}$ did not require that $\mathcal{T}$ be Alexandroff; any topology $\mathcal{T}$ on $X$ produces a quasiorder $\leq_T$ on $X$. We restricted our attention to Alexandroff topologies to obtain the one-to-one correspondence with quasiorders. Quasiorders arising from non-Alexandroff topologies are investigated in Exercise 14.

**Example 8.3.4.** The Hasse diagram for a quasiordered set $(X, \leq)$ is depicted as the Hasse diagram for the partial order on the $\approx$-equivalence classes. Figure 8.3(a) shows the Hasse diagram for a quasiordered set $X = \{a, b, c, d, e, f, g, h, i\}$. For example, we have $a \leq b, a \leq c, b \leq c, c \leq b$, and thus $b \approx c$. The specialization topology from $\leq$ is the collection of increasing sets. A basis for the specialization topology is $\mathcal{B} = \{N(x) : x \in X\} = \{i(x) : x \in X\}$. For example, $i(b) = \{b, c\}$ is the smallest increasing set containing $b$ and thus is the smallest neighborhood of $b$. Some other basic open sets are $N(a) = i(a) = \{a, b, c\}$, $N(e) = i(e) = \{e\}$, $N(d) = i(d) = \{b, c, d, e\}$, and $N(h) = i(h) = X - \{a\}$. The basis $\mathcal{B}$ for the specialization topology is shown in Figure 8.3(b).

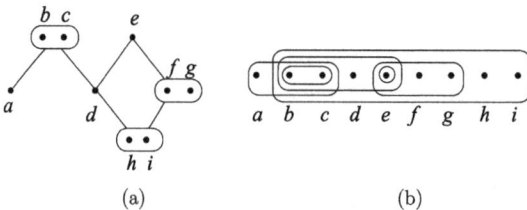

(a)                                        (b)

**Figure 8.3:** (a) The Hasse diagram for a quasiorder and (b) a basis for its specialization topology.

In Chapter 0, we defined several order-theoretic properties, such as upper bounds, minimal elements, and minimum elements, for partially ordered sets. Those definitions apply to quasiordered sets as well. In particular, a function $f : (X, \leq_X) \to (Y, \leq_Y)$ is *increasing* (or *order-preserving*) if for all $a, b \in X$, $a \leq_X b$ implies $f(a) \leq_Y f(b)$.

The following results reinforce the strong link between Alexandroff spaces and quasiordered sets.

**Lemma 8.3.5.** *A function $f : (X, \leq_X) \to (Y, \leq_Y)$ between quasiordered spaces is increasing if and only if for every increasing set $V$ in $Y$, the inverse image $f^{-1}(V)$ is an increasing set in $X$.*

*Proof.* Suppose $(X, \leq_X)$ and $(Y, \leq_Y)$ are quasiordered sets, $f : (X, \leq_X) \to (Y, \leq_Y)$ is an increasing function, and $V$ is an increasing set in $Y$. To see $f^{-1}(V)$ is increasing in $X$, suppose $a \in f^{-1}(V)$ and $a \leq_X b$. Then $f(a) \leq_Y f(b)$. Since $f(a) \in V$ and $V$ is an increasing set, $f(b) \in V$. Thus, $b \in f^{-1}(V)$. This shows $f^{-1}(V)$ is an increasing set in $X$.

Conversely, suppose $f : (X, \leq_X) \to (Y, \leq_Y)$ has the property that $f^{-1}(V)$ is an increasing set in $X$ for every increasing set $V$ in $Y$. To show $f$ is an increasing function, suppose $a \leq_X b$. We want to show $f(a) \leq_Y f(b)$, or $f(b) \in i_{\leq_Y}(f(a))$. Since $i_{\leq_Y}(f(a))$ is an increasing set in $Y$ containing $f(a)$, $f^{-1}(i_{\leq_Y}(f(a)))$ is an increasing set in $X$ containing $a$, and thus containing $b \geq_X a$. Now $b \in f^{-1}(i_{\leq_Y}(f(a)))$ implies $f(b) \in i_{\leq_Y}(f(a))$, as needed. $\qquad\square$

**Theorem 8.3.6.** *Suppose $(X, \mathcal{T}_X)$ and $(Y, \mathcal{T}_Y)$ are Alexandroff spaces having specialization quasiorders $\leq_X$ and $\leq_Y$, respectively, and $f : X \to Y$ is a function. Then $f : (X, \mathcal{T}_X) \to (Y, \mathcal{T}_Y)$ is continuous if and only if $f : (X, \leq_X) \to (Y, \leq_Y)$ is increasing.*

*Proof.* The result follows immediately from Lemma 8.3.5 and the fact (Theorem 8.3.3) that open sets in an Alexandroff topology are the increasing sets in the specialization quasiorder. $\qquad\square$

In an Alexandroff space $X$, $\mathrm{cl}\{x\} = d(x)$, where the decreasing hull is taken in the specialization quasiorder on $X$. In a functionally Alexandroff space, $\mathrm{cl}\{x\} = \{f^n(x)\}_{n=0}^{\infty}$. In particular, $f(x) \in \mathrm{cl}\{x\} = d(x)$, so $f(x) \leq x$. Loosely, this tells us that to draw a Hasse diagram for a functionally Alexandroff topology $\mathcal{T}_f$ determined by $f : X \to X$, we start with $x \in X$, put $f(x) \leq x$, and repeat.

**Example 8.3.7.** Consider the function $f : \{1, 2, \ldots, 7\} \to \{1, 2, \ldots, 7\}$ defined by $f(1) = f(2) = 3, f(3) = 4, f(4) = 5, f(5) = 4, f(6) = 7, f(7) = 7$. In the specialization order for the functionally Alexandroff topology $\mathcal{T}_f$, we have $f(x) \leq x$, and this allows us to generate the Hasse diagram, as shown in Figure 8.4(a). Since the smallest neighborhood of $x \in X$ is $N_f(x) = i(x)$, we take the increasing hulls of singletons to get the basis for $\mathcal{T}_f$ shown in Figure 8.4(b).

Note that the Alexandroff topology depicted in Figure 8.3 is not functionally Alexandroff. If it were $\mathcal{T}_s$ for some function $s$, then intuitively, $s(b)$ should be $a, c$,

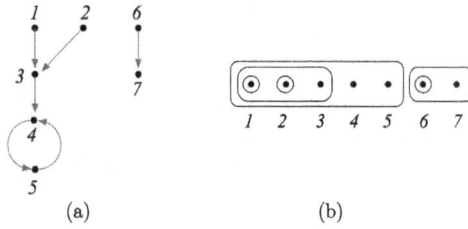

**Figure 8.4:** (a) The function diagram for $f$ and (b) a basis for the functionally Alexandroff topology $\mathcal{T}_f$.

and $d$ simultaneously, which violates the definition of function. Formally, suppose the topology is $\mathcal{T}_s$ for some function $s$. Since $b \in d(c)$ and $c \in d(b)$, we have $b = s^n(c)$ and $c = s^m(b)$ for some $m, n \in \mathbb{N}$, so $b = s^{n+m}(b) = s^0(b)$ and thus $cl\{b\} = \{s^k(b)\}_{k=0}^{\infty} = \{s^k(b)\}_{k=1}^{n+m}$ is a finite cycle which contains $c$. Since the closure of $b$ is the decreasing hull of $b$, we must have $a \in cl\{b\} = \{s^k(b)\}_{k=1}^{n+m}$. But if $a$ is in this cycle with $b$ and $c$, then $b$ and $c$ are each of form $s^j(a)$, and thus $b$ and $c$ should be below $a$ in the diagram. This is false, so the topology of Figure 8.3 cannot be functionally Alexandroff.

The argument of the preceding paragraph really only depended on the points $a, b, c$ from Figure 8.3, which form a subspace as depicted on the left in Figure 8.5. This is essentially half of the proof of the following result. The other half is left to the exercises.

**Theorem 8.3.8.** *If a topology on a finite set $X$ contains either of the subspaces shown in Figure 8.5, then the topology is not functionally Alexandroff.*

**Figure 8.5:** Forbidden subspaces for a finite functionally Alexandroff space.

In fact, the converse of Theorem 8.3.8 holds, so a topology on a finite set is functionally Alexandroff if and only if it contains no subspace as shown in Figure 8.5. Proofs of this may be found in [1] or [37]. Further connections between Alexandroff topologies and quasiorders are surveyed in [43].

## Exercises

1.  In a quasiordered set, show that arbitrary intersections and arbitrary unions of increasing sets are increasing. That is, prove Theorem 8.3.1(b) and (c).

2. A quasiordered set $(X, \leq)$ is totally quasiordered if for every $x, y \in X$, either $x \leq y$ or $y \leq x$. Show that $(X, \leq)$ is totally quasiordered if and only if the associated Alexandroff topology is a nested topology.

3. Suppose $\leq$ is a quasiorder on $X$ which produces the Alexandroff topology $\mathcal{T}$. Describe the Alexandroff topology on $X$ produced by the dual quasiorder $\geq$.

4. Consider the multiples topology on $\mathbb{N}$, having basis $\{M_n : n \in \mathbb{N}\}$ where $M_n = \{kn : k \in \mathbb{N}\}$. Show that this is an Alexandroff topology and find its specialization quasiorder.

5. For $n \in \mathbb{N}$, let $X_n = \{1, 2, 3, \dots, n\}$ and consider the subspace $(X_n, \mathcal{T}_n)$ of the digital line.
   (a) Sketch the Hasse diagram for the specialization quasiorder for $X_1, X_2, X_3$, $X_4, X_5$, and $X_6$.
   (b) List all eight open sets in $(X_4, \mathcal{T}_4)$.
   (c) Show that, for any $n \in \mathbb{N}$, $|\mathcal{T}_n| = F_{n+2}$, where $F_1 = F_2 = 1$ and for $k > 2$, $F_k = F_{k-1} + F_{k-2}$ are the Fibonacci numbers.

6. Shown are bases for some finite (and thus Alexandroff) topologies. For each one, draw the specialization quasiorder.

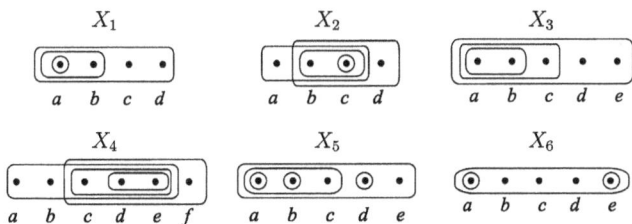

7. Determine which of the topological spaces given in Exercise 6 are functionally Alexandroff and for those that are, give a function $f$ generating them.

8. Suppose $\mathcal{T}$ is a topology on $X$ and $\leq$ is its specialization quasiorder defined by $x \leq y$ if and only if $x \in \mathrm{cl}\{y\}$. Show that $\leq$ is a partial order if and only if $\mathcal{T}$ is $T_0$.

9. Show that every finite $T_0$ space $X$ has at least one closed singleton set $\{x\}$ and at least one isolated point. (You may use Exercise 8.)

10. Suppose $(X, \mathcal{T}_X)$ and $(Y, \mathcal{T}_Y)$ are Alexandroff spaces with specialization orders $\leq_X$ and $\leq_Y$, respectively. Show that $X \times Y$ with the product topology is an Alexandroff space. Describe the specialization order $\leq$ on $X \times Y$ in terms of $\leq_X$ and $\leq_Y$.

11. Define $f : \{2, 3, \dots, 12\} \to \{2, 3, \dots, 12\}$ by taking $f(n)$ to be the largest prime factor of $n$. Draw the quasiorder diagram for the functionally Alexandroff topology $\mathcal{T}_f$ and give a basis for $\mathcal{T}_f$.

12. Define $f : \mathbb{Z} \to \mathbb{Z}$ by $f(n) = n^2$. Consider the functionally Alexandroff topology $\mathcal{T}_f$ and its specialization quasiorder.
    (a) Find all fixed points of $f$.
    (b) Find all isolated points.
    (c) Find all maximal and all minimal points in the quasiorder.

(d) Find $\mathrm{cl}\{-2\}$, $N_f(-16)$, and $N_f(256)$.

(e) Is $(\mathbb{Z}, \mathcal{T}_f)$ connected?

13. Suppose $f : X \to X$, $n \in \mathbb{N}$, and $\mathcal{T}_f^*$ represents the Alexandroff topology of $\mathcal{T}_f$-closed sets.

(a) Show that $\mathcal{T}_f \subseteq \mathcal{T}_{f^n}$.

(b) If $f$ is onto, show that $\mathcal{T}_{f^n} \cap \mathcal{T}_f^* \subseteq \mathcal{T}_f$.

14. Let $\mathcal{T}$ be a topology (not necessarily Alexandroff) on $X$, and let $\leq_{\mathcal{T}}$ be the associated quasiorder defined by $x \leq_{\mathcal{T}} y$ if and only if $x \in \mathrm{cl}\{y\}$. Show that $\mathcal{T}$ is an Alexandroff topology if and only if every $\leq_{\mathcal{T}}$-increasing set is $\mathcal{T}$-open.

15. Given a topological space $X$, define $x \approx y$ if and only if $\mathrm{cl}\{x\} = \mathrm{cl}\{y\}$. Let $q$ denote the quotient map from $X$ to $X/\approx$. With the quotient topology, $X/\approx$ is called the $T_0$-reflection (or, in older literature, the Kolmogorov quotient), denoted $T_0(X)$. Show that $T_0(X)$ is a $T_0$ topological space and if $f : X \to Y$ is any continuous function from $X$ to any arbitrary $T_0$ space $Y$, there exists a continuous function $\hat{f} : T_0(X) \to Y$ such that $\hat{f} \circ q = f$, that is, such that the diagram below commutes. (You may use Exercise 8.)

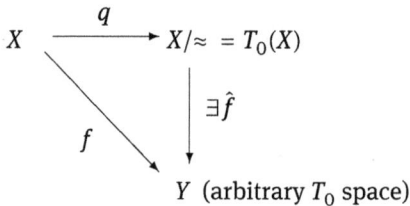

$$
\begin{array}{ccc}
X & \xrightarrow{\ q\ } & X/\approx\ =\ T_0(X) \\
& \searrow_{f} & \downarrow^{\exists \hat{f}} \\
& & Y\ (\text{arbitrary } T_0 \text{ space})
\end{array}
$$

16. Prove that if a topology $\mathcal{T}$ on a finite set $X$ contains a subspace as shown on the right in Figure 8.5, then $\mathcal{T}$ is not functionally Alexandroff.

# 9 Lattice properties

## 9.1 Lattice theory

We recall some order theoretic definitions. Suppose $(X, \leq)$ is a poset. An element $a \in X$ is *maximum* if $x \leq a$ for every $x \in X$, and is *maximal* if $a \leq x$ implies $a = x$. For $A \subseteq X$, $x \in X$ is an *upper bound* of $A$ if $x \geq a$ for all $a \in A$. If the set ub$(A)$ of upper bounds of $A$ has a least element $y$, then $y$ is called the *least upper bound* of $A$ or *supremum* of $A$, denoted lub $A$, sup $A$, or $\bigvee A$. In discussions involving more than one poset, we may add an identifier to the notation by writing $\sup_X A$ or $\bigvee_X A$. If $A = \{x, y\}$ is a two-element set, sup $A = \sup\{x, y\}$ may be written $x \vee y$ and called the *join* of $x$ and $y$. In general, the supremum of a set may not exist.

Any partial order $\leq$ on a set $X$ has a *dual order* $\geq$ defined by $x \geq y$ if and only if $y \leq x$. Any term or statement defined in terms of $\leq$ has an associated dual obtained by replacing the partial order $\leq$ by $\geq$. The dual concepts to those of the preceding paragraph are *minimum element*, *minimal element*, *lower bound of a set*, and *greatest lower bound*. The greatest lower bound or *infimum* of a two-element set $\{x, y\}$ is denoted glb$\{x, y\} = \inf\{x, y\} = \bigwedge\{x, y\} = x \wedge y$, and is called the *meet* of $x$ and $y$.

Elements $a, b$ in a poset $(X, \leq)$ are *incomparable*, denoted $a \| b$ if $a \not\leq b$ and $b \not\leq a$. Infima and suprema may be called infs and sups. A function $f : (X, \leq) \rightarrow (Y, \leq_Y)$ between two posets is *order preserving* or *increasing* if $a \leq b$ implies $f(a) \leq_Y f(b)$. An *order isomorphism* is a bijection $f : (X, \leq) \rightarrow (Y, \leq_Y)$ between two posets with the property that $a \leq b$ if and only if $f(a) \leq_Y f(b)$.

**Definition 9.1.1.** A *lattice* is a poset $(L, \leq)$ in which every pair of elements $x, y$ has a supremum and an infimum.

A *complete lattice* is a poset $(L, \leq)$ in which every nonempty subset has a supremum and an infimum.

If a lattice has a minimum element, it is denoted $\bot$, and called the *bottom element*. If a lattice has a maximum element, it is denoted $\top$ and called the *top element*. A *bounded lattice* is a lattice with $\top$ and $\bot$.

An *inf-semilattice (sup-semilattice)* is a poset $(X, \leq)$ in which every pair of elements $x, y$ has an infimum (supremum). A *complete inf-semilattice (complete sup-semilattice)* is a poset $(X, \leq)$ in which every nonempty subset has an infimum (supremum). An inf-semilattice may also be called a *lower semilattice*, a *meet semilattice*, or an $\wedge$-*semilattice*. Sup-semilattices are defined dually and may be called *upper semilattices*, *join semilattices*, or $\vee$-*semilattices*.

If $(L, \leq)$ is a lattice, every pair of elements has a supremum and an infimum. Iterating, it is easy to show that every finite subset has a supremum and infimum. Thus, finite suprema and infima exist in a lattice, while arbitrary suprema and infima exist in a complete lattice. (By arbitrary suprema and infima, we mean suprema and infima of arbitrary *nonempty* subsets.) It is immediately clear that every totally ordered set $(X, \leq)$ is a lattice, but need not be a complete lattice. For example, the interval $(0, 1)$

https://doi.org/10.1515/9783110686579-010

with the usual order is not a complete lattice since, for example, $\{\frac{1}{n} : n \in \mathbb{N}\}$ has no infimum, or $(0,1)$ has no supremum. Indeed, if $(L, \leq)$ is a complete lattice, then $L$ must have an infimum and a supremum, and thus $L$ must have a least element $\bot$ and a greatest element $\top$. In the literature, 0 and 1 are often used for $\bot$ and $\top$, respectively.

**Example 9.1.2.** Let $X = \{\emptyset, \{a\}, \{b\}, \{c\}, \{a, b\}, \{b, c\}\}$ ordered by set inclusion as depicted in Figure 9.1. $X$ is not a lattice since, for example, $\{a\} \vee \{c\}$ does not exist.

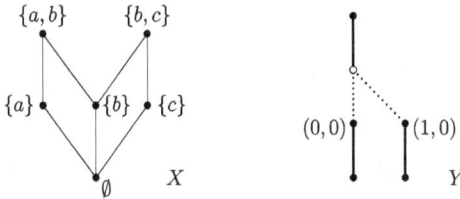

**Figure 9.1:** Posets $X$ and $Y$ are not lattices.

Let $Y = (\{0\} \times ([-1, 0] \cup (1, 2])) \cup (\{1\} \times [-1, 0]) \subseteq \mathbb{R}^2$ have the order $(a, b) \leq (c, d)$ if $(a = c$ and $b \leq d)$ or $(a = 1$ and $d \in (1, 2])$, as suggested in Figure 9.1. $Y$ is not a lattice since $(0, 0) \vee (1, 0)$ does not exist. Note that $(0, 0)$ and $(1, 0)$ have upper bounds, but no least upper bound.

If $(L, \leq)$ is a lattice, then we have two well-defined functions $\vee : L \times L \to L$ and $\wedge : L \times L \to L$. That is, the partial order of a lattice defines two binary operations $\vee$ and $\wedge$ on $L$. Some properties of $\vee$ and $\wedge$ are given in the next theorem.

**Theorem 9.1.3.** *If $(L, \leq)$ is a lattice, then, for all $a, b, c \in L$,*

$$
\begin{array}{lll}
(a \vee b) \vee c = a \vee (b \vee c) & and & (a \wedge b) \wedge c = a \wedge (b \wedge c) & \text{(associativity)} \\
a \vee b = b \vee a & and & a \wedge b = b \wedge a & \text{(commutativity)} \\
a \vee a = a & and & a \wedge a = a & \text{(idempotency)} \\
a \vee (a \wedge b) = a & and & a \wedge (a \vee b) = a. & \text{(absorption)}
\end{array}
$$

*Furthermore, if $\vee$ and $\wedge$ are binary operations on a nonempty set $L$ satisfying the eight properties above, then taking $x \leq y$ if and only if $x \wedge y = x$ gives a partial order $\leq$ on $L$ which makes $(L, \leq)$ a lattice.*

The details of the proof are routine and are omitted. The result says that a lattice $(L, \leq)$ may be viewed as an algebraic structure $(L, \vee, \wedge)$. In particular, when looking at lattice isomorphisms and lattice substructures, rather than considering functions that preserve the order and subsets that inherit the same order, we will require functions that preserve suprema and infima and subsets which inherit the same suprema and infima. Thus, it is not (merely) the order structure which characterizes a lattice, but the sups and infs.

If $(X, \leq)$ is a poset and $A \subseteq X$, then $(A, \leq)$, where $\leq$ is the order inherited from $X$, is a subposet of $(X, \leq)$. If $(L, \leq)$ is a lattice and $A \subseteq L$, then the subposet $(A, \leq)$ need not be a lattice, and if $(A, \leq)$ is a lattice, the suprema and infima in $(A, \leq)$ may not agree with those in $(L, \leq)$. For example, consider the power set lattice $(\mathcal{P}(\{1, 2, 3\}), \subseteq)$. The subset $A = \{\emptyset, \{1, 2\}, \{2, 3\}, \{1, 2, 3\}\}$, given the order $\subseteq$ inherited from $\mathcal{P}(\{1, 2, 3\})$, is a lattice and is a subposet of $\mathcal{P}(\{1, 2, 3\})$, but $\{1, 2\} \wedge \{2, 3\} = \emptyset$ in $A$ while $\{1, 2\} \wedge \{2, 3\} = \{2\}$ in $\mathcal{P}(\{1, 2, 3\})$. Since the infima do not agree, the subposet $(A, \subseteq)$ is not a sublattice of $(\mathcal{P}(\{1, 2, 3\}), \subseteq)$.

**Definition 9.1.4.** A *sublattice* (*complete sublattice*) of a lattice $(L, \leq)$ is a subposet $(A, \leq)$ of $(L, \leq)$ which is a lattice with the property that, for any finite (arbitrary) subsets $S \subseteq A$, the supremum $\bigvee_A S$ and infimum $\bigwedge_A S$ of $S$ taken in $A$ agree with the supremum $\bigvee_L S$ and infimum $\bigwedge_L S$ taken in $L$.

Based on this definition, the following result is immediate.

**Theorem 9.1.5.** *If $L$ is a lattice and $A \subseteq L$, then $A$ is a sublattice (complete sublattice) if and only if $A$ is closed under finite (arbitrary) suprema and infima. That is, $A$ is a sublattice (complete sublattice) of $L$ if and only if for any finite (arbitrary) nonempty subset $S \subseteq A$, $\bigvee_L S \in A$ and $\bigwedge_L S \in A$.*

**Definition 9.1.6.** If $L$ and $M$ are lattices, a *lattice isomorphism* is a bijective function $f : L \to M$ which *preserves* $\vee$ *and* $\wedge$, that is, with $f(x \vee y) = f(x) \vee f(y)$ and $f(x \wedge y) = f(x) \wedge f(y)$ for all $x, y, \in L$.

To check whether a poset $X$ is a complete lattice, we must check for the existence of arbitrary suprema and arbitrary infima. If $X$ is a complete lattice, then it has a bottom element $\bot$. If we know in advance that $X$ has a bottom element, to show $X$ is a complete lattice, the next result shows that it suffices to check only for the existence of arbitrary suprema.

**Theorem 9.1.7.** *If $(X, \leq)$ is a poset with a bottom element $\bot$, then $(X, \leq)$ is a complete lattice if and only if $\sup A$ exists for every nonempty $A \subseteq X$.*

*Proof.* From the definition, if $X$ is a complete lattice, then $\sup A$ exists for every nonempty $A \subseteq X$.

Suppose $(X, \leq)$ has bottom element $\bot$ and arbitrary suprema exist in $X$. To show $X$ is a complete lattice, we must show arbitrary infima exists. Suppose $A$ is a nonempty subset of $X$. The set lb $A$ of lower bounds of $A$ is nonempty, since $\bot$ is a lower bound of $A$. From our hypothesis, $b = \sup(\text{lb } A)$ exists. We will show $b \in \text{lb } A$. Suppose $a \in A$. If $x \in \text{lb } A$, then $x \leq a$. Thus $a$ is an upper bound of lb $A$, so $b = \text{lub}(\text{lb } A) \leq a$. This holds for all $a \in A$, so $b$ is a lower bound of $A$. Since $b = \sup(\text{lb } A)$, clearly $b$ is greater than every other lower bound of $A$. Thus, $b = \sup(\text{lb } A)$ is indeed glb $A = \inf A$. $\qquad\square$

We mention that the dual statement is also true. That is, if $X$ is a poset with top element $\top$ in which arbitrary infima exist, then $X$ is a complete lattice.

Our next result is a fixed-point theorem. If $f : X \rightarrow X$ is a function from a set to itself, a *fixed point* of $f$ is a point $x \in X$ with $f(x) = x$. Fixed-point theorems are an active area of research and have broad applications. Alfred Tarski lectured on the theorem below in 1939 and published it in 1955. He and Bronisław Knaster gave the result in the special case of $(\mathcal{P}(X), \subseteq)$ in 1927, and the result is also called the Knaster–Tarski fixed-point theorem.

**Theorem 9.1.8** (Tarski fixed-point theorem). *If $(L, \leq)$ is a complete lattice and $f :$ $(L, \leq) \rightarrow (L, \leq)$ is an order-preserving function, then $f$ has a fixed point. Furthermore, the set of fixed points of $f$ is a complete lattice.*

*Proof.* To see that the set FP of fixed points of $f$ is a complete lattice, suppose $A \subseteq$ FP. At this point, we do not know that FP $\neq \emptyset$. Let $V = \{x \in L : a \leq f(x) \leq x \, \forall a \in A\}$. Whether $A$ is empty or nonempty, $\top \in V$, so $V \neq \emptyset$. Notice that $V$ is the set of upper bounds of $A$ which satisfy $f(x) \leq x$ (since $a \leq x$ implies $a = f(a) \leq f(x)$ for $a \in A \subseteq$ FP). Let $z = \inf_L V$. We claim that $z \in$ FP and $z = \sup_{FP} A$. For every $v \in V, z \leq v$, so $f(z) \leq f(v) \leq v$, and thus $f(z)$ is a lower bound of $V$, so $f(z) \leq z$. Furthermore, for each $a \in A$, $a$ is a lower bound of $V$, so $a \leq \inf_L V = z$, so $z$ is an upper bound of $A$. As an upper bound of $A$ with $f(z) \leq z$, we have $z \in V$. Now $a \leq z$ for any $a \in A \subseteq$ FP, so $a = f(a) \leq f(z)$ for any $a \in A$, and thus $f(z)$ is an upper bound of $A$. Since $f(z) \leq z$, we have $f(f(z)) \leq f(z)$, so $f(z) \in V$ and $z = \inf_L V \leq f(z) \leq z$. Thus, $z = f(z)$, so $z \in$ FP, and in particular, FP $\neq \emptyset$. It remains to show that $z$ is the least upper bound of $A$ in FP. Suppose $b$ is an upper bound of $A$ in FP. Then, for any $a \in A, a \leq f(b) = b$, so $b \in V$ and thus $z = \inf_L V \leq b$, so $z$ is the least upper bound of $A$ in FP. The dual argument shows that $\inf_{FP} A$ exists, and is $\sup_L \{x \in L : x \leq f(x) \leq a\}$. $\square$

Note that this theorem does not say that the fixed points form a complete sublattice; generally this is not the case. The examples below illustrate this and give an indication of the broad utility of this theorem in areas other than topology.

**Example 9.1.9.** Recall that the power set $\mathcal{P}(X)$ ordered by set inclusion is a complete lattice.

(a) Let $(X, \mathcal{T})$ be a topological space. Let int $: (\mathcal{P}(X), \subseteq) \rightarrow (\mathcal{P}(X), \subseteq)$ be the function that maps $A \subseteq X$ to its interior. Now $A \subseteq B$ implies int $A \subseteq$ int $B$, so int is an order-preserving function on $\mathcal{P}(X)$. The Tarski fixed-point theorem implies that the fixed points of int form a complete lattice. But the fixed points of int are those sets $A$ with $A =$ int $A$, which are precisely the open sets. Thus, if $\mathcal{T}$ is any topology on $X$, $(\mathcal{T}, \subseteq)$ is a complete lattice, called the complete lattice of $\mathcal{T}$-open sets. Given any collection $\{U_i : i \in I\} \subseteq \mathcal{T}, \bigvee_{\mathcal{T}} \{U_i : i \in I\} = \bigcup \{U_i : i \in I\} = \bigvee_{\mathcal{P}(X)} \{U_i : i \in I\}$, but $\bigwedge_{\mathcal{T}} \{U_i : i \in I\}$ need not equal $\bigwedge_{\mathcal{P}(X)} \{U_i : i \in I\} = \bigcap \{U_i : i \in I\}$, since $\bigcap \{U_i : i \in I\}$ need not be open. In $\mathcal{T}, \bigwedge_{\mathcal{T}} \{U_i : i \in I\}$ is the largest open set contained in every $U_i$, which is int$(\bigcap \{U_i : i \in I\})$. Thus, unless $\mathcal{T}$ is an Alexandroff topology, $(\mathcal{T}, \subseteq)$ is not a sublattice of $(\mathcal{P}(X), \subseteq)$.

(b) Let $V$ be a vector space. Recall that $A \subseteq V$ is (vector-space) *convex* if for every $a, b \in A$ and for every scalar $t \in [0,1]$, $a + t(b-a) \in A$. Let $c : (\mathcal{P}(V), \subseteq) \to (\mathcal{P}(V), \subseteq)$ be the map that takes $A \subseteq V$ to the smallest convex set containing $A$, that is, to the *convex hull* of $A$. For example, to find the convex hull of a finite set $F$ in $\mathbb{R}^2$, drive a nail at each point, and stretch a rubber band around the nails. The rubber band encloses the convex hull of $F$. Now clearly $A \subseteq B$ implies $c(A) \subseteq c(B)$, and the fixed points of $c$ are the convex sets. Thus, the convex sets in a vector space form a complete lattice ordered by $\subseteq$.

(c) Let $G$ be a group. Consider the function from $(\mathcal{P}(G), \subseteq)$ to itself which maps $A \subseteq G$ to the smallest subgroup $[A]$ containing $A$. Clearly this map is order-preserving since $A \subseteq B$ implies $[A] \subseteq [B]$, and the fixed points of the map are the subgroups of $G$. Thus, the subgroups of a group $G$ are a complete lattice ordered by $\subseteq$.

If $(X, \leq)$ is a poset, we say $a$ *covers* $b$ and $b$ is *covered by* $a$ if $b < a$ and $b \leq x \leq a$ implies $x \in \{a, b\}$. In drawing the Hasse diagram for a finite poset $(X, \leq)$, the points of $X$ are represented by dots, and a line is drawn downward from $a$ to $b$ if and only if $a$ covers $b$. In a lattice with $\bot$, an *atom* is an element which covers $\bot$. Dually, in a lattice with $\top$, a *coatom* or *dual atom* is an element covered by $\top$. A lattice $L$ is *atomic* if $L$ has a least element $\bot$ and every other element $x \neq \bot$ is greater than or equal to an atom. *Dually atomic lattices* or *coatomic lattices* are defined dually.

If $L$ is a bounded lattice and $a \in L$, a *(lattice) complement* of $a$ is an element $b \in L$ such that $a \vee b = \top$ and $a \wedge b = \bot$. An element of a bounded lattice may have no complement, or may have many complements. A bounded lattice in which every element has a complement is a *complemented lattice*. A bounded lattice in which every element has a unique complement is a *uniquely complemented lattice*.

A lattice $(L, \vee, \wedge)$ is *distributive* if for any $a, b, c \in L$,

$$a \vee (b \wedge c) = (a \vee b) \wedge (a \vee c) \quad \text{and} \quad a \wedge (b \vee c) = (a \wedge b) \vee (a \wedge c).$$

In fact, it is enough to assume only one of the conditions above; each implies the other. The next result is an important characterization of distributive lattices. The proof may be found in any text on lattice theory, such as [13].

**Theorem 9.1.10.** *A lattice $L$ is distributive if and only if it contains no sublattice isomorphic to the "diamond lattice" or the "pentagon lattice" shown in Figure 9.2.*

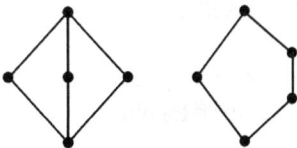

**Figure 9.2:** A lattice is distributive if and only if it contains no sublattice isomorphic to either of these lattices.

## Exercises

1. Suppose $(X, \leq)$ is a poset in which $x \vee y$ exists for every $x, y \in X$. Show that $\bigvee F$ exists for every finite set $F \subseteq X$.

2. Describe arbitrary sups and infs in the posets below.
   (a) $\mathcal{P}(X)$, the power set of $X$, ordered by inclusion.
   (b) $P = \{(a, \infty) : a \in (0, 1) \cup \{2\}\}$, ordered by inclusion.

3. Example 9.1.9 used the Tarski fixed-point theorem to show that a topology $\mathcal{T}$ on a set $X$ is a complete lattice, called the lattice of $\mathcal{T}$-open sets. Show that $(\mathcal{T}, \subseteq)$ is a complete lattice using Theorem 9.1.7.

4. The convex subsets of a vector space $V$ form a complete lattice. (See Example 9.1.9.) Is this a sublattice of $(\mathcal{P}(V), \subseteq)$?

5. A poset $(X, \leq)$ is *directed* if every pair of elements of $X$ has an upper bound. Exhibit a finite directed set which is not a $\vee$-semilattice.

6. For each poset below, determine whether it is a lattice, a complete lattice, a $\wedge$-semilattice, a complete $\wedge$-semilattice, a $\vee$-semilattice, or a complete $\vee$-semilattice. (More than one may apply.)
   (a) $B = \{A \subseteq \mathbb{R}^2 : A \text{ is bounded}\}$, ordered by $\subseteq$.
   (b) $(\mathbb{N}, |)$, where $n|m$ if and only if $m = nk$ for some $k \in \mathbb{N}$.
   (c) The poset $F([0, 1])$ of functions $f : [0, 1] \rightarrow \mathbb{R}$, with the order $f \leq g$ if and only if $f(x) \leq g(x)$ for all $x \in [0, 1]$.

7. With $F([0, 1])$ as in Exercise 6, consider the subposet $C([0, 1]) = \{f \in F([0, 1]) : f \text{ is continuous}\}$. If they exist, find $\bigwedge_{F([0,1])} \{x^n : n \in \mathbb{N}\}$ and $\bigwedge_{C([0,1])} \{x^n : n \in \mathbb{N}\}$.

8. Show that every lattice isomorphism is an order isomorphism.

9. Let $\mathcal{T}$ be a topology on $X$, and let $\mathcal{C} = \{X - U : U \in \mathcal{T}\}$ be the collection of $\mathcal{T}$-closed sets in $X$. Show that $(\mathcal{C}, \supseteq)$ is a complete lattice which is isomorphic to the lattice $(\mathcal{T}, \subseteq)$ of $\mathcal{T}$-open sets.

10. If $X \neq \emptyset$, show that the power set lattice $(\mathcal{P}(X), \subseteq)$ is a uniquely complemented lattice. Identify $\bot$, $\top$, and the unique complement of each $A \subseteq X$.

11. For the diamond and pentagon lattices shown in Figure 9.2, label the points and find all complements of each point.

12. Suppose $L$ is a lattice. Show that $a \vee (b \wedge c) = (a \vee b) \wedge (a \vee c)$ for every $a, b, c \in L$ implies $a \wedge (b \vee c) = (a \wedge b) \vee (a \wedge c)$ for every $a, b, c \in L$.

13. Suppose $P$ is a poset and $\mathcal{F}$ is the collection of all increasing functions from $P$ to $P$ which are strictly below the diagonal. That is $\mathcal{F} = \{f : P \rightarrow P : x \leq y \Rightarrow f(x) \leq f(y) < y \ \forall x, y \in P\}$. Suppose there exist $f, g \in \mathcal{F}$ with $f(x) \leq g \circ g(x)$ for all $x \in P$. Show that $a$ does not cover $f(a)$ for any $a \in P$.

14. Suppose $f : \mathbb{N} \rightarrow \mathbb{N}$ is defined by taking $f(n)$ to be the sum of the digits of $n$. In the quasiorder associated with the functionally Alexandroff topology $\mathcal{T}_f$ on $\mathbb{N}$, describe the set $S$ of integers $m$ which cover 2.

15. Show that the diamond and pentagon lattices shown in Figure 9.2 are not distributive lattices.

16. Show that in a distributive lattice, any element with a complement has a unique complement.
17. Show that every totally ordered set is a distributive lattice.
18. For the bounded lattices below, identify all atoms and coatoms and find all complements of each labeled element.

    $X$ is the finite poset whose Hasse diagram is shown.

    $Y = (\{-1\} \times [0,1]) \cup (\{0\} \times (0,2]) \cup (\{1\} \times [0,1]) \cup \{(0,-1)\}$ with the order $(a,b) \le (c,d)$ if and only if $a = c$ and $b \le d$, together with $(0,-1) \le (a,b) \le (0,2)$ for all $(a,b) \in Y$.

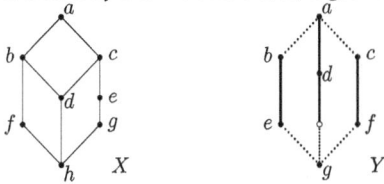

## 9.2 Compact and Hausdorff topologies in $T(X)$

We know several topologies on the plane $\mathbb{R}^2$, including the Euclidean topology $\mathcal{T}_{\mathcal{E}}$, the cofinite topology $\mathcal{T}_{cf}$, the particular point topologies, the discrete topology $\mathcal{T}_D$, and the indiscrete topology $\mathcal{T}_I$. For any topology $\mathcal{T}$ on $\mathbb{R}^2$, we have $\mathcal{T}_I \subseteq \mathcal{T} \subseteq \mathcal{T}_D$. In $\mathbb{R}^2$, every finite subset is $\mathcal{T}_{\mathcal{E}}$-closed, and thus every $\mathcal{T}_{cf}$-closed set is $\mathcal{T}_{\mathcal{E}}$-closed. Taking complements shows $\mathcal{T}_{cf} \subseteq \mathcal{T}_{\mathcal{E}}$. Such considerations are part of the general study of the collection of all topologies on a set (like $\mathbb{R}^2$) and how they are ordered by $\subseteq$.

**Definition 9.2.1.** If $X$ is a nonempty set, $T(X)$ will denote the poset of all topologies on $X$ ordered by set inclusion $\subseteq$.

Recall that if $\mathcal{T}_C, \mathcal{T}_F \in T(X)$ and $\mathcal{T}_C \subseteq \mathcal{T}_F$, we say $\mathcal{T}_C$ is coarser than $\mathcal{T}_F$ and $\mathcal{T}_F$ is finer than $\mathcal{T}_C$.

**Theorem 9.2.2.** $T(X)$ is a complete lattice.

*Proof.* Given a collection of topologies $\{\mathcal{T}_i : i \in I\} \subseteq T(X)$, $\bigcap_{i \in I} \mathcal{T}_i$ is a topology, and is the largest topology contained in every $\mathcal{T}_i$. Thus, $\bigwedge_{i \in I} \mathcal{T}_i = \bigcap_{i \in I} \mathcal{T}_i$. The topology $[\bigcup_{i \in I} \mathcal{T}_i]$ generated by the subbasis $\bigcup_{i \in I} \mathcal{T}_i$ is the smallest topology containing every $\mathcal{T}_i$, and thus is $\bigvee_{i \in I} \mathcal{T}_i$. □

With this result, $T(X)$ will be called the *lattice of topologies on X*.

Note that $\mathcal{T}_1 \vee \mathcal{T}_2$, the smallest topology containing $\mathcal{T}_1$ and $\mathcal{T}_2$, has $\mathcal{T}_1 \cup \mathcal{T}_2$ as a subbasis and thus has a basis of finite intersections of elements of $\mathcal{T}_1 \cup \mathcal{T}_2$. Because $\mathcal{T}_1$ and $\mathcal{T}_2$ are closed under finite intersections, it follows that $\mathcal{T}_1 \vee \mathcal{T}_2$ has a basis $\{U_1 \cap U_2 : U_1 \in \mathcal{T}_1, U_2 \in \mathcal{T}_2\}$.

Suppose $(X, \mathcal{T})$ is Hausdorff (or $T_1$, or $T_0$). Since $\mathcal{T}$ already has enough open sets to achieve the required separation of points, any finer topology $\mathcal{T}_F \supseteq \mathcal{T}$ will also be Hausdorff (or $T_1$, or $T_0$). This says that the Hausdorff (or $T_1$, or $T_0$) topologies on $X$ form an increasing set in the lattice $T(X)$.

Similarly, if $(X, \mathcal{T})$ is compact and $\mathcal{T}_C \subseteq \mathcal{T}$ is a coarser topology on $X$, any $\mathcal{T}_C$-open cover is a $\mathcal{T}$-open cover, and thus has a finite subcover. Thus, the compact topologies on $X$ form a decreasing set in $T(X)$.

If $X$ admits a compact Hausdorff topology, then the decreasing set of compact topologies intersects the increasing set of Hausdorff topologies. The next result shows that the set of compact topologies and the set of Hausdorff topologies can barely intersect; the intersection contains no pair of distinct topologies $\mathcal{T} \subset \mathcal{T}'$, with one strictly finer than the other.

**Theorem 9.2.3.** *If $\mathcal{T}$ is a compact Hausdorff topology on $X$, then $\mathcal{T}$ is maximal in the set of compact topologies on $X$ and is minimal in the set of Hausdorff topologies on $X$.*

*Proof.* Suppose $\mathcal{T}$ is a compact Hausdorff topology on $X$ and $\mathcal{T}_F \supset \mathcal{T}$ is a strictly finer compact topology on $X$. Then there exists a $\mathcal{T}_F$-closed set $A$ which is not $\mathcal{T}$-closed. But since $A$ is closed in the compact space $(X, \mathcal{T}_F)$, it is compact relative to $\mathcal{T}_F$. This implies that $A$ is compact relative to the coarser topology $\mathcal{T}$. As a compact subset of a Hausdorff space $(X, \mathcal{T})$, $A$ must be $\mathcal{T}$-closed. This contradicts our choice of $A$. Thus, there exists no compact topology $\mathcal{T}_F$ strictly finer than a compact Hausdorff topology $\mathcal{T}$.

Suppose $\mathcal{T}$ is a compact Hausdorff topology on $X$ and $\mathcal{T}_C \subset \mathcal{T}$ is a strictly coarser Hausdorff topology on $X$. Then there exists a $\mathcal{T}$-closed set $A$ which is not $\mathcal{T}_C$-closed. As a closed subspace of the compact space $(X, \mathcal{T})$, $A$ is compact. Thus, $A$ is compact as a subspace of the coarser space $(X, \mathcal{T}_C)$. But a compact set in a Hausdorff space is closed, so $A$ is $\mathcal{T}_C$-closed, contrary to our choice of $A$. Thus, there exists no Hausdorff topology $\mathcal{T}_C$ strictly coarser than a compact Hausdorff topology $\mathcal{T}$. $\quad\square$

Figure 9.3 suggests the intersection of the decreasing set of compact topologies and the increasing set of Hausdorff topologies on a set $X$. This figure also suggests that there are maximal compact topologies which are not Hausdorff and minimal Hausdorff topologies which are not compact. This is indeed the case, as seen by the Examples 9.2.7 and 9.2.8 below.

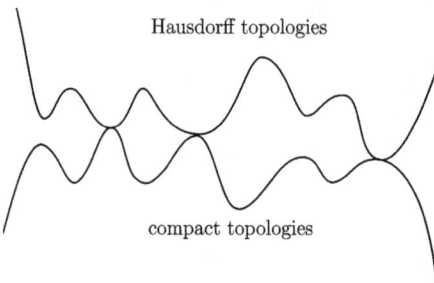

**Figure 9.3:** Compact Hausdorff topologies are maximal compact and minimal Hausdorff topologies in $T(X)$.

To get a topology finer than $T$, we must add open sets. The smallest topology obtained by adding a specified set $A$ is defined below.

**Definition 9.2.4.** If $T$ is a topology on $X$ and $A \subseteq X$, the *simple extension of $T$ by $A$* is the topology $T(A)$ having subbasis $T \cup \{A\}$. Thus, $T(A) = \{U \cup (V \cap A) : U, V \in T\}$.

Intuitively, if $A$ is added to $T$ as a new open set and the result is a compact topology, then open covers of form $\{A\} \cup \mathcal{C}$ should have finite subcovers, which means open covers $\mathcal{C}$ of $X - A$ should have finite subcovers, and thus $X - A$ must be compact. This fact and its converse are our next lemma.

**Lemma 9.2.5.** *If $(X, T)$ is compact, then $(X, T(A))$ is compact if and only if $X - A$ is a compact subspace of $(X, T)$.*

*Proof.* Suppose $(X, T)$ and $(X, T(A))$ are compact. Since $A$ is open in $T(A)$, $X - A$ is closed in the compact space $(X, T(A))$, and thus $X - A$ is compact in $T(A)$. Since $T \subseteq T(A)$, it follows that $X - A$ is compact in $(X, T)$.

Conversely, suppose $(X, T)$ is compact and $X - A$ is compact in $(X, T)$. To show $(X, T(A))$ is compact, suppose $\mathcal{C} = \{U_i \cup (V_i \cap A) : i \in I\}$ is a $T(A)$-open cover of $X$, where $U_i, V_i \in T$ for all $i \in I$. Now $\{U_i : i \in I\}$ and $\{U_i \cup V_i : i \in I\}$, respectively, are $T$-open covers of $X - A$ and $X$, so there exist finite index sets $F_1, F_2 \subseteq I$ such that $\{U_i : i \in F_1\}$ covers $X - A$ and $\{U_i \cup V_i : i \in F_2\}$ covers $X$. In particular, $\{U_i \cup (V_i \cap A) : i \in F_2\}$ covers $A$, so $\{U_i \cup (V_i \cap A) : i \in F_1 \cup F_2\}$ is a finite $T(A)$-subcover of $X$. Thus, $(X, T(A))$ is compact. □

**Theorem 9.2.6.** *If $(X, T)$ is a compact topological space, then the following are equivalent.*
(a) *$(X, T)$ is maximal compact.*
(b) *A subset of $X$ is compact if and only if it is closed.*
(c) *Every continuous bijection from a compact space $(W, T_W)$ to $(X, T)$ is a homeomorphism.*

*Proof.* (a) $\Rightarrow$ (b): Since every closed subset of the compact space $(X, T)$ is compact, we only need to show that (a) implies every compact subset of $X$ is closed. Assuming (a), suppose $A$ is compact but not closed. By Lemma 9.2.5, the simple extension $T(X - A)$ of $T$ by $X - A$ is a strictly finer compact topology on $X$, contrary to the maximality of $T$.

(b) $\Rightarrow$ (c): Assuming (b), suppose $(W, T_W)$ is compact, and $f : (W, T_W) \to (X, T)$ is a continuous bijection. To see that $f$ is a homeomorphism, it suffices to show that $f$ is a closed map. If $K$ is closed in the compact space $W$, then $K$ is compact, so $f(K)$ is compact and, by (b), is closed. Thus $f$ is a closed map, as needed.

(c) $\Rightarrow$ (a): If $(X, T)$ is not maximal compact, then there exists a strictly finer topology $T' \supset T$ which is compact. Now id $: (X, T') \to (X, T)$ is a continuous bijection which is not a homeomorphism. □

If $(X, T)$ is Hausdorff, then both (b) and (c) of Theorem 9.2.6 hold. (See Theorem 5.1.6 and Exercise 14 of Section 5.1.) Thus, this provides an indirect proof that compact Hausdorff spaces are maximal compact spaces, which is half of Theorem 9.2.3.

We conclude this section with examples showing that a maximal compact topology on $X$ need not be Hausdorff, and a minimal Hausdorff topology need not be compact.

**Example 9.2.7.** *Maximal compact topologies need not be Hausdorff.* Let $X = \mathbb{N}^2 \cup \{a, b\}$ where $a, b \notin \mathbb{N}^2$. Given a sequence $(a_n)_{n \in \mathbb{N}}$ of natural numbers, define $U(a_n) = \{a\} \cup \{(n, y) \in \mathbb{N}^2 : n \in \mathbb{N}, y \geq a_n\}$. Thus, $U(a_n)$ consists of the point $a$ and every point of $\mathbb{N}^2$ on or above the graph $\{(n, a_n) : n \in \mathbb{N}\}$ of the sequence $(a_n)_{n \in \mathbb{N}}$. Given $k \in \mathbb{N}$, let $V(k) = \{b\} \cup ([k, \infty) \times \mathbb{N})$. Let $\mathcal{T}$ be the topology on $X$ having basis

$$\mathcal{B} = \{(m, n) : (m, n) \in \mathbb{N}^2\}$$
$$\cup \{U(a_n) : (a_n)_{n \in \mathbb{N}} \text{ is a sequence in } \mathbb{N}\}$$
$$\cup \{V(k) : k \in \mathbb{N}\}.$$

Now $a$ and $b$ cannot be separated by open sets in $X$, so $X$ is not Hausdorff. Any open cover $\mathcal{C}$ of $X$ contains sets $W_a, W_b$ covering $a$ and $b$, and these two sets cover all but finitely many points of $X$, which can be covered by finitely many members of $\mathcal{C}$. Thus, $X$ is compact.

We will show $X$ is maximal compact by showing that every compact set must be closed, or equivalently, every non-closed set is non-compact. Suppose $H \subseteq X$ is not closed. Then $\operatorname{cl} H - H \neq \emptyset$ and, since $X$ is $T_1$, $H$ is infinite. If $(m, n) \in \operatorname{cl} H \cap \mathbb{N}^2$, the neighborhood $\{(m, n)\}$ of $(m, n)$ intersects $H$, so $(m, n) \in H$. Thus, the only possible points in $\operatorname{cl} H - H$ are $a$ and $b$. Suppose $a \in \operatorname{cl} H - H$. There must be some column $\{k\} \times \mathbb{N}$ which contains infinitely many elements of $H$: otherwise, with $a_n = \max(\{1\} \cup (\{n\} \times \mathbb{N}) \cap H)$, the neighborhood $U(1 + a_n)$ of $a$ would not intersect $H$, contrary to $a \in \operatorname{cl} H$. Now $V(k + 1)$ and all the singletons of $\mathbb{N}^2$ give an open cover of $H$ with no finite subcover, so $H$ is non-compact. If $b \in \operatorname{cl} H - H$, then every neighborhood $V(k)$ of $b$ intersects $H$, so we may choose an infinite sequence $((n, a_n))_{n \in J}$ in $H$, where $J \subseteq \mathbb{N}$. For $n \in \mathbb{N} - J$, take $a_n = 1$. This defines a sequence $(a_n)_{n \in \mathbb{N}}$. Now $U(1 + a_n)$ and the singletons in $\mathbb{N}^2$ form an open cover of $H$ with no finite subcover, so in all cases, a non-closed set $H$ is non-compact.

**Example 9.2.8.** *Minimal Hausdorff topologies need not be compact.* Let $X = \mathbb{R} \cup \{a, b\}$ where $a, b \notin \mathbb{R}$. Let $\mathcal{T}$ be the topology on $X$ having basis

$$\mathcal{B} = \{(x - \varepsilon, x + \varepsilon) : x \in \mathbb{R}, \varepsilon > 0\}$$
$$\cup \left\{ \{a\} \cup \bigcup_{|n| \geq M} (2n, 2n + 1) : n \in \mathbb{N}, M \in \mathbb{Z} \right\}$$
$$\cup \left\{ \{b\} \cup \bigcup_{|n| \geq M} (2n - 1, 2n) : n \in \mathbb{N}, M \in \mathbb{Z} \right\}.$$

It is easily seen that $(X, \mathcal{T})$ is Hausdorff. Since $A = \{a, b\} \cup \bigcup\{(n, n + 1) : n \in \mathbb{Z}\} = X - \mathbb{Z}$ is open, $\{A\} \cup \{(n - \frac{1}{4}, n + \frac{1}{4}) : n \in \mathbb{Z}\}$ is an open cover of $X$ with no finite subcover, and thus $(X, \mathcal{T})$ is not compact.

To see that $(X, \mathcal{T})$ is minimal among the Hausdorff topologies on $X$, suppose $\mathcal{T}'$ is a Hausdorff topology on $X$ with $\mathcal{T}' \subseteq \mathcal{T}$. We will show that $\mathcal{B} \subseteq \mathcal{T}'$, and thus $\mathcal{T} \subseteq \mathcal{T}' \subseteq \mathcal{T}$.

Notice that the subspace topology $\mathcal{T}|_{\mathbb{R}}$ on $\mathbb{R}$ as a subspace of $(X, \mathcal{T})$ is the Euclidean topology $\mathcal{T}_{\mathcal{E}}$. Thus, any closed interval $[a, b] \subseteq \mathbb{R} \subseteq X$ is compact in $\mathcal{T}$, and thus is compact in the coarser topology $\mathcal{T}'$. Since the Hausdorff property is hereditary, $[a, b]$ is compact $T_2$ in both $\mathcal{T}'|_{[a,b]}$ and $\mathcal{T}|_{[a,b]}$, and $\mathcal{T}'|_{[a,b]} \subseteq \mathcal{T}|_{[a,b]} = \mathcal{T}_{\mathcal{E}}|_{[a,b]}$. Since the compact $T_2$ topologies are minimal among the Hausdorff topologies on $[a, b]$ (Theorem 9.2.3), $\mathcal{T}'|_{[a,b]} = \mathcal{T}|_{[a,b]} = \mathcal{T}_{\mathcal{E}}|_{[a,b]}$. Given an interval $(c, d)$ in $\mathbb{R}$ we will show $(c, d) \in \mathcal{T}'$. Suppose $x \in (c, d)$. Since $(X, \mathcal{T}')$ is Hausdorff, there exist mutually disjoint $\mathcal{T}'$-open neighborhoods $N_x, N_a$, and $N_b$ of $x, a$, and $b$, respectively. Now $N_x \in \mathcal{T}' \subseteq \mathcal{T}$ implies there exists $m \in \mathbb{Z}$ with $((-\infty, -m) \cup (m, \infty)) \cap N_x \subseteq \mathbb{Z}$. But the restriction of $\mathcal{T}$ to $\mathbb{R}$ is the Euclidean topology, so $N_x$ can have no isolated points, and thus $N_x \subseteq [-m, m]$. On any interval $[z, w]$, the topologies $\mathcal{T}, \mathcal{T}'$ and $\mathcal{T}_{\mathcal{E}}$ agree, and now having $N_x \subseteq [-m, m]$ essentially tells us that nothing can go wrong outside the bounded interval. In particular, $N_x \cup (c, d) \subseteq [z, w] = [-m \wedge c, m \vee d]$, so $(c, d)$ is $\mathcal{T}'|_{[z,w]}$-open, and thus $(c, d) = U \cap [z, w]$ for some $U \in \mathcal{T}'$. Now $U \cap N_x$ is a $\mathcal{T}'$-neighborhood of $x$ contained in $(c, d)$, so $(c, d)$ is $\mathcal{T}'$-open. In particular, the sets of form $(x - \varepsilon, x + \varepsilon) \in \mathcal{B}$ are in $\mathcal{T}'$.

Next, suppose $B = \{a\} \cup \bigcup_{|n| \geq M} (2n, 2n + 1)$ is a basic $\mathcal{T}$-open neighborhood of $a$. To show $B \in \mathcal{T}'$, for any $x \in B$ we must find a $\mathcal{T}'$-neighborhood of $x$ contained in $B$. This is immediate for any $x \in B \cap \mathbb{R}$, since the paragraph above shows $(c, d) \in \mathcal{T}'$ for any $c, d \in \mathbb{R}$. It remains to show the existence of a $\mathcal{T}'$ neighborhood of $a$ contained in $B$. Since $\mathcal{T}'$ is Hausdorff, there exist disjoint $\mathcal{T}'$-open neighborhoods $N_a$ and $N_b$ of $a$ and $b$, respectively. Now $\mathcal{T}' \subseteq \mathcal{T}$ implies that there exists $M \in \mathbb{N}$ such that $N_a$ is disjoint from $\bigcup \{(2n - 1, 2n) : n \in \mathbb{N}, |n| \geq M\}$. Furthermore, the open set $N_a$ cannot contain $2n - 1$ or $2n$ for $|n| \geq M$ and still be disjoint from $\bigcup \{(2n - 1, 2n) : n \in \mathbb{N}, |n| \geq M\}$. Thus, $N_a - B$ must be bounded, say $N_a - B \subseteq [c, d]$. Now $[c, d]$ is compact and Hausdorff in $(X, \mathcal{T}')$, and thus is $\mathcal{T}'$-closed. It follows that $N_a \cap (X - [c, d])$ is a $\mathcal{T}'$-neighborhood of $a$. Since $N_a - B \subseteq [c, d]$ and $N_a$ is the disjoint union of $N_a - B$ and $N_a \cap B$, it follows that $N_a \cap (X - [c, d]) \subseteq N_a \cap B \subseteq B$. Thus, $N_a \cap (X - [c, d])$ is a $\mathcal{T}'$-neighborhood of $a$ contained in $B$, as needed.

A similar argument shows that any basic $\mathcal{T}$-neighborhood of $b$ is $\mathcal{T}'$ open, so $\mathcal{T} = \mathcal{T}'$, and thus $\mathcal{T}'$ is a minimal Hausdorff topology on $X$ which is not compact.

Some of the results of this section are drawn from [7, 9, 10, 23, 24, 30, 32, 34, 44, 46].

## Exercises

1. Given topologies $\mathcal{T}_1, \mathcal{T}_2$ on $X$, it follows from the definitions that $\mathcal{T}_1 \vee \mathcal{T}_2$ has a basis $\{U_1 \cap U_2 : U_1 \in \mathcal{T}_1, U_2 \in \mathcal{T}_2\}$. Suppose $\mathcal{B}_i$ is a basis for $\mathcal{T}_i$ ($i = 1, 2$). Show that $\{B_1 \cap B_2 : B_1 \in \mathcal{B}_1, B_2 \in \mathcal{B}_2\}$ is a basis for some topology $\mathcal{T}_3$ on $X$ and $\mathcal{T}_3 = \mathcal{T}_1 \vee \mathcal{T}_2$.

2. Use the Tarski fixed-point theorem to prove that the collection $T(X)$ of topologies on $X$, ordered by inclusion, is a complete lattice.

3. Suppose $(X, \mathcal{T}_1)$ and $(X, \mathcal{T}_2)$ are connected. Must either of $(X, \mathcal{T}_1 \vee \mathcal{T}_2)$ or $(X, \mathcal{T}_1 \wedge \mathcal{T}_2)$ be connected? Provide proofs or counterexamples.

4. Shown below are bases for three topologies $\mathcal{T}_1$, $\mathcal{T}_2$, and $\mathcal{T}_3$ on $X = \{1, 2, 3, 4, 5\}$. Find bases for $\mathcal{T}_i \vee \mathcal{T}_j$ and $\mathcal{T}_i \wedge \mathcal{T}_j$ for all distinct $i, j \in \{1, 2, 3\}$.

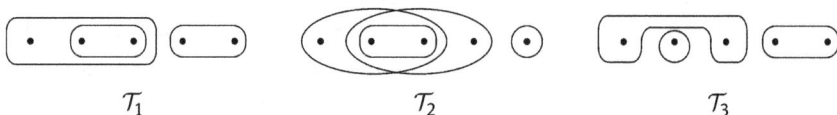

5. Let $(X, \mathcal{T})$ be the interval $[-2, 2]$ with the Euclidean topology. Since this is compact and Hausdorff, any strictly finer topology on $[-2, 2]$ cannot be compact. Let $\mathcal{T}_F$ be the smallest topology on $[-2, 2]$ for which each member of $\mathcal{D} = \{(a, b) \cap [-2, 2] : a, b \in \mathbb{R}, a < b\} \cup \{(-1, 0]\}$ is open. Note that $\mathcal{T} \subset \mathcal{T}_F$.
   (a) Is $\mathcal{D}$ a basis for the topology $\mathcal{T}_F$? If not, describe a basis for $\mathcal{T}_F$.
   (b) Show that $\mathcal{T}_F$ is not compact directly by exhibiting an open cover which has no finite subcover.

6. Suppose $\mathcal{T}, \mathcal{T}_F \in T(X)$ with $\mathcal{T} \subseteq \mathcal{T}_F$.
   (a) Is the complete lattice of $\mathcal{T}$-open sets a sublattice of the complete lattice of $\mathcal{T}_F$-open sets? Give a proof or counterexample.
   (b) Is the complete lattice of $\mathcal{T}$-open sets a complete sublattice of the complete lattice of $\mathcal{T}_F$-open sets? Give a proof or counterexample.

7. Suppose $\mathcal{T}$ is a topology on $X$ and $A \notin \mathcal{T}$. Show that the simple extension $\mathcal{T}(A)$ of $\mathcal{T}$ by $A$ need not cover $\mathcal{T}$ in $T(X)$.

8. Suppose $\mathcal{T}$ is a $T_1$ topology on an infinite set $X$.
   (a) Suppose $A \notin \mathcal{T}$. Pick an element $a \in A \cap \partial A$, and let $\mathcal{B}' = \{\{x\} : x \in X - \{a\}\} \cup \{\{a\} \cup (U - A) : a \in U \in \mathcal{T}\}$. Show that $\mathcal{B}'$ is a basis for a Hausdorff topology $\mathcal{T}'$ on $X$ with $\mathcal{T} \subseteq \mathcal{T}'$ and $A \notin \mathcal{T}'$.
   (b) Show that $\mathcal{T}$ is the intersection of all $T_2$ topologies on $X$ finer than $\mathcal{T}$.

9. Given a topological property $P$, show that the following are equivalent:
   (a) Property $P$ is preserved under continuous bijections. That is, if $(X, \mathcal{T})$ has property $P$ and $f : (X, \mathcal{T}) \to (Y, \mathcal{T}_Y)$ is a continuous bijection, then $(Y, \mathcal{T}_Y)$ has property $P$.
   (b) For any set $X$, $\{\mathcal{T} \in T(X) : (X, \mathcal{T})$ has property $P\}$ is a decreasing set in $T(X)$.
   (Hint: Given a continuous bijection $f : (X, \mathcal{T}) \to (Y, \mathcal{T}_Y)$, consider $\mathcal{T}' = \{f(U) \subseteq Y : U \in \mathcal{T}\}$.)

10. Show that every nonempty set $X$ admits a compact $T_2$ topology.

11. For $X = \mathbb{R} \cup \{a, b\}$ where $a, b \notin \mathbb{R}$, give three non-homeomorphic minimal $T_2$ topologies in $T(X)$ besides the one given in Example 9.2.8.

12. Show that $T(X)$ has a minimum $T_2$ topology if and only if $X$ is a nonempty finite set.

13. Suppose $(X, \mathcal{T})$ is maximal connected (that is, maximal among the connected topologies in $T(X)$). Show that if $A$ and $X - A$ are both connected in $X$, then either $A$ or $X - A$ is open.

14. If $\mathcal{T}$ is a minimal Hausdorff topology in $T(X)$ and $A \subseteq X$ is open and closed, show that the subspace topology $\mathcal{T}|_A$ on $A$ is minimal Hausdorff in $T(A)$.

## 9.3 Sublattices and complements in $T(X)$

The atoms in $T(X)$ are topologies $\{\emptyset, A, X\}$ with exactly three open sets, $\emptyset \subset A \subset X$. Every topology other than the indiscrete topology is above an atom, so $T(X)$ is atomic. Indeed, every $\mathcal{T} \in T(X)$ other than $\perp$ is the supremum of the atoms below it. An atomic lattice with this property is called an *atomistic lattice*.

We will describe some of the coatoms in $T(X)$. For distinct elements $p, q \in X$, let $\mathcal{T}_{\{p,q\}} = \text{Super}(\{p\}) \cup \text{Disjoint}(\{p, q\})$. Thus, the $\mathcal{T}_{\{p,q\}}$-open sets contain $p$ or miss both $p$ and $q$. To see $\mathcal{T}_{\{p,q\}}$ is a coatom, suppose $\mathcal{T}' \supset \mathcal{T}_{\{p,q\}}$ is a strictly finer topology on $X$. Then $\mathcal{T}'$ contains a set $U \notin \mathcal{T}_{\{p,q\}}$, and thus $U$ contains $q$ but not $p$. As the union of two $\mathcal{T}'$-open sets, $(X - \{p, q\}) \cup U = X - \{p\}$ is $\mathcal{T}'$-open. Any sets containing $p$ are $\mathcal{T}'$-open and, intersecting such sets with $X - \{p\}$, we see that any sets not containing $p$ are $\mathcal{T}'$-open. This shows $\mathcal{T}'$ is the discrete topology, so $\mathcal{T}_{\{p,q\}}$ is a coatom.

A full description of the coatoms in $T(X)$ would require a discussion of ultrafilters beyond the scope of our presentation. Omitting the details, we remark that coatoms in $T(X)$ of form $\mathcal{T}_{\{p,q\}}$ are called *principal ultratopologies*, and each is generated by a principal ultrafilter on $X$. All the other coatoms are called *non-principal ultratopologies*, and each is generated by a non-principal ultrafilter.

We have seen that the set of $T_2$ topologies in $T(X)$ is an increasing set whose minimal elements include the compact $T_2$ topologies on $X$. Every infinite set admits many compact $T_2$ topologies (see Exercises 10 and 11 of the previous section), and the infimum of any distinct pair of these in $T(X)$ is not a $T_2$ topology. Thus, the $T_2$ topologies on $X$ do not form a sublattice of $T(X)$. The situation is different for $T_1$ topologies.

**Theorem 9.3.1.** *The set of $T_1$ topologies in $T(X)$ is a complete sublattice of $T(X)$.*

*Proof.* If $\mathcal{T}_i$ is a $T_1$ topology on $X$ for all $i \in I$, $\bigvee_{i \in I} \mathcal{T}_i$ is finer than each of the $T_1$ topologies $\mathcal{T}_i$, and thus is $T_1$. For each $i \in I$ and each $x \in X$, $X - \{x\} \in \mathcal{T}_i$, so $X - \{x\} \in \bigcap_{i \in I} \mathcal{T}_i = \bigwedge_{i \in I} \mathcal{T}_i$, so $\bigwedge_{i \in I} \mathcal{T}_i$ is $T_1$. $\square$

Below, we find some other sublattices of $T(X)$.

**Theorem 9.3.2.** *The collection $A(X)$ of Alexandroff topologies on $X$ is a sublattice of $T(X)$ and a complete lattice, but not a complete sublattice of $T(X)$.*

*Proof.* If $\mathcal{T}_i \in A(X)$ for $i \in I$, then $\bigwedge_{T(X)}\{\mathcal{T}_i : i \in I\} = \bigcap\{\mathcal{T}_i : i \in I\}$. Since $\bigcap\{\mathcal{T}_i : i \in I\} \subseteq \mathcal{T}_{i_0}$ for any $i_0 \in I$, it is closed under arbitrary intersections and thus is in $A(X)$.

If $T_1, T_2 \in A(X)$, then $T_1 \vee T_2 = [T_1 \cup T_2]$ consists of unions of finite intersections of members of $T_1 \cup T_2$ (where $\vee$ is taken in $T(X)$). If $N_i(x)$ is the smallest $T_i$-neighborhood of $x$, then, for every $x \in X$, $N_1(x) \cap N_2(x) \in T_1 \vee T_2$. Since there is no way to obtain a smaller neighborhood of $x$ as a union of finite intersections of members of $T_1 \cup T_2$, $N_1(x) \cap N_2(x)$ is the smallest $T_1 \vee T_2$-neighborhood of $x$, so $T_1 \vee T_2 \in A(X)$. This shows that $A(X)$ is a lattice and a complete $\wedge$-semilattice. Since $A(X)$ has a largest element (the discrete topology), $A(X)$ is a complete lattice by (the dual of) Theorem 9.1.7.

Unless every topology on $X$ is an Alexandroff topology, $T(X)$ is not a complete sublattice: If $T \in T(X)$ is not an Alexandroff topology, then it is the supremum of all the atoms below it, and these atoms are all Alexandroff topologies. Thus, the set of Alexandroff topologies is not closed under arbitrary suprema. $\qquad\square$

Recall that a relation $R$ on $X$ is a subset of $X \times X$, and we write $xRy$ if $(x, y) \in R$. In particular, any set of relations on $X$ can be partially ordered by set inclusion as subsets of $X \times X$. Thus, if $QO(X)$ is the set of quasiorder relations on $X$, then $(QO(X), \subseteq)$ is a subposet of $(\mathcal{P}(X \times X), \subseteq)$. In the light of the one-to-one correspondence between Alexandroff topologies and quasiorders (Theorem 8.3.3), the next result is not surprising.

**Theorem 9.3.3.** *The complete lattice $(A(X), \subseteq)$ of Alexandroff topologies on $X$ is isomorphic to the complete lattice $(QO(X), \supseteq)$ of quasiorders on $X$, ordered by reverse inclusion.*

Recall that every equivalence relation is a quasiorder. The corresponding Alexandroff topologies are the partition topologies. Let $\Pi(X)$ be the set of all partitions of $X$. If $P, R \in \Pi(X)$, recall that $P$ is finer than $R$ if and only if $[x]_P \subseteq [x]_R$ for every $x \in X$, so $[x]_R = \bigcup\{[y]_P : y \in [x]_R\}$. Thus, a finer partition in $\Pi(X)$ produces a finer topology in $A(X)$.

**Theorem 9.3.4.** *The set $\Pi(X)$ of partitions of $X$ is a complete lattice, and is isomorphic to the complete lattice of partition topologies in $T(X)$.*

The $T_0$ Alexandroff topologies correspond to the partial orders on $X$ (see Exercise 8 of Section 8.3). The poset of partial order relations, considered as a subposet of $(\mathcal{P}(X \times X), \subseteq)$ fails to be a lattice. This fact and the proofs of the two previous theorems are left to the exercises.

We now turn to the question of complementation in $T(X)$. A complement of $T \in T(X)$ is a topology $T'$ with $T \vee T' = \top = \mathcal{P}(X)$ and $T \wedge T' = \bot = \{\emptyset, X\}$. Now $T \wedge T' = T \cap T' = \bot$ if and only if the only sets open in both $T$ and $T'$ are $\emptyset$ and $X$. The condition that $T \vee T' = \top = \mathcal{P}(X)$ occurs if and only if for every $x \in X$, there exist $U \in T$ and $U' \in T'$ with $\{x\} = U \cap U'$.

**Example 9.3.5 (The lattice of $T_1$ topologies on an infinite set $X$ is not complemented).** Let $X = A \cup B$, where $A$ and $B$ are infinite disjoint sets, and give $X$ the $T_1$ topology $T = \{U : U \subseteq B \text{ or } A - U \text{ is finite}\}$. Thus, $X$ is the disjoint union of $A$ with the cofinite topology and $B$ with the discrete topology. Suppose $T'$ is a complement of $T$. Now $T'$

restricted to $A$ cannot be discrete, or else $A \in \mathcal{T} \wedge \mathcal{T}' = \perp = \{\emptyset, X\}$. Thus, there exists $a_0 \in A$ such that $\{a_0\} \notin \mathcal{T}'$. Since $\mathcal{T} \vee \mathcal{T}' = \top = P(X)$, there exist $U \in \mathcal{T}$, $V \in \mathcal{T}'$ with $U \cap V = \{a_0\}$. As a $\mathcal{T}$-neighborhood of $a_0$, $U$ contains all but finitely many elements of $A$, so $V$ contains at most finitely many elements of $A$, say $\{a_0, a_1, \dots, a_n\}$. Because $\mathcal{T}'$ is $T_1$, we may separate $a_i$ ($i = 1, \dots, n$) from $\mathcal{T}'$-neighborhoods of $a_0$, and intersecting these with $V$ gives a $\mathcal{T}'$-neighborhood $V' \subseteq V$ of $a_0$ with $V' \cap A = \{a_0\} \notin \mathcal{T}'$. Thus, $V'$ must contain a point $b \in B$. If $V''$ is the intersection of $V'$ with a $\mathcal{T}'$-neighborhood of $b$ excluding $a_0$, then $V'' \in \mathcal{T}'$ is a nonempty subset of $B$. This gives the contradiction that $V'' \in \mathcal{T} \wedge \mathcal{T}' = \{\emptyset, X\}$. Thus, $\mathcal{T}$ has no $T_1$ complement.

Next we will consider complementation in the lattice $A(X)$ of Alexandroff topologies on $X$. We say a quasiordered set $(X, \leq)$ is *connected* if for every pair of points $x, y \in X$, there exists a sequence of points $x = x_0, x_1, \dots, x_n = y \in X$ with $x_{k-1} \leq x_k$ or $x_k \leq x_{k-1}$ for $k = 1, \dots, n$. The maximal connected subsets of $(X, \leq)$ are the connected components. This graph-theoretic definition of connectedness of a quasiordered set is equivalent to topological connectedness in the specialization topology on $X$. The details are left as an exercise.

For the next theorem, we will need two constructions involving quasiordered sets. The first involves splitting points into several clones, each of which has the same position in the poset as the original. Suppose $(Y, \leq)$ is a partially ordered set. For each $y \in Y$, let $D_y$ be an arbitrary nonempty index set, and consider $D[y] = \{(y, i) : i \in D_y\}$ to be the set of duplicates of $y$. Let $Y' = \bigcup\{D[y] : y \in Y\}$, and give $Y'$ the order $(y, i) \leq_D (z, j)$ if and only if $(y, i) = (z, j)$ or $y < z$ in $(Y, \leq)$. It is easy to see that $(Y', \leq_D)$ is a poset. If $(X, \leq)$ is a quasiorder set, consider the associated partial order $\leq$ on the set $Y$ of $\approx$-equivalence classes, where $x \approx y$ if and only if $x \leq y$ and $y \leq x$. For each $y \in Y$, let $D_y = [y]$ be the $\approx$-equivalence class of $y$. Performing the duplication of points as above results in a partially ordered set on $X$ obtained by retaining the order $\leq$ between non-equivalent points and making $x \| y$ for distinct points $x$ and $y$ which are $\approx$-equivalent. (Recall that $x \| y$ means $x \not\leq y$ and $y \not\leq x$.) We call the resulting poset the *parallel division of the quasiorder on $X$ into a partial order on $X$*. (Informally, this process may be called *cloudbusting*.) An example of this is illustrated in Figure 9.4(a) and (b).

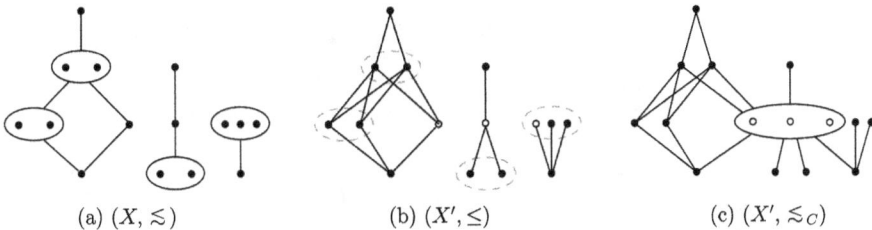

(a) $(X, \lesssim)$      (b) $(X', \leq)$      (c) $(X', \lesssim_C)$

**Figure 9.4:** (a) A quasiordered set $X$. (b) The parallel division of $X$ into a partial order $X'$. (c) The connection of $X'$ at the indicated points from each component.

The second construction involves gluing together quasiordered sets. Given a collection $\{(X_i, \leq_i) : i \in I\}$ of quasiordered sets, for each $i \in I$, pick a point $a_i \in X_i$. We will define a quasiorder $\leq$ on the disjoint union $X = \bigcup\{X_i : i \in I\}$ which puts the points $\{a_i : i \in I\}$ in an $\approx$-equivalence class, and retains transitivity. Specifically, for $x, y \in X$, define $\leq$ by $x \leq y$ if and only if $x \leq_i y$ in some $X_i$ or $x \leq_i a_i$ and $a_j \leq_j y$ for some $i, j \in I$. We call this construction the *connection* of the quasiordered sets $X_i$ at the points $\{a_i\}$. Clearly, this produces a connected quasiordered set. Figure 9.4(b) and (c) illustrate this.

**Theorem 9.3.6.** *The lattice $A(X)$ of Alexandroff topologies on $X$ is a complemented lattice.*

*Proof.* Suppose $\mathcal{T}$ is an Alexandroff topology on $X$ with associated specialization quasiorder $\leq$. Form the parallel division $X'$ of $(X, \leq)$ into a partial order $\leq$ on $X$. Let $\{(X_i, \leq_i)\}_{i \in I}$ be the set of connected components of $(X, \leq)$. For each $i \in I$, pick $a_i \in X_i$ and form the connection $(X', \leq_C)$ of the sets $X_i$ at the points $a_i$, as suggested in Figure 9.4. Let $\mathcal{T}'$ be the Alexandroff topology consisting of the decreasing sets of $(X', \leq_C)$. (This is the specialization topology for $Y$ with the dual order $\geq_C$.) We will show that $\mathcal{T}'$ is a complement of $\mathcal{T}$.

Given $x \in X$, within the connected component of $x$, $i_\leq(x) \cap d_\leq(x)$ is the $\approx$-equivalence class $[x]$ of $x$. $N_{\mathcal{T}}(x) = i_\leq(x)$ is contained in the component of $x$. Since $N_{\mathcal{T}'}(x) = d_{\leq_C}(x)$ and, within the component of $x$, $d_{\leq_C}(x) = \{x\} \cup (d_\leq(x) - [x])$, it follows that $N_{\mathcal{T}}(x) \cap N_{\mathcal{T}'}(x) = \{x\}$ for every $x \in X$.

Now suppose $U \in \mathcal{T} \cap \mathcal{T}'$, $x \in U$, and $x$ is in the connected component $X_i$. For any $y$ in the same component $X_i$, there is a path from $x$ which proceeds up and down repeatedly in $(X, \leq)$, reaching $y$, and this path can be chosen to use no more than one point from each $\approx$-equivalence class. Since $U$ is $\leq$-increasing and $\leq_C$-decreasing, if follows that $y \in U$, and thus $U$ contains the entire component $X_i$ containing $x$. But for any other component $X_j$, since $a_j \leq_C a_i$ and $U$ is $\leq_C$-decreasing, $U$ contains the point $a_j$ and thus contains the entire component $X_j$ of $a_j$. Since this holds for every $j \in I$, $U = X$ and thus $\mathcal{T} \cap \mathcal{T}' = \{\emptyset, X\}$. $\square$

Note that the proof above was constructive: given an Alexandroff topology $\mathcal{T}$ on $X$, it described an explicit algorithm to produce an Alexandroff complement $\mathcal{T}'$. The algorithm required an arbitrary choice of a point $a_i$ from each connected component of $X$. Different choices will yield different complements, so generally, $A(X)$ is not uniquely complemented.

**Corollary 9.3.7.** *If $X$ is a finite set, $T(X)$ is a complemented lattice.*

Juris Hartmanis proved Corollary 9.3.7 directly in 1958. As a generalization, Anne K. Steiner proved Theorem 9.3.6 in 1966. In the same paper, Steiner proved the following important theorem.

**Theorem 9.3.8** (A. K. Steiner). *The lattice $T(X)$ of topologies on $X$ is complemented.*

Her proof used the 1966 result of Haim Gaifman that if every $T_1$ topology on $X$ has a complement in $T(X)$, then every topology on $X$ has a complement, and showed that certain spaces have complements which are Alexandroff topologies. Questions of when a certain kind of topology has a certain kind of complement have been the basis for much subsequent research. Alternate proofs of Theorem 9.3.8 were given by A. C. M. van Rooij in 1968 and Paul Schnare in 1972 (with a minor correction in 1977). Further details of this development are given in [31].

The following theorem provides a remarkable reason for studying sublattices of $T(X)$.

**Theorem 9.3.9.** *Every lattice L is a sublattice of $T(X)$ for some X.*

This is a corollary to the 1946 theorem of Philip Whitman that every lattice is a sublattice of the lattice of partitions $\Pi(X)$, since the lattice of partition topologies is a complete sublattice of $T(X)$.

## Exercises

1. Show that the coatoms in the lattice of $T_1$ topologies are precisely the coatoms in the lattice of $T_2$ topologies.
2. Show that $\mathcal{T} \in T(X)$ is not $T_1$ if and only if $\mathcal{T}$ is below a coatom of form $\mathcal{T}_{\{p,q\}}$.
3. If $X$ is a nonempty set, show that $T(X)$ has a minimum $T_1$ topology, namely, the cofinite topology $\mathcal{T}_{cf}$.
4. If $(L, \leq)$ is a complete lattice and $a \in L$, show that $i(a)$ is a complete sublattice of $L$. Use this to show that the set of $T_1$-topologies on a set $X$ is a complete sublattice of $T(X)$, giving a second proof of Theorem 9.3.1.
5. Prove Theorem 9.3.3.
6. Prove Theorem 9.3.4.
7. Show that the lattice $\Pi(X)$ of partitions of $X$ is not distributive if $|X| \geq 3$.
8. If $PO(X)$ is the collection of partial orders on $X$ ordered by set inclusion of the partial order relations, as subsets of $X \times X$, then $PO(X)$ corresponds to the $T_0$ Alexandroff topologies. Show that $PO(\{a, b\})$ is not a lattice.
9. The proof of Theorem 9.3.2 showed that $A(X)$ is not a complete sublattice of $T(X)$ if $A(X) \neq T(X)$. State and prove a characterization of the sets $X$ for which every topology on $X$ is an Alexandroff topology.
10. Show that a quasiordered set $(X, \leq)$ is connected if and only if the associated specialization topology $\mathcal{T}_\leq$ makes $X$ a connected topological space.
11. Let $X = \{1, 2, 3, 4, 5\}$ and define $f, g : X \to X$ by $f(1) = 2, f(2) = 3, g(1) = 3, g(2) = 1$, and $f(x) = g(x) = x$ for all other points.
    (a) Sketch the functionally Alexandroff topologies $\mathcal{T}_f$ and $\mathcal{T}_g$ and their associated quasiorders.
    (b) Find $\mathcal{T}_f \wedge \mathcal{T}_g$ in $T(X) = A(X)$.

(c) Find $\mathcal{T}_f \wedge \mathcal{T}_g$ in the poset FA($X$) of functionally Alexandroff topologies on $X$.

12. If $\mathcal{T}$ is a minimal $T_0$ topology in $T(X)$ and $A \subseteq X$ is open or closed, show that the subspace topology $\mathcal{T}|_A$ on $A$ is minimal $T_0$ in $T(A)$. (Compare to Exercise 14 of Section 9.2.)

13. Find a complement of the topology whose basis is depicted in Figure 8.3.

14. Find a complement of the topology $\mathcal{T}$ on $X = \{a, b, c, d, e, f, g, h\}$ whose basis is $\{\{a, b\}, \{a, b, c, d\}, \{e, f, g, h\}, \{h\}\}$.

15. Let $\mathcal{T}$ be the topology on $X = \{a, b, c, d\}$ having basis $\{\{a, b, c\}, \{c\}, \{d\}\}$.
    (a) Find all complements of $\mathcal{T}$ which arise from the construction given in the proof of Theorem 9.3.6.
    (b) Find other complements of $\mathcal{T}$ besides those found in part (a).

16. Show that the lattice $A(X)$ of Alexandroff topologies on a nonempty set $X$ is uniquely complemented if and only if $|X| \le 2$.

17. **(The Hartmanis complement)** Suppose $\mathcal{T}$ is a topology on a finite set $X$. Let $\le$ be the specialization quasiorder and let $(X/\approx, \le)$ be the partial order on the $\approx$-equivalence classes, where $a \approx b$ if and only if $a \le b$ and $b \le a$. Let $M_\approx$ be the set of maximal equivalence classes in $(X/\approx, \le)$, let $M \subseteq X$ be the union of the maximal equivalence classes, and let $A$ be a complete set of equivalence class representatives from $M_\approx$. Let $\mathcal{T}'$ be the topology on $X$ having $N(x) = \{x\}$ for $x \notin A$, and $N(a) = A \cup (X - M)$ for $a \in A$. That is, $\mathcal{T}'$ is the Alexandroff topology whose specialization order $\le'$ is $\Delta_X \cup \{(a, x) : a \in A, x \in A \cup (X - M)\}$.
    (a) Show that $\mathcal{T}'$ is a complement of $\mathcal{T}$ in $T(X)$.
    (b) Using the example of $(\mathbb{R}, \mathcal{T})$ where $\mathcal{T} = \{\emptyset, \mathbb{R}\} \cup \{(a, \infty) : a \in \mathbb{R}\} \cup \{[a, \infty) : a \in \mathbb{R}\}$, explain why the Hartmanis complement construction does not generally work for Alexandroff topologies on infinite sets. Find a complement of $(\mathbb{R}, \mathcal{T})$.
    (c) Find a Hartmanis complement of the Alexandroff topology whose specialization order is given in Figure 9.4(a).

# 10 Partially ordered topological spaces

## 10.1 Partially ordered topological spaces

A *partially ordered topological space* is a triple $(X, \mathcal{T}, \leq)$ where $X$ is a nonempty set, $\mathcal{T}$ is a topology on $X$, and $\leq$ is a partial order on $X$. A partial order provides a means to quantify preference, with $a \leq b$ meaning $b$ is preferable to $a$. Thus, partially ordered topological spaces are useful in situations where nearness and preference are required.

Notice than any topological space $(X, \mathcal{T})$ can be viewed as a partially ordered topological space $(X, \mathcal{T}, =)$ ordered by the *trivial order* or *discrete order* of equality. Thus, topological spaces may be viewed as partially ordered topological spaces having the trivial order.

The next example shows that in general, the topology and order of a partially ordered topological space may not interact well. Specifically, it shows that the increasing hull of a closed set need not be closed, and the closure of an increasing set need not be increasing.

**Example 10.1.1.** Let $X$ be the closed upper half-plane with an open ball of radius $1/2$ around the origin deleted, and the origin added, as seen in Figure 10.1. Give $X$ the Euclidean topology as a subspace of plane, and the product order $(a, b) \leq (c, d)$ if and only if $a \leq c$ and $b \leq d$. Let $A = \{(x, -x^{-1}) : x < 0\}$. Now $A$ is closed, but $i(A)$ is the set of all points of $X$ in the open upper half-plane, which is not closed. Also, $i(A)$ is increasing, but its closure is not increasing, since, for example, $(-2, 0) \in \mathrm{cl}(i(A))$, $(-2, 0) \leq (0, 0)$, and $(0, 0) \notin \mathrm{cl}(i(A))$. The set $B = \{(0, 0)\}$ is open, but $i(B)$ is not open: $(0, 5) \in i(B)$ but $i(B)$ contains no neighborhood of $(0, 5)$. Furthermore, $i(B)$ is increasing, but its interior is not: $\mathrm{int}(i(B)) = i(B) - (\{0\} \times [1/2, \infty))$ contains $(0, 0)$ but not $(0, 1) \geq (0, 0)$.

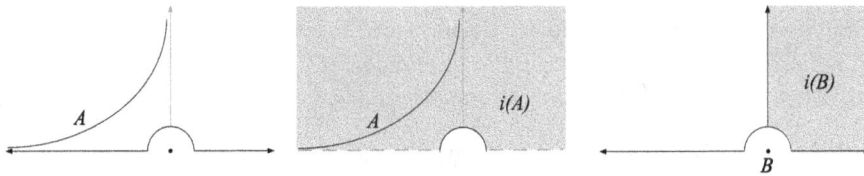

**Figure 10.1:** The interior or closure of an increasing set need not be increasing; the increasing hull of a closed (open) set need not be closed (open).

Since the topology and order of a partially ordered topological space may not interact well, frequently some forms of compatibility between the topology and order are assumed. The first compatibility condition we will consider is convexity.

https://doi.org/10.1515/9783110686579-011

**Definition 10.1.2.** In a poset $(X, \leq)$, a set $A$ is *convex* (or for emphasis, *order convex*) if $a, c \in A$ and $a \leq b \leq c$ imply $b \in A$. Equivalently, $A$ is convex if $A = i(A) \cap d(A)$. A partially ordered topological space $(X, \mathcal{T}, \leq)$ has a *convex topology* if $\mathcal{T}$ has a basis of convex sets.

If $\mathcal{T}$ is an Alexandroff topology on $(X, \leq)$, the convexity of $\mathcal{T}$ would say that if $z$ is in the smallest neighborhood $N(x)$ of $x$ and $x \leq y \leq z$, then $y \in N(x)$. That is, if $z$ is in every neighborhood of $x$ and $y$ is "closer" (in the order sense) to $x$ than $z$, then $y$ must also be close to $x$ in a topological sense.

Any ordered topological space $(X, \mathcal{T}, \leq)$ produces two associated topologies:

$$\mathcal{T}^{\sharp} = \{U \in \mathcal{T} : U = i(U)\} \quad \text{and}$$
$$\mathcal{T}^{\flat} = \{U \in \mathcal{T} : U = d(U)\}.$$

That is, $\mathcal{T}^{\sharp}$ consists of the $\leq$-increasing $\mathcal{T}$-open sets, and $\mathcal{T}^{\flat}$ consists of the $\leq$-decreasing $\mathcal{T}$-open sets. It is easy to show directly that these are both topologies. For an indirect proof that $\mathcal{T}^{\sharp}$ is a topology, recall that the $\leq$-increasing sets form a topology $\mathcal{I}$ on $X$, and note that $\mathcal{T}^{\sharp} = \mathcal{T} \wedge \mathcal{I} = \mathcal{T} \cap \mathcal{I}$.

In a poset $(X, \leq)$, it is immediate from the definitions that the intersection of any increasing set with any decreasing set is a convex set. In particular, every monotone set is convex.

**Theorem 10.1.3.** *If $\mathcal{T}$ is a topology on $(X, \leq)$ and $\mathcal{T} = \mathcal{T}^{\sharp} \vee \mathcal{T}^{\flat}$, then $\mathcal{T}$ is convex.*

*Proof.* If $\mathcal{T} = \mathcal{T}^{\sharp} \vee \mathcal{T}^{\flat}$, then $\mathcal{T}$ has basis $\{U \cap V : U \in \mathcal{T}^{\sharp}, V \in \mathcal{T}^{\flat}\}$. This basis contains only convex sets, so $\mathcal{T}$ is convex. □

Other compatibility conditions between the topology and order of a partially ordered topological space include the ordered separation axioms. Recall that a topological space $(X, \mathcal{T})$ is $T_1$ if and only if for every $x \neq y$ in $X$, there exists a neighborhood of $x$ excluding $y$ (and thus, with $x$ and $y$ interchanged, a neighborhood of $y$ excluding $x$).

**Definition 10.1.4.** A partially ordered topological space $(X, \mathcal{T}, \leq)$ is $T_1$-*ordered* if for every $x \not\leq y$, there exists an increasing neighborhood of $x$ which excludes $y$ and there exists a decreasing neighborhood of $y$ which excludes $x$.

Note that if $x \leq y$, then it would be impossible to find an increasing neighborhood of $x$ which excludes $y$ or a decreasing neighborhood of $y$ which excludes $x$. Thus, the definition says for any distinct points $x$ and $y$, there is a monotone neighborhood of one which excludes the other except when the order absolutely prohibits it.

From the definition of the terms used, Definition 10.1.4 may be rephrased to say $(X, \mathcal{T}, \leq)$ is $T_1$-*ordered* if for every $x \not\leq y$ in $X$, there exists a neighborhood $U_x$ of $x$ with $y \notin i(U_x)$ and there exists a neighborhood $U_y$ of $y$ with $x \notin d(U_x)$. Then $i(U_x)$ is an increasing neighborhood of $x$ and $d(U_y)$ is a decreasing neighborhood of $y$. In general, if $U_x$ is open, $i(U_x)$ need not be open. Thus, as stated, the separating monotone neighborhoods $i(U_x)$ and $d(U_y)$ cannot be assumed to be open neighborhoods. However, the next theorem says that indeed they may be assumed to be open.

**Theorem 10.1.5.** *For a partially ordered topological space* $(X, \mathcal{T}, \leq)$, *the following are equivalent:*
(a) *X is $T_1$-ordered.*
(b) *For every $x \nleq y$ in X, there exists an open increasing neighborhood $U_x$ of x with $y \notin U_x$ and there exists an open decreasing neighborhood $U_y$ of y with $x \notin U_y$.*
(c) *For every $x \in X$, $i(x)$ and $d(x)$ are closed.*
(d) *For every $x \nleq y$ in X, there exists an increasing neighborhood $U_x$ of x with $y \notin U_x$ and there exists a decreasing neighborhood $U_y$ of y with $x \notin U_x$.*

The proof of the theorem is left as an exercise. Thus, for the $T_1$-ordered property, "strengthening" the definition to require separation by monotone *open* neighborhoods does not give a new separation axiom.

A useful characterization of $T_1$ topological spaces is that they are the spaces in which every singleton is closed. Theorem 10.1.5(c) gives the ordered version of this.

If a topological space $(X, \mathcal{T})$ is viewed as a trivially-ordered topological space $(X, \mathcal{T}, =)$, then $i(x) = \{x\} = d(x)$, and thus $(X, \mathcal{T}, =)$ is $T_1$-ordered if and only if $(X, \mathcal{T})$ is $T_1$.

A topological space is $T_2$ if and only if for every $x \neq y$, there exist disjoint neighborhoods of $x$ and $y$. Ordered versions of this are given below.

**Definition 10.1.6.** A partially ordered topological space $(X, \mathcal{T}, \leq)$ is $T_2$-*ordered* if for every $x \nleq y$, there exists an increasing neighborhood $U_x$ of x disjoint from a decreasing neighborhood $U_y$ of y. $(X, \mathcal{T}, \leq)$ is *strongly $T_2$-ordered* if for every $x \nleq y$, there exists an open increasing neighborhood $U_x$ of x disjoint from an open decreasing neighborhood $U_y$ of y.

If $(X, \mathcal{T}, \leq)$ is $T_2$-ordered and $x \neq y$, then both statements $x \leq y$ and $y \leq x$ cannot be true, and the inequality that fails guarantees disjoint neighborhoods $U_x$ and $U_y$ of x and y. Thus, if $(X, \mathcal{T}, \leq)$ is $T_2$-ordered, then $(X, \mathcal{T})$ is $T_2$. Similarly, $T_1$-ordered implies $T_1$. Also, it is clear that strongly $T_2$-ordered implies $T_2$-ordered, which implies $T_1$-ordered.

While the use of monotone neighborhoods or open monotone neighborhoods made no difference in the definition of $T_1$-ordered, the following example shows that a $T_2$-ordered topological space need not be strongly $T_2$-ordered.

**Example 10.1.7.** In the plane, take a copy of the rationals at $y = -1$, a copy of the reals at $y = 0$, and a copy of the irrationals at $y = 1$ with the order $(a, b) \leq (c, d)$ if and only if $a = c$ and $b \leq d$. Remove $(0, 0)$ and add $(0, 1)$, with $(0, 1)$ incomparable to every other element. With the Euclidean subspace topology, call this space $(X, \mathcal{T}, \leq)$.

We will show that this space is $T_2$-ordered, but $(0,-1)$ and $(0,1)$ cannot be separated by oppositely directed open monotone neighborhoods.

If $(x, m) \not\leq (y, n) \in X$ and $x \neq y$, then there exist disjoint neighborhoods $U_x$ and $U_y$ in $\mathbb{R}$ of $x$ and $y$, respectively, and since all the order present is "vertical", there is an increasing neighborhood of $(x, m)$ contained in $U_x \times \mathbb{R}$ and a decreasing neighborhood of $(y, n)$ contained in $U_y \times \mathbb{R}$. Such neighborhoods must be disjoint.

If $(x, m) \not\leq (y, n) \in X$ and $x = y$, then $n < m$, except in the case of $(0, -1) \not\leq (0, 1)$. Except for the case $(0, -1) \not\leq (0, 1)$, if $U$ is a neighborhood of $x$ in $\mathbb{R}$, then $(U \times \{m, m+1\}) \cap X$ is an increasing neighborhood of $(x, m)$ disjoint from the decreasing neighborhood $(U \times \{n, n-1\}) \cap X$ of $(y, n)$.

Now in the case $(0, -1) \not\leq (0, 1)$, let $U$ be a neighborhood of $0$ in $\mathbb{R}$. $(U \cap \mathbb{Q} \times \{-1, 0\}) - \{(0, 0)\}$ is an increasing neighborhood of $(0, -1)$ disjoint from the decreasing neighborhood $(U - \mathbb{Q} \times \{0, 1\}) \cup \{(0, 1)\}$ of $(0, 1)$.

This shows that $X$ is $T_2$-ordered. However, note that any increasing neighborhood of $(0, -1)$ contains $(U \cap \mathbb{Q} \times \{0\}) - \{(0, 0)\}$ and any decreasing neighborhood of $(0, 1)$ contains $(V - \mathbb{Q} \times \{0\})$, where $U$ and $V$ are neighborhoods of $0$ in $\mathbb{R}$. By the denseness of $\mathbb{Q}$ in $\mathbb{R}$, any open increasing neighborhood of $(0, -1)$ must contain irrational points near the origin, and thus must intersect any decreasing neighborhood of $(0, 1)$. Thus, $(0, -1)$ and $(0, 1)$ cannot be separated by disjoint monotone open neighborhoods, so $X$ is not strongly $T_2$-ordered.

It is a standard exercise (Exercise 9 of Section 7.1) to show that a topological space $(X, \mathcal{T})$ is $T_2$ if and only if $\Delta_X = \{(x, x) : x \in X\}$ is a closed subset of the product $X \times X$. Viewing $\Delta_X$ as the graph of the trivial order $=$, this suggests the following result.

**Theorem 10.1.8.** *A partially ordered topological space $(X, \mathcal{T}, \leq)$ is $T_2$-ordered if and only if the graph $G = \{(x, y) \in X^2 : x \leq y\}$ of the partial order is a closed subset of the product $X \times X$.*

*Proof.* Suppose $(X, \mathcal{T}, \leq)$ is $T_2$-ordered. To see $X^2 - G$ is open, suppose $(x, y) \notin G$. Then $x \not\leq y$, so there exists an increasing neighborhood $U$ of $x$ disjoint from a decreasing neighborhood $V$ of $y$. Now $U \times V$ is a neighborhood of $(x, y)$ in $X^2$. If $G \cap (U \times V) \neq \emptyset$, then there exists $(a, b) \in G \cap (U \times V)$, so $a \leq b$, $a \in U = i(U)$, and $b \in V$. Thus, $b \in i(U) = U$, contrary to $U \cap V = \emptyset$. Thus, $U \times V \subseteq X^2 - G$. This shows $X^2 - G$ is open.

Conversely, suppose $X^2 - G$ is open. Then, for any $(x, y) \notin G$, there exists a basic neighborhood $U \times V$ of $(x, y)$ contained in $X^2 - G$. If $z \in i(U) \cap d(V)$, then $u_0 \leq z \leq v_0$ for some $u_0 \in U$, $v_0 \in V$, and thus $u_0 \leq v_0$, so $(u_0, v_0) \in (U \times V) \cap G$, contrary to the choice of $U \times V$. Thus, $i(U) \cap d(V) = \emptyset$, so $i(U)$ is an increasing neighborhood of $x$ disjoint from the decreasing neighborhood $d(V)$ of $y$. This holds for any $(x, y) \notin G$, that is, for any $x \not\leq y$, so $X$ is $T_2$-ordered. □

**Corollary 10.1.9.** *In a $T_2$-ordered topological space $(X, \mathcal{T}, \leq)$, if $x_n \leq y_n$ for all $n \in \mathbb{N}$ and the sequence $(x_n, y_n)_{n \in \mathbb{N}}$ converges to $(x, y)$ in $X \times X$, then $x \leq y$.*

*Proof.* Suppose $x_n \le y_n$ for all $n \in \mathbb{N}$ and $(x_n, y_n)_{n \in \mathbb{N}}$ converges to $(x, y)$. This says $(x_n, y_n)_{n \in \mathbb{N}}$ is a convergent sequence in the closed set $G \subseteq X^2$, and thus $G$ contains the limit $(x, y)$. $\qquad\square$

The theorem below is an ordered version of the fact that a compact subset of a Hausdorff space is closed.

**Theorem 10.1.10.** *Suppose $(X, \mathcal{T}, \le)$ is a $T_2$-ordered space. If $K \subseteq X$ is compact, then $i(K)$ and $d(K)$ are closed.*

*Proof.* Suppose $K$ is compact. To show $i(K)$ is closed, suppose $a \notin i(K)$. Then $x \not\le a$ for all $x \in K$. By the $T_2$-ordered condition, for each $x \in K$, there exist open neighborhoods $N_x$ of $x$ and $M_x$ of $a$ with $i(N_x) \cap d(M_x) = \emptyset$. Let $\{N_{x_i} : i = 1, \dots, n\}$ be a finite subcover of the open cover $\{N_x : x \in K\}$ of $K$. Now $M = \bigcap \{M_{x_i} : i = 1, \dots, n\}$ is a neighborhood of $a$. If there exists $z \in M \cap i(K)$, then $z \ge k$ for some $k \in K \subseteq \bigcup \{N_{x_i} : i = 1, \dots, n\}$. Thus, $k \in N_{x_j}$ for some $j \in \{1, \dots, n\}$, so $z \in i(N_{x_j}) \cap M_{x_j}$, contrary to the choice of $M_{x_j}$ and $N_{x_j}$. Thus, for an arbitrary $a \notin i(K)$, we have found a neighborhood $M$ of $a$ contained in $X - i(K)$, so $X - i(K)$ is open and thus $i(K)$ is closed. A similar argument shows $d(K)$ is closed. $\qquad\square$

**Theorem 10.1.11.** *Suppose $(X, \mathcal{T}, \le)$ is a compact $T_2$-ordered space. For any increasing set $A \subseteq X$ and any open set $U$ containing $A$, there exists an open increasing set $V$ with $A \subseteq V \subseteq U$. In other words, in a compact $T_2$-ordered space $(X, \mathcal{T}, \le)$, every open neighborhood of an increasing set $A$ contains a $\mathcal{T}^{\#}$-neighborhood of $A$.*

*Proof.* Suppose $X$ is compact $T_2$-ordered, $A \subseteq X$ is increasing, and $U$ is an open set containing $A$. Now $X - U$ is closed and thus compact in $X$. By Theorem 10.1.10, $d(X - U)$ is closed and decreasing, so $V = X - d(X - U)$ is open and increasing. Now $X - U \subseteq d(X - U)$ implies $V = X - d(X - U) \subseteq U$. To see $A \subseteq V$, suppose not. Then there exists $a \in A \cap d(X - U)$, so there exists $x \in X - U$ with $a \le x$. Since $A$ is increasing, $x \in A \subseteq U$, which contradicts $x \in X - U$. $\qquad\square$

The previous theorem has a natural application to ordered separation axioms.

**Theorem 10.1.12.** *Every compact $T_2$-ordered space is strongly $T_2$-ordered. That is, if $x \not\le y$ in a compact $T_2$-ordered space, then there exists an open increasing neighborhood of $x$ disjoint from an open decreasing neighborhood of $y$.*

*Proof.* Suppose $x \not\le y$ in a compact $T_2$-ordered space $(X, \mathcal{T}, \le)$. Then $i(x)$ and $d(y)$ are disjoint, and since $X$ is $T_1$-ordered, these sets are also closed. Now $(X, \mathcal{T})$ is compact and $T_2$, and thus is normal, so there exist disjoint open sets $U_x$ and $U_y$ with $i(x) \subseteq U_x$ and $d(y) \subseteq U_y$. Now applying Theorem 10.1.11 and its dual, there exist an open increasing set $V_x$ and an open decreasing set $V_y$ with $i(x) \subseteq V_x \subseteq U_x$ and $d(y) \subseteq V_y \subseteq U_y$. Thus, $V_x$ and $V_y$ are the desired oppositely directed open monotone neighborhoods of $x$ and $y$. $\qquad\square$

Now we give another nice property of compact $T_2$-ordered spaces.

**Theorem 10.1.13.** *If $(X, \mathcal{T}, \leq)$ is a compact $T_2$-ordered topological space, then $\mathcal{T}$ is a convex topology.*

*Proof.* Suppose $(X, \mathcal{T}, \leq)$ is compact and $T_2$-ordered. To show $\mathcal{T}$ is convex, it suffices to show that every open neighborhood of $a \in X$ contains a convex neighborhood of $a$. Suppose $U$ is an open neighborhood of $a$. Then $X - U$ is compact. For any $x \in X - U$, either $a \not\leq x$ or $x \not\leq a$, so there exist disjoint sets $M_x$ and $N_x$, one in $\mathcal{T}^{\sharp}$ and one in $\mathcal{T}^{\flat}$, with $a \in M_x$ and $x \in N_x$. The open cover $\{N_x : x \in X - U\}$ of $X - U$ has a finite subcover $\{N_x : x \in F\}$ where $F \subseteq X - U$ is finite. Now $W = \bigcap\{M_x : x \in F\}$ is disjoint from $\bigcup\{N_x : x \in F\}$, and since $X - U \subseteq \bigcup\{N_x : x \in F\}$, we have $a \in W \subseteq U$. But $W$ is a finite intersection of open monotone neighborhoods of $X$, and thus is a convex neighborhood of $a$ contained in $U$. $\qquad\square$

## Exercises

1. Suppose $(X, \mathcal{T}, \leq)$ is a partially ordered topological space. Show that $\mathcal{T}$ has a basis of convex open sets if and only if for every $x \in X$ and for every neighborhood $U$ of $x$, there exists a convex open neighborhood $W$ of $x$ with $W \subseteq U$. That is, $\mathcal{T}$ is a convex topology on $(X, \leq)$ if and only if it is a *locally convex topology* on $(X, \leq)$.

2. Given a partially ordered topological space $(X, \mathcal{T}, \leq)$, let $\mathcal{T}'$ be the topology on $X$ whose closed sets are the $\leq$-increasing sets. In Example 10.1.1, we noted that in a partially ordered topological space, the increasing hull of a closed set need not be closed and the closure of an increasing set need not be increasing. Interpret this statement in terms of $\mathcal{T}'$-closures and $\mathcal{T}$-closures, and in terms of how $\mathcal{T}$ and $\mathcal{T}'$ compare in the lattice $T(X)$ of topologies on $X$.

3. Let $(X, \leq)$ be $[0, 1]$ with the usual order. For $\varepsilon > 0$, define $U(0, \varepsilon) = \{0\} \cup (1 - \varepsilon, 1)$, $U(1, \varepsilon) = \{1\} \cup (0, \varepsilon)$, and for $0 < x < 1$, $U(x, \varepsilon) = (x - \varepsilon, x + \varepsilon)$. Let $\mathcal{T}$ be the topology on $X$ having basis $\{U(x, \varepsilon) : x \in [0, 1], \varepsilon > 0\}$. Determine whether $(X, \mathcal{T}, \leq)$ is $T_1$-ordered or $T_2$-ordered, and whether $\mathcal{T}$ is convex.

4. Suppose $\leq$ is a partial order on $X$, and $\mathcal{T}_{\leq}$ is the associated Alexandroff (specialization) topology on $X$. Must the partially ordered topological space $(X, \mathcal{T}_{\leq}, \leq)$ be $T_1$-ordered? Must $\mathcal{T}_{\leq}$ be convex?

5. If $A$ is a subset of a partially ordered topological space $(X, \mathcal{T}, \leq)$, let $I(A)$ denote the smallest closed increasing set containing $A$. $I(A)$ is called the *closed increasing hull* of $A$. The *closed decreasing hull* $D(A)$ of $A \subseteq X$ is defined dually. We have seen examples to show that $I(A) \neq \mathrm{cl}(i(A))$ and $I(A) \neq i(\mathrm{cl}(A))$. Describe $I(A)$ and $D(A)$ in terms of $\mathcal{T}^{\sharp}$ and $\mathcal{T}^{\flat}$.

6. Prove Theorem 10.1.5.

7. Give a proof or counterexample: For points $x, y$ in an ordered topological space, there exists an open increasing neighborhood of $x$ excluding $y$ if and only if there exists a neighborhood $U$ of $x$ with $y \notin i(U)$.

8. Show that $(X, \mathcal{T}, \leq)$ is $T_1$-ordered if and only if $i(x) = \bigcap\{U \in \mathcal{T}^\sharp : x \in U\}$ and $d(x) = \bigcap\{U \in \mathcal{T}^\flat : x \in U\}$ for all $x \in X$.

9. Determine whether the $T_1$-ordered and $T_2$-ordered properties are hereditary and productive (where a product $\prod\{X_i : i \in I\}$ carries the product topology and product order $\langle x_i \rangle_{i \in I} \leq \langle y_i \rangle_{i \in I}$ if and only if $x_i \leq y_i$ in $X_i$ for every $i \in I$).

10. The definitions for ordered separation axioms in a partially ordered topological space may also be used to define the $T_2$-ordered and $T_1$-ordered properties for any quasiordered topological space $(X, \mathcal{T}, \leq)$. Of the following implications which hold for partially ordered topological spaces, which still hold for quasiordered topological spaces? Justify your answers.

$$T_2\text{-ordered} \quad \Rightarrow \quad T_1\text{-ordered}$$
$$\Downarrow \qquad\qquad\qquad \Downarrow$$
$$T_2 \qquad \Rightarrow \qquad T_1$$

11. Determine whether each partially ordered topological space $(X, \mathcal{T}, \leq)$ below is $T_2$-ordered. If it is $T_2$-ordered, prove it. If it is not $T_2$-ordered, then (i) find $x \not\leq y$ in $X$ which cannot be separated by monotone neighborhoods, (ii) find a sequence $(x_n, y_n)_{n \in \mathbb{N}}$ converging to $(x, y)$ with $x_n \leq y_n$ for every $n \in \mathbb{N}$, but $x \not\leq y$, and (iii) sketch the graph of the partial order in $X^2$ and show that it is not closed.

   (a) $X = \mathbb{R}$, $\mathcal{T}$ is the Euclidean topology, and $\leq$ is the usual order on the positive reals and is equality on $(-\infty, 0]$, with positive numbers and nonpositive numbers not related. (Thus, $x \leq y$ if and only if $x = y$ or $0 < x \leq y$.)

   (b) $X = \mathbb{R}$, $\mathcal{T}$ is the Euclidean topology, and $\leq$ is the usual order on the nonnegative reals and is equality on the negative reals, with negative numbers and nonnegative numbers not related. (Thus, $x \leq y$ if and only if $x = y$ or $0 \leq x \leq y$.)

12. A *bitopological space* is a triple $(X, \mathcal{T}_1, \mathcal{T}_2)$ where $X$ is a set and $\mathcal{T}_1$ and $\mathcal{T}_2$ are topologies on $X$. A bitopological space $(X, \mathcal{T}_1, \mathcal{T}_2)$ is *pairwise* $T_1$ if for any pair of distinct points $x, y \in X$, either there exists a $\mathcal{T}_1$-neighborhood of $x$ excluding $y$ or a $\mathcal{T}_2$-neighborhood of $x$ excluding $y$.

   (a) If $(X, \mathcal{T}, \leq)$ is a $T_1$-ordered topological space, show that $(X, \mathcal{T}^\sharp, \mathcal{T}^\flat)$ is pairwise $T_1$.

   (b) Let $(X, \mathcal{T}, \leq)$ be the interval $[0, 1]$ with the usual topology and order from the real line, except with $0$ noncomparable to all other points. Show that $(X, \mathcal{T}^\sharp, \mathcal{T}^\flat)$ is pairwise $T_1$ but $(X, \mathcal{T}, \leq)$ is not $T_1$-ordered.

13. Exercise 12 shows that the definition given for the $T_1$-ordered property is not aligned with the bitopological concept of pairwise $T_1$. Here is an alternate definition of $T_1$-ordered which remedies this: $(X, \mathcal{T}, \leq)$ is $T_1^K$-*ordered* if for every $x \in X$, $\{x\} = \bigcap\{U \in \mathcal{T}^\sharp \cup \mathcal{T}^\flat : x \in U\}$.

   (a) Show that $(X, \mathcal{T}, \leq)$ is $T_1^K$-ordered if and only if $(X, \mathcal{T}^\sharp, \mathcal{T}^\flat)$ is pairwise $T_1$.

   (b) Show that $T_1$-ordered implies $T_1^K$-ordered.

(c)  Show that $(X, \mathcal{T}, \leq)$ is a $T_1^K$-ordered if and only if, for every $x \in X$, $\{x\} = I(x) \cap D(x)$, where $I(x)$ and $D(x)$ are as defined in Exercise 5.

(d)  Show that if the order on $(X, \mathcal{T}, \leq)$ is a total order, then $X$ is $T_1$-ordered if and only if $X$ is $T_1^K$-ordered.

(For other shortcomings of the original definition of $T_1$-ordered, see [28].)

## 10.2  Normally ordered topological spaces

Recall that a topological space is $T_4$ if disjoint closed sets can be separated by open sets, and is normal if it is $T_1$ and $T_4$. The ordered versions of $T_4$ and normality involve separating disjoint monotone closed sets.

**Definition 10.2.1.** A partially ordered topological space $(X, \mathcal{T}, \leq)$ is $T_4$-*ordered* if, for any closed increasing set $A$ and any closed decreasing set $B$ disjoint from $A$, there exist open sets $U_A$ and $U_B$ with $A \subseteq U_A$, $B \subseteq U_B$, and $i(U_A) \cap d(U_B) = \emptyset$. A partially ordered topological space $(X, \mathcal{T}, \leq)$ is *strongly* $T_4$-*ordered* if, for any closed increasing set $A$ and any closed decreasing set $B$ disjoint from $A$, there exists an open increasing set $U_A$ containing $A$ disjoint from an open decreasing set $U_B$ containing $B$. A partially ordered topological space is *normally ordered (strongly normally ordered)* if it is $T_1$-ordered and (strongly) $T_4$-ordered.

Thus, a $T_4$-ordered space is one in which disjoint oppositely directed monotone closed sets can be separated by disjoint oppositely directed monotone neighborhoods, and is strongly $T_4$-ordered if in addition, the separating monotone neighborhoods may be chosen to be open. (Here we use the convention that an open neighborhood of a set $A$ is an open set containing $A$, and a neighborhood of $A$ is any set containing an open neighborhood of $A$.)

Since any compact $T_2$ space is normal, the following result is not surprising.

**Theorem 10.2.2.** *Any compact $T_2$-ordered topological space is strongly normally ordered.*

This theorem shows that the compact $T_2$-ordered spaces are among the best behaved ordered topological spaces. The proof is analogous to the proof of the corresponding result Theorem 7.1.8 for (trivially ordered) topological spaces.

Total orders are particularly nice partial orders, and with the right topology defined in terms of the order, they give another class of strongly normally ordered spaces. First, we present a natural way to define a topology from any partial order.

**Definition 10.2.3.** If $(X, \leq)$ is a partially ordered set, the *order topology* on $X$ is the topology generated by the subbasis of open rays $(a, \rightarrow) = \{x \in X : a < x\} = i(a) - \{a\}$ and $(\leftarrow, b) = \{x \in X : x < b\} = d(b) - \{b\}$. If $\leq$ is a total order on $X$, then $(X, \leq)$ with the order topology is called a *linearly ordered topological space*, or a *LOTS*.

Though a linear order is another name for a total order, note that a LOTS is more than an ordered topological space whose order is linear: a LOTS must carry the order topology.

Clearly the collection $\mathcal{S} = \{(a, \rightarrow) : a \in X\} \cup \{(\leftarrow, b) : b \in X\}$ of all open rays is a subbasis for a topology on $X$. If $X = \{x\}$ has only one element, then the empty intersection gives $\{x\}$ and the empty union gives $\emptyset$. If $X$ has a smallest element $a$ and at least one other element $x > a$, then $(a, \rightarrow), (\leftarrow, x) \in \mathcal{S}$ and thus $\bigcup \mathcal{S} = X$. If $X$ has no smallest element, then, for any $x \in X$, there exists $x^- < x$, so $x \in (x^-, \rightarrow) \in \mathcal{S}$ and thus $\bigcup \mathcal{S} = X$.

The intersection of two oppositely directed open rays $(a, \rightarrow) \cap (\leftarrow, b)$ in $(X, \leq)$ will be denoted $(a, b)$ and called the open interval with endpoints $a$ and $b$. Open rays and the whole space $X = (\leftarrow, \rightarrow)$ are also considered to be open intervals which have fewer than two endpoints. In general, the open intervals in a poset $(X, \leq)$ are not closed under finite intersections and do not form a basis for the order topology. The example below illustrates this.

**Example 10.2.4.** Let $X = \{a, b, c, d, e, f\}$ with the partial order depicted in Figure 10.2. Now $(a, e) = \{b, c\}$ and $(a, f) = \{c, d\}$, but $(a, e) \cap (a, f) = \{c\}$ is not an open interval and contains no open interval around $c$. Thus, the open intervals do not form a basis for any topology on $X$.

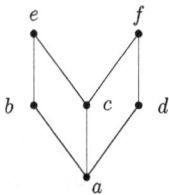

**Figure 10.2:** The open intervals are not a basis for a topology.

If it is surprising to find that the intersection of two intervals need not be an interval, it is perhaps because most of our experience with intervals occurs in $\mathbb{R}$, which is a LOTS. The next theorem addresses this situation.

**Theorem 10.2.5.** *The order topology on a totally ordered set $(X, \leq)$ has a basis $\mathcal{B} = \{(a, b) : a, b \in X\} \cup \{(a, \rightarrow) : a \in X\} \cup \{(\leftarrow, b) : b \in X\} \cup \{X\}$ consisting of the open intervals (with 0, 1, or 2 endpoints) in $X$.*

The proof is left as an exercise. Due to this result, the order topology on a totally ordered set (that is, the topology on a LOTS) is sometimes called the *interval topology*. The proof of the next result is also left to the exercises.

**Theorem 10.2.6.** *Every LOTS is strongly normally ordered.*

We now turn our attention to another large class of normally ordered spaces introduced in [42].

**Definition 10.2.7.** A *(partially) ordered metric space* is a triple $(X, m, \leq)$ where $X$ is a set, $m$ is a metric on $X$, and $\leq$ is a partial order on $X$. If $n$ is a natural number, the ordered metric space $(X, m, \leq)$ is $\frac{1}{n}$-*ball transitive* if $x \leq y$ implies $B(x, \frac{\varepsilon}{n}) \subseteq d(B(y, \varepsilon))$ and $B(y, \frac{\varepsilon}{n}) \subseteq i(B(x, \varepsilon))$ for all $\varepsilon > 0$. We will say a metric space $X$ with a partial order is *ball transitive* if it is $\frac{1}{n}$-ball transitive for some natural number $n$.

**Example 10.2.8.** Let $X = \mathbb{R} \times \{0, 1\}$ with the Euclidean metric $m$. Define $\leq$ by $(x, y) \leq (z, w)$ if and only if $(x, y) = (z, w)$ or $y = 0, w = 1$, and $z = 2x$. Thus, besides $(x, y) \leq (x, y)$ for every $(x, y) \in X$, we have only $(x, 0) \leq (2x, 1)$ for all $x \in \mathbb{R}$, as suggested in Figure 10.3. The increasing hull of a ball $B((x, 0), \varepsilon)$ on the lower segment contains the ball $B((2x, 1), 2\varepsilon)$, and thus $B((2x, 1), \frac{\varepsilon}{n})$ for any natural number $n$. The decreasing hull of a ball $B((x, 1), \varepsilon)$ on the upper segment contains the ball $B((\frac{x}{2}, 0), \frac{\varepsilon}{2})$, and thus $(X, m, \leq)$ is $\frac{1}{2}$-ball transitive (and not 1-ball transitive).

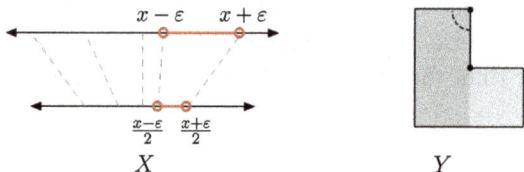

**Figure 10.3:** $X$ is $\frac{1}{2}$-ball transitive. $Y$ is not ball transitive.

Let $Y = [-1, 1]^2 - (0, 1]^2$ be the subset of $\mathbb{R}^2$ shown in Figure 10.3, with the Euclidean metric $m$ and the product order $(a, b) \leq (x, y)$ if and only if $a \leq x$ and $b \leq y$. Now $(0, 0) \leq (0, 1)$, but for any $\varepsilon < 1$, $d(B((0, 1), \varepsilon))$ contains no ball of positive radius around $(0, 0)$. Thus, $Y$ is not ball-transitive. Note that $Y$ is compact, $T_2$-ordered (and hence normally ordered), and convex, so none of these properties implies ball transitivity.

Next, we will show that with the natural product construction, a product of ball transitive spaces is ball transitive.

**Theorem 10.2.9.** *If $(X, m_X, \leq_X)$ is a $\frac{1}{n}$-ball transitive space and $(Y, m_Y, \leq_Y)$ is a $\frac{1}{k}$-ball transitive space, then $X \times Y$ with the metric $m_X + m_Y$ and the product order is a $\frac{1}{n+k}$-ball transitive space.*

*Proof.* Suppose $(a, b) \leq (s, t)$ in $X \times Y$ and $(x, y) \in B((a, b), \varepsilon)$. Then $m_X(x, a) + m_Y(y, b) < \varepsilon$, so $x \in B_X(a, \varepsilon)$ and $y \in B_Y(b, \varepsilon)$. From the hypotheses, it follows that $x \in d(B_X(s, n\varepsilon))$ and $y \in d(B_Y(t, k\varepsilon))$. Thus, there exist points $s' \in B_X(s, n\varepsilon)$ and $t' \in B_Y(t, k\varepsilon)$ with $(x, y) \leq (s', t')$. Since $(s', t') \in B_{X \times Y}((s, t), (n + k)\varepsilon)$, it follows that

$$B((a, b), \varepsilon) \subseteq d(B((s, t), (n + k)\varepsilon)) \quad \text{in } X \times Y.$$

With the dual argument, this proves $X \times Y$ is $\frac{1}{n+k}$-ball transitive. $\square$

Given the usual metric and order, it is easy to see that the real line $\mathbb{R}$ is 1-ball transitive. It may be geometrically clear that the Euclidean spaces $\mathbb{R}^2$ and $\mathbb{R}^3$ with the product order are also 1-ball transitive. While our visualization is limited for higher dimensions, the previous theorem shows that, for any $n \in \mathbb{N}$, the space $\mathbb{R}^n$ with the taxicab metric $m((x_1, \ldots, x_n), (y_1, \ldots, y_n)) = \sum_{i=1}^{n} |x_i - y_i|$ and the product order is ball transitive. It is left as an exercise to show that this implies $\mathbb{R}^n$ with the Euclidean topology and product order is ball transitive.

Since $\mathbb{R}^2$ is ball transitive, the space $Y$ of Example 10.2.8 shows that ball transitive is not hereditary, even to convex compact $T_2$-ordered subspaces. In that space $Y$, with $\varepsilon < 1$ note that $B((0,1), \varepsilon)$ is open but $d(B((0,1), \varepsilon))$ is not open. This cannot happen in a ball transitive space.

**Theorem 10.2.10.** *If $(X, m, \leq)$ is a ball transitive ordered metric space and $U$ is open in $X$, then $i(U)$ and $d(U)$ are open in $X$.*

*Proof.* Suppose $U$ is open in $X$ and $x \in i(U)$. Then there exists $y \in U$ such that $y \leq x$. Choose $\varepsilon > 0$ so that $B(y, \varepsilon) \subseteq U$. Since $X$ is ball transitive, there exists a natural number $n$ such that $B(x, \frac{\varepsilon}{n}) \subseteq i(B(y, \varepsilon)) \subseteq i(U)$. Thus, $i(U)$ is open. The dual argument shows that $d(U)$ is open. $\qquad\square$

If $F$ is closed in a ball transitive space, $i(F)$ and $d(F)$ need not be closed, as seen by considering $F = \{(-x, \frac{1}{x}) \in \mathbb{R}^2 : x > 0\}$ in $\mathbb{R}^2$. However, we have the following result.

**Theorem 10.2.11.** *In a ball transitive space, the closures and interiors of decreasing sets are decreasing, and the closures and interiors of increasing sets are increasing.*

*Proof.* Suppose $A$ is a decreasing set in a ball transitive space $X$, $y \in \mathrm{cl}(A)$, and $z \leq y$. We want to show $z \in \mathrm{cl}(A)$. Suppose there exists $\varepsilon > 0$ such that $B(z, \varepsilon) \cap A = \emptyset$. Noting that $A = d(A)$, it follows that $i(B(z, \varepsilon)) \cap A = \emptyset$. But $y \in i(B(z, \varepsilon))$, and, by Theorem 10.2.10, $i(B(z, \varepsilon))$ is open. Thus, $i(B(z, \varepsilon))$ is an open neighborhood of $y$ that does not intersect $A$, contrary to $y \in \mathrm{cl}(A)$. This shows that the closure of a decreasing set $A$ is decreasing. The dual argument shows the closure of an increasing set is increasing.

If $A$ is decreasing, $X - A$ is increasing, and by the previous paragraph, $\mathrm{cl}(X - A)$ is also increasing. Thus, $\mathrm{int}(A) = X - \mathrm{cl}(X - A)$ is decreasing. The dual argument completes the proof. $\qquad\square$

**Theorem 10.2.12.** *Every ball transitive ordered metric space $(X, m, \leq)$ is strongly $T_4$-ordered.*

*Proof.* Suppose $A$ and $B$ are disjoint closed subsets of a $\frac{1}{n}$-ball transitive space $X$, with $A$ increasing and $B$ decreasing. For any $a \in A$, $a$ is an element of the open increasing set $X - B$, so there exists $\varepsilon_a$ such that $i(B(a, \varepsilon_a)) \cap B = \emptyset$. Similarly, for any $b \in B$, there exists $\varepsilon_b$ such that $d(B(b, \varepsilon_b)) \cap A = \emptyset$. Put

$$U_A = \bigcup_{a \in A} i\left( B\left( a, \frac{\varepsilon_a}{2n} \right) \right), \quad \text{and}$$

$$U_B = \bigcup_{b \in B} d\left( B\left( b, \frac{\varepsilon_b}{2n} \right) \right).$$

By Theorem 10.2.10 and properties of open and monotone sets, $U_A$ is an open increasing neighborhood of $A$ and $U_B$ is an open decreasing neighborhood of $B$. To show that $U_A$ and $U_B$ are disjoint, suppose to the contrary that $x \in U_A \cap U_B$. Then there exist points $a \in A$ and $b \in B$ such that $x \in i(B(a, \frac{\varepsilon_a}{2n})) \cap d(B(b, \frac{\varepsilon_b}{2n}))$. Thus, there exist points $a' \in B(a, \frac{\varepsilon_a}{2n})$ and $b' \in B(b, \frac{\varepsilon_b}{2n})$ with $a' \le x \le b'$. In the case $\varepsilon_b \ge \varepsilon_a$, we have

$$a \in B\left( a', \frac{\varepsilon_a}{2n} \right) \subseteq B\left( a', \frac{\varepsilon_b}{2n} \right) \subseteq d\left( B\left( b', \frac{\varepsilon_b}{2} \right) \right) \subseteq d(B(b, \varepsilon_b)),$$

contrary to $A \cap d(B(b, \varepsilon_b)) = \emptyset$. In the case $\varepsilon_a \ge \varepsilon_b$, we have

$$b \in B\left( b', \frac{\varepsilon_b}{2n} \right) \subseteq B\left( b', \frac{\varepsilon_a}{2n} \right) \subseteq i\left( B\left( a', \frac{\varepsilon_a}{2} \right) \right) \subseteq i(B(a, \varepsilon_a)),$$

contrary to $B \cap i(B(a, \varepsilon_a)) = \emptyset$. Now $U_A$ and $U_B$ are the desired monotone open sets separating $A$ and $B$. □

This gives the following immediate corollary.

**Corollary 10.2.13.** $\mathbb{R}^n$ *with the Euclidean topology and product order is strongly normally ordered.*

*Proof.* It is an easy exercise to show that $\mathbb{R}^n$ with the Euclidean topology is $T_1$-ordered. Since the Euclidean topology is generated by the taxicab metric $m$, $(\mathbb{R}^n, m, \le)$ is $T_1$-ordered and, by the previous theorem, strongly $T_4$-ordered. Thus, $(\mathbb{R}^n, m, \le)$ is strongly normally ordered. But strongly normally ordered is a topological property, independent of the metric, and since $m$ generates the Euclidean topology, the Euclidean space $\mathbb{R}^n$ is strongly $T_4$-ordered. □

## Exercises

1. Prove Theorem 10.2.2.
2. Show that every (strongly) normally ordered space is (strongly) $T_2$-ordered.
3. Which of the properties $T_4$-ordered, strongly $T_4$-ordered, normally ordered, and strongly normally ordered are satisfied by the space of Example 10.1.7? Justify your answers.
4. (A subspace of a LOTS need not be a LOTS.) Consider the subspace $Y = [0, 1) \cup [2, 4]$ of the LOTS $\mathbb{R}$. Show that $Y$ is not a LOTS, even though $Y$ inherits the total order from $\mathbb{R}$. A subspace of a LOTS is called a *generalized ordered space*, or a *GO-space*.
5. Prove Theorem 10.2.5: the order topology on a totally ordered set $(X, \le)$ has a basis of open intervals (including intervals with 0, 1 or 2 endpoints).
6. Prove Theorem 10.2.6: every LOTS is strongly normally ordered.

7. An immediate consequence of Theorem 10.2.9 is that $\mathbb{R}^n$ with the product order and taxicab metric $m((x_1,\ldots,x_n),(y_1,\ldots,y_n)) = \sum_{i=1}^n |x_i - y_i|$ is $\frac{1}{n}$-ball transitive. Give a detailed proof that $\mathbb{R}^n$ with the product order and Euclidean metric $m_\varepsilon$ is ball transitive.

8. Which of the properties $T_1$-ordered, $T_2$-ordered, and ball transitive are satisfied by the spaces below? Justify your answers.

   (a) Let $X = [0,1]$ with the usual Euclidean metric and the usual order on $[0,1)$, with $1 \le 1$ and $1 || x$ for all $x \in [0,1)$.

   (b) Let $Y = [0,1] \times \{0,1\} \subseteq \mathbb{R}^2$ with the Euclidean metric and the order $\Delta_Y \cup \{((x,0),(x,1)) : x \in (0,1]\}$. That is, besides $(x,y) \le (x,y)$ for every $(x,y) \in Y$, we have $(x,0) \le (x,1)$ for $x \in (0,1]$.

   (c) Let $Z = [1,\infty) \times \{0,1\} \subseteq \mathbb{R}^2$ with the Euclidean metric and the order $\Delta_Z \cup \{((x,0),(x^2,1)) : x \in [1,\infty)\}$. That is, besides $(x,y) \le (x,y)$ for every $(x,y) \in Y$, we have $(x,0) \le (x^2,1)$ for $x \in [1,\infty)$.

9. Show that $\mathbb{R}^n$ with the Euclidean metric and product order is $T_1$-ordered.

10. Let $X = \mathbb{R} - (0,1)$ with the Euclidean metric and the usual order, except $0 \nleq 1$. Show that $X$ is ball transitive but not $T_1$-ordered.

11. Let $X$ be a copy of $\mathbb{R}$ embedded in $\mathbb{R}^2$ as shown, with $(-\infty,-1)$ bent to lie on the graph of $y = 1/x$ and $(1,\infty)$ lying on $y = 0$. Keep the linear order from $\mathbb{R}$ and give $X$ the Euclidean metric from $\mathbb{R}^2$. Show that this linearly ordered metric space is not ball transitive.

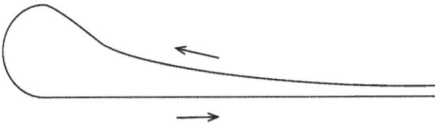

12. A lattice $L$ is a *metric lattice* if there exists a strictly increasing function $f : L \to \mathbb{R}$ with $f(x) + f(y) = f(x \vee y) + f(x \wedge y)$ for all $x,y \in L$; then $m(x,y) = f(x \vee y) - f(x \wedge y)$ defines a metric on $L$. Metric lattices are discussed in Garrett Birkhoff's *Lattice Theory* [6], where it is shown that, for any $a,x,y$ in a metric lattice, $m(a \vee x, a \vee y) + m(a \wedge x, a \wedge y) \le m(x,y)$. Use this to prove that every metric lattice $(L,m,\le)$ is ball transitive.

# 11 Variations on metric spaces

## 11.1 Pseudometrics

In this chapter, we will consider distance functions which may not meet all of the requirements of a metric. We have seen that on the set $C[a, b]$ of continuous functions from $[a, b]$ to $\mathbb{R}$, we may define a metric $d(f, g) = \int_a^b |f(x) - g(x)|\, dx$, with the distance between $f$ and $g$ being the area between them. This function $d$ is not a metric on the set of integrable functions over $[a, b]$. For example, if $f(x) = x$ on $[a, b] = [0, 1]$, $g(x) = x$ on $[0, 1)$, and $g(1) = 5$, then $f$ and $g$ are integrable, and $d(f, g) = \int_0^1 |f(x) - g(x)|\, dx = 0$, but $f \neq g$. Thus, the property of a metric that $d(f, g) = 0$ implies $f = g$ fails. If we relax the definition of a metric to omit this condition, then we obtain a *pseudometric*.

**Definition 11.1.1.** A *pseudometric* on a set $X$ is a function $d : X \times X \to [0, \infty)$ such that
(a)  $p(x, y) \geq 0$ for all $x, y \in X$ (nonnegativity);
(b)  $p(x, x) = 0$ for all $x \in X$;
(c)  $p(x, y) = p(y, x)$ for all $x, y \in X$ (symmetry);
(d)  $p(x, y) \leq p(x, z) + p(z, y)$ for all $x, y, z \in X$ (triangle inequality).

The pair $(X, p)$ is a pseudometric space. As in a metric space, $B(x, \varepsilon) = \{y \in X : p(x, y) < \varepsilon\}$ is the *ball* of radius $\varepsilon$ centered at $x$, and the collection $\mathcal{B} = \{B(x, \varepsilon) : x \in X, \varepsilon > 0\}$ of all balls is a basis for the *pseudometric topology* on $X$.

The proof that the collection of balls in a metric space forms a basis for a topology remains valid for the collection of balls in a pseudometric space.

When we drop the metric condition that $d(x, y) = 0$ implies $x = y$ to get a pseudometric, we should ask what we lose. One consequence of that condition was that every metric space was Hausdorff: if $x \neq y$, then with $\varepsilon = \frac{1}{2} d(x, y)$, the balls $B(x, \varepsilon)$ and $B(y, \varepsilon)$ were disjoint neighborhoods separating $x$ and $y$. This proof of the Hausdorff property is not valid in a pseudometric space, and for good reason: pseudometric spaces need not be Hausdorff. The next example illustrates this.

**Example 11.1.2.** On a nonempty set $X$, define $p : X \times X \to [0, \infty)$ by $p(x, y) = 0$ for all $x, y \in X$. It is easy to see that $p$ satisfies all the conditions of a pseudometric. For any $x \in X$ and any $\varepsilon > 0$, we have $B(x, \varepsilon) = X$, so the pseudometric topology is the indiscrete topology on $X$. This pseudometric is called the *trivial pseudometric*.

This example shows that pseudometric spaces need not be $T_2, T_1$, nor even $T_0$. However, pseudometric spaces satisfy the higher separation axioms which do not require $T_1$. We first present a useful technique. If we can measure the distance between points, there is a natural way to measure the distance between a point and a set.

https://doi.org/10.1515/9783110686579-012

**Definition 11.1.3.** Suppose $(X,p)$ is a pseudometric space, $\emptyset \neq A \subseteq X$, and $x \in X$. The *distance from $x$ to $A$* is

$$d_A(x) = \inf\{p(x,a) : a \in A\}.$$

**Theorem 11.1.4.** *If $(X,p)$ is a pseudometric space and $A$ is a nonempty subset of $X$, then $d_A(x) = 0$ if and only if $x \in \mathrm{cl}\, A$. Furthermore, if $A \subseteq X$ is nonempty and closed, then $d_A : X \to \mathbb{R}$ is a continuous function.*

*Proof.* The first statement follows since

$$\begin{aligned}
d_A(x) = 0 &\iff \inf\{p(a,x) : a \in A\} = 0 \\
&\iff \forall \varepsilon > 0 \quad \exists a \in A, p(x,a) < \varepsilon \\
&\iff \forall \varepsilon > 0 \quad B(x,\varepsilon) \cap A \neq \emptyset \\
&\iff x \in \mathrm{cl}\, A.
\end{aligned}$$

Suppose $A$ is a nonempty closed subset of $X$. To see that $d_A$ is continuous, for any $\varepsilon > 0$, we will show that, for $\delta = \varepsilon$, $y \in B(x,\delta)$ implies $|d_A(x) - d_A(y)| < \varepsilon$. Suppose $y \in B(x,\delta)$. By the triangle inequality,

$$p(a,x) \leq p(a,y) + p(y,x) \quad \text{and} \quad p(a,x) \geq p(a,y) - p(x,y).$$

Taking the infimum over $a \in A$, and noting that $\inf\{p(a,y) + c : a \in A\} = \inf\{p(a,y) : a \in A\} + c$ for the constant $c = \pm p(x,y)$, we have

$$d_A(x) \leq d_A(y) + p(x,y) \quad \text{and} \quad d_A(x) \geq d_A(y) - p(x,y).$$

Now $y \in B(x,\delta) = B(x,\varepsilon)$ implies $p(x,y) < \varepsilon$, which leads to

$$d_A(x) - d_A(y) < \varepsilon \quad \text{and} \quad d_A(x) - d_A(y) > -\varepsilon.$$

Together, this gives $|d_A(x) - d_A(y)| < \varepsilon$, as needed. $\square$

If $A$ is closed in a pseudometric space $X$ and $x \notin A$, the result above implies that $d_A$ is a continuous real-valued function which maps $A$ to $0$ and $x$ to $d_A(x) > 0$. Thus, $X$ is $T_{3.5}$, and therefore is $T_3$. The proof given in Theorem 7.1.7 that every metric space is $T_4$ remains valid for pseudometric spaces. This gives the following result.

**Theorem 11.1.5.** *Every pseudometric space is $T_3$, $T_{3.5}$, and $T_4$.*

**Example 11.1.6.** Consider the set $\mathbb{R}^{\mathbb{R}}$ of functions $f : \mathbb{R} \to \mathbb{R}$. Define $p : \mathbb{R}^{\mathbb{R}} \times \mathbb{R}^{\mathbb{R}} \to [0,\infty)$ by $p(f,g) = |f(0)-g(0)|$, so the distance between functions $f$ and $g$ is the vertical distance between their graphs at $x = 0$. In other words, for $f,g \in \prod_{j \in \mathbb{R}} \mathbb{R}$, we take $p(f,g)$ to be $d_\varepsilon(\pi_0(f), \pi_0(g))$ where $d_\varepsilon$ is the Euclidean metric on $\pi_0(\prod_{j \in \mathbb{R}} \mathbb{R}) = \mathbb{R}$. Since the distance between functions is measured only in the 0th coordinate by a metric on that factor, $p$ clearly satisfies the conditions of a pseudometric. It is not a metric, however, since $p(\sin(x), x^2) = 0$ even though the functions $\sin(x)$ and $x^2$ are not equal.

The crux of the example above is applied in the next example.

**Example 11.1.7.** Consider the set $X = C(\mathbb{R}) \subseteq \mathbb{R}^{\mathbb{R}}$ of continuous real-valued functions $f : \mathbb{R} \to \mathbb{R}$ with domain $\mathbb{R}$, and let $[a, b]$ be a compact subset of $\mathbb{R}$. Now the set $C[a, b]$ of continuous real-valued functions $f : [a, b] \to \mathbb{R}$ is obtained from $C(\mathbb{R})$ essentially by discarding part of the graph of each $f \in C(\mathbb{R})$. Now we know metrics on $C[a, b]$, such as the sup-metric or the areawise metric. For $f, g \in C(\mathbb{R})$, we may use the metric distance between their restrictions to $[a, b]$ in any specified metric on $C[a, b]$ as the pseudo-metric distance between $f$ and $g$ as functions on $\mathbb{R}$. Again, because the distances are defined in terms of an existing metric, this gives a pseudometric $p$. But again, because there are distinct functions $f, g$ on $\mathbb{R}$ which agree on $[a, b]$, $p(f, g) = 0$ does not imply $f$ and $g$ are equal.

Recalling that an equivalence relation on a set $X$ specifies equality of some at-tribute of the elements of $X$, the previous two examples suggest a connection between pseudometrics, equivalence relations, and metrics. In Example 11.1.6, the distance be-tween two functions in $\mathbb{R}^{\mathbb{R}}$ is completely determined by their values at 0, so for the purpose of finding distances, we may ignore all other values of the functions, and de-clare two functions to be equivalent if and only if they agree at 0. Now, the metric on $\mathbb{R} = \pi_0(\prod_{j \in \mathbb{R}} \mathbb{R})$ gives a metric on the equivalence classes from this equivalence relation. In terms of the pseudometric, the following are equivalent: (a) $p(f, g) = 0$, (b) $d_{\mathcal{E}}(f(0), g(0)) = 0$, and (c) $f \equiv g$.

Similar remarks apply to the Example 11.1.7, using the equivalence relation on $C(\mathbb{R})$ defined by $f \equiv g$ if and only if $f(x) = g(x)$ for all $x \in [a, b]$.

We now formalize this link.

**Theorem 11.1.8.** *Suppose $p$ is a pseudometric on $X$ with pseudometric topology $\mathcal{T}_p$.*
(a) *The relation on $X$ defined by $x \equiv y$ if and only if $p(x, y) = 0$ is an equivalence relation.*
(b) *If $Y = \{[x] : x \in X\}$ is the set of $\equiv$-equivalence classes, then the function $d : Y \times Y \to [0, \infty)$ defined by $d([x], [y]) = p(x, y)$ is a metric on $Y$.*
(c) *The metric topology $\mathcal{T}_d$ on $Y$ is the quotient topology $\mathcal{T}_f$ where $f : (X, p) \to Y = \{[x] : x \in X\}$ is the function defined by $f(x) = [x]$.*

*Proof.* (a) The reflexivity and symmetry of $\equiv$ follow from $p(x, x) = 0$ and $p(x, y) = p(y, x)$ for all $x, y \in X$. Transitivity follows from the other two defining properties of a pseu-dometric, for if $p(x, y) = 0$ and $p(y, z) = 0$, then $0 \le p(x, z) \le p(x, y) + p(y, z) = 0 + 0$, so $p(x, z) = 0$.

(b) Suppose $Y$ and $d$ are as in the statement of (b). First, we must show that $d$ is well-defined. Suppose $[x] = [x']$ and $[y] = [y']$. Then $p(x, x') = 0 = p(y, y')$. To see $d([x], [y]) = d([x'], [y'])$, we must show $p(x, y) = p(x', y')$. Now $p(x, y) \le p(x, x') + p(x', y') + p(y', y) = p(x', y')$ and $p(x', y') \le p(x', x) + p(x, y) + p(y, y') = p(x, y)$, so $p(x, y) = p(x', y')$, and thus $d$ is well-defined.

Because $d([x], [y]) = p(x, y)$, $d$ clearly is a pseudometric on $X$; the defining properties hold for $d$ since they hold for $p$. To show $d$ is a metric, it only remains to show that $d([x], [y]) = 0$ implies $[x] = [y]$. But $d([x], [y]) = p(x, y) = 0$ if and only if $x \equiv y$, so $[x] = [y]$.

(c) Recalling that $V \in \mathcal{T}_f$ if and only if $f^{-1}(V) \in \mathcal{T}_p$, we want to show $V \in \mathcal{T}_d$ if and only if $f^{-1}(V) \in \mathcal{T}_p$. Now $V \in \mathcal{T}_d$ if and only if $V$ can be written as a union $\bigcup_{i \in I} B([x_i], \varepsilon_i)$ of $d$-balls and $f^{-1}(V) \in \mathcal{T}_p$ if and only if it can be written as a union of $p$-balls. Now

$$z \in f^{-1}(B([x], \varepsilon)) \iff f(z) = [z] \in B([x], \varepsilon)$$
$$\iff d([x], [z]) = p(x, z) < \varepsilon$$
$$\iff z \in B_p(x, \varepsilon),$$

so $f^{-1}(B([x], \varepsilon)) = B_p(x, \varepsilon)$. Since $f^{-1}(\bigcup_{i \in I} B([x_i], \varepsilon_i)) = \bigcup_{i \in I} f^{-1}(B([x_i], \varepsilon_i)) = \bigcup_{i \in I} B_p(x_i, \varepsilon_i)$, it follows that $V \in \mathcal{T}_d$ if and only if $f^{-1}(V) \in \mathcal{T}_p$. □

By the theorem above, every pseudometric on $X$ gives a metric on a partition of $X$. Our next result shows the converse: every metric on a partition of $X$ gives a pseudometric on $X$. Furthermore, these operations are "inverses" in the sense that iterating them will return the original structure.

**Theorem 11.1.9.** *Suppose $\approx$ is an equivalence relation on $X$, $Y = \{[x] : x \in X\}$ is the set of $\approx$-equivalence classes, and $d$ is a metric on $Y$. Then $p : X \times X \to \mathbb{R}$ defined by $p(x, y) = d([x], [y])$ is a pseudometric on $X$, and if $f : X \to Y$ is the map $f(x) = [x]$, then the d-metric topology on $X$ is the $\mathcal{T}_f$ quotient topology.*

*Proof.* The pseudometric properties of $p$ follow easily from the metric properties of $d$, and the equivalence of the topologies follows from the last argument of the proof of Theorem 11.1.8 that $f^{-1}(B([x], \varepsilon)) = B_p(x, \varepsilon)$. □

Combined, the two previous results say that pseudometrics on $X$ are precisely metrics on equivalence classes (or partitions) of $X$. Doing mathematics on equivalence classes is an important recurring theme that arises in loss-of-resolution situations. If the Euclidean plane is represented by a computer screen consisting of $m \times n$ pixels, the overlaid pixel grid provides a partition of the plane, so working with the pixels is an example of working with equivalence classes. Similarly, if demographic data is collected from a sampling of locations in the USA but is then summarized by state, in effect we have a loss of resolution from the entire USA to a "pixelated" version consisting of equivalence classes Alabama, Alaska, Arizona, Arkansas, ..., Wisconsin, Wyoming.

This situation is analogous to quasiorders on $X$ being partial orders on equivalence classes. Pseudometrics are metrics on equivalence classes. Both may be used to model a loss of resolution.

Sometimes it is of interest to consider more than one pseudometric on a set $X$.

**Definition 11.1.10.** A *gauge* on a set $X$ is a collection $\mathcal{P} = \{p_i : i \in I\}$ of pseudometrics $p_i$ on $X$. A gauge $\mathcal{P}$ on $X$ generates a topology $\mathcal{T}_\mathcal{P}$ having a subbasis consisting of all balls from all the pseudometrics $p_i$ ($i \in I$). That is, $\mathcal{T}_\mathcal{P} = \bigvee_{p_i \in \mathcal{P}} \mathcal{T}_{p_i}$ where $\mathcal{T}_{p_i}$ is the pseudometric topology on $X$ generated by $p_i$. A *gauge space* is a topological space $(X, \mathcal{T})$ where $\mathcal{T} = \mathcal{T}_\mathcal{P}$ is generated by some gauge $\mathcal{P}$ on $X$. A gauge space (or a gauge) is *separating* if for every pair of distinct points $x \neq y$ in $X$, there exists $p_i \in \mathcal{P}$ with $p_i(x, y) \neq 0$. Depending on the emphasis, we may denote a gauge space by $(X, \mathcal{P})$, $(X, \mathcal{T}_\mathcal{P})$, or $(X, \mathcal{T})$.

**Theorem 11.1.11.** *Suppose $\mathcal{P} = \{p_i : i \in I\}$ is a gauge on $X$.*
(a) *$\mathcal{P}$ is separating if and only if $\mathcal{T}_\mathcal{P}$ is a Hausdorff topology on $X$.*
(b) *$(X, \mathcal{T}_\mathcal{P})$ is $T_{3.5}$.*
(c) *If $A \subseteq X$, then $\{p_i|_{A \times A} : i \in I\}$ is a gauge on $A$ which generates the subspace topology on $A$, as a subspace of $(X, \mathcal{T}_\mathcal{P})$.*

(a) and (c) follow easily from the definitions. (b) follows from the fact that $\mathcal{T}_\mathcal{P} = \bigvee_{p_i \in \mathcal{P}} \mathcal{T}_{p_i}$, and the supremum of $T_{3.5}$ topologies in the lattice of topologies must be $T_{3.5}$. Details are left to the exercises. While (c) shows that the property of being a gauge space is hereditary, the next result shows that it is also productive.

**Theorem 11.1.12.** *Suppose $(X_j, \mathcal{P}_j)$ is a gauge space for each $j$ in an index set $J$. Then $\prod_{j \in J} X_j$ with the product topology is a gauge space.*

*Proof.* Suppose $(X_j, \mathcal{P}_j)$ is a gauge space for each $j \in J$, with $\mathcal{P}_j = \{p_{i,j} : i \in I_j\}$. Each pseudometric $p_{i,j}$ on $X_j$ defines a pseudometric $\hat{p}_{i,j}$ on $\prod_{j \in J} X_j$ by taking the $\hat{p}_{i,j}$-distance between two vectors in $\prod_{j \in J} X_j$ to be the $p_{i,j}$-distance between the projections of those vectors onto the $j$th coordinate $X_j$. Now $\hat{\mathcal{P}} = \{\hat{p}_{i,j} : i \in I_j, j \in J\}$ is a family of pseudometrics on $\prod_{j \in J} X_j$ which generates the gauge topology $\mathcal{T}_{\hat{\mathcal{P}}}$ having subbasis $\{B_{\hat{p}_{i,j}}(\langle x_j \rangle_{j \in J}, \varepsilon) : i \in I_j, j \in J, \varepsilon > 0\}$. But for a fixed $(i, j) \in I_j \times J$, observe that $B_{\hat{p}_{i,j}}(\langle x_j \rangle_{j \in J}, \varepsilon) = \prod_{k \in J} U_k$, where $U_k = X_k$ for $k \neq j$ and $U_j = B_{p_{i,j}}(x_j, \varepsilon)$, is a $\mathcal{P}_j$-subbasic neighborhood of $x_j$. Thus, the gauge topology $\mathcal{T}_{\hat{\mathcal{P}}}$ has the same subbasis as the product topology on $\prod_{j \in J} (X_j, \mathcal{T}_{\mathcal{P}_j})$. $\square$

**Theorem 11.1.13.** *$(X, \mathcal{T})$ is a separating gauge space if and only if $\mathcal{T}$ is completely regular.*

*Proof.* By Theorem 11.1.11(a) and (b), a separating gauge space is $T_2$ and $T_{3.5}$, and thus is completely regular. Suppose $(X, \mathcal{T})$ is completely regular. Then $X$ may be embedded in its Stone–Čech compactification $\beta X$. Recall that $\beta X$ is a subspace of $\prod\{I_f : f \in C^*(X)\}$, where $C^*(X)$ is the collection of bounded continuous real-valued functions on $X$ and $I_f$ is a compact interval in $\mathbb{R}$ containing $f(X)$. Now $\mathbb{R}$ is $T_2$ and pseudometrizable, and thus is a $T_2$ gauge space. Since subspaces, products, and homeomorphic images of $T_2$ gauge spaces are $T_2$ gauge spaces, it follows that $\beta X$ and $X$ are $T_2$ gauge spaces. $\square$

## Exercises

1. Define $p : \mathbb{R}^2 \times \mathbb{R}^2 \to \mathbb{R}$ by $p((a, b), (x, y)) = |a - x|$. Show that $p$ is a pseudometric. Discuss the associated metric space on the equivalence classes of $\mathbb{R}^2$ from the equivalence relation $(a, b) \approx (x, y)$ if and only if $p((a, b), (x, y)) = 0$.

2. Suppose $p$ is a pseudometric on $X$ generating the topology $\mathcal{T}_p$. Prove that $p$ is a metric if and only if $(X, \mathcal{T}_p)$ is $T_0$. (Thus, any non-metrizable pseudometric space is $T_3$, $T_{3.5}$, and $T_4$, but is neither regular, completely regular, nor normal)

3. Let $\mathbb{F} = \{f : \mathbb{R} \to \mathbb{R} : \int_{-\infty}^{\infty} f(x)\,dx$ exists and is finite$\}$. For $f, g \in \mathbb{F}$, taking $p(f, g) = \int_{-\infty}^{\infty} |f(x) - g(x)|\,dx$ defines a pseudometric on $\mathbb{F}$.
   (a) If $(f_n)_{n \in \mathbb{N}}$ is a sequence in $\mathbb{F}$ which converges pointwise to $f$, must $(f_n)_{n \in \mathbb{N}}$ converge to $f$ in $(\mathbb{F}, \mathcal{T}_p)$?
   (b) For $n \in \mathbb{N}$, let $g_n(x) = \sin^n(x)$ for $x \in [-2\pi, 2\pi]$ and $g_n(x) = 0$ otherwise. In $(\mathbb{F}, \mathcal{T}_p)$, does $(g_n)_{n \in \mathbb{N}}$ converge? If so, find all of its limits.

4. Suppose $X$ is a set and $f : X \to \mathbb{R}$ is a function. Define $p : X \times X \to \mathbb{R}$ by $p(x, y) = |f(x) - f(y)|$ for every $x, y \in X$. Show that $p$ is a pseudometric on $X$ and find necessary and sufficient conditions on $f$ to make $p$ a metric.

5. Determine whether the topological spaces below are pseudometrizable.
   (a) $(X, \mathcal{T})$ where $X = \{1, 2, 3, 4\}$ and $\mathcal{T} = \{\emptyset, \{1, 2\}, X\}$.
   (b) $(X, \mathcal{T})$ where $X = \{1, 2, 3, 4\}$ and $\mathcal{T} = \{\emptyset, \{1, 2\}, \{3, 4\}, X\}$.
   (c) $(\mathbb{R}, \mathcal{T})$ where $\mathcal{T} = \{(a, \infty) : a \in \mathbb{R}\} \cup \{\emptyset, \mathbb{R}\}$.

6. Show that every partition topology on $X$ is pseudometrizable.

7. If $d$ is the Euclidean metric on $\mathbb{R}^2$, we may define the distance between two nonempty subsets $A, B \subseteq \mathbb{R}^2$ by $D(A, B) = \inf\{d(a, b) : a \in A, b \in B\}$.
   (a) Show that $D(A, B) = 0$ does not imply $A = B$.
   (b) Show that $D$ does not satisfy the triangle inequality.
   (c) Give an example of two disjoint nonempty closed sets $A, B$ in $\mathbb{R}^2$ with $D(A, B) = 0$.

8. Prove Theorem 11.1.11(a): A gauge space $(X, \mathcal{P})$ is separating if and only if $\mathcal{T}_\mathcal{P}$ is a Hausdorff topology on $X$.

9. In the lattice $T(X)$ of topologies on $X$, if $\mathcal{T}_i$ is $T_{3.5}$ for each $i \in I$, show that $\bigvee \{\mathcal{T}_i : i \in I\}$ is $T_{3.5}$. (Note that this proves Theorem 11.1.11(b).)

10. Prove Theorem 11.1.11(c): A subspace of a gauge space is a gauge space.

## 11.2 Quasi-metrics

It is not uncommon to hear an answer to "How far is it?" given in hours rather than in kilometers or miles. Measured in hours or in energy exerted, the distance by boat from St. Louis downstream to New Orleans will be different from the distance upstream from New Orleans to St. Louis. In a city with one-way streets, the driving distance from $x$ to $y$ may be different from the driving distance from $y$ to $x$. These examples can be modeled by distance functions which are not symmetric. A *quasi-metric* drops the symmetry condition.

**Definition 11.2.1.** A *quasi-metric* on a set $X$ is a function $q : X \times X \to [0, \infty)$ such that
(a)  $q(x, y) \geq 0$ for all $x, y \in X$ (nonnegativity);
(b)  $x = y$ if and only if $q(x, y) = q(y, x) = 0$;
(c)  $q(x, y) \leq q(x, z) + q(z, y)$ for all $x, y, z \in X$ (triangle inequality).

The pair $(X, q)$ is a quasi-metric space. As in a metric space, $B(x, \varepsilon) = \{y \in X : q(x, y) < \varepsilon\}$ is the *ball* of radius $\varepsilon$ centered at $x$, and the collection $\mathcal{B} = \{B(x, \varepsilon) : x \in X, \varepsilon > 0\}$ of all balls is a basis for the *quasi-metric topology* on $X$.

Since distances need not be symmetric in a quasi-metric space, generally $y \in B(x, \varepsilon)$ does not imply $x \in B(y, \varepsilon)$. The example below is a prototypical quasi-metric.

**Example 11.2.2.** On $\mathbb{R}$, define

$$q(x, y) = \begin{cases} y - x & \text{if } x \leq y, \\ 0 & \text{if } y < x. \end{cases}$$

It is routine to confirm that $q$ is a quasi-metric. Now $q(x, y)$ tells how far to the right of $x$ the point $y$ is, where "how far" is measured in the usual Euclidean metric. Considering $\mathbb{R}$ as a vertical line, $q$ may be used to model the distance over which an external force must be applied to move from $x$ to $y$. If $y < x$, then no external force must be applied since gravity will move the object downward; only upward distances require external force. Now $B_q(x, \varepsilon) = (-\infty, x + \varepsilon)$, so the topology on $\mathbb{R}$ generated by this quasi-metric is the left ray topology $\mathcal{T}_q = \{(-\infty, a) : a \in \mathbb{R}\} \cup \{\emptyset, \mathbb{R}\}$.

If $q$ is a quasi-metric on $X$, then the *opposite* of $q$ is the function $q^{\mathrm{op}}$ defined by $q^{\mathrm{op}}(x, y) = q(y, x)$. It is easy to see that $q^{\mathrm{op}}$ is a quasi-metric. In the example above, $q(x, y)$ tells how far $y$ is to the right of $x$ and $q^{\mathrm{op}}(x, y)$ tells how far $y$ is to the left of $x$. Since at least one of these is zero, their sum will give the Euclidean distance between $x$ and $y$. The next theorem shows that this is typical.

**Theorem 11.2.3.** *If $q$ is a quasi-metric on $X$, then $d : X \times X \to [0, \infty)$ defined by $d(x, y) = q(x, y) + q^{\mathrm{op}}(x, y)$ is a metric on $X$ called the* symmetrization metric *from $q$. The symmetrization metric topology $\mathcal{T}_d$ on $X$ is finer than the quasi-metric topology $\mathcal{T}_q$ on $X$.*

*Proof.* Clearly $d(x, y) \geq 0$, $d(x, y) = d(y, x)$, and $d(x, x) = 0$. If $d(x, y) = 0$, then $q(x, y) = q(y, x) = 0$, so $x = y$. If $x, y, z \in X$, then $d(x, y) = q(x, y) + q^{\mathrm{op}}(x, y) \leq q(x, z) + q(z, y) + q^{\mathrm{op}}(x, z) + q^{\mathrm{op}}(z, y) = d(x, z) + d(z, y)$. From $q(x, y) \leq d(x, y)$, it follows that, for any $x \in X$ and any $\varepsilon > 0$, $B_d(x, \varepsilon) \subseteq B_q(x, \varepsilon)$, and thus $\mathcal{T}_d$ is finer than $\mathcal{T}_q$.  □

In the quasi-metric $q$ of Example 11.2.2, we had $q^{\mathrm{op}}(x, y) = 0$ if and only if $x \leq y$ in the usual order on $\mathbb{R}$. Again, this is no coincidence, as seen in the next theorem. This theorem illustrates another striking connection between topology and order.

**Theorem 11.2.4.**
(a) *Every quasi-metric $q$ on $X$ defines a partial order $\leq_q$ on $X$ by taking $x \leq_q y$ if and only if $q(x,y) = 0$.*
(b) *Every partial order $\leq$ on $X$ defines a quasi-metric $q_\leq$ on $X$ by $q_\leq(x,y) = 0$ if $x \leq y$ and $q_\leq(x,y) = 1$ if $x \not\leq y$.*

The straightforward proof is left as an exercise. Starting with a quasi-metric $q$, iteratively applying parts (a) and (b) of the theorem may not return the original quasi-metric. Indeed, applying (b) only returns a quasi-metric with values 0 or 1, while the original quasi-metric was arbitrary. However, starting with a partial order $\leq$, iteratively applying (b) and (a) will return the original partial order.

If $\leq$ is a partial order and $q_\leq$ is the associated quasi-metric given in (b) of the theorem above, then $B(x,\varepsilon) = X$ if $\varepsilon > 1$ and $B(x,\varepsilon) = i(x)$ if $\varepsilon \leq 1$. Thus, the $\mathcal{T}_{q_\leq}$-open sets are the increasing sets in $(X,\leq)$. That is, $\mathcal{T}_{q_\leq}$ is the Alexandroff specialization topology generated by the partial order $\leq$.

Given the partial order of equality on a set $X$, the quasi-metric $q_=$ on $X$ is $q_=(x,y) = 0$ if $x = y$ and $q_=(x,y) = 1$ if $x \neq y$. This is the discrete metric on $X$. For an arbitrary partial order $\leq$, the quasi-metric $q_\leq$ may be viewed as a generalization of the discrete metric.

We now introduce some topological properties involving cardinalities of sets.

**Definition 11.2.5.** Suppose $(X,\mathcal{T})$ is a topological space.

A *neighborhood base* for $x \in X$ is a collection $\mathcal{B}_x$ of neighborhoods of $x$ such that every neighborhood of $x$ contains a neighborhood from $\mathcal{B}_x$.

$(X,\mathcal{T})$ is *first countable* if every point has a countable neighborhood base.

$(X,\mathcal{T})$ is *second countable* if there is a countable basis $\mathcal{B}$ for the topology $\mathcal{T}$.

$(X,\mathcal{T})$ is *separable* if there exists a countable dense subset $D \subseteq X$.

In any metric, pseudometric, or quasi-metric space $X$, if $U$ is a neighborhood of $x \in X$, then $U$ must contain $B(x,\frac{1}{n})$ for some $n \in \mathbb{N}$. That is, $\mathcal{B}_x = \{B(x,\frac{1}{n}) : n \in \mathbb{N}\}$ is a countable neighborhood base at any point $x \in X$. Thus, every metric, pseudometric, or quasi-metric space is first countable. The crucial properties guaranteeing that such spaces are first countable are that the topologies are generated by the $\varepsilon$-balls $B(x,\varepsilon)$ where the $\varepsilon$'s are from $(0,\infty)$, and $(\frac{1}{n})_{n \in \mathbb{N}}$ is a countable sequence in $(0,\infty)$ converging to 0.

In general, metric spaces need not be second countable nor separable. For example, the discrete metric on an uncountable set $X$ is neither second countable nor separable.

**Theorem 11.2.6.** *If $(X,\mathcal{T})$ is metrizable and separable, then it is second countable.*

*Proof.* Suppose $(X,\mathcal{T})$ is metrizable and separable. Let $D$ be a countable dense subset of $X$. For every $x \in X$, $\mathcal{B}_x = \{B(x,\frac{1}{n}) : n \in \mathbb{N}\}$ is a countable neighborhood base for $x$. Let $\mathcal{B} = \bigcup\{\mathcal{B}_x : x \in D\}$. Now as a countable union of countable collections, $\mathcal{B}$ is countable. To show $\mathcal{B}$ is a basis for $\mathcal{T}$, suppose $U \in \mathcal{T}$. For any $x \in U$, there exists $B(x,\frac{1}{m}) \subseteq U$.

Since $D$ is dense in $X$, there exists $a \in B(x, \frac{1}{2m}) \cap D$. Now $x \in B(a, \frac{1}{2m})$, and we will show $B(a, \frac{1}{2m}) \subseteq B(x, \frac{1}{m})$. Suppose $y \in B(a, \frac{1}{2m})$. Then $d(x,y) \le d(x,a) + d(a,y) \le \frac{1}{2m} + \frac{1}{2m} = \frac{1}{m}$, so $y \in B(x, \frac{1}{m})$. Thus, for each $x \in U$, we can find $B(a, \frac{1}{2m}) \in \mathcal{B}$ with $x \in B(a, \frac{1}{2m}) \subseteq U$, so $\mathcal{B}$ is a countable basis for $\mathcal{T}$. $\qquad\square$

We will now present a familiar Hausdorff space which is quasi-metrizable but not metrizable. Recall that the lower limit topology on $\mathbb{R}$ is the topology generated by the basis $\{[x, x + \varepsilon) : x \in \mathbb{R}, \varepsilon > 0\}$.

**Example 11.2.7.** Let $X = \mathbb{R}$ and define

$$q(x,y) = \begin{cases} y - x & \text{if } x \le y, \\ 5 & \text{if } y < x. \end{cases}$$

It is routine to confirm that $q$ is a quasi-metric. If $y$ is to the right of $x$, then $q(x,y)$ gives the usual Euclidean distance between them. Nothing to the left of $x$ is close to $x$, if we take "close" to mean less than 5 units away. For $\varepsilon \le 5$, $B_q(x,\varepsilon) = [x, x + \varepsilon)$ and if $\varepsilon > 5$, $B_q(x,\varepsilon) = (-\infty, x + \varepsilon)$. Since the rays $(-\infty, x + \varepsilon)$ and intervals $[x, x + \varepsilon)$ with $\varepsilon > 5$ are unions of sets $[y, y + \varepsilon')$ with radii less than 5, the collection $\{[x, x + \varepsilon) : x \in X, \varepsilon > 0\}$ is a basis for the topology $\mathcal{T}_q$ on $\mathbb{R}$. Thus, $\mathcal{T}_q$ is the lower limit topology on $\mathbb{R}$.

Since $q(3, \pi) \approx .14159$ and $q(\pi, 3) = 5$, we have $\pi \in B(3, 1)$ but $3 \notin B(\pi, 1)$.

Every nonempty open set in $(\mathbb{R}, \mathcal{T}_q)$ contains a rational number, so $\mathbb{Q}$ is a countable dense subset of $(\mathbb{R}, \mathcal{T}_q)$, and thus the space is separable. Suppose $\mathcal{B}$ is a basis for $\mathcal{T}_q$. For each $x \in \mathbb{R}$, pick a neighborhood $B_x \in \mathcal{B}$ with $x \in B_x \subseteq [x, x + 1)$. Now consider the collection $\mathcal{B}' = \{B_x : x \in \mathbb{R}\} \subseteq \mathcal{B}$. If $a < b$ then $a \notin B_b \subseteq [b, b + 1)$ so $B_a \ne B_b$, and thus $x \ne y$ implies $B_x \ne B_y$. Now $\mathcal{B}'$ is indexed by an uncountable set $\mathbb{R}$ and contains no duplicates, so $\mathcal{B}'$, and therefore the arbitrarily chosen basis $\mathcal{B}$, are uncountable. This shows $\mathcal{T}_q$ has no countable basis, and thus is not second countable. Now if the separable space $(\mathbb{R}, \mathcal{T}_q)$ were metrizable, by the previous theorem, it would be second countable. Since it is not second countable, the lower limit topology on $\mathbb{R}$ is not metrizable.

**Hemimetrics.** If we drop the condition that $q(x,y) = q(y,x) = 0$ implies $x = y$ from the definition of a quasi-metric, we get a hemimetric.

**Definition 11.2.8.** A *hemimetric* on a set $X$ is a function $h : X \times X \to [0, \infty)$ such that
(a) $h(x,y) \ge 0$ for all $x, y \in X$ (nonnegativity);
(b) $h(x,x) = 0$ for all $x \in X$;
(c) $h(x,y) \le h(x,z) + h(z,y)$ for all $x, y, z \in X$ (triangle inequality).

The pair $(X, h)$ is a hemimetric space. As in a metric space, $B(x, \varepsilon) = \{y \in X : h(x,y) < \varepsilon\}$ is the *ball* of radius $\varepsilon$ centered at $x$, and the collection $\mathcal{B} = \{B(x, \varepsilon) : x \in X, \varepsilon > 0\}$ of all balls is a basis for the *hemimetric topology* on $X$.

Recall that a pseudometric dropped the metric condition that $d(x,y) = d(y,x) = 0$ implies $x = y$, and a quasi-metric dropped the symmetry condition. Since a hemimetric drops both of these conditions, in the literature, hemimetrics are also called *quasi-pseudometrics* or *pseudo-quasi-metrics*.

The theorem below was given by W. A. Wilson in 1931.

**Theorem 11.2.9.** *Every second countable topological space is hemimetrizable.*

*Proof.* Suppose $(X, \mathcal{T})$ is second countable. Let $\mathcal{B}$ be a countable basis for $\mathcal{T}$. To set up the idea of the proof, we will first consider the case that $\mathcal{B}$ is finite. For $x, y \in X$, define $h(x,y)$ to be the number of elements of $\mathcal{B}$ which contain $x$ but not $y$. (Loosely speaking, $h(x,y)$ is the vote count when the elements $B \in \mathcal{B}$ are asked if they think $y$ is not close to $x$.) Clearly $h(x,y) \geq 0$ for all $x, y \in X$ and $h(x,x) = 0$. Suppose $x, y, z \in X$ are given. To see $h(x,y) \leq h(x,z) + h(z,y)$, suppose $B$ is one of the elements of $\mathcal{B}$ counted by $h(x,y)$; that is, suppose $x \in B$ and $y \notin B$. Now either $z \in B$ or $z \notin B$. If $z \in B$ then $B$ will be counted by $h(z,y)$, and if $z \notin B$ then $B$ will be counted by $h(x,z)$. Thus, every $B \in \mathcal{B}$ counted by $h(x,y)$ is counted either by $h(x,z)$ or $h(z,y)$, so $h(x,y) \leq h(x,z) + h(z,y)$.

Now suppose the countable basis $\mathcal{B} = \{B_n : n \in \mathbb{N}\}$ is countably infinite. Defining $h(x,y)$ to be the number of elements of $\mathcal{B}$ which contain $x$ but not $y$ may not give a function $h : X \times X \to [0, \infty)$, since $h(x,y)$ might be infinite. We will perform a weighted count to ensure that the result is finite. Given $x, y \in X$, let $M_{x,y} = \{n \in \mathbb{N} : x \in B_n, y \notin B_n\}$, and define

$$h(x,y) = \begin{cases} \sum_{n \in M_{x,y}} \frac{1}{2^n} & \text{if } x \neq y, \\ 0 & \text{if } x = y. \end{cases}$$

It is easy to see that, for any $x, y \in X$, $0 \leq h(x,y) \leq \sum_{n \in \mathbb{N}} 2^{-n} = 1$. For any given $x, y, z \in X$, if $n \in M_{x,y}$, then $x \in B_n$ and $y \notin B_n$. If $z \in B_n$, then $n \in M_{z,y}$ and if $z \notin B_n$ then $n \in M_{x,z}$. Thus, $M_{x,y} \subseteq M_{x,z} \cup M_{z,y}$, so the sums associated with $h(x,z) + h(z,y)$ contain all the terms included in the sum for $h(x,y)$, so $h(x,y) \leq h(x,z) + h(z,y)$. □

The proof of the following theorem is left to the exercises.

**Theorem 11.2.10.** *A hemimetric $h$ on $X$ is a quasi-metric if and only if the hemimetric topology is $T_0$.*

The following corollary gives a sufficient condition and a necessary condition for a topological space to be quasi-metrizable.

**Corollary 11.2.11.** *Every second countable $T_0$ space is quasi-metrizable.*
*Every quasi-metrizable space is $T_0$ and first countable.*

*Proof.* By Theorem 11.2.9, second countable implies hemimetrizable, and by Theorem 11.2.10, the additional $T_0$ hypothesis implies the hemimetric is in fact a quasi-metric. Any quasi-metrizable space $X$ is clearly first countable, and Theorem 11.2.10 shows that it is also $T_0$. □

Theorem 11.2.10 shows that the quasi-metrics are the $T_0$ hemimetrics. Alexandroff topologies arise from quasiorders $\leq$, and the $T_0$ Alexandroff topologies arise from partial orders. Loosely speaking, a partial order may be thought of as a $T_0$ quasiorder. The connections between quasi-metrics, partial orders, and their Alexandroff specialization topologies given in Theorem 11.2.4 and the discussion following it are thus connections between the $T_0$ hemimetrics, $T_0$ quasiorders, and $T_0$ Alexandroff topologies. Dropping the $T_0$ hypothesis gives the analogous connection between hemimetrics, quasiorders, and Alexandroff topologies, stated below.

**Theorem 11.2.12.**

(a) *Every hemimetric h on X defines a quasiorder $\leq_h$ on X by taking $x \leq_h y$ if and only if $h(x,y) = 0$.*

(b) *Every quasiorder $\leq$ on X defines a hemimetric $h_\leq$ on X by $h_\leq(x,y) = 0$ if $x \leq y$ and $h_\leq(x,y) = 1$ if $x \not\leq y$. Furthermore, the topology generated by the hemimetric $h_\leq$ is the Alexandroff specialization topology $\mathcal{T}_\leq$.*

Further results on quasi-metrics and hemimetrics can be found in [19, 20, 21, 29].

## Exercises

1. Suppose $k$ is a nonnegative real number. Define $q : \mathbb{R} \times \mathbb{R} \rightarrow [0, \infty)$ by

$$q(x,y) = \begin{cases} y - x & \text{if } x \leq y, \\ k(x - y) & \text{if } y \leq x. \end{cases}$$

The function $q$ models the time required to travel between two points on the real line if there is a constant headwind which slows travel in one direction. Show that $q$ is a quasi-metric on $\mathbb{R}$. (Note that the quasi-metric of Example 11.2.2 is of this form, for $k = 0$.)

2. Suppose $q$ and $r$ are quasi-metrics on $X$ and $a, b$ are nonnegative real numbers, not both zero. Show that $m(x,y) = aq(x,y) + br(x,y)$ defines a quasi-metric on $X$.

3. Suppose $q$ is a quasi-metric on $X$ and $a, b$ are positive real numbers. Find necessary and sufficient conditions for the quasi-metric $m = aq(x,y) + bq^{\text{op}}(x,y)$ to be a metric.

4. Suppose the points of $\mathbb{Z}^2$ are connected by the following grid of one-way streets: for $n \in \mathbb{Z}$, streets at $y = 2n$ run east, streets at $y = 2n + 1$ run west, streets at $x = 2n$ run north, and streets at $x = 2n + 1$ run south. Let $d((m, n), (a, b))$ be the length of the shortest route along these streets from $(m, n)$ to $(a, b)$. Find $d((0, 0), (a, b))$ for all $(a, b) \in \mathbb{Z}^2$.

5. Prove Theorems 11.2.4 and 11.2.12.

6. Is the continuous image of a first countable topological space first countable? Is the continuous image of a second countable topological space second countable? Give a proof or counterexample.

7. Let $q$ be the quasi-metric generating the lower limit topology on $\mathbb{R}$, as given in Example 11.2.7.
   (a) Confirm that $q$ satisfies the triangle inequality.
   (b) Identify the metric topology from the symmetrization metric $d(x,y) = q(x,y) + q^{op}(x,y)$.

8. The topology $\mathcal{T}$ on $\mathbb{R}$ generated by the basis $\mathcal{B} = \{[x, x+\varepsilon) : x \in \mathbb{Q}, \varepsilon \in (0,1)\} \cup \{(x - \varepsilon, x] : x \in \mathbb{R} - \mathbb{Q}, \varepsilon \in (0,1)\}$ suggests the distance function $q : \mathbb{R} \times \mathbb{R} \to \mathbb{R}$ given by

$$q(x,y) = \begin{cases} y - x & \text{if } y \geq x, \ x \in \mathbb{Q}, \\ x - y & \text{if } x \geq y, \ x \notin \mathbb{Q}, \\ 1 & \text{otherwise.} \end{cases}$$

   (a) With $\varepsilon \in (0,1)$ and $B(x,\varepsilon) = \{y \in \mathbb{R} : q(x,y) < \varepsilon\}$, find $B(x,\varepsilon)$ if $x \in \mathbb{Q}$ and if $x \notin \mathbb{Q}$.
   (b) Is $q$ a quasi-metric that generates $\mathcal{T}$? Prove your answer.

9. If $q$ is a quasi-metric on $X$ such that $\leq_q$ is a total order on $X$ and $d(x,y) = q(x,y) + q^{op}(x,y)$ is the symmetrization metric, show that

$$q(x,y) = \begin{cases} 0 & \text{if } x \leq_q y, \\ d(x,y) & \text{if } x \not\leq_q y, \end{cases}$$

   and furthermore, $x \leq_q y \leq_q z$ implies $d(x,y) \vee d(y,z) \leq d(x,z)$.

10. If $q$ is a quasi-metric on $X$, show that the closed ball $\bar{B}_q(x,\varepsilon) = \{y \in X : q(x,y) \leq \varepsilon\}$ is closed in the topology generated by $q^{op}$. Give an example to show that $\bar{B}_q(x,\varepsilon)$ need not be closed in the topology generated by $q$.

11. Suppose $q$ is a quasi-metric on $X$. Show that $m(x,y) = q(x,y) \vee q^{op}(x,y)$ is a metric on $X$ which generates the same topology on $X$ as $d(x,y) = q(x,y) + q^{op}(x,y)$.

12. Suppose $q$ is a quasi-metric on $X$ and $m(x,y) = q(x,y) \vee q^{op}(x,y)$ is the metric on $X$ defined in Exercise 11. Show that every ball $B_m(x,\varepsilon)$ is order convex with respect to the order $\leq_q$.

13. Prove Theorem 11.2.10: a hemimetric $h$ on $X$ is a quasi-metric if and only if the hemimetric topology is $T_0$.

14. Suppose $h$ is a hemimetric on $X$. Show that the hemimetric topology $\mathcal{T}_h$ is $T_1$ if and only if for $x, y \in X$, $h(x,y) = 0$ implies $x = y$.

15. Let $\mathcal{F}$ be the collection of finite subsets of a set $X$. Determine whether or not the functions below are metrics, pseudometrics, quasi-metrics, or hemimetrics.
    (a) For $A, B \in \mathcal{F}$, define $r(A, B) = |A - B|$.
    (b) For $A, B \in \mathcal{F}$, define $s(A, B) = |A - B| + |B - A|$. (Thus, $s(A, B)$ is the cardinality of the symmetric difference $(A - B) \cup (B - A)$ of $A$ and $B$.)

16. Determine whether the topological spaces below are quasi-metrizable or not. For those that are quasi-metrizable, exhibit a quasi-metric generating the topology.
    (a) The digital line topology $\mathcal{T}_{dl}$, generated by the basis $\{\{2n + 1\} : n \in \mathbb{Z}\} \cup \{\{2n - 1, 2n, 2n + 1\} : n \in \mathbb{Z}\}$, on $\mathbb{Z}$.
    (b) The cofinite topology on $\mathbb{R}$.

(c) The single transmitter topology $\mathcal{T}$ on $X = [-1, 1]$, where $\mathcal{T}$ has basis $\mathcal{B} = \{\{x\} : x \in X - \{0\}\} \cup \{(-1, 1)\}$.

17. Prove that every compact metric space is separable (and thus, second countable, by Theorem 11.2.6). Explain whether or not the proof you give would hold if "metric" is replaced by "pseudometric", "quasi-metric", or "hemimetric".

18. For $i = 1, 2$, suppose $\leqslant_i$ is a quasiorder on $X$, $\mathcal{T}_{\leqslant_i}$ is the associated specialization topology, and $h_i$ is the associated hemimetric defined in Theorem 11.2.12(b).
    (a) Show that $\leqslant_1 \cap \leqslant_2$ is a quasiorder on $X$.
    (b) Show that $\mathcal{T}_{\leqslant_1 \cap \leqslant_2} = \mathcal{T}_{\leqslant_1} \vee \mathcal{T}_{\leqslant_2}$.
    (c) Show that $h = h_1 \vee h_2$ is a hemimetric on $X$, and the associated quasiorder, as in Theorem 11.2.12(a), is $\leqslant_1 \cap \leqslant_2$.

19. It is widely known that if $(X, \|\cdot\|)$ is a normed vector space, $d(x, y) = \|x - y\|$ defines a metric on $X$. An *asymmetric norm* on a vector space $X$ is a function $\|\cdot\| : X \to [0, \infty)$ with
    (a) $\|x\| \geq 0$ for all $x \in X$
    (b) $\|x\| = \|-x\| = 0$ implies $x = 0$
    (c) $\|ax\| = a\|x\|$ for all $x \in X$ and all positive $a \in [0, \infty)$
    (d) $\|x + y\| \leq \|x\| + \|y\|$ for all $x, y \in X$.
    (See the book *Functional Analysis in Asymmetric Normed Spaces* [11] for more on asymmetric norms.) If $\|\cdot\|$ is an asymmetric norm on the vector space $X$, show that $q(x, y) = \|x - y\|$ is a quasi-metric on $X$.

20. One the collection $F = \{f \in \mathbb{R}^{[0,1]} : f$ is continuous and $\int_0^1 f(x)\, dx = 0\}$, define $\|f\| = \max\{f(x) : x \in [0, 1]\}$.
    (a) Show that $\|\cdot\|$ is a asymmetric norm, as defined in Exercise 19.
    (b) Show that it is not true that $\|af\| = |a|\,\|f\|$ for all $a \in \mathbb{R}, f \in F$, so $\|\cdot\|$ is not a norm on $F$.

## 11.3 Partial metrics

In Euclidean geometry and the theory of metric spaces, a point $x$ has no length or width, and the distance from $x$ to $x$ is zero. In practice, particularly in computer applications, we must use representations of points which are not exact, and one approximation (whether a single pixel on a screen, or a truncated decimal such as 3.14) may represent many different exact values. The distances between these exact values represented by a single approximation $a$ suggest the consideration of metrics allowing nonzero distances from $a$ to $a$. The self-distance $d(a, a)$ essentially gives a measure of the ambiguity of the point $a$. In 1994, Steven Matthews [36] quantified these notions by introducing partial metrics which relax the metric restriction that $d(x, x) = 0$ and adjust the triangle inequality accordingly.[1]

---

[1] Reprinted with permission from [22].

**Definition 11.3.1.** A *partial metric* on a set $X$ is a function $\rho : X \times X \to [0, \infty)$ such that
(a) $\rho(x, y) \geq \rho(x, x) \geq 0$ for all $x, y \in X$ (small self-distances);
(b) $x = y$ if and only if $\rho(x, x) = \rho(x, y) = \rho(y, y)$;
(c) $\rho(x, y) = \rho(y, x)$ for all $x, y \in X$ (symmetry);
(d) $\rho(x, y) \leq \rho(x, z) + \rho(z, y) - \rho(z, z)$ for all $x, y, z \in X$ (triangle inequality).

The pair $(X, \rho)$ is a partial metric space. As in a metric space, $B(x, \varepsilon) = \{y \in X : \rho(x, y) < \varepsilon\}$ is the *ball* of radius $\varepsilon$ centered at $x$, and the collection $\mathcal{B} = \{B(x, \varepsilon) : x \in X, \varepsilon > 0\}$ of all balls is a basis for the *partial metric topology* on $X$.

Note that $B(x, \rho(x, x)) = \{y : \rho(x, y) < \rho(x, x)\} = \emptyset$, and in particular, $x \in B(x, \varepsilon)$ if and only if $\varepsilon > \rho(x, x)$. Thus, $\{B(x, \varepsilon) : \varepsilon > 0\}$ may not be a collection of neighborhoods of $x$, and thus may not be a neighborhood base at $x$. To ensure that $x \in B(x, \varepsilon)$ for all $\varepsilon > 0$, we could redefine the $\varepsilon$-ball centered at $x$ to be $B^+(x, \varepsilon) = \{y \in X : \rho(x, y) < \rho(x, x) + \varepsilon\}$.

Another important point that should immediately be addressed is that the collection $\mathcal{B}$ of all balls really is a basis for a topology. In the topology generated by a metric (or quasi-metric, or pseudometric) $d$, the proof that the balls form a basis depends on the standard triangle inequality to show that if $z \in B(x, \varepsilon)$, then $B(z, \varepsilon - d(x, z)) \subseteq B(x, \varepsilon)$; then, for $z \in B(x, \varepsilon) \cap B(y, \varepsilon')$ and $\delta = \min\{\varepsilon - d(x, z), \varepsilon' - d(y, z)\}$, $z \in B(z, \delta) \subseteq B(x, \varepsilon) \cap B(y, \varepsilon')$. In a partial metric space, the triangle inequality has been altered and these arguments will also need to be altered. To this end, we will show that if $\rho$ is a partial metric on $X$,

$$\text{for } z \in B(x, \varepsilon) \quad \text{and} \quad \delta = \varepsilon - \rho(x, z) + \rho(z, z), \quad z \in B(z, \delta) \subseteq B(x, \varepsilon).$$

If $z \in B(x, \varepsilon)$, then $\rho(x, z) < \varepsilon$, so $\delta = \varepsilon - \rho(x, z) + \rho(z, z) > \rho(z, z)$ and thus $z \in B(z, \delta)$. If $w \in B(z, \delta)$, then $\rho(w, z) < \delta = \varepsilon - \rho(z, x) + \rho(z, z)$. Rearranging terms gives $\rho(w, z) + \rho(z, x) - \rho(z, z) < \varepsilon$, and by the triangle inequality for partial metrics, this gives $\rho(w, x) < \varepsilon$. Thus, $B(z, \delta) \subseteq B(x, \varepsilon)$. Now as in the metric case, if $z \in B(x, \varepsilon) \cap B(y, \varepsilon')$, then $B(z, \mu) \subseteq B(x, \varepsilon) \cap B(y, \varepsilon')$ for $\mu = \min\{\varepsilon - \rho(x, z) + \rho(z, z), \varepsilon' - \rho(y, z) + \rho(z, z)\}$.

Another consequence of the fact that $B(x, \varepsilon)$ contains a $\delta$-ball around each of its points is that $\{B(x, \varepsilon) : \varepsilon > \rho(x, x)\}$ is a neighborhood base at $x$.

A partial metric in which all self-distances $\rho(x, x)$ are zero is easily seen to be a metric.

We present some examples of partial metrics.

**Example 11.3.2.** On $[0, \infty)$, define $\rho(x, y) = x \vee y$. To confirm the triangle inequality, note that for any $x, y, z \in [0, \infty)$ we have $x \leq (x \vee z)$ and $(z \vee y) - (z \vee z) \geq 0$, so $x \leq (x \vee z) + (z \vee y) - (z \vee z)$. Similarly, $y \leq (x \vee z) + (z \vee y) - (z \vee z)$, so $(x \vee y) \leq (x \vee z) + (z \vee y) - (z \vee z)$, as needed. It is easily seen that $\rho$ satisfies the other properties of a partial metric. In this example, the self-distances are $\rho(x, x) = x$. The balls are

$$B(x, \varepsilon) = \{y : x \vee y < \varepsilon\} = \begin{cases} \emptyset & \text{if } x \geq \varepsilon, \\ [0, \varepsilon) & \text{if } x < \varepsilon. \end{cases}$$

To a mathematician, $\pi$ and 4/7 are exact numbers. Since the decimal expansions of $\pi$ and 4/7 do not terminate, computers can only approximate these numbers. If $\pi$ and 4/7 are approximated by 3.14 and 0.571, respectively, then allowing for rounding, this really tells us that $\pi$ is a number in the interval [3.135, 3.145] and 4/7 is a number in the interval [0.5705, 0.5715]. With only these approximations, we could conclude that the distance between $\pi$ and 4/7 is at most $3.145 - 0.5705$, which is the greatest distance between two points of $[0.5705, 0.5715] \cup [3.135, 3.145]$. Thus, machine numbers actually represent intervals, so it is important to be able to measure the distance between two intervals. This situation, described more carefully in the next example, was one of the initial examples motivating partial metrics.

**Example 11.3.3.** Let $\mathcal{I}$ be the collection of nonempty compact intervals $[a, b]$ in $\mathbb{R}$. Define a partial metric on $\mathcal{I}$ by $\rho([a, b], [c, d]) = \max\{b, d\} - \min\{a, c\}$. Thus, $\rho([a, b], [c, d])$ is the length of the smallest interval containing $[a, b] \cup [c, d]$ (that is, the length of the *convex hull* of $[a, b] \cup [c, d]$). The verification that $\rho$ is a partial metric is left as an exercise. Note that the self-distance $\rho([a, b], [a, b])$ is the length of the interval $[a, b]$. For example, when rounding all decimals to the nearest hundredth, $\pi \approx 3.14$, so $\pi$ is identified as a number in an interval $[3.135, 3.145]$ with self-distance 1/100.

The ball $B([a, b], \varepsilon)$ consists of all the intervals $[x, y]$ such that the distance $x$ extends to the left of $a$ plus the distance $y$ extends to the right of $b$ is less than $\varepsilon$. Note that $\rho([a, b], [x, y]) = \rho([a, b], [a, b])$ if and only if the length of the convex hull of $[a, b] \cup [x, y]$ equals the length of $[a, b]$. This occurs if and only if $[x, y] \subseteq [a, b]$. In particular, the partial order $\subseteq$ on the collection of nonempty compact intervals is characterized by $[x, y] \subseteq [a, b]$ if and only if $\rho([a, b], [x, y]) = \rho([a, b], [a, b])$.

Partial metrics are richly connected to many of the other structures we have studied. We have noted that in a partial metric space $(X, \rho)$, the basis $\{B(x, \varepsilon) : x \in X, \varepsilon > 0\}$ generates the same topology as $\{B(x, \rho(x, x) + \varepsilon) : x \in X, \varepsilon > 0\}$, since the latter collection consists of the nonempty balls from the first collection. This suggests considering $q(x, y) = \rho(x, y) - \rho(x, x)$; then $q(x, x) = 0$ but $q(x, y)$ may not equal $q(y, x)$, and $q$ is in fact a quasi-metric.

**Theorem 11.3.4.** *Suppose $\rho$ is a partial metric on $X$.*
(a) *$q(x, y) = \rho(x, y) - \rho(x, x)$ is a quasi-metric on $X$ which generates the same topology as $\rho$.*
(b) *$d(x, y) = 2\rho(x, y) - \rho(x, x) - \rho(y, y)$ is a metric on $X$.*
(c) *Taking $x \leq_\rho y$ if and only if $\rho(x, y) = \rho(x, x)$ defines a partial order $\leq_\rho$ on $X$.*

*Proof.* The routine verification of (a) is left to the exercises. Since $q$ is a quasi-metric, $q(x, y) + q^{op}(x, y) = d(x, y) = 2\rho(x, y) - \rho(x, x) - \rho(y, y)$ is the symmetrization metric for $q$. The relation $\leq_\rho$ is the partial order $\leq_q$ defined from the quasi-metric $q$ by $x \leq_q y$ if and only if $0 = q(x, y) = \rho(x, y) - \rho(x, x)$. ☐

Notice that in Example 11.3.3, we had $\rho([a, b], [x, y]) = \rho([a, b], [a, b])$ if and only if $[x, y] \subseteq [a, b]$, so $\leq_\rho$ is reverse inclusion $\supseteq$. If these intervals represent approximations of a fixed real number, a smaller interval ("$\subseteq$") is a better approximation ("$\geq_\rho$").

Partial metrics are also linked with *weighted metrics*.

**Definition 11.3.5.** A weighted metric space is a metric space $(X, d)$ together with a weight function $|\cdot|$ satisfying $|x| \geq 0$ for all $x \in X$, and $d(x, y) \geq |x| - |y|$ for all $x, y \in X$. If $(X, d, |\cdot|)$ is a weighted metric space, we may simply say $d$ is a *weighted metric*.

**Theorem 11.3.6.** *Every partial metric $\rho$ on $X$ gives a weighted metric space $(X, d, |\cdot|)$ where $d(x, y) = 2\rho(x, y) - \rho(x, x) - \rho(y, y)$ and $|x| = \rho(x, x)$. Every weighted metric space $(X, d, |\cdot|)$ gives a partial metric $\rho(x, y) = (|x| + |y| + d(x, y))/2$ with $\rho(x, x) = |x|$.*

*Proof.* Suppose $\rho$ is a partial metric on $X$. By the previous theorem, $d(x, y) = 2\rho(x, y) - \rho(x, x) - \rho(y, y)$ is a metric. With $|x| = \rho(x, x)$, clearly $|x| \geq 0$ for all $x \in X$. Since $\rho(x, y) \geq \rho(x, x)$, we have $d(x, y) \geq 2\rho(x, x) - \rho(x, x) - \rho(y, y) = |x| - |y|$. Thus, $(X, d, |\cdot|)$ is a weighted metric space.

Conversely, if $(X, d, |\cdot|)$ is a weighted metric space and $\rho(x, y) = (|x| + |y| + d(x, y))/2$, then clearly $\rho(x, x) = |x|$ and $\rho(y, x) = \rho(x, y) \geq \rho(x, x) = |x| \geq 0$ for any $x, y \in X$. Suppose $\rho(x, y) = \rho(x, x) = \rho(y, y)$. The first of these equations implies $|x| + |y| + d(x, y) = |x| + |x|$, so $d(x, y) = |x| - |y|$. Similarly, interchanging $x$ and $y$ shows $d(x, y) = -(|x| - |y|)$, and it follows that $d(x, y) = 0$, so $x = y$. Finally, for any $x, y, z \in X$, we have $2(\rho(x, z) + \rho(z, y) - \rho(z, z)) = d(x, z) + |x| + |z| + d(z, y) + |z| + |y| - 2|z| = d(x, z) + d(z, y) + |x| + |y| \geq d(x, y) + |x| + |y| = 2\rho(x, y)$, so the partial metric triangle inequality holds. $\square$

We remark that the transitions from partial metric to weighted metric and from weighted metric to partial metric given in the theorem above are inverses; iterating them will return the original structure. This follows since both transitions have $|x| = \rho(x, x)$, and then the defining equations $d(x, y) = 2\rho(x, y) - \rho(x, x) - \rho(y, y)$ and $\rho(x, y) = (|x| + |y| + d(x, y))/2$ are equivalent. Thus, the weighted metric spaces are precisely the partial metric spaces.

## Exercises

1. If $d$ is a metric on $X$ and $k \in [0, \infty)$, show that $\rho = k + d$ is a partial metric and describe the associated partial order $\leq_\rho$.
2. Let $\rho$ be the partial metric on $[0, \infty)$ defined by $\rho(x, y) = x \vee y$.
   (a) Describe all sequences which converge to 2.
   (b) Find all limits of the constant sequence $(a)_{n \in \mathbb{N}}$.
   (c) Find all limits of the sequence $(2 + (-1)^n)_{n \in \mathbb{N}}$.
   (d) Describe the non-convergent sequences in $([0, \infty), \rho)$.
   (e) If $(x_n, y_n)_{n \in \mathbb{N}}$ converges to $(x, y)$ in $([0, \infty), \rho) \times ([0, \infty), \rho)$, does $(\rho(x_n, y_n))_{n \in \mathbb{N}}$ converge to $\rho(x, y)$ in $\mathbb{R}$ with the Euclidean metric?

3. Verify that $\rho$ defined on the set $\mathcal{I}$ of nonempty compact intervals in $\mathbb{R}$ by $\rho([a, b], [c, d]) = \max\{b, d\} - \min\{a, c\}$ (as in Example 11.3.3) satisfies the partial metric triangle equality.

4. Let $\rho$ be the partial metric on the set $\mathcal{I}$ of nonempty compact intervals in $\mathbb{R}$ defined by $\rho([a, b], [c, d]) = \max\{b, d\} - \min\{a, c\}$. Describe the subspace $\{[a, a] : a \in \mathbb{R}\}$.

5. If $C[0, 1]$ is the collection of continuous real-valued functions on $[0, 1]$, define $\bar{\rho}$ : $C[0, 1] \times C[0, 1] \to \mathbb{R}$ by taking $\bar{\rho}(f, g)$ to be the area of the smallest rectangle $[0, 1] \times [a, b]$ which contains the graphs of $f$ and $g$. Which properties of a partial metric does $\bar{\rho}$ satisfy?

6. Prove Theorem 11.3.4(a): If $\rho$ is a partial metric on $X$, then $q(x, y) = \rho(x, y) - \rho(x, x)$ is a quasi-metric on $X$ which generates the same topology as $\rho$.

7. Let $\rho$ be a partial metric on $X$ and let $\leq_\rho$ be the associated partial order defined by $x \leq_\rho y$ if and only if $\rho(x, y) = \rho(x, x)$. Let $\mathcal{T}_{\leq_\rho}$ be the specialization topology, consisting of the $\leq_\rho$-increasing subsets of $X$, and let $\mathcal{T}_\rho$ be the partial metric topology generated by $\rho$.
   (a) Show $\mathcal{T}_\rho \subseteq \mathcal{T}_{\leq_\rho}$.
   (b) Show $\mathcal{T}_{\leq_\rho} \subseteq \mathcal{T}_\rho$ if and only if for every $x \in X$, there exists $\varepsilon > \rho(x, x)$ with $i(x) = B(x, \varepsilon)$.

8. If $(X, \rho)$ is a partial metric space and $k \in [0, \infty)$, show that $A = \{x \in X : \rho(x, x) \leq k\}$ and $B = \{x \in X : \rho(x, x) < k\}$ are increasing sets in $(X, \leq_\rho)$.

9. From Exercise 8, if $(X, \rho)$ is a partial metric space then $M = \{x \in X : \rho(x, x) = 0\}$ is an increasing set in $(X, \leq_\rho)$. Show that every point of $M$ is a maximal point in $(X, \leq_\rho)$.

10. Consider the function $p : \mathbb{R}^2 \to [0, \infty)$ defined by $p(x, y) = \max\{|x - y|, |x + y|\}$. (See Exercise 6 of Section 6.1.) Determine whether $p$ is a partial metric.

## 11.4 Other variations of metrics

We end this chapter with a brief listing of some other forms of metrics.

Recall that a metric on $X$ is a function $d : X \times X \to [0, \infty)$ satisfying

(M1) $d(x, y) \geq 0$ for all $x, y \in X$ (nonnegativity);
(M2) $d(x, x) = 0$ for all $x \in X$;
(M3) $d(x, y) = 0$ implies $x = y$;
(M4) $d(x, y) = d(y, x)$ for all $x, y \in X$ (symmetry);
(M5) $d(x, y) \leq d(x, z) + d(z, y)$ for all $x, y, z \in X$ (triangle inequality).

A pseudometric dropped condition (M3). A quasi-metric dropped condition : (M4) and then revised condition (M3) to $d(x, y) = 0 = d(y, x)$ implies $x = y$. Dropping both conditions (M3) and (M4) gives a hemimetric (also known as a quasi-pseudometric or pseudo quasi-metric). There is not universal agreement on these definitions. Some authors require quasimetrics to be $T_1$, as characterized by Exercise 14 of Section 11.2;

what we call a quasi-metric, they would call a $T_0$ quasi-pseudometric. Our terminology follows that most commonly used by computer scientists.

All of the forms of metrics mentioned so far have included the triangle inequality. Dropping the triangle inequality (M5) gives a *semimetric*.

A partial metric dropped condition (M2), which then required rewording of conditions (M1), (M3), and (M5).

The condition (M1) is actually redundant, since a metric is a function $d$ with codomain $[0, \infty)$. In most practices, metrics are used to measure distances in some sense, and distances should not be negative. However, taking distances only from $[0, \infty)$ imposes the notable restriction that any metric topology has a countable neighborhood base $\{B(x, \frac{1}{n}) : n \in \mathbb{N}\}$ at each point. This follows since in the codomain $[0, \infty)$ with the Euclidean topology, 0 has a countable neighborhood base. If we wish to obtaining any topology which is not first countable, then we must allow the distances to come from some set $A$ other than $[0, \infty)$. What properties should $A$ have? The triangle inequality requires that $A$ have an additive structure. Many metric arguments require forcing things to be less than $\varepsilon/2$, or choosing $\varepsilon$ to be the infimum of two values. Thus, $A$ should be an algebraic structure with a set of positive elements $P$ (to serve as radii) which is closed under finite infima and halving, among other technical properties. Replacing the condition (M1) in the definition of a metric by $d(x, y) \in A$ for a suitable structure $A$ (called a value semigroup, or value quantale), we get a *generalized metric*. The significance of generalized metrics was made clear in Ralph Kopperman's 1988 paper "All topologies come from generalized metrics" [27].

Thus, besides the vast amount of mathematics devoted to metric spaces, meaningful research has been done in spaces in which any one of the conditions (M1)–(M5) have been dropped. Furthermore, conditions are sometimes added, as well. We mention only one.

An *ultrametric* on $X$ is a metric $d$ on $X$ which satisfies the following strengthened version of the triangle inequality:

$$d(x, y) \leq \max\{d(x, z), d(z, y)\} \quad \text{for all } x, y, z \in X.$$

Thus, in an ultrametric space, the triangle inequality $d(x, y) \leq d(x, z) + d(z, y)$ does not require the sum of both terms on the right; the larger term on the right already suffices.

The prefix metric on the set of binary sequences (Exercise 16 of Section 6.1) and the discrete metric on any set $X$ are ultrametrics. Some interesting properties of ultrametric spaces are given in the exercises.

## Exercises

1. Prove these properties of ultrametric spaces:
   (a) Every triangle has at least two equal sides.
   (b) If $y \in B(x, \varepsilon)$, then $B(x, \varepsilon) = B(y, \varepsilon)$.

(c) If two balls intersect, then they are nested, with the ball of smaller radius contained in the other.

(d) Every ball $B(x, \varepsilon)$ is closed.

2. Define $d : \mathbb{R} \times \mathbb{R} \to \mathbb{R}$ by $d(x, x) = 0$ for all $x \in \mathbb{R}$, and $d(x, y) = \max\{|x|, |y|\}$ for all $x \neq y$ in $\mathbb{R}$. Show that $d$ is an ultrametric.

3. Suppose $(x_n)$ is a sequence in an ultrametric space $(X, u)$. Show that the following are equivalent.

(a) There exists $N$ such that $m, n > N$ imply $u(x_n, x_m) < \varepsilon$.

(b) There exists $N$ such that $n > N$ implies $u(x_n, x_{n+1}) < \varepsilon$.

4. Define $m : C[0, 1] \times C[0, 1] \to \mathbb{R}$ by $m(f, g) = \min\{|f(x) - g(x)| : x \in [0, 1]\}$. Which of the properties (M1)–(M5) of a metric are satisfied by $m$?

5. Let $X$ be the closed upper half-plane $\mathbb{R} \times [0, \infty)$ and let $p = (-1, 0)$. Let $d$ be the Euclidean metric on $\mathbb{R}^2$. Define $m : X \times X \to \mathbb{R}$ by $m(x, y) = \max\{d(x, y), d(x, p), d(y, p)\}$. Which of the properties (M1)–(M5) of a metric are satisfied by $m$? Is $m$ any of the named versions of metrics we have encountered?

6. Let $\boldsymbol{\theta}$ represent the point $(\cos \theta, \sin \theta)$ on the unit circle, and let $X = \{\boldsymbol{\theta} : \theta \in [0, \pi]\}$ be the upper half of the unit circle in $\mathbb{R}^2$. Define the "surplus length" $s : X \times X \to \mathbb{R}$ by taking $s(\boldsymbol{\theta}, \boldsymbol{\alpha})$ to be the circular arc length in $X$ between $\boldsymbol{\theta}$ and $\boldsymbol{\alpha}$ minus the length of the line segment in $\mathbb{R}^2$ between $\boldsymbol{\theta}$ and $\boldsymbol{\alpha}$. Which of the properties (M1)–(M5) of a metric are satisfied by $s$? Is $s$ any of the named versions of metrics we have encountered?

7. Let $\mathcal{G}$ be the collection of bounded nonempty open subsets of the Euclidean plane. Define $d : \mathcal{G} \times \mathcal{G} \to \mathbb{R}$ by taking $d(A, B)$ to be the percentage of $B$ not covered by $A$. (Technically, $d(A, B) = m(B - A)/m(B)$ where $m$ represents the Lebesgue measure.) Which of the properties (M1)–(M5) of a metric are satisfied by $d$?

8. A function $u : X \times X \to \mathbb{R}$ satisfying the properties (M1)–(M4) of a metric is a $k$-ultrametric on $X$ if there exists $k \geq 1$ such that, for any $x, y, z \in X$, $u(x, y) \leq k \max\{u(x, z), u(z, y)\}$. Show that if $d$ is a metric on $X$, then $d$ is a 2-ultrametric and in general, for $n \in \mathbb{N}$, $d^n$ is a $2^n$-ultrametric.

9. A large city $C \subseteq \mathbb{R}^2$ has subway stations at the points of $S = \{s_1, s_2, \dots, s_n\} \subseteq C$. Let $d$ be the Euclidean metric on $\mathbb{R}^2$. Determine which of the defining properties (M1)–(M5) of a metric are satisfied by the following functions. Justify your answers.

(a) The function $D : \mathcal{F}(X) \times \mathcal{F}(X) \to \mathbb{R}$, where $\mathcal{F}(X)$ is the collection of all nonempty finite subsets of $\mathbb{R}^2$, defined by $D(A, B) = \min\{d(a, b) : a \in A, b \in B\}$.

(b) Define an equivalence relation on $\mathbb{R}^2$ by $x \approx y$ if and only if $x = y$ or $\{x, y\} \subseteq S$, and let $\mathbb{R}^2/\approx$ be the set of equivalence classes. Define $m : \mathbb{R}^2/\approx \to \mathbb{R}$ by $m([x], [y]) = D([x], [y])$.

(c) The *walking distance* $w$ on $\mathbb{R}^2$ defined by $w(x, y) = \min\{D(x, S) + D(S, y), d(x, y)\}$ (where $D(z, S) = D(S, z)$ is understood to be $D(\{z\}, S)$).

# 12 Uniform structures

## 12.1 Uniform continuity in metric spaces

Recall that if $X$ and $Y$ are metric spaces, a function $f : X \rightarrow Y$ is continuous if and only if

$$\forall a \in X \; \forall \varepsilon > 0 \; \exists \delta_{a,\varepsilon} > 0 \quad \text{such that} \quad x \in B(a, \delta_{a,\varepsilon}) \Rightarrow f(x) \in B(f(a), \varepsilon),$$

or equivalently,

$$\forall a \in X \; \forall \varepsilon > 0 \; \exists \delta_{a,\varepsilon} > 0 \quad \text{such that} \quad d(x, a) < \delta_{a,\varepsilon} \Rightarrow d(f(x), f(a)) < \varepsilon.$$

This is the standard pointwise definition of continuity given in Section 3.1. The pointwise approach says that, for each point $a$ in the domain, given any tolerance level $\varepsilon$ to specify the closeness of $f(x)$ to $f(a)$, we may find a $\delta$ (specifying closeness of $x$ to $a$) to guarantee that if $x$ is within $\delta$ of $a$, then $f(x)$ is within $\varepsilon$ of $f(a)$. For a given tolerance level $\varepsilon$, the choice of $\delta$ clearly depends on $\varepsilon$, but because we are verifying the continuity of $f$ pointwise, it also depends on the point $a$ in the domain at which we are testing the continuity. This dependence is emphasized in the notation $\delta_{a,\varepsilon}$ above.

For certain continuous functions, given an $\varepsilon > 0$ to specify the tolerance level, there is a single $\delta$, dependent on $\varepsilon$ alone, which ensures that fluctuating the input by less than $\delta$ will always guarantee that the output fluctuates less than $\varepsilon$, independent of the location $a$ of the input in the domain. Such functions are called *uniformly continuous*.

**Definition 12.1.1.** If $X$ and $Y$ are metric spaces, a function $f : X \rightarrow Y$ is *uniformly continuous* if and only if

$$\forall \varepsilon > 0 \; \exists \delta_\varepsilon > 0 \quad \text{such that} \quad \forall a \in X, \; x \in B(a, \delta_\varepsilon) \Rightarrow f(x) \in B(f(a), \varepsilon).$$

Equivalently, $f$ is uniformly continuous if and only if

$$\forall \varepsilon > 0 \; \exists \delta_\varepsilon > 0 \quad \text{such that} \quad \forall x, y \in X, \; d(x, y) < \delta_\varepsilon \Rightarrow d(f(x), f(y)) < \varepsilon.$$

In the first definition above, notice the placement of the quantified statement "$\forall a \in X$", which distinguishes continuity from uniform continuity. The subscripts on $\delta$ emphasize the dependence of $\delta$ only on $\varepsilon$ for uniform continuity, and on both $\varepsilon$ and the point $a$ in the domain for continuity. Generally, these subscripts will be omitted. Clearly, uniform continuity implies continuity. The examples below will illustrate the distinction between continuity and uniform continuity.

Give $\mathbb{R}$ and the interval $[0, \infty)$ the usual Euclidean metric $d(x, y) = |x - y|$ and consider the function $f : [0, \infty) \rightarrow \mathbb{R}$ defined by $f(x) = x^2$. We will show that $f$ is continuous, but not uniformly continuous.

https://doi.org/10.1515/9783110686579-013

To show that $f$ is continuous, given $a \in [0, \infty)$ and $\varepsilon > 0$, we must find $\delta > 0$, which may depend on both $a$ and $\varepsilon$, such that $x \in B(a, \delta)$ implies $f(x) \in B(f(a), \varepsilon)$, that is, such that

$$|x - a| < \delta \Rightarrow |f(x) - f(a)| = |x^2 - a^2| < \varepsilon.$$

Because we get to choose the $\delta$, we control the size of $|x - a|$. Thus, we hope to find a copy of $|x - a|$ hiding in the term $|x^2 - a^2|$ we wish to force to be small. By simple factoring, $|x^2 - a^2| = |x + a||x - a|$. To know how small we should make $|x - a|$, we need a bound on the remaining factor $|x + a|$. Recall that our choice of $\delta$ specifies closeness of $x$ to $a \in [0, \infty)$, and thus specifies closeness of $|x + a|$ to $|2a| = 2a$. To get an exact bound on $|x + a|$, let us stipulate that $\delta \le 1$. Since $|x - a| < \delta$, we are stipulating that $x$ will always be within 1 unit of $a$, and thus $|x + a|$ is within 1 unit of $2a$. Formally,

$$|x - a| < \delta \le 1 \Rightarrow \quad -1 < x - a < 1$$
$$\Rightarrow \quad 2a - 1 < x + a < 2a + 1$$
$$\Rightarrow -2a - 1 < x + a < 2a + 1 \quad \text{since} -2a \le 2a$$
$$\Rightarrow \quad |x + a| < 2a + 1.$$

Now if $|x - a| < \delta$, we have

$$|f(x) - f(a)| = |x^2 - a^2| = |x + a||x - a| < (2a + 1)\delta.$$

Since we wish to have $|f(x) - f(a)| < \varepsilon$, from the previous line we simply choose $\delta$ so that $(2a + 1)\delta = \varepsilon$; that is, we choose $\delta = \frac{\varepsilon}{2a+1}$, subject to the stipulation that $\delta \le 1$. That is, we choose $\delta = \min\{\frac{\varepsilon}{2a+1}, 1\}$.

We have worked backwards from the desired conclusion $|x^2 - a^2| < \varepsilon$ to find a suitable choice of $\delta$. While this is the difficult part, one may now write a concise direct proof working forwards, as Carl F. Gauss would have preferred. Gauss believed that just as scaffolding used to construct a building would be unsightly after the completion, so too are the constructive tools unsightly to a proof. Niels Abel said of Gauss, "He is like the fox, who effaces his tracks in the sand with his tail." Historian W. W. Rouse Ball said of Gauss, "he removes every trace of the analysis by which he reached his results, and studies to give a proof which while rigorous shall be as concise and synthetical as possible." Such a proof that $f : [0, \infty) \to \mathbb{R}$ defined by $f(x) = x^2$ is continuous is given below.

Given $a \in [0, \infty)$ and $\varepsilon > 0$, let $\delta = \min\{\frac{\varepsilon}{2a+1}, 1\}$. Now $|x - a| < \delta \le 1$ implies $-1 < x-a < 1$, so $-2a-1 \le 2a-1 < x+a < 2a+1$, and in particular, $|x+a| < 2a+1$. Thus, $|x-a| < \delta$ implies $|f(x)-f(a)| = |x^2-a^2| = |x+a||x-a| < (2a+1)\delta \le (2a+1)\frac{\varepsilon}{2a+1} = \varepsilon$, as needed.

For $f$ to be uniformly continuous, for any $\varepsilon > 0$ we would need to find a $\delta$ independent of $a$ which gives the desired inequalities. The dependence of $\delta$ on $a$ was evident in the proof given, so our proof fails to show that $f$ is uniformly continuous. Logically, we have not shown that there is no way to choose $\delta$ independent of $a$; we have merely shown that our first proof was not sufficient. To show that $f$ is not uniformly continuous, we need more.

We will show $f(x) = x^2$ is not uniformly continuous by showing for $\varepsilon = 1$, for any $\delta > 0$, there exist $x, a \in [0, \infty)$ with $|x - a| < \delta$, but with $|f(x) - f(a)| \geq \varepsilon = 1$. Given $\delta > 0$, let $x = \frac{1}{\delta} + \frac{\delta}{2}$ and $a = \frac{1}{\delta}$. Now $|x - a| < \delta$ yet

$$|f(x) - f(a)| = \left|\left(\frac{1}{\delta} + \frac{\delta}{2}\right)^2 - \left(\frac{1}{\delta}\right)^2\right| = 1 + \frac{\delta^2}{4} > 1 = \varepsilon.$$

Figure 12.1 illustrates that, for a fixed tolerance level $\varepsilon$ in the codomain, we may find a $\delta_{a,\varepsilon}$-ball around any $a \in [0, \infty)$ which maps into the $\varepsilon$-ball around $f(a)$. Observe that, as the values of $a$ increase to infinity and the slopes $f'(a)$ increase to infinity, the widths of the $\delta_{a,\varepsilon}$-balls must decrease to zero. Thus, there is no $\delta > 0$ which will serve for every $a \in [0, \infty)$.

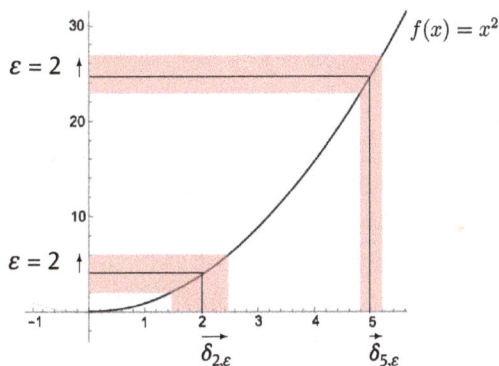

**Figure 12.1:** For a fixed $\varepsilon$, the value of $\delta_{a,\varepsilon}$ shrinks to 0 as $a \to \infty$.

The geometric interpretation of this example should suggest that, for calculus-type functions, that is, functions whose domain and codomain are subsets of the Euclidean line $\mathbb{R}$, there should be some connection between unbounded derivatives and uniform continuity. Loosely speaking, a steeper slope will require a smaller $\delta$. If the slopes are bounded by $M$, then perhaps the $\delta$ which works for the steepest slopes will work for all cases. The next theorem confirms this.

**Theorem 12.1.2.** *If $D$ is an interval in the Euclidean line $\mathbb{R}$, $f : D \to \mathbb{R}$ is a differentiable function, and $|f'(x)| < M$ for all $x \in D$, then $f$ is uniformly continuous.*

*Proof.* By the mean value theorem from calculus, for any $a, x \in D$, there exists $c$ between $a$ and $x$ in $D$ with

$$\frac{|f(x) - f(a)|}{|x - a|} = |f'(c)| < M,$$

so $|f(x) - f(a)| < M|x - a|$. Take $\delta_\varepsilon = \frac{\varepsilon}{M}$. Now

$$|x - a| < \delta_\varepsilon = \frac{\varepsilon}{M} \Rightarrow |f(x) - f(a)| < M \cdot \frac{\varepsilon}{M} = \varepsilon. \qquad \square$$

The converse is not true. That is, a uniformly continuous differentiable function from an interval $D$ to $\mathbb{R}$ may have an unbounded derivative. The function $g(x) = \sqrt{x}$ on the interval $(0, \infty)$ is such a function. As $x$ approaches zero through positive values, $g'(x)$ approaches infinity. However, the problematic unbounded derivatives occur in a very restricted location, and a given tolerance level $\varepsilon$ will allow enough fluctuation to encompass the location of the unbounded derivatives. The details are left as an exercise.

The theorem above gives a way to recognize certain functions as uniformly continuous, but they must necessarily be differentiable real-valued functions of a real variable. That is, the theorem above was more of a real analysis result than a topological one. The next theorem is an important topological result.

**Theorem 12.1.3.** *Suppose $X$ and $Y$ are metric spaces and $f : X \rightarrow Y$ is a continuous function. If $X$ is compact, then $f$ is uniformly continuous.*

*Proof.* Suppose $\varepsilon > 0$ is given. Under the hypotheses, the continuity of $f$ guarantees that, for every $c \in X$, there exists $\delta_c > 0$ such that $x \in B(c, \delta_c)$ implies $f(x) \in B(f(c), \varepsilon/2)$. Now $\{B(c, \delta_c/2) : c \in X\}$ is an open cover of the compact space $X$, so there exists a finite subcover $\mathcal{F} = \{B(c_i, \delta_{c_i}/2) : i = 1, \ldots, n\}$ of $X$. Take $\delta = \min\{\delta_{c_i}/2 : i = 1, \ldots, n\}$.

Suppose $a, x \in X$ and $x \in B(a, \delta)$. Since $\mathcal{F}$ covers $X$, there exists $j \in \{1, \ldots, n\}$ with $a \in B(c_j, \delta_{c_j}/2)$, so $d(a, c_j) < \delta_{c_j}/2$. Since $d(x, a) < \delta \le \delta_{c_j}/2$, the triangle inequality gives

$$d(x, c_j) \le d(x, a) + d(a, c_j) < \frac{\delta_{c_j}}{2} + \frac{\delta_{c_j}}{2} = \delta_{c_j}.$$

Also, $d(a, c_j) < \delta_{c_j}/2 < \delta_{c_j}$. Thus, both $x$ and $a$ are in $B(c_j, \delta_{c_j})$. This implies $f(x)$ and $f(a)$ are both in $B(f(c_j), \varepsilon/2)$. Again applying the triangle inequality, we get

$$d(f(x), f(a)) \le d(f(x), f(c_j)) + d(f(c_j), f(a)) < \frac{\varepsilon}{2} + \frac{\varepsilon}{2} = \varepsilon.$$

This shows that, for any $a \in X$, $x \in B(a, \delta)$ implies $f(x) \in B(f(a), \varepsilon)$, and thus $f$ is uniformly continuous. $\qquad \square$

We close this section with a necessary condition for uniform continuity. Recall that a sequence $(x_n)$ in a metric space $(X, d)$ is Cauchy if for any $\varepsilon > 0$ there exists $n \in \mathbb{N}$ such that $d(x_j, x_k) < \varepsilon$ for all $j, k > n$.

**Theorem 12.1.4.** *If $X$ and $Y$ are metric spaces, $(x_n)$ is a Cauchy sequence in $X$, and $f : X \rightarrow Y$ is uniformly continuous, then $(f(x_n))$ is a Cauchy sequence in $Y$.*

*Proof.* Suppose $f$ is uniformly continuous and $(x_n)$ is a Cauchy sequence in $X$. Given $\varepsilon > 0$, the condition on $f$ guarantees there exists a $\delta > 0$ such that $d(f(x_j), f(x_k)) < \varepsilon$ for any $x_j, x_k \in X$ with $d(x_j, x_k) < \delta$. The condition on $(x_n)$ guarantees an $n \in \mathbb{N}$ such that $d(x_j, x_k) < \delta$ for $j, k > n$. Now we have

$$j, k > n \Rightarrow d(x_j, x_k) < \delta \Rightarrow d(f(x_j), f(x_k)) < \varepsilon,$$

and since $\varepsilon > 0$ was arbitrary, $(f_n)$ is a Cauchy sequence in $Y$. $\qquad\square$

## Exercises

1. Consider the function $f : (0, \infty) \to \mathbb{R}$ defined by $f(x) = \frac{1}{x}$, where the domain and codomain have the Euclidean metric.
   (a) Show that $f$ is continuous.
   (b) Show that $f$ is not uniformly continuous.

2. Let $g : (0, \infty) \to \mathbb{R}$ be defined by $g(x) = \sqrt{x}$, where the domain and codomain have the Euclidean metric. Prove that $g(x)$ is uniformly continuous, even though the derivative $g'(x)$ is not bounded on $(0, \infty)$.

3. Show that $f : \mathbb{R} \to \mathbb{R}$ defined by $f(x) = \sqrt[3]{x}$ is uniformly continuous. Is $f$ differentiable on $\mathbb{R}$? Is the derivative $f'(x)$ bounded?

4. Which of the following functions are uniformly continuous? Prove your answers using the $\varepsilon$–$\delta$ definition (rather than any theorems).
   (a) $f : \mathbb{R} \to \mathbb{R}$ given by $f(x) = \begin{cases} x & \text{if } x < 0 \\ 3x & \text{if } x \geq 0 \end{cases}$
   (b) $g : \mathbb{R} \to \mathbb{R}$ given by $g(x) = \begin{cases} x & \text{if } x < 1 \\ 3x & \text{if } x \geq 1 \end{cases}$
   (c) $h : [1, \infty) \to [1, \infty)$ where $h$ is the piecewise linear function connecting $(n, \Delta_n)$ to $(n + 1, \Delta_{n+1})$ for each $n \in \mathbb{N}$, where $\Delta_n = 1 + \cdots + n = n(n + 1)/2$ is the $n$th *triangular number*.

5. Suppose $X$ is a metric space and $f$ and $g$ are uniformly continuous functions from $X$ to $\mathbb{R}$.
   (a) Show that $f + g : X \to \mathbb{R}$ is uniformly continuous.
   (b) Show that, for any constant $k \in \mathbb{R}$, $kf : X \to \mathbb{R}$ is uniformly continuous.

6. Suppose $X$, $Y$, and $Z$ are metric spaces, and $f : X \to Y$ and $g : Y \to Z$ are uniformly continuous. Show that $g \circ f : X \to Z$ is uniformly continuous.

7. For each part below, give an example satisfying the conditions, or prove that no such example can exist.
   (a) Continuous functions $f, g : \mathbb{R} \to \mathbb{R}$ which are not uniformly continuous, but whose product $fg$ is uniformly continuous.
   (b) Continuous functions $f, g : \mathbb{R} \to \mathbb{R}$ which are not uniformly continuous, but whose sum $f + g$ is uniformly continuous.

(c)  Uniformly continuous functions $f, g : \mathbb{R} \to \mathbb{R}$ whose product $fg$ is not uniformly continuous.

(d)  A continuous function $f : [0,1] \to \mathbb{R}$ such that $g : [0,2] \to \mathbb{R}$ defined by $g(x) = f(2x)$ is not uniformly continuous.

8.  Show that if $f : \mathbb{R} \to \mathbb{R}$ is continuous and is uniformly continuous on $A \subseteq \mathbb{R}$, then $f$ is uniformly continuous on cl $A$.

9.  Suppose $f : \mathbb{R} \to \mathbb{R}$ is continuous, $\mathbb{R} = A \cup B$, and the restrictions $f|_A$ and $f|_B$ are uniformly continuous. Must $f$ be uniformly continuous? Give a proof or counterexample.

10.  Suppose $f : \mathbb{R} \to \mathbb{R}$ is a continuous nonnegative function with $\int_{-\infty}^{\infty} f(x)\, dx = a < \infty$.

(a)  Show that if $f$ is uniformly continuous, then $\lim_{x \to \infty} f(x) = 0$.

(b)  Show by example that if $f$ is not uniformly continuous, then $\lim_{x \to \infty} f(x)$ need not be 0.

11.  Find a continuous function $f$ between two metric spaces $X$ and $Y$ and a Cauchy sequence $(x_n)$ in $X$ such that $(f(x_n))$ is not a Cauchy sequence. That is, show that uniform continuity is required for the result of Theorem 12.1.4.

12.  Suppose $f : \mathbb{R} \to \mathbb{R}$ is defined by $f(x) = x^2$. Show that if $(x_n)$ is a Cauchy sequence in $\mathbb{R}$, then $(f(x_n))$ is a Cauchy sequence. Since we have seen that $f$ is not uniformly continuous, this shows that the converse of Theorem 12.1.4 fails.

13.  Prove that $f : (a, b) \to \mathbb{R}$ is uniformly continuous if and only if $f$ can be extended to a continuous function $f^* : [a, b] \to \mathbb{R}$.

## 12.2 Uniformities

In this section, we will define a uniformity on a set $X$. A set equipped with a uniformity is called a uniform space. We will see that uniform spaces fall between metric spaces and topological spaces. More precisely, every metric gives a uniformity, and every uniformity gives a topology. One motivation for the development of uniform spaces was to extend the notion of uniform continuity to a larger class of topological spaces than the metric spaces.

Uniform continuity was defined for functions between metric spaces $X$ and $Y$. The fundamental idea that made the continuity *uniform* was that, for a given $\varepsilon > 0$, there was a single $\delta > 0$ for which the balls $B(x, \delta)$ were of interest, for every $x \in X$. That is, instead of considering balls around each individual point $x \in X$, uniform continuity involves determining neighborhoods around every point of $X$ in one fell swoop. One way to graphically depict the $\delta$-neighborhood of $x \in X$ for every $x \in X$, as shown in Figure 12.2(a), is to draw the neighborhood vertically over a copy of $x$ on a horizontal axis. (Of course, such graphical depictions work well when $X \subseteq \mathbb{R}$; for other sets $X$, such figures are only suggestive tools.)

The concept of uniform continuity involved giving a neighborhood of every $x \in X$ all at once. If we relax our metric condition that the neighborhood of every point $x \in X$ is a ball of the same radius $\delta$, we might consider sets such as $U$ depicted in Figure 12.2(b). Such a set $U$ is a subset of $X \times X$ and thus is a relation on $X$. Topologies can be defined by the collection of neighborhoods of every point. Uniformities will be defined by a collection of relations $U$ on $X$, each of which provides a neighborhood of every point $x \in X$. We start with some terminology.

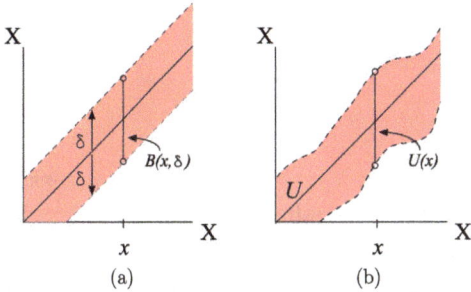

**Figure 12.2:** The shaded sets specify, in one fell swoop, a neighborhood of every point $x \in X$.

### 12.2.1 Relations

Recall that a *relation* from a set $X$ to a set $Y$ is a subset of $X \times Y$. If $R \subseteq X \times Y$ is a relation from $X$ to $Y$ and $(x, y) \in R$, we say $x$ is related to $y$ by $R$ and write $xRy$. A relation from $X$ to $X$ is called a relation *on* $X$. The *diagonal relation* $\Delta_X$ on $X$ is $\{(x, x) : x \in X\}$. The diagonal relation is sometimes called the identity relation or equality: $x$ is related to $y$ by $\Delta_X$ if and only if $x = y$. Every relation $R$ on $X$ has an inverse relation $R^{-1} = \{(y, x) \in X \times X : (x, y) \in R\}$.

A function $f : X \to Y$ may be viewed as the relation $G_f = \{(x, y) \in X \times Y : y = f(x)\} = \{(x, f(x)) \in X \times Y : x \in X\}$ from $X$ to $Y$. The relation $G_f$ may be called the graph of the function. Just as we may compose functions $f : X \to Y$ and $g : Y \to Z$ to get $g \circ f : X \to Z$ with $G_{g \circ f} = \{(x, z) \in X \times Z : \exists y \in Y \text{ with } (x, y) \in G_f, (y, z) \in G_g\}$, we may compose relations. Specifically, if $R$ is a relation from $X$ to $Y$ and $S$ is a relation from $Y$ to $Z$, then their *composition* $S \circ R$ is the relation from $X$ to $Z$ defined by

$$S \circ R = \{(x, z) \in X \times Z : \exists y \in Y \text{ with } xRy, ySz\}.$$

As suggested by function terminology, if $R$ is a relation from $X$ to $Y$ and $A \subseteq X$, the *image* of $A$ under $R$ is $R(A) = \{y \in Y : \exists a \in A \text{ with } aRy\}$. If $A$ is a singleton $\{x\}$, the image of $\{x\}$ under $R$ is called the *slice* of $R$ determined by $x$, and is denoted $R(x)$. In Figure 12.2(b), a relation $U$, a point $x$, and the slice $U(x)$ determined by $x$ are labeled. If $R$ is a relation on $X$, then we denote $R \circ R$ by $R^2$, $R \circ R \circ R$ by $R^3$, and so on.

Familiar properties of relations may be phrased using this terminology.

**Definition 12.2.1.** Let $R$ be a relation on $X$.

$R$ is reflexive if and only if $\Delta_X \subseteq R$.

$R$ is symmetric if and only if $R = R^{-1}$.

$R$ is antisymmetric if and only if $R \cap R^{-1} \subseteq \Delta_X$.

$R$ is transitive if and only if $R \circ R \subseteq R$.

Recall that a reflexive, transitive, symmetric relation is an *equivalence relation* and a reflexive transitive, antisymmetric relation is a *partial order*.

### 12.2.2 Filters

Just as a topology on $X$ is a collection of subsets of $X$ satisfying certain properties, a filter on $X$ is a collection of subsets of $X$ satisfying certain properties. Filters may be used to study convergence and are particularly useful in topological spaces in which some point has no countable neighborhood base, so convergence to the point cannot be accomplished with sequences (which are countable).

**Definition 12.2.2.** A *filter* on $X$ is a nonempty collection $\mathcal{F}$ of subsets of $X$ such that

(F1) $\emptyset \notin \mathcal{F}$;

(F2) $A \in \mathcal{F}$ and $A \subseteq B$ imply $B \in \mathcal{F}$, and

(F3) $A_1, \ldots, A_n \in \mathcal{F}$ implies $A_1 \cap \cdots \cap A_n \in \mathcal{F}$.

That is, a filter on $X$ is a collection of nonempty subsets of $X$ which is closed under the formation of supersets and is closed under finite intersections.

The most common filters encountered in topology are the *neighborhood filters*.

**Definition 12.2.3.** If $X$ is a topological space and $x \in X$, the *neighborhood filter* at $x$ is the collection

$$\mathcal{V}_x = \{V \subseteq X : x \in \text{int}(V)\}$$

of all neighborhoods of $x$.

We have noted that some introductory textbooks require neighborhoods to be open sets. If neighborhoods are required to be open, then the (open) neighborhoods of $x$ do not form a filter, since supersets of an open neighborhood of $x$ need not be an open neighborhood of $x$. Since the open neighborhoods of a point $x$ form a neighborhood basis for $x$, in many cases open neighborhoods are adequate. However, for filter considerations, the more general definition of neighborhoods is required.

The traditional use of the letter $\mathcal{V}$ for neighborhood filters arises from the French word *voisinage* for neighborhood.

**Example 12.2.4.**

(a) If $\emptyset \neq A \subseteq X$, then the collection $\mathcal{F} = \{S \subseteq X : A \subseteq S\}$ of supersets of $A$ is easily seen to be a filter.

(b) In $\mathbb{R}$, let $\mathcal{G} = \{C \subseteq \mathbb{R} : \mathbb{R} - C \text{ is finite}\}$ be the collection of cofinite subsets of $\mathbb{R}$. Since $\mathbb{R} - \emptyset$ is infinite, $\emptyset \notin \mathcal{G}$. If $C$ is any cofinite set, any superset of $C$ will have a smaller complement, and thus is also cofinite. Since $\mathcal{G}$ is a subcollection of the cofinite topology on $\mathbb{R}$ (indeed, $\mathcal{G} \cup \{\emptyset\}$ is the cofinite topology), $\mathcal{G}$ is closed under finite intersections. Thus, $\mathcal{G}$ is a filter on $\mathbb{R}$.

(c) Let $\mathcal{H} = \{A \cap \mathbb{Q} : A \subseteq \mathbb{R}, \pi \in \mathrm{int}_{\mathbb{R}} A\}$ be the collection of all neighborhoods of $\pi$ in the Euclidean line $\mathbb{R}$ restricted to $\mathbb{Q}$. It is easy to check that $\mathcal{H}$ is a filter on $\mathbb{Q}$.

A filter $\mathcal{F}$ on $X$ is *free* if $\bigcap \mathcal{F} = \emptyset$, and is *fixed* otherwise. In the example above, $\mathcal{F}$ is fixed, while $\mathcal{G}$ and $\mathcal{H}$ are free filters. Filters must be closed under finite intersections; free filters show that they need not be closed under arbitrary intersections.

### 12.2.3 Uniformities

Now let us return to the concept of a uniformity on a set $X$. The idea is that we will use a single relation like $U$ in Figure 12.2(b) to give a neighborhood of every point of $x \in X$. Such a set $U$ will be called an *entourage*. In most topologies, every point will have many different neighborhoods, and similarly, in most uniform space there will be a large collection of different entourages $U$. The properties of neighborhoods of a point will dictate the properties needed for the collection of entourages. The neighborhoods of $x$ form a filter on $X$, so the entourages should form a filter on $X \times X$. A neighborhood of $x$ must contain $x$, so any entourage $U$ should contain $\Delta_X$. That is, each entourage is a reflexive relation on $X$. This motivates the topological aspects of a uniformity, which we now define.

**Definition 12.2.5.** A *uniformity* (or *uniform structure*) on a nonempty set $X$ is a collection $\mathcal{U}$ of subsets of $X \times X$ such that

(U1) $\mathcal{U}$ is a filter on $X \times X$;
(U2) $\Delta_X \subseteq U$ for every $U \in \mathcal{U}$;
(U3) $U \in \mathcal{U}$ implies $U^{-1} \in \mathcal{U}$;
(U4) For every $U \in \mathcal{U}$, there exists $V \in \mathcal{U}$ with $V^2 \subseteq U$.

The members $U$ of a uniformity are called *entourages*. If $\mathcal{U}$ is a uniformity on $X$, the pair $(X, \mathcal{U})$ is a *uniform space*.

An entourage $U$ may be loosely thought of as a collection of neighborhoods $U(x)$ of each point $x \in X$, or as a "neighborhood of the diagonal" in $X \times X$.

While conditions $(U1)$ and $(U2)$ arose from topological properties we would expect of neighborhoods, conditions $(U3)$ and $(U4)$ arise as generalizations of metric space properties. If $(x, y) \in U$ for some entourage $U \in \mathcal{U}$, we say $x$ is $U$-*close* to $y$. Thus, (U3) says $x$ is $U$-close to $y$ if and only if $y$ is $U^{-1}$-close to $x$. A metric version of this would say $x$ is $\varepsilon$-close to $y$ if and only if $y$ is $\varepsilon'$-close to $x$ for some $\varepsilon'$. (Indeed, by the metric symmetric condition that $d(x, y) = d(y, x)$, we may take $\varepsilon' = \varepsilon$, that is, $x \in B(x, \varepsilon)$ if and only if $y \in B(x, \varepsilon)$.) Many proofs in metric spaces involve showing a distance, say from $x$ to $z$, is less than $\varepsilon$ by showing the distances between the pairs $(x, y)$ and $(y, z)$ are each less than $\varepsilon/2$ and then using the triangle inequality. The condition (U4) is a generalization of splitting a closeness measure ($\varepsilon$ for metric spaces, $U$ for uniform spaces) in half and linking two parts (by the triangle inequality for metrics, by $V^2$ for uniformities) to get something as close or closer. Condition (U4) says that given a closeness measure $U \in \mathcal{U}$, we can find a $V \in \mathcal{U}$ such that if $x$ is $V$-close to $y$ and $y$ is $V$-close to $z$, then $x$ is $V^2$-close, and thus $U$-close, to $z$.

Next, we show that every uniformity produces a topology.

**Definition 12.2.6.** Suppose $\mathcal{U}$ is a uniformity on $X$. The *uniform topology* on $X$ induced by $\mathcal{U}$ is

$$\mathcal{T}_{\mathcal{U}} = \{A \subseteq X : \forall x \in A \; \exists U \in \mathcal{U} \text{ with } U(x) \subseteq A\}.$$

If $(X, \mathcal{T})$ is a topological space whose topology $\mathcal{T}$ equals the uniform topology $\mathcal{T}_{\mathcal{U}}$ from some uniformity on $X$, then we say $X$ is *uniformizable* and $\mathcal{U}$ is *compatible* with the topology $\mathcal{T}$.

To verify that $\mathcal{T}_{\mathcal{U}}$ is indeed a topology on $X$, we first note that clearly $\emptyset$ and $X$ are members of $\mathcal{T}_{\mathcal{U}}$. Suppose $A_i \in \mathcal{T}_{\mathcal{U}}$ for $i \in I$, where $I$ is an arbitrary index set. For $x \in \bigcup_{i \in I} A_i$, there exists $i_0 \in I$ with $x \in A_{i_0}$. Since $A_{i_0}$ is open, there exists $U \in \mathcal{U}$ with $U(x) \subseteq A_{i_0} \subseteq \bigcup_{i \in I} A_i$, and thus $\bigcup_{i \in I} A_i \in \mathcal{T}_{\mathcal{U}}$. Finally, suppose $A_1, \dots, A_n$ are $\mathcal{T}_{\mathcal{U}}$-open and $x \in \bigcap_{i=1}^n A_i$. For each $i = 1, \dots, n$, there exists $U_i \in \mathcal{U}$ with $U_i(x) \subseteq A_i$. Let $U = \bigcap_{i=1}^n U_i$. Since $\mathcal{U}$ is a filter, $U \in \mathcal{U}$. Now for each $i = 1, \dots, n$, we have $U \subseteq U_i$, so $U(x) \subseteq U_i(x) \subseteq A_i$, and thus $U(x) \subseteq \bigcap_{i=1}^n A_i$. This shows that $\mathcal{T}_{\mathcal{U}}$ is closed under finite intersections.

From the definition, it is easy to see that every open set $A$ in $\mathcal{T}_{\mathcal{U}}$ is a union of slices $U(x) \subseteq A$. It does not follow that the collection $\mathcal{B} = \{U(x) : x \in X, U \in \mathcal{U}\}$ of all slices of entourages is a basis for $\mathcal{T}_{\mathcal{U}}$, since the slices $U(x)$ are not all $\mathcal{T}_{\mathcal{U}}$-open.

Next, we show that every metric produces a uniformity.

**Example 12.2.7.** Suppose $(X, d)$ is a metric space. For $\delta > 0$, let

$$U_\delta = \{(x, y) \in X \times X : d(x, y) < \delta\}$$
$$= \bigcup_{x \in X} \{x\} \times B(x, \delta).$$

Note that Figure 12.2(a) depicts $U_\delta$. Then $\mathcal{U}_d = \{W \subseteq X \times X : \exists \delta > 0 \text{ with } U_\delta \subseteq W\}$ is a uniformity on $X$ called the *metric uniformity* induced by $d$. Note that $\mathcal{U}_d$ consists of the supersets of the sets $U_\delta$ ($\delta > 0$). It is easily seen that conditions (U1), (U2), and (U3) from the definition of a uniformity are satisfied by $\mathcal{U}_d$. For (U4), suppose $W \in \mathcal{U}_d$. Then there exists $\delta > 0$ with $U_\delta \subseteq W$. Let $V = U_{\delta/2}$. For any $(x, z) \in V^2 = U_{\delta/2}^2$, there exists $y \in X$ with $(x, y) \in U_{\delta/2}$ and $(y, z) \in U_{\delta/2}$. Thus, $d(x, y) < \delta/2$ and $d(y, z) < \delta/2$. By the triangle inequality, $d(x, z) < \delta$, so $(x, z) \in U_\delta \subseteq W$. This shows $V^2 = U_{\delta/2}^2 \subseteq W$, so (U4) holds. It is immediate from the definitions that the uniform topology on $X$ generated by the metric uniformity induced by $d$ is simply the metric topology on $X$ induced by $d$.

### 12.2.4 Base for a uniformity

The metric uniformity was defined to be all the supersets of the $\delta$-neighborhoods $U_\delta$ of the diagonal $\Delta_X$ in $X \times X$. The neighborhood filter $\mathcal{V}_x$ at $x$ was defined to be all the supersets of *open* neighborhoods of $x$. Such constructions using supersets are common and prompt the following definitions.

**Definition 12.2.8.** If $\mathcal{F}$ is a filter on $X$, a subcollection $\mathcal{B}$ of $\mathcal{F}$ is a *base* (or *filter base*) for the filter $\mathcal{F}$ if $\mathcal{F}$ consists of all supersets of elements of $\mathcal{B}$.

If $\mathcal{U}$ is a uniformity on $X$, a subcollection $\mathcal{B}$ of $\mathcal{U}$ is a *base* for the uniformity $\mathcal{U}$ if $\mathcal{U}$ consists of all supersets of elements of $\mathcal{B}$.

Thus, the collection of open neighborhoods of $x$ is a filter base for the neighborhood filter $\mathcal{V}_x$, and the collection $\{U_\delta : \delta > 0\}$ is a base for the metric uniformity $\mathcal{U}_d$.

**Theorem 12.2.9.** *Every uniformity $\mathcal{U}$ has a base of symmetric entourages.*

*Proof.* Suppose $\mathcal{U}$ is a uniformity on set $X$. For $U \in \mathcal{U}$, let $U^s = U \cap U^{-1}$. Note that $(x, y) \in U^s = U \cap U^{-1}$ if and only if $(y, x) \in U^{-1} \cap U = U^s$, so $U^s$ is symmetric. $U^s$ is called the *symmetrization* of the entourage $U$. Let $\mathcal{U}^s = \{U^s : U \in \mathcal{U}\}$. For $U \in \mathcal{U}$, $U^{-1} \in \mathcal{U}$ by (U3), and $\mathcal{U}$ is closed under finite intersections by (U1), so $U^s \in \mathcal{U}$, and thus $\mathcal{U}^s \subseteq \mathcal{U}$. Furthermore, every entourage $U \in \mathcal{U}$ is a superset of $U^s \in \mathcal{U}^s$. Thus, $\mathcal{U}^s$ is a base for $\mathcal{U}$ which consists of symmetric entourages. Indeed, $\mathcal{U}^s$ is the collection of all symmetric entourages in $\mathcal{U}$. $\quad\square$

When dealing with bases for topologies, you may recall results to recognize when $\mathcal{B}$ is a basis for a given topology $\mathcal{T}$ (namely, $\mathcal{B} \subseteq \mathcal{T}$ and every element of $\mathcal{T}$ is a union of elements of $\mathcal{B}$), and results of a different flavor to determine whether a collection $\mathcal{B}$ of subsets of $X$ is a basis for some unknown topology on $X$ (namely, $\bigcup \mathcal{B} = X$, and for any $B_1, B_2 \in \mathcal{B}$ and any $x \in B_1 \cap B_2$, there exists $B_3 \in \mathcal{B}$ with $x \in B_3 \subseteq B_1 \cap B_2$). Similar situations apply for filter bases and bases for uniformities.

**Theorem 12.2.10.** *A nonempty collection $\mathcal{B}$ of subsets of a set $X$ is a filter base if and only if $\emptyset \notin \mathcal{B}$ and whenever $B_1, \ldots, B_n \in \mathcal{B}$, there exists $B \in \mathcal{B}$ with $B \subseteq B_1 \cap \cdots \cap B_n$.*

*Proof.* If $\mathcal{B}$ is a base for a filter $\mathcal{F}$, then since $\mathcal{B} \subseteq \mathcal{F}$, $\emptyset \notin \mathcal{B}$. Whenever $B_1, \ldots, B_n \in \mathcal{B}$, $B = B_1 \cap \cdots \cap B_n \in \mathcal{F}$, so we have $B \in \mathcal{B}$ with $B \subseteq B_1 \cap \cdots \cap B_n$.

Conversely, suppose $\mathcal{B}$ satisfies the conditions given. To show that $\mathcal{B}$ is the base for some filter $\mathcal{F}$, from the definition of filter base, $\mathcal{F} = \{F \subseteq X : \exists B \in \mathcal{B}, B \subseteq F\}$. Thus, it only remains to show that $\mathcal{F} = \{F \subseteq X : \exists B \in \mathcal{B}, B \subseteq F\}$ is indeed a filter. Clearly $\emptyset \notin \mathcal{F}$. Given $F_1, \ldots, F_n \in \mathcal{F}$, for each $i = 1, \ldots, n$, there exist $B_i \in \mathcal{B}$ with $B_i \subseteq F_i$. From the assumptions, there exists $B \in \mathcal{B}$ with $B \subseteq B_1 \cap \cdots \cap B_n$, so $B \subseteq F_1 \cap \cdots \cap F_n$, and thus $F_1 \cap \cdots \cap F_n \in \mathcal{F}$. Thus, $\mathcal{F}$ is a filter. □

Recall that a uniformity on $X$ is a filter on $X \times X$ and filters (uniformities) are generated from a filter base (uniformity base) by taking all supersets of the elements of the base. Thus, it should be expected that a filter base which satisfies enough of the properties of a uniformity should be a base for a uniformity.

**Theorem 12.2.11.** *If $X$ is a nonempty set, a collection $\mathcal{B}$ of subsets of $X \times X$ is a base for some uniformity on $X$ if and only if*
(B1) *$\mathcal{B}$ is a filter base on $X \times X$.*
(B2) *$\Delta_X \subseteq B$ for every $B \in \mathcal{B}$.*
(B3) *$B \in \mathcal{B}$ implies there exists $C \in \mathcal{B}$ with $C \subseteq B^{-1}$.*
(B4) *For every $B \in \mathcal{B}$, there exists $C \in \mathcal{B}$ with $C^2 \subseteq B$.*

The conditions (B1)–(B4) above should be carefully compared to the defining conditions (U1)–(U4) of a uniformity. The essential differences are that a base for a uniformity need only be a filter base, and in (B3), the symmetric relation $B^{-1}$ need not be in the base to imply $B^{-1}$ is in the uniformity induced by the base. The proof of this theorem is left to the exercises.

We end this section with some examples.

**Example 12.2.12.** Given $X \neq \emptyset$, let $\mathcal{B} = \{\Delta_X\}$. Now $\mathcal{B}$ is a base for the uniformity $\mathcal{U}$ of all supersets of the diagonal. For every $x \in X$, the slice $\Delta_X(x) = \{x\} \subseteq \{x\}$, so from Definition 12.2.6, $\{x\} \in \mathcal{T}_{\mathcal{U}}$. Thus, the induced topology $\mathcal{T}_{\mathcal{U}}$ is the discrete topology on $X$. Another base for this uniformity is $\mathcal{C} = \{U_\varepsilon : \varepsilon > 0\}$ where $U_\varepsilon = \{(x, y) \in X \times X : d(x, y) < \varepsilon\}$ and $d$ is the discrete metric. In particular, $\mathcal{B}$ and $\mathcal{C} = \{\Delta_X, X \times X\}$ are distinct bases for the same uniformity.

**Example 12.2.13.** Given $n \in \mathbb{N}$, let $U_n = \{(a, b) \in \mathbb{Z}^2 : a \equiv b \bmod 3^n\}$. We will show that $\mathcal{B} = \{U_n : n \in \mathbb{N}\}$ is a base for a uniformity on $\mathbb{Z}$. Clearly $(a, a) \in U_n$ for every $a \in \mathbb{Z}$ and every $n \in \mathbb{N}$, so (B2) holds and $\emptyset \notin \mathcal{B}$. If $m > n$ in $\mathbb{N}$ and $3^m$ divides $b - a$, then $3^n$ divides $b - a$, so $U_m \subseteq U_n$. Thus, $\mathcal{B}$ is a nested collection, and is therefore closed under finite intersections and is a filter base. Clearly $U_n = U_n^{-1}$ for each $n \in \mathbb{N}$, so (B3) holds. For (B4), note that if $(a, c) \in U_n^2$, then there exists $b \in \mathbb{Z}$ with $(a, b) \in U_n$ and $(b, c) \in U_n$, so $a \equiv b \equiv c \bmod 3^n$, and thus $(a, c) \in U_n$. Thus, $U_n^2 \subseteq U_n$ for every $U_n \in \mathcal{B}$. The prime base 3 in this example may be replaced by any prime $p$; the resulting uniformity is called the *p-adic uniformity* on $\mathbb{Z}$.

## Exercises

1. Suppose $U$ and $V$ are relations on $X$ with $V \subseteq U$. Show (a) $V^2 \subseteq U^2$ and (b) for any $x \in X$, $V(x) \subseteq U(x)$.

2. Show that if $U$ is a reflexive relation on $X$, then $U \subseteq U^n$ for any $n \in \mathbb{N}$.

3. If $U$ and $V$ are relations on $X$ and $a \in X$, show that $(U \cap V)(a) = U(a) \cap V(a)$.

4. Suppose $U \subseteq \mathbb{R} \times \mathbb{R}$ is a relation on $\mathbb{R}$ and $x \in \mathbb{R}$. Describe a visual algorithm to find the slice $U^2(x)$.

5. Suppose $V$ is a relation on a topological space $(X, \mathcal{T})$. Consider the statements:
   (a) $V$ is open in the product $(X, \mathcal{T}) \times (X, \mathcal{T})$.
   (b) For every $x \in X$, the slice $V(x)$ is open in $(X, \mathcal{T})$.
   Are (a) and (b) equivalent, or does either statement imply the other? Justify your answers with proofs or counterexamples.

6. Suppose $(X, \mathcal{T})$ is a topological space, $V$ is a relation on $X$, and $V$ is open in $(X, \mathcal{T})^2$. Show that $V \circ V = V^2$ is open in $(X, \mathcal{T})^2$.

7. In a poset $(X, \leq)$, an *order filter* is a nonempty increasing subset $F$ such that every pair of elements in $F$ has a lower bound in $F$. Show that $\mathcal{F}$ is a filter on $X$ if and only if $\mathcal{F}$ is an order filter in the poset $(\mathcal{P}(X) - \{\emptyset\}, \subseteq)$.

8. If $\mathcal{U}$ is a uniformity on $X$ and $U \in \mathcal{U}$, show that the symmetrization $U^s = U \cap U^{-1}$ of $U$ is the largest symmetric entourage contained in $U$.

9. Suppose $\mathcal{U}$ is a uniformity on $X$. Show that $\bigcap \mathcal{U}$ is an equivalence relation on $X$.

10. Let $X$ be the unit interval $[0, 1]$ and $B = (0, 1]^2 \cup \{(0, 0)\}$. Show that $\mathcal{B} = \{B\}$ is a base for a uniformity $\mathcal{U}$. Describe the topology induced by $\mathcal{U}$.

11. For $a \in \mathbb{R}$, let $U_a = \Delta_{\mathbb{R}} \cup [a, \infty)^2$. Show that $\mathcal{B} = \{U_a : a \in \mathbb{R}\}$ is a base for a uniformity $\mathcal{U}$. Describe the topology induced by $\mathcal{U}$.

12. Find all possible uniformities on a two-point set $X = \{a, b\}$. List all uniformizable topologies and all non-uniformizable topologies on $X = \{a, b\}$.

13. Suppose $\mathcal{U}$ is a uniformity on $X$ and $U \in \mathcal{U}$. Show that, for every $n \in \mathbb{N}$, there exists $V \in \mathcal{U}$ with $V^n \subseteq U$.

14. Suppose $X$ is a topological space and $x \in X$ is not an isolated point. Show that the collection $\mathcal{B}_x^{\text{del}} = \{U - \{x\} : U \in \mathcal{V}_x\}$ of deleted neighborhoods of $x$ is a filter base. Is $\mathcal{B}_x^{\text{del}}$ every a filter? The filter generated by $\mathcal{B}$ is called the *deleted neighborhood filter* at $x$.

15. Show that the collection of open intervals $\mathcal{B} = \{(a, 1) : 0 \leq a < 1\}$ is a filter base on $X = (-1, 1) \subseteq \mathbb{R}$. Is the filter generated by $\mathcal{B}$ a neighborhood filter of any $x \in X$? Is the filter generated by $\mathcal{B}$ a deleted neighborhood filter (see Exercise 14) for any $x$ in the compactification $Y = [-1, 1]$ of $(-1, 1)$?

16. Suppose $\mathcal{B}$ is a base for a uniformity $\mathcal{U}$ on a set $X$.
    (a) If $\mathcal{B} \subseteq \mathcal{C} \subseteq \mathcal{U}$, must $\mathcal{C}$ be a base for $\mathcal{U}$?
    (b) If $\mathcal{B} \subseteq \mathcal{C} \subseteq \mathcal{P}(X)$, must $\mathcal{C}$ be a base for $\mathcal{U}$?
    Prove your answers.

17. Find two uniformities $\mathcal{U}$ and $\mathcal{W}$ on a set $X$ such that $\mathcal{U} \cup \mathcal{W}$ is neither a uniformity nor a base for a uniformity on $X$.

18. A nonempty collection $\mathcal{S}$ of subsets of $X \times X$ is a *subbase* for a uniformity on $X$ if and only if the collection of finite intersections of members of $\mathcal{S}$ forms a base for a uniformity on $X$. Show that if $\mathcal{U}$ and $\mathcal{W}$ are uniformities on a set $X$, then $\mathcal{U} \cup \mathcal{W}$ is a subbase for a uniformity on $X$.

19. For $\varepsilon > 0$, let $W_\varepsilon$ be the relation on $\mathbb{R}$ whose slices are given by

$$
W_\varepsilon(x) = \begin{cases} (x - \varepsilon, x + \varepsilon/2) \cup (x + \varepsilon, x + 2\varepsilon) & \text{if } 1 < x < 2 \\ (x - 3\varepsilon, x - 2\varepsilon] \cup (x - \varepsilon, x + 2\varepsilon) & \text{if } 3 < x < 4 \\ (x - \varepsilon, x + 2\varepsilon) & \text{otherwise,} \end{cases}
$$

as suggested in the figure below.

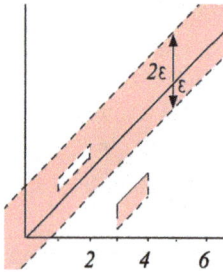

Show that $\mathcal{W} = \{W_\varepsilon : \varepsilon > 0\}$ is a base for a uniformity on $\mathbb{R}$ which induces the usual topology on $\mathbb{R}$.

20. Prove Theorem 12.2.11.

21. **(Filter convergence)** A filter $\mathcal{F}$ on a topological space $X$ is said to *converge to* $x \in X$, denoted $\mathcal{F} \to x$, if and only if $\mathcal{V}_x \subseteq \mathcal{F}$. In $\mathbb{R}$ with the Euclidean topology, let $\mathcal{F} = \{A \subseteq \mathbb{R} : \text{there exists } m \in \mathbb{N} \text{ with } \{\frac{1}{n} : n \geq m\} \subseteq A\}$ be the collection of supersets of tails of the sequence $(\frac{1}{n})_{n=1}^{\infty}$. Show that $\mathcal{F}$ is a filter converging to $0$.

## 12.3 The uniform topology

In this section, we investigate properties of the topology induced by a uniformity.

### 12.3.1 Closure and interior

**Theorem 12.3.1.** *Suppose $\mathcal{U}$ is a uniformity on $X$ and $A \subseteq X$. Then in the induced topology $\mathcal{T}_\mathcal{U}$ on $X$,*

$$
\operatorname{int} A = \{a \in A : U(a) \subseteq A \text{ for some } U \in \mathcal{U}\}.
$$

*Proof.* Let $B = \{a \in A : U(a) \subseteq A$ for some $U \in \mathcal{U}\}$.

Since int $A \in \mathcal{T}_{\mathcal{U}}$, the definition of $\mathcal{T}_{\mathcal{U}}$ tells us that if $x \in$ int $A$, there exists $U \in \mathcal{U}$ with $U(x) \subseteq$ int $A \subseteq A$, and thus $x \in B$. This shows int $A \subseteq B \subseteq A$. To show int $A = B$, we will show $B$ is open.

To see that $B$ is open, from the definition of $\mathcal{T}_{\mathcal{U}}$ we must show for any $a \in B$, there exists $V \in \mathcal{U}$ with $V(a) \subseteq B$. Given $a \in B$, there exists $U \in \mathcal{U}$ with $U(a) \subseteq A$. Pick $V \in \mathcal{U}$ with $V^2 \subseteq U$. We will show $V(a) \subseteq B$. Suppose $y \in V(a)$. For any $z \in V(y)$, we have $(a, y), (y, z) \in V$, so $(a, z) \in V^2$, or $z \in V^2(a)$. Thus, $V(y) \subseteq V^2(a)$. Now $y \in V(y) \subseteq V^2(a) \subseteq U(a) \subseteq A$, and from the definition of $B$, $y \in B$. Since $y$ was an arbitrary element of $V(a)$, this shows $V(a) \subseteq B$, as needed. $\qquad\square$

From Theorem 12.3.1, it is easy to see that $x \in$ int $U(x)$ for any $x \in X$ and $U \in \mathcal{U}$, so $U(x)$ is a neighborhood of $x$. Furthermore, if $N$ is any open neighborhood of $x$, $N \in \mathcal{T}_{\mathcal{U}}$ implies that, for any $y \in N$, there exists $U \in \mathcal{U}$ with $U(y) \subseteq N$. In particular, since $x \in N$, there exists $U \in \mathcal{U}$ with $U(x) \subseteq N$. Thus, every neighborhood $N$ of $x$ contains a neighborhood of form $U(x)$ for some $U \in \mathcal{U}$. This formally proves the entirely expected result that $\{U(x) : U \in \mathcal{U}\}$ is a base for the neighborhood filter $V_x$ at $x$. Indeed, our motivation for the definition of a uniformity was that the slices $U(x)$ should form a basis for the neighborhoods at $x$. We state this formally, in a strengthened form.

**Corollary 12.3.2.** *If $\mathcal{U}$ is a uniformity on $X$, then, for each $x \in X$, the $\mathcal{T}_{\mathcal{U}}$-neighborhood filter at $x$ is $V_x = \{U(x) : U \in \mathcal{U}\}$.*

*Proof.* In the discussion above, we saw that $\{U(x) : U \in \mathcal{U}\}$ is a basis for the neighborhood filter $V_x$, so every neighborhood $V$ of $x$ contains a basic neighborhood $U(x)$ for some $U \in \mathcal{U}$. But since $\mathcal{U}$ is a filter, $U' = U \cup (\{x\} \times V) \in \mathcal{U}$, and $U'(x) = V$. Thus, every neighborhood $V$ of $x$ has form $U'(x)$ for some $U' \in \mathcal{U}$. $\qquad\square$

Note that $\mathcal{B} = \{U(x) : x \in X, U \in \mathcal{U}\}$ is the union of neighborhood bases for each $x \in X$, but $\mathcal{B}$ is not generally a basis for the topology $\mathcal{T}_{\mathcal{U}}$. The problem is that members of the neighborhood filter need not be open, while members of the topology, and thus of any basis, must be open. We note that the converse of Corollary 12.3.2 fails. Exercise 1 of Section 12.5 presents a collection $\mathcal{U}$ such that, for each $x \in X$, $\{U(x) : U \in \mathcal{U}\}$ is a base for the $\mathcal{T}$-neighborhood filter $V_x$ at $x$, where $\mathcal{T}$ is a uniformizable topology but $\mathcal{U}$ is not a uniformity.

With these results, the following theorem about subspaces should be expected.

**Theorem 12.3.3.** *If $\mathcal{U}$ is a uniformity on $X$ and $Y$ is a nonempty subset of $X$, then $\mathcal{U}_Y = \{U \cap (Y \times Y) : U \in \mathcal{U}\}$ is a uniformity on $Y$, and the uniform topology $\mathcal{T}_{\mathcal{U}_Y}$ from $\mathcal{U}_Y$ is the subspace topology which $Y$ inherits from $(X, \mathcal{T}_{\mathcal{U}})$.*

The proof is left to the exercises.

**Theorem 12.3.4.** *Suppose $\mathcal{U}$ is a uniformity on $X$ and $A \subseteq X$. Then in the induced topology $\mathcal{T}_{\mathcal{U}}$ on $X$,*

$$\operatorname{cl} A = \bigcap \{U(A) : U \in \mathcal{U}\}.$$

*Proof.* From the definition of a uniformity, note that $U \in \mathcal{U}$ if and only if $U^{-1} \in \mathcal{U}$, so for any $x \in X$, $\{U^{-1}(x) : U \in \mathcal{U}\}$ is a base for the neighborhood filter $\mathcal{V}_x$. Now

$$
\begin{aligned}
x \in \operatorname{cl} A &\iff \text{every neighborhood of } x \text{ intersects } A \\
&\iff \forall U \in \mathcal{U}, U^{-1}(x) \cap A \neq \emptyset \\
&\iff \forall U \in \mathcal{U}, \exists a \in A \text{ such that } (x, a) \in U^{-1} \\
&\iff \forall U \in \mathcal{U}, \exists a \in A \text{ such that } (a, x) \in U \\
&\iff \forall U \in \mathcal{U}, x \in U(A),
\end{aligned}
$$

which gives the desired equality. □

### 12.3.2 The square of a uniform space

Suppose $\mathcal{U}$ is a uniformity on $X$. The entourages $U \in \mathcal{U}$ are subsets of $X \times X$. Since $\mathcal{U}$ gives a topology $\mathcal{T}_\mathcal{U}$ on $X$, it is natural to consider the product space $(X, \mathcal{T}_\mathcal{U}) \times (X, \mathcal{T}_\mathcal{U})$. With this product topology, $X \times X$ is called the *square* of the uniform space $X$. If $X$ is a uniform space, we will assume $X \times X$ carries this product topology if not explicitly stated otherwise.

**Theorem 12.3.5.** *If $\mathcal{U}$ is a uniformity on $X$ and $R \subseteq X \times X$ is a relation on $X$, then the closure of $R$ in the product space $(X, \mathcal{T}_\mathcal{U}) \times (X, \mathcal{T}_\mathcal{U})$ is*

$$
\operatorname{cl} R = \bigcap \{U \circ R \circ U : U \in \mathcal{U}\}.
$$

*Proof.* First we note that if $\mathcal{B}$ is any base for $\mathcal{U}$, then

$$
\bigcap \{U \circ R \circ U : U \in \mathcal{U}\} = \bigcap \{B \circ R \circ B : B \in \mathcal{B}\}.
$$

Indeed, "$\supseteq$" holds since every $U \in \mathcal{U}$ contains a $B \in \mathcal{B}$, and "$\subseteq$" holds since $\mathcal{B} \subseteq \mathcal{U}$.

By Theorem 12.2.9, the collection $\mathcal{U}^s$ of symmetric entourages in $\mathcal{U}$ is a base for $\mathcal{U}$. Now for $R \subseteq X \times X$, we have

$$
\begin{aligned}
(x, y) \in \operatorname{cl} R &\iff \forall U \in \mathcal{U}^s, U(x) \times U(y) \text{ intersects } R \\
&\iff \forall U \in \mathcal{U}^s, \exists (r, s) \in (U(x) \times U(y)) \cap R \\
&\iff \forall U \in \mathcal{U}^s, \exists r, s \in X \text{ with } (x, r) \in U, (y, s) \in U, (r, s) \in R \\
&\iff \forall U \in \mathcal{U}^s, \exists r, s \in X \text{ with } (x, r) \in U, (r, s) \in R, (s, y) \in U^{-1} \\
&\iff \forall U \in \mathcal{U}^s, (x, y) \in U^{-1} \circ R \circ U = U \circ R \circ U \\
&\iff (x, y) \in \bigcap \{U \circ R \circ U : U \in \mathcal{U}^s\},
\end{aligned}
$$

and the result follows from the observation above. □

**Theorem 12.3.6.** *Suppose $\mathcal{U}$ is a uniformity on $X$. Then the collections*

$$\mathcal{V} = \{U \in \mathcal{U} : U \text{ is open in } X \times X\} \quad \text{and}$$
$$\mathcal{C} = \{U \in \mathcal{U} : U \text{ is closed in } X \times X\}$$

*of open entourages and of closed entourages are each a base for the uniformity $\mathcal{U}$.*

*Proof.* Suppose $U \in \mathcal{U}^s$ is a symmetric entourage in $\mathcal{U}$. We claim $U \subseteq \text{int}(U^3)$. Suppose $(a, b) \in U$. Since $(a, a), (a, b), (b, b) \in U$, $(a, b) \in U^3$. To see $(a, b) \in \text{int}(U^3)$, we will show that the neighborhood $U(a) \times U(b)$ of $(a, b)$ is contained in $U^3$. Suppose $(x, y) \in U(a) \times U(b)$. Then $(a, x), (b, y) \in U = U^{-1}$, so $(x, a), (a, b), (b, y) \in U$, and thus $(x, y) \in U^3$. This shows $U \subseteq \text{int}(U^3)$.

   To show that the collection $\mathcal{V}$ of open entourages is a base for $\mathcal{U}$, given an arbitrary $U \in \mathcal{U}$, we must find $V \in \mathcal{V}$ with $V \subseteq U$. Given $U \in \mathcal{U}$, there exists $W \in \mathcal{U}$ with $W^3 \subseteq U$ (see Exercise 13 of Section 12.2). We may further assume $W$ is symmetric, for if not, replace $W$ by $W^s = W \cap W^{-1}$ and note that $W^s \subseteq W$ implies $(W^s)^3 \subseteq W^3 \subseteq U$. Now by the preceding paragraph, $W \subseteq \text{int}(W^3) \subseteq \text{int } U \subseteq U$. Now since $\mathcal{U}$ is closed under the formation of supersets and $W \in \mathcal{U}$, we have int $U \in \mathcal{U}$. Thus, every entourage in $\mathcal{U}$ contains an open entourage, so $\mathcal{V}$ is a base for $\mathcal{U}$.

   To show that the collection $\mathcal{C}$ of closed entourages is a base for $\mathcal{U}$, suppose $U \in \mathcal{U}$ and let $W \in \mathcal{U}$ be an entourage with $W^3 \subseteq U$. By Theorem 12.3.5, $W \subseteq \text{cl } W \subseteq W^3 \subseteq U$, and as a superset of $W \in \mathcal{U}$, cl $W$ is a closed entourage contained in $U$. Thus, $\mathcal{C}$ is a base for $\mathcal{U}$. $\square$

   We have two important corollaries which follow immediately.

**Corollary 12.3.7.** *If $(X, \mathcal{U})$ is a uniform space, every entourage $U \in \mathcal{U}$ is a (not necessarily open) neighborhood of the diagonal. That is, for $U \in \mathcal{U}$, $\Delta_X \subseteq \text{int } U$ and int $U \in \mathcal{U}$.*

**Corollary 12.3.8.** *If $(X, \mathcal{U})$ is a uniform space, then $(X, \mathcal{T}_\mathcal{U})$ is $T_3$.*

*Proof.* Recall (Theorem 7.1.9) that $X$ is $T_3$ if and only if for every $x \in X$, every neighborhood of $x$ contains a closed neighborhood. (That is, every neighborhood filter $\mathcal{V}_x$ has a filter base of closed neighborhoods, or every point has a neighborhood base of closed neighborhoods.) The crux of the proof is that $\mathcal{V}_x = \{U(x) : U \in \mathcal{U}\}$, and since $\mathcal{U}$ has a base of closed entourages, $\mathcal{V}_x$ will have a base of slices $C(x)$ of closed entourages $C \in \mathcal{U}$, and these slices will be closed. Formally, any $\mathcal{T}_\mathcal{U}$-neighborhood of $x \in X$ has form $U(x)$ where $U \in \mathcal{U}$. We may view the neighborhood $U(x)$ as $U \cap (\{x\} \times X)$. By Theorem 12.3.6, there exists a closed entourage $C \in \mathcal{U}$ with $C \subseteq U$. In particular, $C(x)$ is a neighborhood of $x$ and $C(x) \subseteq U(x)$. Furthermore, since $C$ is closed in $X \times X$, the intersection $C \cap (\{x\} \times X) = C(x) \subseteq U(x)$ is closed in the copy $\{x\} \times X$ of $X$ embedded in $X \times X$. $\square$

   The next theorem provides further connections between separation axioms in a uniform space.

**Theorem 12.3.9.** *Suppose* $(X, \mathcal{U})$ *is a uniform space. Then the following are equivalent.*
(a) $(X, \mathcal{T}_{\mathcal{U}})$ *is* $T_1$.
(b) $(X, \mathcal{T}_{\mathcal{U}})$ *is Hausdorff.*
(c) $(X, \mathcal{T}_{\mathcal{U}})$ *is regular.*
(d) $\bigcap \mathcal{U} = \Delta_X$.

*Proof.* Since $X$ is regular if and only if it is $T_3$ and $T_1$, by Corollary 12.3.8 we have $(a) \Rightarrow$ $(c) \Rightarrow (b) \Rightarrow (a)$, showing the equivalence of $(a)$, $(b)$, and $(c)$. The proof that $(d)$ is also equivalent is left to the exercises. $\qquad\square$

The final theorem of this section is an elegant result. The proof is more intricate than the proofs of our other results on uniform spaces.

**Theorem 12.3.10.** *Every compact Hausdorff topological space* $(X, \mathcal{T})$ *is uniformizable, and there is a unique uniformity compatible with* $\mathcal{T}$, *namely the collection* $\mathcal{U}$ *of all neighborhoods of the diagonal* $\Delta_X$.

*Proof.* Suppose $(X, \mathcal{T})$ is a compact Hausdorff topological space. First, we show that the collection $\mathcal{U} = \{U \subseteq X \times X : \Delta_X \subseteq \text{int } U\}$ of neighborhoods of the diagonal is a uniformity compatible with $\mathcal{T}$. Clearly, $\mathcal{U}$ satisfies conditions (U1) and (U2) of the definition of a uniformity. Since the function $h : X^2 \to X^2$ defined by $h(x, y) = (y, x)$ is a homeomorphism which maps $U$ to $U^{-1}$, (U3) is satisfied.

It remains to check (U4). Suppose $U \in \mathcal{U}$ is a neighborhood of $\Delta_X$. Then, for any $(x, x) \in \Delta_X$, there exists a basic open neighborhood $N_x \times N_x$ of $(x, x)$ contained in $U$. Now $\{N_x : x \in X\}$ is an open cover of the compact space $X$, so there exists a finite subcover $\{N_i : i \in F\}$, where $F = \{x_1, x_2, \dots, x_n\}$ is a finite subset of $X$. With $U' = \bigcup\{N_i \times N_i : i \in F\}$, we have $U' \subseteq U$, so it suffices to show the existence of $V \in \mathcal{U}$ with $V^2 \subseteq U'$. For any subset $J \subseteq F$, set $X_J = \bigcup\{N_i : i \in J\}$ and $Y_J = \bigcup\{N_i : i \in F - J\}$. Now $X_J$ and $Y_J$ are open and $X_J \cup Y_J = \bigcup\{N_i : i \in F\} = X$. Thus, $X - X_J$ and $X - Y_J$ are disjoint closed subsets of $X$. Since every compact Hausdorff space is normal, there exist disjoint open sets $G_J, H_J$ in $X$ with $X - X_J \subseteq G_J$ and $X - Y_J \subseteq H_J$. Taking complements, we see that $X_J \supseteq X - G_J \supseteq H_J$ and similarly $Y_J \supseteq G_J$. Thus, for every $J \subseteq F$, $C_J = \{G_J, H_J, X_J \cap Y_J\}$ is an open cover of $X$. Consider the collection

$$\mathcal{D} = \left\{ \bigcap_{J \subseteq F} Q_J : Q_J \in C_J = \{G_J, H_J, X_J \cap Y_J\} \right\}.$$

Every element $D \in \mathcal{D}$ has form $\bigcap\{Q_J : J \subseteq F\}$, and as a finite intersection of open sets, is open. Furthermore, given any $x \in X$ and any $J \subseteq F$, by choosing $Q_J$ to be an element of the cover $C_J$ which contains $x$, we see that there exists an element $D = \bigcap\{Q_J : J \subseteq F\}$ which contains $x$, so $\mathcal{D}$ is a finite open cover of $X$. Let $V = \bigcup\{D \times D : D \in \mathcal{D}\}$. Now $V$ is an open neighborhood of $\Delta_X$ (so $V \in \mathcal{U}$), and we claim $V^2 \subseteq U'$. Suppose $(x, z) \in V^2$. To see $(x, z) \in U' = \bigcup\{N_i \times N_i : i \in F\}$, suppose to the contrary $\{x, z\} \not\subseteq N_i$ for any $i \in F$. Since $(x, z) \in V^2$, there exists $y \in X$ with $(x, y) \in V$ and $(y, z) \in V$. Thus, there exist

$D_1, D_2 \in \mathcal{D}$ with $(x, y) \in D_1 \times D_1$ and $(y, z) \in D_2 \times D_2$. In particular, $y \in D_1 \cap D_2$. For $J_0 = \{k \in F : x \in N_k\}$, we have $X_{J_0} = \bigcup\{N_i : x \in N_i\}$, so $x \in X_{J_0} - Y_{J_0}$ and $z \in Y_{J_0} - X_{J_0}$. Now $x \notin X_{J_0} \cap Y_{J_0}$ and $x \notin G_{J_0} \subseteq Y_{J_0}$, so $x \in D_1 = \bigcap\{Q_J^1 : J \subseteq F\}$ implies $Q_{J_0}^1 = H_{J_0}$. Similarly, $y \in D_2 = \bigcap\{Q_J^2 : J \subseteq F\}$ implies $Q_{J_0}^2 = G_{J_0}$. Since $D_1 \subseteq Q_{J_0}^1 = H_{J_0} \subseteq X_{J_0}$, $D_2 \subseteq Q_{J_0}^2 = G_{J_0} \subseteq Y_{J_0}$, and $X_{J_0} \cap Y_{J_0} = \emptyset$, we have $D_1 \cap D_2 = \emptyset$, contrary to $y \in D_1 \cap D_2$. This completes the proof that $V^2 \subseteq U' \subseteq U$, and thus $\mathcal{U}$ is a uniformity.

Having shown that $\mathcal{U}$ is a uniformity, we next show that it is compatible with $\mathcal{T}$, that is, $\mathcal{T}_{\mathcal{U}} = \mathcal{T}$. Suppose $A \in \mathcal{T}$ and $x \in A$. Then $U = X^2 - (\{x\} \times (X - A))$ is an open neighborhood of $\Delta_X$ in $(X, \mathcal{T})^2$ with $U(x) = A$. Since $U \in \mathcal{U}$, this shows that $A \in \mathcal{T}_{\mathcal{U}}$. Conversely, suppose $A \in \mathcal{T}_{\mathcal{U}}$. Then, for any $x \in A$, there exists $U \in \mathcal{U}$ with $U(x) \subseteq A$. Without loss of generality (by Theorem 12.3.6), we may take $U$ to be open in $(X, \mathcal{T})^2$. Now $U(x) = U \cap (\{x\} \times X)$ is open in the subspace $\{x\} \times X$ of $(X, \mathcal{T})^2$, and since $\{x\} \times X$ is homeomorphic to $(X, \mathcal{T})$, it follows that $B = U(x)$ is open in $\mathcal{T}$. This shows that, for any $x \in A$, there exists $B = U(x) \in \mathcal{T}$ with $x \in B \subseteq A$, so $A \in \mathcal{T}$.

Finally, we show that $\mathcal{U}$ is the unique uniformity compatible with $\mathcal{T}$. Suppose that $\mathcal{W}$ is a uniformity with $\mathcal{T}_{\mathcal{W}} = \mathcal{T}$. By Corollary 12.3.7, $\mathcal{W} \subseteq \mathcal{U}$. To see $\mathcal{U} \subseteq \mathcal{W}$, by Theorem 12.3.6, it suffices to show that every open member of $\mathcal{U}$ is an element of $\mathcal{W}$. Suppose $U \in \mathcal{U}$ is an open neighborhood of the diagonal. Then $U$ and $X^2 - \Delta_X$ cover $X^2$. Since $(X, \mathcal{T}) = (X, \mathcal{T}_{\mathcal{W}})$ is Hausdorff, we have $\bigcap \mathcal{W} = \Delta_X$, and furthermore, since $\mathcal{W}$ has a base of its closed entourages (again Theorem 12.3.6), we have $\Delta_X = \bigcap\{\text{cl } W : W \in \mathcal{W}\}$, so $X^2 - \Delta_X = \bigcup\{X^2 - \text{cl } W : W \in \mathcal{W}\}$. Thus, $\{U\} \cup \{X^2 - \text{cl } W : W \in \mathcal{W}\}$ is an open cover of the compact space $X \times X$, so there exist $W_1, \ldots, W_n \in \mathcal{W}$ with

$$X^2 = U \cup (X^2 - \text{cl } W_1) \cup (X^2 - \text{cl } W_2) \cup \cdots \cup (X^2 - \text{cl } W_n).$$

It follows that

$$X^2 - U \subseteq (X^2 - \text{cl } W_1) \cup (X^2 - \text{cl } W_2) \cup \cdots \cup (X^2 - \text{cl } W_n)$$
$$U \supseteq \text{cl } W_1 \cap \text{cl } W_2 \cap \cdots \cap \text{cl } W_n$$
$$\supseteq W_1 \cap W_2 \cap \cdots \cap W_n \in \mathcal{W}.$$

As a superset of an entourage in $\mathcal{W}$, we have $U \in \mathcal{W}$, as needed. $\square$

An important corollary follows.

**Corollary 12.3.11.** *Every completely regular space is uniformizable.*

*Proof.* This follows from the theorem since uniformizability is a hereditary property, and every completely regular space $X$ is a subspace of its Stone-Čech compactification $\beta X$, which is compact and Hausdorff. $\square$

In Section 12.4, we will see that, for $T_1$ spaces, the converse of the corollary above holds. This gives the significant result that a $T_1$ space is completely regular if and only if it is uniformizable.

**Example 12.3.12.** Consider the topology $\mathcal{T}$ on $\mathbb{R}$ generated by the basis $\mathcal{B} = \{[x, x + \varepsilon) : x \in \mathbb{Q}, \varepsilon > 0\} \cup \{(x - \varepsilon, x] : x \in \mathbb{R} - \mathbb{Q}, \varepsilon > 0\}$. It is easy to verify that $(\mathbb{R}, \mathcal{T})$ is completely regular (Exercise 6 of Section 7.2), and thus is uniformizable. Furthermore, $(\mathbb{R}, \mathcal{T})$ is separable but not second countable (Exercise 13 below). By Theorem 11.2.6, metrizable and separable imply second countable, so this uniformizable space $(\mathbb{R}, \mathcal{T})$ is not metrizable. (Exercise 1 of Section 12.5 also addresses this example.)

### 12.3.3 Uniform continuity

Suppose $(X, d_X)$ is a metric space with metric uniformity $\mathcal{U}$ having a base $\{U_\varepsilon : \varepsilon > 0\}$ of $\varepsilon$-neighborhoods of the diagonal, and $(Y, d_Y)$ is a metric space with metric uniformity $\mathcal{V}$ having a base $\{V_\varepsilon : \varepsilon > 0\}$ of $\varepsilon$-neighborhoods of the diagonal. Recall that $(x, y) \in U_\delta$ if and only if $d_X(x, y) < \delta$. Now the definition of $f : (X, d_X) \to (Y, d_Y)$ being uniformly continuous between the metric spaces,

$$\forall \varepsilon > 0 \, \exists \delta > 0 \quad \text{such that } d_X(x, y) < \delta \Rightarrow d_Y(f(x), f(y)) < \varepsilon,$$

can be restated as

$$\forall \varepsilon > 0 \, \exists \delta > 0 \quad \text{such that } (x, y) \in U_\delta \Rightarrow (f(x), f(y)) \in V_\varepsilon.$$

Since every entourage in a uniformity contains a basic entourage, these statements are equivalent to

$$\forall V \in \mathcal{V} \, \exists U \in \mathcal{U} \quad \text{such that } (x, y) \in U \Rightarrow (f(x), f(y)) \in V.$$

We will take this last statement as the definition of uniform continuity between two uniform spaces.

**Definition 12.3.13.** A function $f : (X, \mathcal{U}) \to (Y, \mathcal{V})$ between two uniform spaces is *uniformly continuous* if for every $V \in \mathcal{V}$, there exists $U \in \mathcal{U}$ such that $(x, y) \in U$ implies $(f(x), f(y)) \in V$.

Our discussion above shows that if $X$ and $Y$ are metric spaces with the metric uniformity, the concept of uniform continuity of $f : X \to Y$ is identical whether $f$ is viewed as a function between two metric spaces or as a function between two uniform spaces. However, since not all uniform spaces are metrizable, the definition extends the metric definition of uniform continuity.

A function between topological spaces is continuous if and only if inverse images of open sets are open. Our next result will show that a function between uniform spaces is uniformly continuous if and only if inverse images of entourages are entourages. However, if $f : X \to Y$, the entourages in question are subsets of $Y \times Y$ and $X \times X$, so we must adjust the function $f$ before we can discuss images or inverse images of entourages. For a function $f : X \to Y$, we define

$$(f \times f) : X \times X \to Y \times Y \quad \text{by } (f \times f)(x_1, x_2) = (f(x_1), f(x_2)).$$

**Theorem 12.3.14.** $f : (X,\mathcal{U}) \rightarrow (Y,\mathcal{V})$ *is uniformly continuous if and only if for every* $V \in \mathcal{V}, (f \times f)^{-1}(V) \in \mathcal{U}.$

*Proof.* The result follows from the equivalence of the statements

$$(f \times f)^{-1}(V) \in \mathcal{U} \Longrightarrow \exists U \in \mathcal{U} \quad \text{such that } U \subseteq (f \times f)^{-1}(V)$$
$$\Longleftrightarrow \exists U \in \mathcal{U} \quad \text{such that } (f \times f)U \subseteq V$$
$$\Longleftrightarrow \exists U \in \mathcal{U} \quad \text{such that } (x,y) \in U \Rightarrow (f(x),f(y)) \in V. \qquad \square$$

## Exercises

1. Prove Theorem 12.3.3.
2. Suppose $\mathcal{B}$ is a base for a uniformity $\mathcal{U}$ on a set $X$. In the induced topology, the neighborhood filter of $x \in X$ is $V_x = \{U(x) : U \in \mathcal{U}\}$. Is $\{B(x) : B \in \mathcal{B}\}$ a filter base for $V_x$?
3. Suppose $(X,\mathcal{U})$ is a uniform space. Show that the collection $\mathcal{V}^s = \{U \in \mathcal{U} : U = U^{-1} \text{ and } U \text{ is open in } X \times X\}$ of open symmetric entourages and the collection $\mathcal{C}^s = \{U \in \mathcal{U} : U = U^{-1} \text{ and } U \text{ is closed in } X \times X\}$ of closed symmetric entourages are both bases for $\mathcal{U}$.
4. Suppose $(X,\mathcal{U})$ is a uniform space, $U \in \mathcal{U}$, and $x \in X$.
   (a) Show that $\operatorname{cl}(U(x)) \subseteq U^2(x)$.
   (b) If $U$ is symmetric, show that $\operatorname{cl}(U(x)) \subseteq U^4(y)$ for any $y \in U(x)$.
5. Suppose $(X,\mathcal{U})$ is a uniform space, $U \in \mathcal{U}$, and $x \in X$. Must $\operatorname{cl}(U(x)) = (\operatorname{cl} U)(x)$? Does either inclusion hold?
6. Suppose $(X,\mathcal{U})$ is a uniform space and $R$ is a relation on $X$. Show that $R \subseteq \bigcap\{\operatorname{int}(U \circ R \circ U) : U \in \mathcal{U}^s\}$.
7. If $(X,\mathcal{U})$ is a uniform space, Corollary 12.3.7 shows that every entourage $U \in \mathcal{U}$ is a neighborhood of the diagonal. Show that the converse fails by exhibiting a neighborhood of the diagonal $\Delta_{\mathbb{R}}$ which is not an entourage in the metric uniformity on $\mathbb{R}$ induced by the Euclidean metric.
8. Suppose $(X,\mathcal{U})$ is a uniform space. Prove that $(X, \mathcal{T}_{\mathcal{U}})$ is $T_2$ if and only if $\bigcap \mathcal{U} = \Delta_X$.
9. Suppose $(X,\mathcal{U})$ is a uniform space. Without appealing to Theorem 12.3.9 and without using the defining condition (U4) of a uniformity, prove that $(X, \mathcal{T}_{\mathcal{U}})$ is $T_1$ if and only if $\bigcap \mathcal{U} = \Delta_X$.
10. Let $D_\varepsilon$ be the diamond $\{(x, y) \in \mathbb{R}^2 : |x| + |y| < \varepsilon\}$, $U_\varepsilon = \{(x, y) \in \mathbb{R}^2 : |x - y| < \varepsilon\}$ and $V_\varepsilon = D_{1+\varepsilon} \cup U_\varepsilon$. Show that $\mathcal{B} = \{V_\varepsilon : \varepsilon > 0\}$ is a base for a uniformity on $\mathbb{R}$ which generates a topology which is not $T_1$.
11. Consider $(\mathbb{R}, \mathcal{T}_c)$, where $\mathcal{T}_c$ is the cofinite topology. Find a uniformity $\mathcal{U}$ on $\mathbb{R}$ which induces $\mathcal{T}_c$ or prove that $\mathcal{T}_c$ is not uniformizable.
12. Give a proof or counterexample to this statement: Every compact $T_1$ topological space is uniformizable.

13. Show that the topological space $(\mathbb{R}, \mathcal{T})$ of Example 12.3.12 is separable but not second countable.

14. Suppose $\mathcal{B}$ is a basis for a uniformity $\mathcal{V}$ on $Y$. Prove $f : (X, \mathcal{U}) \to (Y, \mathcal{V})$ is uniformly continuous if and only if for every $B \in \mathcal{B}$, $(f \times f)^{-1}(B) \in \mathcal{U}$.

15. Let $\mathcal{U}$ be the Euclidean metric uniformity on $X = (0, \infty)$. Show that $f : (X, \mathcal{T}_{\mathcal{U}}) \to (X, \mathcal{T}_{\mathcal{U}})$ defined by $f(x) = 1/x$ is not uniformly continuous by finding $V_\varepsilon \in \mathcal{U}$ such that $(f \times f)^{-1}(V_\varepsilon) \notin \mathcal{U}$.

## 12.4 Uniformities and pseudometrics

Given a pseudometric $p$ on $X$ and $\varepsilon > 0$, we may define the "$\varepsilon$-neighborhood of the diagonal"

$$U_\varepsilon = \{(x, y) \in X \times X : p(x, y) < \varepsilon\} = \bigcup_{x \in X} \{x\} \times B(x, \varepsilon),$$

just as we did for a metric. Exactly as for the metric case given in Example 12.2.7, $\{U_\varepsilon : \varepsilon > 0\}$ is a base for a uniformity on $X$, called the *pseudometric uniformity* generated by $p$. The pseudometric uniformity induces the pseudometric topology on $X$. We have seen in Section 6.1 that every metric $d : X \times X \to [0, \infty)$ is continuous, where $X$ carries the metric topology, $X \times X$ the product topology, and $[0, \infty)$ the Euclidean topology. Those proofs remain valid for pseudometrics.

If $\mathcal{U}$ and $\mathcal{W}$ are uniformities on $X$ and $\mathcal{U} \subseteq \mathcal{W}$, then we say $\mathcal{W}$ is *finer* than $\mathcal{U}$ and $\mathcal{U}$ is *coarser* than $\mathcal{W}$. Clearly, a finer uniformity has more entourages and thus produces a finer uniform topology.

If $f : (X, \mathcal{T}) \to Y$ is continuous and $\mathcal{T}_f$ is a topology on $X$ finer than $\mathcal{T}$, then $f : (X, \mathcal{T}_f) \to Y$ is continuous. Combining these observations gives the following result.

**Theorem 12.4.1.** *Suppose $p$ is a pseudometric on $X$ and $\mathcal{U}$ is a uniformity on $X$. If the uniform topology $\mathcal{T}_{\mathcal{U}}$ is finer than the pseudometric topology $\mathcal{T}_p$ on $X$, then $p : (X, \mathcal{T}_{\mathcal{U}})^2 \to [0, \infty)$ is continuous. In particular, if $\mathcal{U}$ is finer than the pseudometric uniformity generated by $p$, then $p : (X, \mathcal{T}_{\mathcal{U}})^2 \to [0, \infty)$ is continuous.*

The hypotheses of the last sentence above are commonly used and are named: given a uniformity $\mathcal{U}$ on $X$, a pseudometric $p$ on $X$ is *uniform with respect to $\mathcal{U}$* if and only if $\mathcal{U}$ is finer than the pseudometric uniformity generated by $p$. That is, $p$ is uniform with respect to $\mathcal{U}$ if and only if for every $\varepsilon > 0$, $U_\varepsilon \in \mathcal{U}$.

Our next goal will be to show that every uniformizable space is $T_{3.5}$. We will use the following lemma.

**Lemma 12.4.2.** *Suppose $\mathcal{U}$ is a uniformity on $X$ and $V_0, V_1, V_2, \ldots$ is a nested sequence of symmetric entourages in $\mathcal{U}$ such that*

$$\cdots \subseteq V_{n+1} \subseteq V_{n+1}^3 \subseteq V_n \subseteq \cdots \subseteq V_2 \subseteq V_2^3 \subseteq V_1 \subseteq V_1^3 \subseteq V_0 = X \times X.$$

Then there exists a pseudometric $p$ on $X$ such that

$$V_n \subseteq \{(x,y) \in X^2 : p(x,y) \le 2^{-n}\} \subseteq V_{n-1} \quad \text{for every } n \in \mathbb{N}.$$

*Proof.* For $x, y \in X$, put

$$f(x,y) = \begin{cases} 0 & \text{if } (x,y) \in \bigcap_{n=1}^{\infty} V_n, \\ 2^{-n} & \text{if } (x,y) \in V_n - V_{n+1}. \end{cases}$$

In particular, if $f(x,y) = 2^{-n}$, then $(x,y) \in V_0, V_1, \dots, V_n$ but $(x,y) \notin V_{n+1}$. Thus, $(x,y) \in V_n$ if and only if $f(x,y) \le 2^{-n}$. Now for $x, y \in X$, define

$$p(x,y) = \bigwedge_{\substack{x_0, x_1, \dots, x_m \in X \\ x = x_0, x_m = y}} \sum_{i=1}^{m} f(x_{i-1}, x_i).$$

It is easy to see that $p(x,x) = 0$ and $p(x,y) = p(y,x)$. The triangle inequality follows immediately from the definition of $p(x,y)$ as an infimum over all paths from $x$ to $y$. Thus, $p$ is a pseudometric. Next, we will prove

$$\frac{1}{2}f(x,y) \le p(x,y) \le f(x,y) \quad \text{for all } x, y \in X. \tag{12.1}$$

The result will follow quickly from these inequalities: Assuming (12.1), $(x,y) \in V_n$ implies $f(x,y) \le 2^{-n}$, so $p(x,y) \le f(x,y) \le 2^{-n}$. Also, if $p(x,y) \le 2^{-n}$, then $\frac{1}{2}f(x,y) \le p(x,y)$ implies $f(x,y) \le 2^{-(n-1)}$, so $(x,y) \in V_{n-1}$.

The second inequality in (12.1) is immediate from the definition of $p(x,y)$ as an infimum of a set containing $f(x,y)$. For the first inequality, we will show that given any path $x = x_0, x_1, \dots, x_m = y$ of length $m$ from $x$ to $y$ in $X$,

$$\frac{1}{2}f(x,y) \le a \quad \text{where } a = \sum_{i=1}^{m} f(x_{i-1}, x_i). \tag{12.2}$$

If $a \ge \frac{1}{2}$, then (12.2) holds since $f(x,y) \le 1$.

Next, we consider the case $a = 0$. Note that

$$a = 0 \Rightarrow f(x_{i-1}, x_i) = 0 \qquad \text{for } i = 1, \dots, m,$$
$$\Rightarrow (x_{i-1}, x_i) \in \bigcap_{n=1}^{\infty} V_n \quad \text{for } i = 1, \dots, m,$$
$$\Rightarrow (x,y) \in V_n^m \qquad \text{for all } n \in \mathbb{N}.$$

By the nested assumption $V_j \subseteq V_j^3 \subseteq V_{j-1}$, it follows that $V_j^{3k} \subseteq V_{j-k}$, and if $m = 3k - r$ for $r \in \{0, 1, 2\}$, then

$$(x,y) \in V_n^m = V_n^{3k-r} \subseteq V_n^{3k} \subseteq V_{n-k} \subseteq V_{n-k-1} \subseteq \cdots \subseteq V_1 \subseteq V_0 \quad \text{for all } n \ge k.$$

Thus, $(x,y) \in V_n$ for all $n \in \mathbb{N}$, so $f(x,y) = 0$ and (12.2) holds.

Finally, suppose $a \in (0, \frac{1}{2})$. We will use induction on $m$. For $m = 1$, we have $\frac{1}{2}f(x,y) \le f(x,y)$ so the inequality holds. For $m = 2$ we have $a = f(x,x_1) + f(x_1,y)$, so both terms are less than or equal to $a$. Let $k \in \mathbb{N}$ be the smallest natural number with $2^{-k} \le a$. If $f(x,x_1) = 2^{-n} \le a < \frac{1}{2}$, then $2 \le k \le n$. The definition of $f$ implies $(x,x_1) \in V_j$ for all $j \le n$, so in particular, $(x,x_1) \in V_k$. If $f(x,x_1) = 0$, then $(x,x_1) \in V_k$. Thus, in both cases $(x,x_1) \in V_k$. The same argument shows $(x_1,y) \in V_k$. Thus, $(x,y) \in V_k^2 \subseteq V_k^3 \subseteq V_{k-1}$, so $f(x,y) \le 2^{-(k-1)}$, and thus $\frac{1}{2}f(x,y) \le 2^{-k} \le a$, proving (12.2) when $m = 2$.

Now suppose (12.2) holds for chains of length $1, 2, \ldots m - 1$ and suppose $x = x_0, x_1, \ldots, x_m = y$ is a chain of length $m \ge 3$ from $x$ to $y$. In this case, we will break the chain into three subchains $x_0, \ldots, x_j$; $x_j, x_{j+1}$; and $x_{j+1}, \ldots x_m$ where $j$ is the smallest natural number which makes the partial sum $\sum_{i=1}^{j+1} f(x_{i-1}, x_i)$ exceed half of the total sum $\sum_{i=1}^{m} f(x_{i-1}, x_i) = a$. Now $\sum_{i=1}^{j} f(x_{i-1}, x_i) \le a/2$ and $\sum_{i=j+2}^{m} f(x_{i-1}, x_i) \le a/2$. Applying (12.2) to these subchains of length less than $m$ gives $\frac{1}{2}f(x,x_j) \le \frac{a}{2}$ and $\frac{1}{2}f(x_{j+1}, x_m) \le \frac{a}{2}$. Thus $f(x,x_j) \le a$ and $f(x_{j+1}, x_m) \le a$. Clearly $f(x_j, x_{j+1}) \le a$, since the sum $a$ contains $f(x_j, x_{j+1})$ as one term. Now (mirroring the case for $m = 2$), let $k \in \mathbb{N}$ be the smallest natural number with $2^{-k} \le a$. If $f(x,x_j) = 2^{-n} \le a < \frac{1}{2}$, then $2 \le k \le n$ and $(x,x_j) \in V_k$. If $f(x,x_j) = 0$, then $(x,x_j) \in V_k$. In either case, we have $(x,x_j) \in V_k$. Similarly, $(x_j, x_{j+1}), (x_{j+1}, y) \in V_k$. Thus, $(x,y) \in V_k^3 \subseteq V_{k-1}$, so $f(x,y) \le 2^{-(k-1)}$, and thus $\frac{1}{2}f(x,y) \le 2^{-k} \le a$, proving (12.2) for paths of length $m$. By mathematical induction, (12.2) holds for every $m \in \mathbb{N}$. □

The lemma above allows us to prove the following important theorem.

**Theorem 12.4.3.** *A $T_1$ topological space $(X, \mathcal{T})$ is uniformizable if and only if it is completely regular.*

*Proof.* Corollary 12.3.11 showed that every completely regular space is uniformizable. Suppose $(X, \mathcal{T})$ is uniformizable and $T_1$. To see $X$ is completely regular, we must show it is $T_{3.5}$. Suppose $A$ is a closed set in $X$ not containing the point $b$. Let $\mathcal{U}$ be a uniformity on $X$ inducing $\mathcal{T}$, and let $V \in \mathcal{U}$ be an open symmetric entourage with $V(b) \cap A = \emptyset$. Take $V_0 = X^2$ and $V_1 = V$. From the properties of a uniformity (see Exercise 13 of Section 12.2), we can construct a sequence of symmetric entourages $V_2, V_3, \ldots$ satisfying the hypotheses of Lemma 12.4.2. Let $p$ be the pseudometric guaranteed by Lemma 12.4.2. Given any $\varepsilon > 0$, for $2^{-n} < \varepsilon$, we have $V_n \subseteq \{(x,y) : p(x,y) \le 2^{-n}\} \subseteq U_\varepsilon = \{(x,y) : p(x,y) < \varepsilon\}$, so $U_\varepsilon \in \mathcal{U}$ for every $\varepsilon > 0$. Thus, by Theorem 12.4.1, $p$ is continuous. Now $\{(x,y) : p(x,y) \le 2^{-n}\} \subseteq V_{n-1}$ for any $n \in \mathbb{N}$, and with $n = 2$ we have $\{(x,y) : p(x,y) < 1/4\} \subseteq \{(x,y) : p(x,y) \le 1/4\} \subseteq V_1 = V$. Thus, $4p(x,y) < 1$ implies $(x,y) \in V$, and thus $y \notin V(b)$ implies $4p(b,y) \ge 1$. Define $g : X \to \mathbb{R}$ by $g(y) = \min\{4p(b,y), 1\}$. Now $g$ is continuous, and $y \in A$ implies $y \notin V(b)$, so $g(y) = 1$. Furthermore, $g(b) = 0$. Thus, $g$ is a continuous function separating $A$ and $b$. □

The proof above actually shows that any uniformizable topological space is $T_{3.5}$, without the use of the $T_1$ property. Our proof that completely regular implies uniformizable used the Stone–Čech compactification, which required the $T_2$ property implied by complete regularity. Thus, in that proof, we cannot drop the $T_1$ condition to show that $T_{3.5}$ implies uniformizable. This is a shortcoming of the proof; other proofs can be used to show that $T_{3.5}$ implies uniformizable, so the uniformizable spaces are precisely the $T_{3.5}$ spaces.

Collecting some previous results, we have the following characterization of completely regular spaces.

**Theorem 12.4.4.** *Let $(X, \mathcal{T})$ be a topological space. The following are equivalent.*
(a) *$(X, \mathcal{T})$ is completely regular.*
(b) *$(X, \mathcal{T})$ is a subspace of a compact Hausdorff space.*
(c) *$(X, \mathcal{T})$ is a separating gauge space.*
(d) *$(X, \mathcal{T})$ is uniformizable by a uniformity $\mathcal{U}$ with $\bigcap \mathcal{U} = \Delta_X$.*

*Proof.* See Corollary 7.3.6 and Theorems 11.1.13, 12.3.9, and 12.4.3. □

## Exercises

1. Suppose $\mathcal{U}$ and $\mathcal{W}$ are uniformities on $X$ inducing the uniform topologies $\mathcal{T}_{\mathcal{U}}, \mathcal{T}_{\mathcal{W}}$ on $X$. We have noted that if $\mathcal{W}$ is finer than $\mathcal{U}$, then $\mathcal{T}_{\mathcal{W}}$ is finer than $\mathcal{T}_{\mathcal{U}}$. Give proofs or counterexamples for the following statements. (Hint: Consider the uniformity of Exercise 11 of Section 12.2.)
   (a) If $\mathcal{W}$ is strictly finer than $\mathcal{U}$, then $\mathcal{T}_{\mathcal{W}}$ is strictly finer than $\mathcal{T}_{\mathcal{U}}$.
   (b) If $\mathcal{T}_{\mathcal{W}}$ is finer than $\mathcal{T}_{\mathcal{U}}$, then $\mathcal{W}$ is finer than $\mathcal{U}$.
2. The function $p((x, y), (a, b)) = |x - a|$ is a pseudometric on $\mathbb{R}^2$. Determine whether it is uniform with respect to the uniformities given below.
   (a) The metric uniformity $\mathcal{U}^{\mathcal{E}}$ from the Euclidean metric on $\mathbb{R}^2$.
   (b) The metric uniformity $\mathcal{U}^T$ from the taxicab metric on $\mathbb{R}^2$.
   (c) The metric uniformity $\mathcal{U}^D$ from the discrete metric on $\mathbb{R}^2$.
3. Let $X$ be the set of integrable functions from $[0, 1]$ to $\mathbb{R}$. The function $p(f, g) = \int_0^1 |f(x) - g(x)| \, dx$ is a pseudometric on $X$. Determine whether $p$ is uniform with respect to the sup-metric uniformity on $X$.
4. Let $X$ be the set of integrable functions from $[0, 1]$ to $\mathbb{R}$. The function $r(f, g) = |f(0) - g(0)|$ is a pseudometric on $X$. Determine whether $r$ is uniform with respect to (a) the sup-metric uniformity on $X$, and (b) the pseudometric uniformity from the pseudometric $p(f, g) = \int_0^1 |f(x) - g(x)| \, dx$.
5. If $K = \{1/n : n \in \mathbb{N}\}$, the collection $\mathcal{B} = \{(a, b) \subseteq \mathbb{R} : a < b\} \cup \{(a, b) - K \subseteq \mathbb{R} : a < b\}$ is a basis for a topology on $\mathbb{R}$ called the *K-topology*. Determine whether or not this topology is uniformizable.
6. Let $X = [0, 1]^2$ with the topology having basis $\mathcal{T}_{\mathcal{E}} \cup \{\{(0, 0)\} \cup (0, \varepsilon)^2 : \varepsilon > 0\}$ where $\mathcal{T}_{\mathcal{E}}$ is the Euclidean topology on $[0, 1]^2$. Determine whether or not this topology is uniformizable.

7.  Suppose $\mathcal{U}$ and $\mathcal{W}$ are uniformities on $X$. If $\bigcap \mathcal{U} = \bigcap \mathcal{W}$, is $\mathcal{T}_{\mathcal{U}} = \mathcal{T}_{\mathcal{W}}$?
8.  Suppose $\mathcal{U}$ and $\mathcal{W}$ are uniformities on $X$. If $\mathcal{T}_{\mathcal{U}} = \mathcal{T}_{\mathcal{W}}$, is $\bigcap \mathcal{U} = \bigcap \mathcal{W}$?

## 12.5 Quasi-uniformities

We have seen applications requiring asymmetric distances. Quasi-metrics dropped the symmetry condition from the definition of a metric. Quasi-metrics are clearly a part of the study of *asymmetric topology*. The $T_1$ property has a certain symmetry to it: given $x \neq y$, there is a neighborhood of $x$ excluding $y$ and a neighborhood of $y$ excluding $x$. The $T_0$ property did not have this symmetry: there may be a neighborhood of $x$ excluding $y$ but no neighborhood of $y$ excluding $x$. Thus, $T_0$ spaces and generally spaces which are not $T_1$ are also a part of asymmetric topology. For classical applications of topology to real analysis, spaces are usually considered to be $T_2$. In application to computer science, spaces of machine numbers or pixels are finite, and the non-discrete topologies on finite sets are not $T_1$. Thus, applications of topology to computer science constitute a significant motivation for and a significant part of asymmetric topology.

Uniform spaces have a built-in symmetry condition. If $U$ is in a uniformity $\mathcal{U}$, then so is $U^{-1}$. If we drop this condition, we get a *quasi-uniformity*. Quasi-uniformities are also important for modeling asymmetric situations.

**Definition 12.5.1.** A *quasi-uniformity* on a nonempty set $X$ is a collection $\mathcal{U}$ of subsets of $X \times X$ such that
(Q1)  $\mathcal{U}$ is a filter on $X \times X$.
(Q2)  $\Delta_X \subseteq U$ for every $U \in \mathcal{U}$.
(Q3)  For every $U \in \mathcal{U}$, there exists $V \in \mathcal{U}$ with $V^2 \subseteq U$.

The *quasi-uniform topology* on $X$ induced by a quasi-uniformity $\mathcal{U}$ is

$$\mathcal{T}_{\mathcal{U}} = \{A \subseteq X : \forall x \in A \ \exists U \in \mathcal{U} \text{ with } U(x) \subseteq A\}.$$

A collection $\mathcal{B}$ is a base for the quasi-uniformity $\mathcal{U}$ if $\mathcal{B} \subseteq \mathcal{U}$ and every $U \in \mathcal{U}$ contains some $B \in \mathcal{B}$.

The proof of Corollary 12.3.2 (and Theorem 12.3.1) did not use the symmetric property of the uniformity, so the result holds for quasi-uniform spaces. For reference, we state it here.

**Theorem 12.5.2.** *If $\mathcal{U}$ is a quasi-uniformity on $X$, then, for each $x \in X$, the $\mathcal{T}_{\mathcal{U}}$-neighborhood filter at $x$ is $\mathcal{V}_x = \{U(x) : U \in \mathcal{U}\}$.*

Exercise 1 shows that the converse of Theorem 12.5.2 fails.
The proof of the following result is straightforward and is left to the exercises.

**Theorem 12.5.3.** *A collection $\mathcal{B}$ of subsets of $X^2$ is a base for some quasi-uniformity if and only if $\mathcal{B}$ is filter base and for every $B \in \mathcal{B}$, there exists $C \in \mathcal{B}$ with $\Delta_X \subseteq C^2 \subseteq B$.*

Recall that a quasiorder dropped the antisymmetry condition from the definition of a partial order. That is, a quasiorder on $X$ is a reflexive, transitive relation $R \subseteq X^2$. Every quasiorder $\leq$ on $X$ gives an Alexandroff topology $\mathcal{T}_\leq$ on $X$ consisting of the increasing sets, and (see Exercise 8 of Section 8.3.) $\mathcal{T}_\leq$ is $T_0$ if and only if $\leq$ is a partial order.

The next result includes some interesting connections between quasiorders and quasi-uniformities.

**Theorem 12.5.4.**
(a) *If $R$ is a quasiorder on $X$, then $\{R\}$ is a base for a quasi-uniformity on $X$.*
(b) *If $\mathcal{U}$ is a quasi-uniformity on $X$, then $\bigcap \mathcal{U}$ is a quasiorder on $X$.*
(c) *A quasi-uniform topology $\mathcal{T}_\mathcal{U}$ is $T_0$ if and only if $\bigcap \mathcal{U}$ is a partial order on $X$.*
(d) *A quasi-uniform topology $\mathcal{T}_\mathcal{U}$ is $T_1$ if and only if $\bigcap \mathcal{U} = \Delta_X$.*

*Proof.* (a) follows directly from the definitions and Theorem 12.5.3.

(b) If $\mathcal{U}$ is a quasi-uniformity on $X$, then $\Delta_X \subseteq U$ for any $U \in \mathcal{U}$, so $\Delta_X \subseteq \bigcap \mathcal{U}$, so $\bigcap \mathcal{U}$ is reflexive. To see it is transitive, suppose $(x, y), (y, z) \in \bigcap \mathcal{U}$. For any $U \in \mathcal{U}$, pick $V \in \mathcal{U}$ with $V^2 \subseteq U$. Now $(x, y), (y, z) \in V$, so $(x, z) \in V^2 \subseteq U$. Thus, $(x, y) \in \bigcap \mathcal{U}$.

(c) Suppose $\mathcal{T}_\mathcal{U}$ is $T_0$ and $(x, y) \in \bigcap \mathcal{U} - \Delta_X$. Then $x \neq y$ and $y \in U(x)$ for every $U \in \mathcal{U}$. By the $T_0$ condition, there must exist $V \in \mathcal{U}$ giving a basic neighborhood $V(y)$ with $x \notin V(y)$. Thus, $(y, x) \notin V$, so $(y, x) \notin \bigcap \mathcal{U}$. Thus, the quasiorder $\bigcap \mathcal{U}$ is antisymmetric and therefore a partial order.

Conversely, suppose $\bigcap \mathcal{U}$ is a partial order and $x \neq y$. If $(x, y) \notin \bigcap \mathcal{U}$, then there exists $U \in \mathcal{U}$ with $(x, y) \notin U$, so $U(x)$ is a $\mathcal{T}_\mathcal{U}$-neighborhood of $x$ which excludes $y$. If $(x, y) \in \bigcap \mathcal{U}$, then antisymmetry implies $(y, x) \notin \bigcap \mathcal{U}$, and, as above, there exists a neighborhood $U(y)$ of $y$ which excludes $x$. Thus, $\mathcal{T}_\mathcal{U}$ is $T_0$.

(d) Suppose $\mathcal{T}_\mathcal{U}$ is $T_1$. Given $x \neq y$, there exists $U \in \mathcal{U}$ with $y \notin U(x)$. Thus, $(x, y) \notin U$, so $(x, y) \notin \bigcap \mathcal{U}$. Since $\Delta_X \in \bigcap \mathcal{U}$, it follows that $\bigcap \mathcal{U} = \Delta_X$.

Conversely, suppose $\bigcap \mathcal{U} = \Delta_X$. Now if $x \neq y$, then $(x, y) \notin \bigcap \mathcal{U}$, so there exists $U \in \mathcal{U}$ with $(x, y) \notin U$, and $U(x)$ is a neighborhood of $y$ which excludes $x$. □

Recall that every quasi-metric $q$ on $X$ generates a $T_0$ topology on $X$ and a partial order $\leq$ on $X$ defined by $x \leq y$ if and only if $q(x, y) = 0$. The connection between quasi-metrics and quasi-uniformities is given in the theorem below. The proof is left to the exercises.

**Theorem 12.5.5.** *Suppose $q$ is a quasi-metric on $X$. For $\varepsilon > 0$, let $U_\varepsilon = \{(x, y) \in X \times X : q(x, y) < \varepsilon\}$. Then the collection $\mathcal{B} = \{U_\varepsilon : \varepsilon > 0\}$ is a base for a quasi-uniformity $\mathcal{U}$ on $X$, and $\bigcap \mathcal{U}$ is the partial order $\leq$ defined by $x \leq y$ if and only if $q(x, y) = 0$.*

Recall that a topology is uniformizable if and only if it is $T_{3.5}$. The next theorem, proven by William Pervin in 1962, gives us a remarkable result on which topologies are quasi-uniformizable: Every topology is quasi-uniformizable. The quasi-uniformity introduced in the proof is called the *Pervin quasi-uniformity*.

**Theorem 12.5.6.** *Every topology is induced by a quasi-uniformity.*

*Proof.* If $\mathcal{T}$ is a topology on $X$, for $A \in \mathcal{T}$, let $U_A = (A \times A) \cup (X - A \times X)$, as shown in Figure 12.3, and let $\mathcal{S} = \{U_A : A \in \mathcal{T}\}$. We will show that $\mathcal{S}$ is a subbasis for a quasi-uniformity $\mathcal{U}$; that is, the collection $\mathcal{B}$ of finite intersections of elements of $\mathcal{S}$ is a basis for a quasi-uniformity $\mathcal{U}$, and then we will see that $\mathcal{T}_{\mathcal{U}} = \mathcal{T}$.

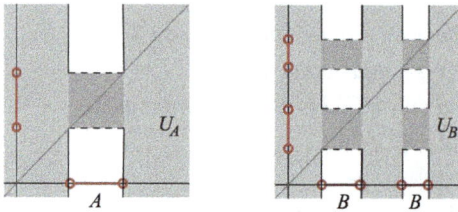

**Figure 12.3:** Subbasic elements of the Pervin quasi-uniformity.

Clearly $\Delta_X \subseteq U_A$ for every $U_A \in \mathcal{S}$, and thus $\Delta_X \subseteq B$ for every $B \in \mathcal{B}$, and $\emptyset \notin \mathcal{B}$. $\mathcal{B}$ is closed under finite intersections, since a finite intersection of finite intersections of elements of $\mathcal{S}$ is a finite intersection of elements of $\mathcal{S}$. Thus, $\mathcal{B}$ is a filter base. To see $\mathcal{B}$ is a base for a quasi-uniformity, by Theorem 12.5.3, it only remains to show that, for any $B \in \mathcal{B}$, there exists $C \in \mathcal{B}$ with $C^2 \subseteq B$. We first show that if $A \in \mathcal{T}$, then $U_A^2 = U_A$. Every reflexive relation $U$ has $U \subseteq U^2$, so we only need to show $U_A^2 \subseteq U_A$. Suppose $(x,y) \in U_A^2$. If $x \in X - A$, then $\{x\} \times X \subseteq U_A$, so $(x,y) \in U_A$. If $x \in A$, then $(x,y) \in U_A^2$ implies $(x,z),(z,y) \in U_A$ for some $z \in X$. But $x \in A$ and $(x,z) \in U_A$ implies $z \in A$, and then $z \in A$ and $(z,y) \in U_A$ implies $y \in A$. Thus, $(x,y) \in A \times A \subseteq U_A$, and this completes the proof that $U_A^2 = U_A$. Now if $B \in \mathcal{B}$, then $B = U_{A_1} \cap \cdots \cap U_{A_n}$ for $A_1, \ldots, A_n \in \mathcal{T}$, and

$$B^2 \subseteq U_{A_1}^2 \cap \cdots \cap U_{A_n}^2 = U_{A_1} \cap \cdots \cap U_{A_n} = B.$$

Thus, for $C = B \in \mathcal{B}$, we have $C^2 \subseteq B$.

Having shown that $\mathcal{B}$ is a base for some quasi-uniformity $\mathcal{U}$ (called the Pervin quasi-uniformity), it remains to show that the topology $\mathcal{T}_{\mathcal{U}}$ induced by this quasi-uniformity is $\mathcal{T}$. Given any nonempty $A \in \mathcal{T}$ and any $a \in A$, $U_A \in \mathcal{U}$ so the slice $U_A(a) = A$ is in $\mathcal{T}_{\mathcal{U}}$. Thus, $\mathcal{T} \subseteq \mathcal{T}_{\mathcal{U}}$. Now suppose $A \in \mathcal{T}_{\mathcal{U}}$ and $a$ is an arbitrary point in $A$. To see $A \in \mathcal{T}$, we must find a $\mathcal{T}$-neighborhood of $a$ contained in $A$. In $\mathcal{T}_{\mathcal{U}}$, $a$ has a neighborhood base $\{B(a) : B \in \mathcal{B}\}$, so there exists $B \in \mathcal{B}$ with $a \in B(a) \subseteq A$. Now $B = U_{A_1} \cap \cdots \cap U_{A_n}$ where $A_1, \ldots, A_n \in \mathcal{T}$, so $B(a) = U_{A_1}(a) \cap \cdots \cap U_{A_n}(a)$. For each $i = 1, \ldots, n$, the slice $U_{A_i}(a)$ is either $X$ or $A_i$, and thus is $\mathcal{T}$-open. It follows that $B(a)$ is a $\mathcal{T}$-neighborhood of $a$ contained in $A$. Since $a \in A$ was arbitrary, $A \in \mathcal{T}$, so $\mathcal{T}_{\mathcal{U}} \subseteq \mathcal{T}$. □

Quasi-uniformities are explored in depth in [17]. There are many excellent sources on uniform spaces, including [8, 16].

## Exercises

1. Consider the topology $\mathcal{T}$ on $\mathbb{R}$ generated by the basis $\mathcal{B} = \{[x, x + \varepsilon) : x \in \mathbb{Q}, \varepsilon > 0\} \cup \{(x-\varepsilon, x] : x \in \mathbb{R} - \mathbb{Q}, \varepsilon > 0\}$. In Example 12.3.12, we saw that this topology is uniformizable. Finding a uniformity (or even a quasi-uniformity) which generates $\mathcal{T}$ is not as simple as one might expect. For $\varepsilon > 0$, define

$$U_\varepsilon = \left(\bigcup_{x\in\mathbb{Q}} \{x\} \times [x, x + \varepsilon)\right) \cup \left(\bigcup_{x\in\mathbb{R}-\mathbb{Q}} \{x\} \times (x - \varepsilon, x]\right),$$

and let $\mathcal{U} = \{W \subseteq \mathbb{R}^2 : \exists \varepsilon > 0 \text{ with } U_\varepsilon \subseteq W\}$. Note that $\{U(x) : U \in \mathcal{U}\}$ is a basis for $\mathcal{T}$.
   (a) Show that, for any $\varepsilon > 0$, $U_\varepsilon^2$ contains the Euclidean $\varepsilon/2$ neighborhood of the diagonal $\bigcup_{x\in X}\{x\} \times (x - \varepsilon/2, x + \varepsilon/2)$.
   (b) Show that $\mathcal{U}$ is not a quasi-uniformity.
   (Compare to Exercise 8 of Section 11.2.)

2. Prove Theorem 12.5.3: A collection $\mathcal{B}$ of subsets of $X^2$ is a base for some quasi-uniformity if and only if $\mathcal{B}$ is filter base and for every $B \in \mathcal{B}$, there exists $C \in \mathcal{B}$ with $\Delta_X \subseteq C^2 \subseteq B$.

3. Describe a quasi-uniformity on $\mathbb{R}$ whose induced topology is the lower limit topology $\mathbb{R}_l$. Verify that this really is a quasi-uniformity.

4. Suppose $\mathcal{U}$ is a quasi-uniformity on $X$.
   (a) Show that $\mathcal{U}^{-1} = \{U^{-1} : U \in \mathcal{U}\}$ is also a quasi-uniformity on $X$ (called the *conjugate quasi-uniformity*).
   (b) Show that the quasi-uniformity $U$ is a uniformity if and only if $\mathcal{U} = \mathcal{U}^{-1}$.
   (c) Show that $\mathcal{U} \cup \mathcal{U}^{-1}$ is a subbasis for a uniformity $\mathcal{U}^s$ on $X$.

5. Exercise 9 of Section 12.2 showed that if a quasi-uniformity $\mathcal{U}$ is a uniformity, then $\bigcap \mathcal{U}$ is an equivalence relation. Does the converse hold? That is, give a proof or counterexample to the statement: If $\mathcal{U}$ is a quasi-uniformity and $\bigcap \mathcal{U}$ is an equivalence relation, then $\mathcal{U}$ is in fact a uniformity.

6. Prove Theorem 12.5.5.

7. Given a topology $\mathcal{T}$ on $X$, with $U_A = (A \times A) \cup (X - A \times X)$, the collection $\mathcal{S} = \{U_A : A \in \mathcal{T}\}$ is a subbasis for the Pervin quasi-uniformity on $X$. If $\mathcal{B}_\mathcal{T}$ is a basis for $\mathcal{T}$, is $\mathcal{S} = \{U_B : B \in \mathcal{B}_\mathcal{T}\}$ a subbasis for the Pervin quasi-uniformity?

8. For each topology below, draw some typical subbasic elements $U_A$ for the Pervin quasi-uniformity, as suggested in Figure 12.3.
   (a) The cofinite topology $\mathcal{T}_{cf}$ on $[0, \infty)$.
   (b) The discrete topology on $\mathbb{R}$.
   (c) The right ray topology on $\mathbb{R}$.

# 13 Continuous deformation of sets and curves

## 13.1 Continuous deformation of closed sets in $\mathbb{R}^2$

Using some suggestive terminology we will define below, Figure 13.1 shows two dynamic closed sets $A$ and $B$ in the plane which move from being disjoint to *meeting* and then to *overlapping*. Would it be possible for dynamic closed sets $A$ and $B$ in the plane to move from being disjoint directly to overlapping, without first meeting? Such questions require a careful definition of meeting, overlapping, and continuous deformation, and the answers may depend on what kinds of sets (closed, compact, connected, etc.) are permitted.

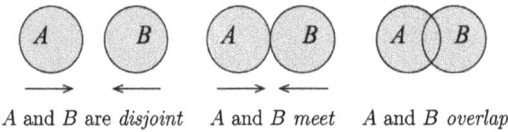

$A$ and $B$ are *disjoint*    $A$ and $B$ *meet*    $A$ and $B$ *overlap*

**Figure 13.1:** A transition from disjoint to meets to overlaps.

Applications to geography have motivated much research on how pairs of planar sets may be related. In geographical studies, pairs of sets may represent the habitats of a predator and of its prey, or a wheat growing region and a drought region. Keeping with these geographical motivations, throughout this section we will mainly consider pairs of closed sets of the plane, although many of our definitions are stated in more generality.

The nature of the intersection between two sets $A$ and $B$ may be classified according to whether their boundaries intersect, their interiors intersect, and whether the boundary of either set intersects the interior of the other. (See [4, 15].)

**Definition 13.1.1.** For sets arbitrary $X_1, X_2, X_3, X_4$, define $\chi(X_1, X_2, X_3, X_4) = (x_1, x_2, x_3, x_4)$ where $x_i = 0$ if $X_i = \emptyset$ and $x_i = 1$ if $X_i \neq \emptyset$ $(i = 1, 2, 3, 4)$. Given two sets $A$, $B$ in a topological space, their *4-intersection value* is the binary 4-tuple

$$\chi(\partial A \cap \partial B, \text{int } A \cap \text{int } B, \partial A \cap \text{int } B, \text{int } A \cap \partial B).$$

The 4-intersection values realized by subsets of the Euclidean plane which are homeomorphic to closed disks have suggestive names, given in Table 13.1.

For arbitrary closed sets $A$ and $B$ in the plane, all 16 possible 4-intersection values may be realized. The 4-intersection value $(0, 1, 1, 1)$ is realized by connected sets $A = [0, 2] \times \mathbb{R}$ and $B = [1, 3] \times \mathbb{R}$, and by disconnected sets $C = [-2, 2]^2 \cup ([5, 7] \times [-1, 1])$ and $D = \bar{B}((6, 0), 2) \cup \bar{B}((0, 0), 1)$, where $\bar{B}(x, \varepsilon) = \{z \in \mathbb{R}^2 : d(x, z) \leq \varepsilon\}$, as depicted in Figure 13.2. The sets $E = F = \mathbb{R}^2$ have 4-intersection value $(0, 1, 0, 0)$.

https://doi.org/10.1515/9783110686579-014

**Table 13.1:** 4-intersection values realized by homeomorphic copies of closed disks.

| $\chi(\partial A \cap \partial B, \text{int } A \cap \text{int } B, \partial A \cap \text{int } B, \text{int } A \cap \partial B)$ | | | |
|---|---|---|---|
| Disjoint $(0,0,0,0)$ | Meet $(1,0,0,0)$ | Overlap $(1,1,1,1)$ | Equal $(1,1,0,0)$ |
| $A$ $B$ | $A$ $B$ | $A$ $B$ | $A=B$ |
| Covers $(1,1,0,1)$ | Covered by $(1,1,1,0)$ | Contains $(0,1,0,1)$ | Inside $(0,1,1,0)$ |
| $A$ $B$ | $A$ $B$ | $A$ $B$ | $A$ $B$ |

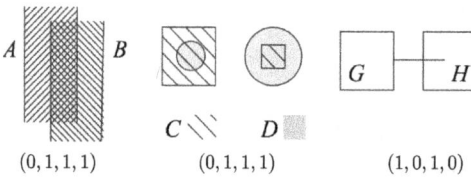

$$(0,1,1,1) \qquad (0,1,1,1) \qquad (1,0,1,0)$$

**Figure 13.2:** More 4-intersection values.

The 4-intersection value $(1,0,1,0)$ is realized with $G = [-1,1]^2 \cup ([1,3] \times \{0\})$ and $H = [2,4] \times [-1,1]$, as seen in Figure 13.2. Note that the set $G$ has a "whisker" which introduces boundary points of $G$ which are not boundary points of int $G$. We may wish to prohibit sets with whiskers.

**Definition 13.1.2.** A subset $A$ of a topological space $X$ is a *regular closed set* if $A = \text{cl}(\text{int } A)$.

A regular closed set in the plane can have no "whiskers". It is left to the exercises to show that, for a regular closed set $A$, $\partial A = \partial(\text{int } A)$ and $(1,0,1,0)$ and several other 4-intersections values cannot be realized with regular closed sets $A$ and $B$.

Let us return to the idea of dynamically changing closed sets $A(t)$ and $B(t)$ in the Euclidean plane. Figure 13.1 shows that it is possible for disjoint sets $A$ and $B$ (with value $(0,0,0,0)$) to morph to sets which meet (with value $(1,0,0,0)$) without passing through any other intermediate values. In this situation, we will say the values $(0,0,0,0)$ and $(1,0,0,0)$ are *adjacent*. Explicitly, 4-intersection values $(x,y,z,w)$ and $(x',y',z',w')$ are *adjacent* if there exist dynamic sets $A(t)$ and $B(t)$ moving continuously over a time interval $(a,b)$ containing $t_0$, with $A(t)$, $B(t)$ having value $(x,y,z,w)$ for $t \in (a,t_0)$, having value $(x',y',z',w')$ for $t \in (t_0,b)$, and either value $(x,y,z,w)$ or $(x',y',z',w')$ at $t_0$. Figure 13.1 also shows that the 4-intersection value $(1,0,0,0)$ (meets) is adjacent to $(1,1,1,1)$ (overlaps). The question presented there was whether disjoint $(0,0,0,0)$ is adjacent to overlaps $(1,1,1,1)$.

Before we formally define what is meant by sets $A(t)$ and $B(t)$ moving continuously, we will present some motivating examples and questions.

First, we present some dynamic sets $A(t)$ and $B(t)$ which move directly from disjoint to overlapping, skipping the meets stage.

**Example 13.1.3.** (a) Let $A$ be the stationary set $[-1,1]^2$. For $t < 0$, let $B(t) = [2,4]^2$. For $t \geq 0$, let $B(t) = [0,2]^2$. Now $A(t)$ and $B(t)$ are disjoint for $t < 0$ and overlap for $t \geq 0$, but the set $B(t)$ hardly appears to be moving continuously.

(b) Let $A$ be the stationary set $\{(x,y) \in \mathbb{R}^2 : x \geq 1, y \geq 1/x\}$ and for time $t \in \mathbb{R}$, let $B(t) = \mathbb{R} \times (-\infty, t]$. In any reasonable definition of continuous motion, we would expect that $A(t) = A$ and $B(t)$ are moving continuously. For $t \leq 0$, $A(t)$ and $B(t)$ are disjoint. For $t > 0$, $A(t)$ and $B(t)$ overlap. There is no instant when the boundaries of $A(t)$ and $B(t)$ intersect but their interiors do not.

(c) For all $t \in \mathbb{R}$, let $A(t) = A = [2,4] \times [-1,1]$, and let $B(t) = [-1,1]^2 \cup ([1,3] \times \mathrm{cl}(-t,t))$. Note that, for $t \leq 0$, the interval $(-t,t)$ is empty, so $B(t) = [-1,1]^2$ is disjoint from $A(t)$. For $t > 0$, $B(t)$ contains $[1,3] \times [-t,t]$, which causes the boundary and interior of $B$ to intersect the boundary and interior of $A$. Thus, $A(t)$ and $B(t)$ are disjoint for $t \leq 0$ and overlap for $t > 0$. Note the importance of using $\mathrm{cl}(-t,t)$ instead of $[-t,t]$ in this construction. At $t = 0$, $\mathrm{cl}(-t,t)$ is empty, while $[-t,t]$ is not. Had we used $[-t,t]$ in the definition of $B(t)$, then at $t = 0$, $B(t)$ would have a whisker and $A(0)$ and $B(0)$ would meet. Using $\mathrm{cl}(-t,t)$, however, $A(t)$ and $B(t)$ are regular closed sets at all times.

One might try to use continuity of the area of closed sets $A(t)$ and $B(t)$ to help quantify continuous motion. However, note that in Example 13.1.3(a), the area $a(t)$ of $A(t)$ and $b(t)$ of $B(t)$ are constantly 4 and thus are continuous functions. However, the area $ab(t)$ of $A(t) \cap B(t)$ is not continuous. It jumps from 0 for $t < 0$ to 1 for $t \geq 0$.

Example 13.1.3(b) clearly depends on the asymptotic unbounded nature of the sets. That is, this example clearly depends on the non-compactness of $A(t)$ and $B(t)$, and brings up the question of whether examples exist using compact sets. (Not only are $A$ and $B$ unbounded, but they have infinite area.)

The sets $A(t)$ and $B(t)$ of Example 13.1.3(c) are regular closed and compact for all $t \in \mathbb{R}$, and not only are the areas $a(t)$ and $b(t)$ of $A(t)$ and $B(t)$ continuous, but the area $ab(t)$ of $A(t) \cap B(t)$ is also continuous. While these areas are continuous, the sudden appearance of the extension $[1,3] \times \mathrm{cl}(-t,t)$ for $t > 0$ suggests some form of discontinuous morphing. For $t = 0$, $B(t)$ is contained in the open set $(-2,2)^2$, but there is no neighborhood of $t = 0$ over which $B(t)$ remains in $(-2,2)^2$. This motivates our next definition.

**Definition 13.1.4.** A function $B : \mathbb{R} \rightarrow \mathcal{P}(\mathbb{R}^2)$ is *upper semicontinuous* (or *u.s.c.*, or *upper Vietoris continuous*) at $t_0$ if for every open set $U \subseteq \mathbb{R}^2$ with $B(t_0) \subseteq U$, there exists $\delta > 0$ such that $|t - t_0| < \delta$ implies $B(t) \subseteq U$.

A function $B : \mathbb{R} \to \mathcal{P}(\mathbb{R}^2)$ is *lower semicontinuous* (or *l. s. c.*, or *lower Vietoris continuous*) at $t_0$ if for every open set $U \subseteq \mathbb{R}^2$ with $B(t_0) \cap U \neq \emptyset$, there exists $\delta > 0$ such that $|t - t_0| < \delta$ implies $B(t) \cap U \neq \emptyset$.

If $B : \mathbb{R} \to \mathcal{P}(\mathbb{R}^2)$ is both u. s. c. and l. s. c. at $t_0$, then it is said to be *Vietoris continuous* at $t_0$. A function $B$ is *upper semicontinuous* (*lower semicontinuous, Vietoris continuous*) if it is upper semicontinuous (lower semicontinuous, Vietoris continuous) at each point of its domain.

Now from our discussion above, the set $B(t)$ of Example 13.1.3(c) is not u. s. c. at $t = 0$. Upper semicontinuity prevents a set from expanding suddenly outside any given neighborhood of the set. Lower semicontinuity prevents a set from suddenly shrinking. For example, the function

$$B(t) = \begin{cases} [-3,3]^2 & \text{if } t \leq 0, \\ [-1,1]^2 & \text{if } t > 0, \end{cases}$$

is u. s. c. at every point $t \in \mathbb{R}$, but is not l. s. c. at $t = 0$. The open set $U = (2,4)^2$ intersects $B(0)$ but there is no neighborhood of $t = 0$ over which $B(t)$ always intersects $U$.

Our next example shows that even if $A(t)$ and $B(t)$ are Vietoris continuous, they may go directly from disjoint to overlapping without first meeting.

**Example 13.1.5.** $A(t)$ will be a comb space, starting (at $t = 1$) with just the base of the comb $A(1) = [0,1] \times \{0\}$. As $t$ decreases from 1 to $\frac{1}{2}$, $A(t)$ grows three teeth from the base of the comb at $x = 0$, $x = \frac{1}{2}$, and $x = 1$. The heights of the teeth grow continuously from 0 to 1, so $A(\frac{1}{2}) = A(1) \cup \{0, \frac{1}{2}, 1\} \times [0,1]$. This initial step of tooth growth grows not only the interior tooth at $x = \frac{1}{2}$, but also the two outer teeth at $x = 0$ and $x = 1$. As $t$ decreases from $t = \frac{1}{2}$ to $\frac{1}{4}$, $A(t)$ grows two new teeth from the base, at the midpoints of the bases of the existing teeth. As $t$ decreases from $t = \frac{1}{4}$ to $\frac{1}{8}$, $A(t)$ grows four new teeth from the midpoints of bases of existing teeth, and so on. When $t \leq 0$, set $A(t) = [0,1]^2$.

Formally, $A(t)$ is given by

$$A(t) = [0,1] \times \{0\} \quad \text{for } t \geq 1,$$
$$A(2^{-n}) = A(1) \cup (\{2^{-n}m : m = 0,1,\ldots,2^n\} \times [0,1]) \quad \text{for } n \in \mathbb{N},$$
$$A(t) = A(2^{-n}) \cup (\{2^{-n-1}m : m = 0,1,\ldots,2^{n+1}\} \times [0, 2 - 2^{n+1}t])$$
$$\text{for } t \in [2^{-n-1}, 2^{-n}], n \in \mathbb{N} \cup \{0\},$$
$$A(t) = [0,1]^2 \quad \text{for } t \leq 0.$$

It is easy to visualize that $A(t)$ is both u. s. c. and l. s. c., but has a discontinuous jump in area at $t = 0$.

Let $B(t)$ be the reflection of $A(t)$ over the line $y = \frac{7}{8}$ translated to the left by $s(t)$ where $s(t)$ is the piecewise linear function with $s(t) = 0$ for $t \leq 0$, $s(t) = \frac{1}{2}$ for $t \geq 1$, and $s(2^{-n}) = 2^{-n-1} =$ half the distance between existing teeth at time $t = 2^{-n}$, for $n \in \mathbb{N}$. Now $A(t)$ and $B(t)$ are compact-valued u. s. c. and l. s. c. functions with $A(t) \cap B(t) = \emptyset$

for $t > 0$ and $A(t) \cap B(t) = [0,1] \times [\frac{3}{4}, 1]$ for $t \le 0$. In particular, $A(t)$ and $B(t)$ transform from disjoint to overlapping without "meeting" first.

For $t > 0$, the sets $A(t)$ and $B(t)$ are never regular closed, but it is easy to visualize that by adding positive shrinking width to the base and teeth of the combs, we can achieve the same results with regular closed sets, each homeomorphic to a closed disk.

Another illustrative example can be constructed using spirals.

**Example 13.1.6.** For $t \in [0,1)$, let $A(t) = \{(r, \theta) : r = (1 - t)\theta, 0 \le \theta \le (1 - t)^{-1}\}$ and let $B(t) = \{(r, \theta) : r = (1 - t)(\theta + \pi), 0 \le \theta \le (1 - t)^{-1} - \pi\}$. For $t \ge 1$, put $A(t) = B(t) = \{(r, \theta) : r \le 1\}$. For $t \in [0,1)$, $A(t)$ and $B(t)$ are disjoint Archimedian spirals, winding tighter around the origin with an increasing number of coils in the unit circle as $t$ approaches 1, as suggested in Figure 13.3. It is easy to see that the functions $A(t)$ and $B(t)$ are u. s. c. and l. s. c. and assume compact values for $t \ge 0$, with $A(t)$ and $B(t)$ disjoint for $t \in [0,1)$ and equal for $t \ge 1$. Thus, disjoint and equal are adjacent using u. s. c. and l. s. c. compact-valued functions. Again, these sets are not regular closed for $t \in [0,1)$, but it is easy to see that the spirals could be thickened slightly to make each homeomorphic to a closed disk.

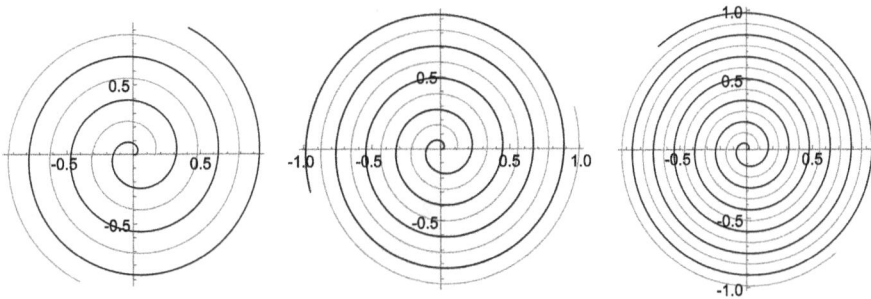

**Figure 13.3:** Spirals $A(t)$ and $B(t)$ for $t = .95, .965$, and $.975$.

In the transformations from disjoint to overlapping or equal, we pass instantly from int $A \cap$ int $B = \emptyset$ to int $A \cap$ int $B \ne \emptyset$. Two ways this may occur are that the interior of $B$ may instantly appear inside int $A$, as in Example 13.1.3(c), or int $A$ may engulf int $B$ as in Examples 13.1.5 and 13.1.6. In the former case, $B(t)$ was not upper semicontinuous, and in the latter, the areas of $A(t), B(t)$, and $A(t) \cap B(t)$ were not continuous. If $A(t)$ and $B(t)$ are disjoint, upper semicontinuity will prevent int $B$ from instantly appearing inside int $A$, and continuity of the areas will prevent int $B$ from being engulfed by int $A$. This is our next theorem.

**Theorem 13.1.7.** *Suppose $A$ and $B$ are u. s. c. functions on $\mathbb{R}$, with values $A(t)$ and $B(t)$ which are closed subsets of $\mathbb{R}^2$ with finite areas, and suppose the area of $A(t) \cap B(t)$ is* *a continuous function of t. Then the 4-intersection value disjoint* $(0,0,0,0)$ *is not adjacent to 4-intersection values of form* $(x,1,z,w)$. *That is, the 4-intersection value disjoint* $(0,0,0,0)$ *is not adjacent to any 4-intersection value with* int $A \cap$ int $B \neq \emptyset$.

*Proof.* If $(0,0,0,0)$ is adjacent to $(x,1,z,w)$ by closed subsets $A(t), B(t)$ changing upper semicontinuously with the area of $A(t) \cap B(t)$ continuous, then, by restricting, rescaling, and possibly reversing the time interval, we may assume that one of the following two cases hold:

(a) int $A(t) \cap$ int $B(t) \neq \emptyset$ for $t < 0$ and $A(t) \cap B(t) = \emptyset$ for $t \geq 0$, or
(b) int $A(t) \cap$ int $B(t) \neq \emptyset$ for $t \leq 0$ and $A(t) \cap B(t) = \emptyset$ for $t > 0$.

That is, either there is a first instant of being disjoint or there is a last instant of int $A(t) \cap$ int $B(t) \neq \emptyset$.

In case (a), since $A(0)$ and $B(0)$ are disjoint closed sets in the normal space $\mathbb{R}^2$, there exist disjoint open sets $G_A$ and $G_B$ with $A \subseteq G_A$ and $B \subseteq G_B$. By the u. s. c. condition, there exists $\delta > 0$ such that $t \in (-\delta, 0]$ implies $A(t) \subseteq G_A$ and $B(t) \subseteq G_B$. This contradicts the hypotheses of (a), so (a) cannot occur.

In case (b), if $a(t)$ is the area of $A(t) \cap B(t)$, then we have $a(t) = 0$ for $t > 0$ and $a(0) \neq 0$. Thus the inverse image $a^{-1}(\{0\})$ of the closed set $\{0\}$ contains $(0, \infty)$ but not 0, and therefore is not closed. This contradicts the continuity of $a$, and thus case (b) cannot occur. □

## Exercises

1. Find all 4-intersection values which are possible between $A = \mathbb{R}^2$ and $B \subseteq \mathbb{R}^2$. Prove that your list is complete and give examples for each 4-intersection value which is realized.

2. If $A$ is a subset of a topological space, show that $\partial(\text{int } A) \subseteq \partial A$, and equality holds if and only if $A \subseteq \text{cl}(\text{int } A)$. In particular, note that equality holds if $A$ is open or is regular closed.

3. Show that 4-intersection values of form $(x, 0, 1, w)$ and $(x, 0, z, 1)$ (where $x, z, w \in \{0, 1\}$) cannot be realized with regular closed sets $A$ and $B$.

4. Illustrate pairs of sets in the plane which have 4-intersection values $(0,0,0,1)$, $(0,0,1,0)$, $(0,0,1,1)$, $(1,0,0,1)$, and $(1,0,1,1)$.

5. Describe modifications of the spiral spaces given in Example 13.1.6 to show the 4-intersection value $(0,0,0,0)$ (disjoint) is adjacent to the 4-intersection values $(1,1,0,1)$ (covers) and $(0,1,0,1)$ (contains) using compact sets $A(t)$ and $B(t)$ which are Vietoris continuous.

6. Show that if $A(t)$ and $B(t)$ are regular closed sets for all $t \in \mathbb{R}$ and the areas of $A(t), B(t)$, and $A(t) \cap B(t)$ are continuous, positive functions, then the 4-intersection values $(0,0,0,0)$ (disjoint) and $(1,1,0,1)$ (covers) are not adjacent.

7. In any topological space $X$, show that if $C$ is connected and $C \cap \partial A = \emptyset$, then $C \subseteq \mathrm{int}\, A$ or $C \cap \mathrm{cl}\, A = \emptyset$.

8. Use Exercise 7 to show that if $A$ and $B$ are nonempty connected subsets of the plane, then they have 4-intersection value $(0, 1, 0, 1)$ if and only if $B \subset \mathrm{int}\, A$.

9. (a) Use Exercise 7 to show that if $A$ and $B$ are subsets of $\mathbb{R}^2$ which are homeomorphic to the closed unit disk and have 4-intersection value $(1, 1, 0, 0)$, then $A = B$.

   (b) Give an example of distinct closed sets $A, B$ of $\mathbb{R}^2$ with 4-intersection value $(1, 1, 0, 0)$.

## 13.2 Continuous deformation of planar curves

In Examples 13.1.3(c) and 13.1.5 of dynamically changing sets which go directly from disjoint to overlapping, the lengths of the boundaries of the sets are discontinuous. The length of the boundary of a set may not be a good measure of the convergence of the sets. The triangular region $A = \{(x, y) \in [0, 1]^2 : y \le x\}$ may be approximated by the region $A_n$ obtained by replacing the diagonal side of $A$ by a staircase of $n$ steps with height and width $1/n$. It is reasonable to say that in some sense, $(A_n)_{n=1}^{\infty}$ converges to $A$, but the length of the boundary of $A_n$ is 4 for every $n \in \mathbb{N}$, while the length of the boundary of $A$ is $2 + \sqrt{2}$.

We now introduce a method for describing continuous deformation of curves in the plane.

**Definition 13.2.1.** The interval $[0, 1]$ with the Euclidean topology will be denoted by $I$. In a topological space $(X, \mathcal{T})$, a *path* (or *curve*) from $a$ to $b$ is a continuous function $f : I \to X$ with $f(0) = a$ and $f(1) = b$. A path from $a$ to $a$ is called a *closed curve* or *loop* based at $a$. A curve $f$ is a *simple curve* if $f$ is one-to-one. A closed curve $f$ is a *simple closed curve* if $f$ is one-to-one on $[0, 1)$. A simple closed curve in the Euclidean plane is a *Jordan curve*.

In practice, a curve or path in $(X, \mathcal{T})$ may be thought of either as the function $f$ from $I$ to $X$ or as the image $f(I)$ as a subset of $X$. Simple curves do not intersect themselves. An important theorem about simple closed planar curves was given by Camille Jordan in 1887. While the result is entirely expected, the proof is surprisingly difficult and is omitted.

**Theorem 13.2.2** (Jordan curve theorem). *A simple closed curve in $\mathbb{R}^2$ divides the plane into three regions: the curve $C$, the bounded region $\mathrm{In}(C)$ inside the curve, and the unbounded region $\mathrm{Out}(C)$ outside the curve. Furthermore, the region $\mathrm{In}(C)$ is homeomorphic to an open ball in $\mathbb{R}^2$.*

Jordan only gave the first sentence of the theorem above. The second sentence was shown by Arthur Schönflies in 1906. The combined results are often called the

Jordan–Schönflies theorem. In particular, note that the region inside a Jordan curve is open and path connected. A Jordan curve together with the region inside (or outside) of it is a regular closed set.

If $F : I^2 \to \mathbb{R}^2$ is a continuous function, then, for each fixed $t_0 \in I$, $F(t_0, x) : I \to \mathbb{R}^2$ is continuous, and thus is a path in $\mathbb{R}^2$. As $t_0$ ranges from 0 to 1, the paths $F(t_0, x)$ move continuously in the plane, since $F$ is continuous. Thus, a function $F : I^2 \to \mathbb{R}^2$ provides a continuous morphing of the path $F(0, x)$ to the path $F(1, x)$. Such a function $F$ is a *homotopy* between the paths.

**Definition 13.2.3.** Suppose $Y$ is a topological space. Two paths $f, g : I \to Y$ in $Y$ are *path homotopic* if there exists a continuous function $F : I^2 \to Y$ with $F(0, x) = f(x)$ and $F(1, x) = g(x)$ for all $x \in I$. The function $F$ is a *path homotopy* from $f$ to $g$. More generally, if $X$ and $Y$ are topological spaces and $f$ and $g$ are continuous functions from $X$ to $Y$, then $f$ is *homotopic* to $g$ if there exists a continuous function $F : I \times X \to Y$ (called a *homotopy*) with $F(0, x) = f(x)$ and $F(1, x) = g(x)$ for all $x$.

Clearly, path homotopies are special cases of homotopies; since we will only consider path homotopies, we refer to them as simply as homotopies. Figure 13.4 depicts a homotopy between two paths.

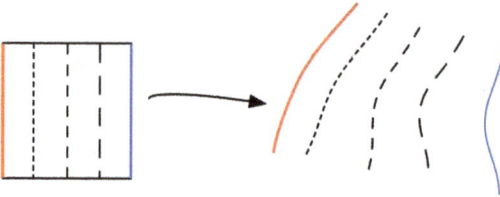

**Figure 13.4:** A homotopy.

In the plane, any paths $f$ and $g$ are homotopic. This can be shown by the *straight-line homotopy* $H : I^2 \to \mathbb{R}^2$ defined by $H(t, x) = (1-t)f(x)+tg(x)$. Note that, for a fixed $x_0 \in I$, $H(t, x_0)$ traces the straight line from $f(x_0)$ to $g(x_0)$ as $t$ goes from 0 to 1. Figure 13.5 shows the straight-line homotopy from the unit circle $f(x) = (\cos(2\pi x), \sin(2\pi x))$ to the line segment $g(x) = (2 - 4x, 2)$ for $x \in I$.

The next theorem is a fundamental result on homotopies.

**Theorem 13.2.4.** *On the set $C$ of paths in a topological space $(Y, \mathcal{T})$, the relation $f \approx g$ if and only if $f$ is homotopic to $g$ is an equivalence relation.*

*Proof.* Note that $f \in C$ if and only if $f : I \to (Y, \mathcal{T})$ is continuous. Define $F : I^2 \to Y$ by $F(t, x) = f(x)$. For any open set $V \subseteq Y$, $F^{-1}(V) = I \times f^{-1}(V)$, which is open in $I^2$, so $F$ is a homotopy from $f$ to $f$, and thus $\approx$ is reflexive.

If $F : I^2 \to Y$ is a homotopy with $F(0, x) = f(x)$ and $F(1, x) = g(x)$, then $G(t, x) = F(1 - t, x)$ is a homotopy with $G(0, x) = g(x)$ and $G(1, x) = f(x)$, so $\approx$ is symmetric.

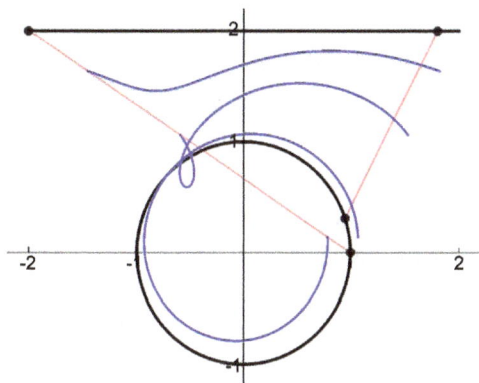

**Figure 13.5:** The straight-line homotopy from the unit circle to a line segment.

Suppose $f \approx g$ and $g \approx h$. Then there exist homotopies $F$ from $f$ to $g$ and $G$ from $g$ to $h$. To get a homotopy $H : I^2 \to Y$ from $f$ to $h$, we should trace the homotopy $F$ for $t \in [0, 0.5]$ and trace $G$ for $t \in [0.5, 1]$. The function that does this is

$$H(t, x) = \begin{cases} F(2t, x) & \text{if } t \in [0, 0.5], \\ G(2t - 1, x) & \text{if } t \in [0.5, 1]. \end{cases}$$

As the composition of continuous functions, $F(2t, x)$ and $G(2t - 1, x)$ are continuous, and $H$ is continuous by the pasting lemma. Thus, $f \approx h$, so $\approx$ is transitive.  □

With the "is homotopic to" equivalence relation, the equivalence class of a curve $f : I \to Y$ in $Y$ is called the *homotopy class* of $f$.

If $F : I^2 \to \mathbb{R}^2$ is a homotopy from $f(x) = F(0, x)$ to $g(x) = F(1, x)$ and $F(t, 0) = F(t, 1)$ for all $t \in I$, then each slice $t = t_0$ defines a closed curve $F(t_0, x)$, which bounds a region $A(t_0)$ in the plane. If additionally $F$ is one-to-one on each $\{t_0\} \times [0, 1)$, then each slice $t = t_0$ defines a simple closed curve, which by the Jordan curve theorem, bounds a well-defined region homeomorphic to an open ball. The union of the bounding curve and the enclosed region gives a closed and bounded region in the plane. Such sets are Lebesgue measurable, so they have a well-defined finite area. Thus, such a homotopy provides a continuous deformation of planar sets whose boundaries are simple closed curves.

**Definition 13.2.5.** If $A : I \to \mathcal{P}(\mathbb{R}^2)$ and $A(t)$ is compact and $\partial A(t)$ is a simple closed curve for every $t \in I$, we say the sets $A(t)$ are *changing homotopically* if there is a homotopy $F : I^2 \to \mathbb{R}^2$ with $F(t, x) = \partial A(t)$ for every $t \in I$.

Our next result says that if sets $A(t)$ are changing homotopically, $x$ is in the interior of $A(0)$, and $x$ never intersects the boundary of any $A(t)$, then $x$ is in the interior of all the sets $A(t)$.

**Lemma 13.2.6.** *If $A : I \to \mathcal{P}(\mathbb{R}^2)$ is changing homotopically, $x$ is inside $A(t_0)$, and $x \notin \partial A(t)$ for $t \in (t_0 - \delta, t_0 + \delta)$, then $x$ is inside $A(t)$ for all $t \in (t_0 - \delta, t_0 + \delta)$. Thus, if a set $S$ is inside $A(t_0)$ and $S \cap \partial A(t) = \emptyset$ for $t \in (t_0 - \delta, t_0 + \delta)$, then $S$ is inside $A(t)$ for all $t \in (t_0 - \delta, t_0 + \delta)$.*

*Proof.* A transition from $x$ inside $A(t_0)$ to outside $A(t)$ on an interval is possible if and only if a transition from outside to inside is possible on an interval, so it suffices to show that a point starting outside $A(t_0)$ remains outside. Suppose $F : I^2 \to \mathbb{R}^2$ is a homotopy with $F(t, x) = \partial A(t)$ for all $t \in I$, $x$ is outside $A(t_0)$, and $x \notin F(t, I) = \partial A(t)$ for $t \in (t_0 - \delta, t_0 + \delta)$. Let $U$ be a neighborhood of $x$ contained in $\mathrm{Out}(\partial A(t_0))$. Now $\mathbb{R}^2$ is a subspace of its one-point compactification $\mathbb{R}^2 \cup \{\infty\}$, which, by stereographic projection, is homeomorphic to the sphere $S^2 = \{(x, y, z) \in \mathbb{R}^3 : x^2 + y^2 + z^2 = 1\}$, where $\infty$ corresponds to the north pole $p = (0, 0, 1) \in S^2$. Thus, we may view $F$ as a map from $I^2$ to $S^2$. Compose $F : I^2 \to S^2$ with a homeomorphism $h$ on $S^2$ which translates $x$ to the pole $p$ and note that $h$ preserves the set of outside points of each boundary curve. Now the neighborhood $h(U)$ of the point $p$ at infinity avoids the boundaries $h(\partial A(t))$ for $t \in (t_0 - \delta, t_0 + \delta)$, and thus is contained in the unbounded outside of those curves. Translating back through $h^{-1}$ shows that $x$ remains outside all the curves $\partial A(t) = F(t, I)$ for $t \in (t_0 - \delta, t_0 + \delta)$. $\square$

For the following result, we will use the fact that if $C$ is a closed subset of the plane $\mathbb{R}^2$ and $\varepsilon > 0$ is given, then there exists an open set $G$ with $C \subseteq G$ such that the area of $G - C$ is less than $\varepsilon$. This believable result follows from the fact that every closed set in $\mathbb{R}^2$ is Lebesgue measurable, and a set $C \subseteq \mathbb{R}^2$ is Lebesgue measurable if and only if for every $\varepsilon > 0$, there exist a closed set $F$ and an open set $G$ with $F \subseteq C \subseteq G$ such that the measure of $G - F$ is less than $\varepsilon$. We may apply this result to any Jordan curve $C$, since the continuous image $C \subseteq \mathbb{R}^2$ of the compact set $I$ is compact and thus closed.

**Theorem 13.2.7.** *If sets $A(t)$ in the plane are changing homotopically, then $A$ is u. s. c. and l. s. c., and has continuous area.*

*Proof.* Suppose $F : I^2 \to \mathbb{R}^2$ is a homotopy with $F(t, x) = \partial A(t)$ for every $t \in I$. To see $A$ is u. s. c., suppose $U \subseteq \mathbb{R}^2$ is an open set containing $A(t_0)$. By the hypotheses, $A$ is compact-valued, so we may assume $U$ is bounded, for otherwise, we may replace $U$ by $U \cap B((0, 0), N)$ where $A \subseteq B((0, 0), N)$. Now $F(t_0, I) = \partial A(t_0) \subseteq A(t_0) \subseteq U$, so $V = F^{-1}(U)$ is an open set in $I^2$ containing the slice $\{t_0\} \times I$. By the compactness of $I$, the tube lemma implies $V$ contains an open tube $(t_0 - \varepsilon, t_0 + \varepsilon) \times I$ around the slice $\{t_0\} \times I$. Thus, for $t \in (t_0 - \varepsilon, t_0 + \varepsilon)$, $F(t, I) = \partial A(t) \subseteq U$. Since $\partial A(t) \subseteq U$ is bounded and $\mathrm{Out}(\partial A(t))$ is unbounded, it follows that $\mathrm{In}(\partial A(t)) \subseteq U$. That is, if the boundary of $A(t)$ remains in the bounded open set $U$, then the closed bounded region $A(t)$ remains inside $U$ for $t \in (t_0 - \varepsilon, t_0 + \varepsilon)$. Thus, $A$ is u. s. c.

The proof that $A$ is l. s. c. is left to the exercises.

To show that the area $a(t)$ of $A(t)$ is continuous at each $t_0 \in I$, suppose a neighborhood $(a(t_0) - \varepsilon, a(t_0) + \varepsilon)$ of $a(t_0)$ is given. Let $U_\varepsilon$ be an open set of area less than

$\varepsilon$ which contains $\partial A(t_0) = F(t_0, I)$. Now $F^{-1}(U_\varepsilon)$ is open in $I^2$ and, as above, contains a tube $(t_0 - \delta, t_0 + \delta) \times I$. For $t \in (t_0 - \delta, t_0 + \delta)$, we have $\partial A(t) = F(t, I) \subseteq U_\varepsilon$, so $A(t_0) - U_\varepsilon \subseteq A(t) \subseteq A(t_0) \cup U_\varepsilon$, and thus $|a(t) - a(t_0)| < \varepsilon$. $\qquad\square$

Our next result shows that sets moving homotopically cannot go directly from disjoint to overlapping.

**Theorem 13.2.8.** *If $A$ and $B$ are changing homotopically, then the 4-intersection value $(0, 0, 0, 0)$ (disjoint) is not adjacent to any 4-intersection value of form $(x, 1, z, w)$.*

*Proof.* The proof is similar to that of Theorem 13.1.7. Suppose to the contrary that disjoint is adjacent to $(x, 1, z, w)$. With a restriction, rescaling, and possible reversing of the time interval, we reduce the problem to one of the two following cases:
(a) $\operatorname{int} A(t) \cap \operatorname{int} B(t) \neq \emptyset$ for $t \in [0, 1/2)$ and $A(t) \cap B(t) = \emptyset$ for $t \in [1/2, 1]$
(b) $\operatorname{int} A(t) \cap \operatorname{int} B(t) \neq \emptyset$ for $t \in [0, 1/2]$ and $A(t) \cap B(t) = \emptyset$ for $t \in (1/2, 1]$.

That is, either there is a first instant $t = 1/2$ when $A(t)$ and $B(t)$ are disjoint or there is a last instant $t = 1/2$ of $\operatorname{int} A \cap \operatorname{int} B \neq \emptyset$. The case (a) follows as in the proof of Theorem 13.1.7, using the normality of $\mathbb{R}^2$ and the upper semicontinuity of $A$ and $B$. In case (b), since $\operatorname{int} A(1/2) \cap \operatorname{int} B(1/2) \neq \emptyset$, there exists $x \in \operatorname{int} A(1/2) \cap \operatorname{int} B(1/2)$. Since $x$ is not in the closed set $\partial A(1/2) \cup \partial B(1/2)$, by regularity of the plane, there exist disjoint open sets $U$ and $V$ with $x \in U$ and $\partial A(1/2) \cup \partial B(1/2) \subseteq V$. If $F$ is a homotopy with $F(t, I) = \partial A(t)$, since $V$ is an open neighborhood of $\partial A(1/2) = F(1/2, I)$, $F^{-1}(V)$ is an open set in $I^2$ containing $\{1/2\} \times I$, and thus containing a tube $(1/2 - \delta, 1/2 + \delta) \times I$. Hence, for $t \in (1/2 - \delta, 1/2 + \delta)$, $F(t, I) = \partial A(t) \subseteq V$ and it follows that $\partial A(t)$ is disjoint from $U$. Since $U$ starts in the interior of $A(1/2)$ and does not cross $\partial A(t)$ for $t \in [1/2, 1/2 + \delta)$, $U$ remains in the interior of $A(t)$ for $t \in [1/2, 1/2 + \delta)$. The same argument applies for $B$, showing that $U$ remains in the interior of $B(t)$ for $t \in [1/2, 1/2 + \delta')$ for some $\delta' > 0$. Now for all $t$ strictly between $1/2$ and $1/2 + \min\{\delta, \delta'\}$, we have $\emptyset \neq U \subseteq \operatorname{int} A(t) \cap \operatorname{int} B(t)$, contrary to the choice of $t = 1/2$ as the last instant when the interiors intersected. $\qquad\square$

## Exercises

1.  The depiction of the straight-line homotopy in Figure 13.5 suggests that the point $(1, 0)$ is mapped by straight lines to both $(-2, 2)$ and $(2, 2)$. Explain why this does not contradict the definition of the homotopy being a function.

2.  Consider the unit circle $f(x) = (\cos(2\pi x), \sin(2\pi x))$ and the line segment $g(x) = (2 - 4x, 2)$ for $x \in I$, as depicted in Figure 13.5. If $f$ and $g$ are considered to be paths in $\mathbb{R}^2 - \{(0, 1.5)\}$, explain why the straight-line homotopy from $f$ to $g$ in $\mathbb{R}^2$ is not a homotopy $F : I^2 \to \mathbb{R}^2 - \{(0, 1.5)\}$ from $f$ to $g$ in $\mathbb{R}^2 - \{(0, 1.5)\}$. Prove that $f$ and $g$ are homotopic in $\mathbb{R}^2 - \{(0, 1.5)\}$ by exhibiting a homotopy.

3. Show that the boundary of the square $[0,1]^2$ is homotopic to the boundary of the triangle having vertices $(0,0), (1,0), (1,1)$.

4. Suppose $f$ and $g$ are paths in $(X, \mathcal{T})$ from $a$ to $b$. Show that $f$ and $g$ are homotopic. (Compare with Exercise 4 of Section 13.3.)

5. Suppose $F : I^2 \to \mathbb{R}$ is a homotopy.
   (a) Define $G : I^2 \to I \times \mathbb{R}$ by $G(t,x) = (x, F(t,x))$. Thus, $G(t_0, x)$ maps $I$ onto the graph of $F(t_0, x)$. Show that $G$ is continuous.
   (b) Show that the sequence of functions $f_n = F(1/n, x)$ converges uniformly to $F(0, x)$.

6. Suppose $F : I^2 \to \mathbb{R}$ is a homotopy and for $t_0 \in I$, $A(t_0) = \int_0^1 F(t_0, x)\, dx$. Show that this defines a continuous function $A : I \to \mathbb{R}$.

7. Suppose $C$ is a rectifiable Jordan curve of length $L$ parametrized by arc length, so $C = f([0, L])$ for a continuous length-preserving function $f : [0, L] \to \mathbb{R}^2$. Without appealing to Lebesgue measure, show that, for any $\varepsilon > 0$, there exists an open set $G$ of area less than $\varepsilon$ with $C \subseteq G$.

8. Exhibit homotopies which show that the sets $A(t)$ and $B(t)$ of Figure 13.1 are moving homotopically.

9. Suppose a unit circle is moving in the plane so that its center traces a semicircular arc of radius 4 units. Exhibit a homotopy showing that the unit circles are moving homotopically.

10. Prove that if sets $A(t)$ are changing homotopically, then $A : [0,1] \to \mathcal{P}(\mathbb{R}^2)$ is lower semicontinuous.


## 13.3 The fundamental group

In this section, we will associate to any given topological space a group, called the fundamental group. Homeomorphic topological spaces will produce isomorphic fundamental groups, so the fundamental group is a topological property. In particular, two topological spaces whose fundamental groups are not isomorphic cannot be homeomorphic. The fundamental group is based on curves in the topological space.

Suppose $X$ is a topological space and $a \in X$. Recall that a *loop* in $X$ based at $a$ is a continuous function $f : I \to X$ with $f(0) = f(1) = a$. We say two loops $f, g$ based at $a$ are *homotopic with fixed base point* if there is a homotopy $F : I^2 \to X$ from $f$ to $g$ with $F(t, 0) = F(t, 1) = a$ for all $t \in I$. As in Theorem 13.2.4, it is easy to show that among the loops in $X$ based at $a$, "homotopic with fixed base" is an equivalence relation. An equivalence class in this relation will be called a *homotopy class of loops based at $a$*, or simply a *homotopy class*. The collection of all homotopy classes of loops in $X$ based at $a$ will be denoted $\pi_1(X, a)$.

If $f, g : I \to X$ are loops in $X$ based at $a$, we define the *product $fg : I \to X$* by

$$fg(t) = \begin{cases} f(2t) & \text{if } t \in [0, 0.5], \\ g(2t - 1) & \text{if } t \in [0.5, 1]. \end{cases} \tag{13.1}$$

Thus, $fg$ traces $f$ at double speed for $t \in [0, 0.5]$, then traces $g$ at double speed for $t \in [0.5, 1]$. By the pasting lemma, $fg$ is continuous, and thus is a loop in $X$ based at $a$. It is easy to show that, for any loops $f, g, h$ based at $a$, $(fg)h$ is homotopic (with the same fixed base) to $f(gh)$.

Recall that a *group* is a set $G$ with an operation $\cdot$ such that

(a) $(a \cdot b) \cdot c = a \cdot (b \cdot c)$ for all $a, b, c \in G$ (associativity);

(b) there exists $e \in G$ such that $a \cdot e = e \cdot a = a$ for all $a \in G$ (existence of an *identity element*);

(c) for every $a \in G$, there exists $a^{-1} \in G$ such that $a \cdot a^{-1} = a^{-1} \cdot a = e$ (existence of inverses).

Two groups $(G, \cdot)$ and $(H, *)$ are *isomorphic*, denoted $G \cong H$, if there exists a bijection $h : G \to H$ which *preserves the operation*, that is, such that $h(a \cdot b) = h(a) * h(b)$ for all $a, b \in G$. The *trivial group* is the group $\{e\}$ consisting of one element. A group $(G, \cdot)$ is *abelian* if and only if the operation is commutative (that is, $a \cdot b = b \cdot a$ for all $a, b \in G$).

**Theorem 13.3.1.** *The collection $\pi_1(X, a)$ of homotopy classes of loops in $X$ based at $a$, with the operation $[f] \cdot [g] = [fg]$, is a group, called the* fundamental group *of $X$ based at $a$.*

*Proof.* First we will show that the operation $[f] \cdot [g] = [fg]$ is a well-defined operation on $\pi_1(X, a)$. Suppose $[f] = [f']$ and $[g] = [g']$. Then there is a homotopy $F$ based at $a$ from $f$ to $f'$ and a homotopy $G$ based at $a$ from $g$ to $g'$. Define $H : I^2 \to X$ by

$$H(t, x) = \begin{cases} F(t, 2x) & \text{if } x \in [0, 0.5], \\ G(t, 2x - 1) & \text{if } x \in [0.5, 1]. \end{cases}$$

Note that $H$ restricted to $I \times [0, 0.5]$ traces the homotopy $F$ with each path $F(t_0, x)$ traced at double speed, while $H$ restricted to $I \times [0.5, 1]$ similarly provides a copy of $G$. Since $H$ is a homotopy based at $a$ from $fg$ to $f'g'$, we have $[fg] = [f'g']$. Now $([f] \cdot [g]) \cdot [h]$ traces $f$ and $g$ for $t \in [0, 0.5]$, and $h$ for $t \in [0.5, 1]$, while $[f] \cdot ([g] \cdot [h])$ trace $f$ for $t \in [0, 0.5]$, then $g$ and $h$ for $t \in [0.5, 1]$. It is straightforward to rescale the time parameters to show that the two resulting loops based at $a$ are homotopic, and thus determine the same homotopy class in $\pi_1(X, a)$.

The identity element in $\pi_1(X, a)$ is the homotopy class of the constant path $e(x) = a$. Given $[f] \in \pi_1(X, a)$, $fe$ traces $f$ in the first half second and remains constant for the last half second. The homotopy

$$F(t, x) = \begin{cases} f(\frac{2x}{1+t}) & \text{if } x \in [0, \frac{1+t}{2}], \\ a & \text{if } x \in [\frac{1+t}{2}, 1], \end{cases}$$

traces $f$ for the first half second and $a$ for the last half second when $t = 0$, and as $t$ increases, it traces $f$ on a longer interval and $a$ on a shorter interval until $t = 1$, when it traces $f$ once for $t \in [0,1]$. This homotopy based at $a$ shows $[fe] = [f]$. Similarly, $[ef] = [f]$.

Given a loop $f : I \rightarrow X$ based at $a$, we claim that the inverse of $[f] \in \pi_1(X, a)$ is the homotopy class of $f^{\leftarrow}(t) = f(1 - t)$. Note that $f^{\leftarrow}$ is simply $f$ traced in reverse. Now $ff^{\leftarrow}$ traces the loop $f$ from $a$ to $a$ then retraces that path in reverse direction back to $a$. To see the constant function $e(x) = a$ is homotopic to $ff^{\leftarrow}$, we use the continuous function $F : I^2 \rightarrow X$ such that, for $t \in I$, the loop $F(t, I)$ traces the first 100t % of $f$ for $x \in [0, 0.5]$, then retraces that path in reverse for $x \in [0.5, 1]$, as suggested in Figure 13.6. The homotopy that does this is

$$F(t,x) = \begin{cases} f(2tx) & \text{if } x \in [0, 0.5], \\ f(2t(1-x)) & \text{if } x \in [0.5, 1]. \end{cases} \qquad \square$$

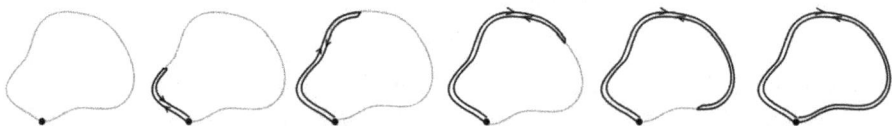

**Figure 13.6:** A homotopy from $a$ to $ff^{\leftarrow}$.

If $f \in \pi_1(X, a)$ is a loop in $X$ based at $a$, observe that the straight-line "homotopy" from $ff^{\leftarrow}$ to $a$ will not generally show that $ff^{\leftarrow}$ is homotopic to $a$. If $f$ is as shown in Figure 13.6 and $X$ is a subset of $\mathbb{R}^2$ which excludes some points inside the loop $f$, then the straight lines from $f(x)$ to $a$ are not all contained in $X$, and thus the straight-line "homotopy" is not a function from $I^2$ to $X$, and thus not a homotopy into $X$. Since $[f^{\leftarrow}] = [f]^{-1}$ in $\pi_1(X, a)$, $f^{\leftarrow}$ is sometimes denoted $f^{-1}$.

Paths, such as $ff^{\leftarrow}$ for $f \in \pi_1(X, a)$, which can be continuously shrunk to a point have a special name.

**Definition 13.3.2.** A path in $(X, \mathcal{T})$ which is homotopic to a constant path $e(x) = a$ is said to be *null homotopic*. In particular, $f \in \pi_1(X, a)$ is null homotopic if and only if $[f]$ is the identity element $[e]$ in $\pi_1(X, a)$.

**Example 13.3.3.**
(a) Any two paths in $\mathbb{R}^2$ are homotopic by the straight-line homotopy, so every loop based at $a \in \mathbb{R}^2$ is null homotopic, and thus $\pi_1(\mathbb{R}^2, a)$ is the trivial group $\{[e]\}$.
(b) If $A = \{(x, y) \in \mathbb{R}^2 : 1 \le x^2 + y^2 \le 3\}$, and $a = (2, 0) \in X$, it is not the case that every loop in $A$ based at $a$ is null homotopic. The path $f_1(x) = (2\cos(2\pi x), 2\sin(2\pi x))$ loops once around the hole of the annulus $A$, and cannot be shrunk to the base point $a$, staying in $A$. Figure 13.7 shows three members of $[f_1]$, that is, three loops in $A$ based at $a = (2, 0)$ which are homotopic to $f_1$. Any two loops in $A$ which circle the origin exactly $n$ times in the same orientation are homotopic. The homotopy

classes of $\pi_1(A, a)$ are characterized by the orientation (positive is counterclockwise, negative is clockwise) and how many times the loops circles around the origin. Thus, $\pi_1(A, a)$ is the additive group of integers $(\mathbb{Z}, +)$. For any choice of the base point $a$ in the annulus, the fundamental group based at $a$ will be $\mathbb{Z}$. We can conclude that the annulus $A$ is not homeomorphic to $\mathbb{R}^2$ since, for every base point $a \in \mathbb{R}^2$, $\pi_1(\mathbb{R}^2, a)$ is the trivial group and is not $\mathbb{Z}$.

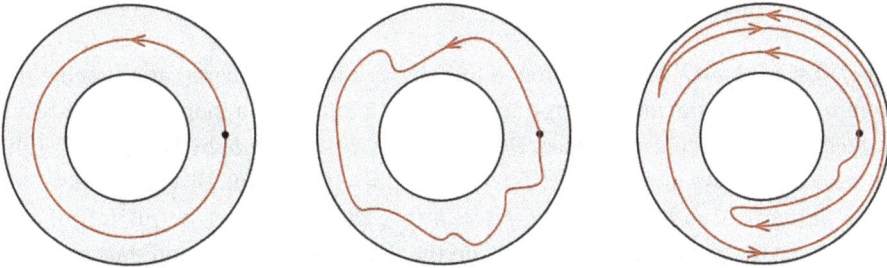

**Figure 13.7:** The loop $(2\cos(2\pi x), 2\sin(2\pi x))$ and two paths homotopic to it in the annulus.

(c) If $S^1 = \{(x, y) \in \mathbb{R}^2 : x^2 + y^2 = 1\}$, then as in (b), for any $a \in S^1$, $\pi_1(S^1, a) = (\mathbb{Z}, +)$. Clearly $S^1$ is not homeomorphic to the annulus, even though they have the same fundamental group.

(d) Recall that a torus is a surface homeomorphic to $S^1 \times S^1$. If $a$ is any point on a torus $T$, then the homotopy classes of paths based at $a$ are determined by how may radial and how many axial loops are made, and in which orientation. To see that $\pi_1(T, a)$ is an abelian group, it suffices to show that an axial loop $a$ and a radial loop $r$ commute. A homotopy from $ar$ to $ra$ is easily seen when viewing $T$ as a quotient of $[0, 2] \times [0, 1]$, as suggested in Figure 13.8. Thus, $\pi_1(T, a) = \mathbb{Z} \times \mathbb{Z}$.

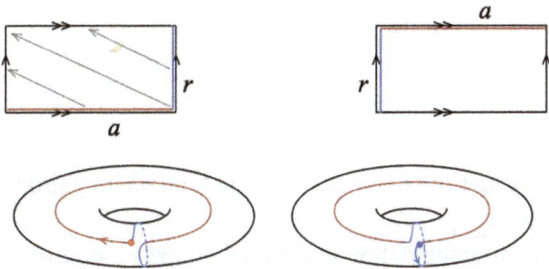

**Figure 13.8:** The fundamental group of the torus is Abelian.

In the examples above, the base point of the fundamental group was not important: any base point gave the same fundamental group. This is not always the case. For example, let $X$ be the union of an annulus $A$ and a unit disk $D$ in the plane, with $A \cap D = \emptyset$. Now for any $a \in A$, loops based at $A$ must be contained in $A$, and $\pi_1(X, a) = (\mathbb{Z}, +)$. However, for any $d \in D$, $\pi_1(X, d)$ is the trivial group. The base point chosen is significant in this example since loops based at $a \in A$ cannot reach points in $D$ and loops based at $d \in D$ cannot reach points in $A$. This suggests the following result.

**Theorem 13.3.4.** *If there is a path from $a$ to $b$ in the topological space $X$, then $\pi_1(X, a) \cong \pi_1(X, b)$.*

*Proof.* Let $p : I \rightarrow X$ be a path from $a$ to $b$. If $f$ is a loop in $X$ based at $a$, tracing $p^{\leftarrow}$ from $b$ to $a$, then tracing $f$, then tracing $p$ from $a$ to $b$ gives a loop based at $b$. This suggests the isomorphism needed. Define $h : \pi_1(X, a) \rightarrow \pi_1(X, b)$ by $h([f]) = [p^{\leftarrow}(fp)]$. By associativity, we may define $h([f]) = [p^{\leftarrow}][f][p] = [p]^{-1}[f][p]$. Given $[g] \in \pi_1(X, b)$, $[g] = h([pgp^{\leftarrow}])$, so $h$ is onto. If $h([f]) = h([g])$, then $[p]^{-1}[f][p] = [p]^{-1}[g][p]$, and multiplying by $[p]$ on the left and $[p]^{-1}$ on the right gives $[f] = [g]$ and thus $h$ is one-to-one. To see that $h$ preserves the operation, note that $h([f][g]) = [p]^{-1}[f][g][p] = [p]^{-1}[f][p][p]^{-1}[g][p] = h([f])h([g])$. $\square$

**Corollary 13.3.5.** *If $(X, \mathcal{T})$ is path connected and $a, b \in X$, then $\pi_1(X, a)$ is isomorphic to $\pi_1(X, b)$.*

As a consequence, in a path connected topological space $(X, \mathcal{T})$, we may speak of *the fundamental group* $\pi_1(X)$, omitting any reference to the base point for the loops.

The following definition is frequently used in complex analysis.

**Definition 13.3.6.** A topological space $(X, \mathcal{T})$ is *simply connected* if it is path connected and every loop is null homotopic.

Thus, a simply connected space is a path connected space with trivial fundamental group. The trivial fundamental group condition is loosely described as "having no holes".

## Exercises

1. Provide explicit details to show that the operation $\cdot$ on the fundamental group $\pi_1(X, a)$ is associative.
2. For an arbitrary $n \in \mathbb{N}$, give a parametrization of a loop $f_n(\theta)$ based at $(2, 0)$ in the annulus $X = \{(x, y) \in \mathbb{R}^2 : 1 \le x^2 + y^2 \le 3\}$ which circles the origin $n$ times counterclockwise and crosses itself no more than $n$ times.
3. Does the converse of Theorem 13.3.4 hold? That is, if $a$ and $b$ are points in a topological space $(X, \mathcal{T})$ with $\pi_1(X, a) \cong \pi_1(X, b)$, then is there a path from $a$ to $b$ in $X$? Prove your answer.

4. Paths $f$ and $g$ in $(X, \mathcal{T})$ from $a$ to $b$ are said to be *homotopic by based paths* if there is a homotopy $F$ from $f$ to $g$ with $F(t, 0) = a$ and $F(t, 1) = b$ for all $t \in I$. Show that if for every $a, b \in X$, every pair of paths $f, g$ from $a$ to $b$ in $(X, \mathcal{T})$ are homotopic by based paths, then every loop in $(X, \mathcal{T})$ is null homotopic. Does the result hold if for every $a \neq b$ in $X$, every pair of paths $f, g$ from $a$ to $b$ in $(X, \mathcal{T})$ are homotopic by based paths? (Compare to Exercise 4 of Section 13.2.)

5. Is the fundamental group of the figure-eight space $Y = \{(x, y) : x^2 + y^2 = 1\} \cup \{(x, y) : (x - 2)^2 + y^2 = 1\}$ isomorphic to the fundamental group of the torus? Justify your answer.

6. Discuss the fundamental groups of the following path connected subsets of the Euclidean plane.

    (a) $X = \{(x, y) : x^2 + y^2 \le 1\} \cup \{(x, y) : (x - 2)^2 + y^2 \le 1\}$
    (b) $Y = \{(x, y, z) \in \mathbb{R}^3 : x^2 + y^2 + z^2 = 1\}$
    (c) $Z = \{(x, y, z) \in \mathbb{R}^3 : x^2 + y^2 + z^2 \le 1\}$
    (d) $W = \{(x, y, z) \in \mathbb{R}^3 : 1 \le x^2 + y^2 + z^2 \le 4\}$
    (e) $C = \{(x, y, z) \in \mathbb{R}^3 : x^2 + y^2 = 1, 0 \le z \le 1\}$

7. Suppose $X$ and $Y$ are topological spaces, $x \in X$, and $y \in Y$. Show that $\pi_1(X \times Y, (x, y)) \cong \pi_1(X, x) \times \pi_1(Y, y)$. Recall that, for groups $(G, \cdot)$ and $(H, *)$, $G \times H$ has group operation $(g_1, h_1)(g_2, h_2) = (g_1 \cdot g_2, h_1 * h_2)$.

## 13.4 The Vietoris topology and the Hausdorff metric

In Section 13.2, we discussed ways to quantify continuous deformation of nonempty closed sets $A(t) \subseteq \mathcal{P}(\mathbb{R}^2)$. A dynamically changing set $A(t)$ is a function $A : I \to \mathcal{P}(\mathbb{R}^2)$, where the time interval $I \subseteq \mathbb{R}$ carries the Euclidean topology. To formally define continuity of the function $A$, we must define a topology on the codomain $\mathcal{P}(\mathbb{R}^2)$ of the function $A$.

Dynamically changing sets are examples of *set-valued functions* or *multifunctions*, that is, functions $f : X \to \mathcal{P}(Y)$ for some sets $X, Y$. For example, if $\mathcal{U}$ is a uniformity on $X$, then every entourage $U \in \mathcal{U}$ defines a set-valued function $U : X \to \mathcal{P}(X)$ where $U(x) = \{y \in X : (x, y) \in U\}$; that is, $U(x)$ is the slice of $U$ determined by $x$.

If $(X, \mathcal{T})$ is a topological space and $\mathcal{C} \subseteq \mathcal{P}(X)$ is a collection of subsets of $X$, a topology on $\mathcal{C}$ is called a *hyperspace topology*.

Here we introduce one of the most common hyperspace topologies, the Vietoris topology. Extensions of the results in this section can be found in the books *Theory of Correspondences* by E. Klein and A. C. Thompson [26], *Topologies on Closed and Closed Convex Sets* by G. Beer [5], and *Hyperspaces of Sets* by S. Nadler [39].

**Definition 13.4.1.** Given a set $X$, $\mathcal{P}_0(X)$ is the collection $\mathcal{P}(X) - \{\emptyset\}$ of nonempty subsets of $X$. If $(X, \mathcal{T})$ is a topological space, $\mathcal{F}_0(X)$ is the set of nonempty closed subsets of $X$, and $\mathcal{K}_0(X)$ is the collection of nonempty compact subsets of $X$.

Recall that a function $A : \mathbb{R} \to \mathcal{P}(\mathbb{R}^2)$ is Vietoris continuous if and only if it is simultaneously upper Vietoris continuous and lower Vietoris continuous (that is, simultaneously upper semicontinuous and lower semicontinuous). This suggests that the Vietoris topology on $\mathcal{P}_0(X)$ will be defined in terms of an upper topology and a lower topology.

$A : \mathbb{R} \to \mathcal{P}(\mathbb{R}^2)$ is upper Vietoris continuous at $t_0 \in \mathbb{R}$ if for every open set $U \subseteq \mathbb{R}^2$ with $A(t_0) \subseteq U$, there exists a neighborhood of $t_0$ over which the values of $A$ remain inside $U$. This suggests that the collection $\{B \subseteq \mathbb{R}^2 : \emptyset \neq B \subseteq U\}$ of nonempty subsets of $\mathbb{R}^2$ which are contained in an open set $U \subseteq \mathbb{R}^2$ should be an open set in the upper Vietoris topology on $\mathcal{P}_0(\mathbb{R}^2)$.

Similarly, the definition of lower Vietoris continuity suggests that collections $\{B \subseteq \mathbb{R}^2 : B \cap U \neq \emptyset\}$ for any open set $U \subseteq \mathbb{R}^2$ should be open sets in the lower Vietoris topology on $\mathcal{P}_0(\mathbb{R}^2)$.

This motivates the following definition.

**Definition 13.4.2.** If $(X, \mathcal{T})$ is a topological space, the *upper Vietoris topology* $\mathcal{T}_\mathcal{U}$ on $\mathcal{P}_0(X)$ is the coarsest topology on $\mathcal{P}_0(X)$ in which each collection

$$\{B \in \mathcal{P}_0(X) : B \subseteq U\}, \quad U \in \mathcal{T},$$

is $\mathcal{T}_\mathcal{U}$-open. The *lower Vietoris topology* $\mathcal{T}_\mathcal{L}$ on $\mathcal{P}_0(X)$ is the coarsest topology on $\mathcal{P}_0(X)$ in which each collection

$$\{B \in \mathcal{P}_0(X) : B \cap U \neq \emptyset\}, \quad U \in \mathcal{T},$$

is $\mathcal{T}_\mathcal{L}$-open. The *Vietoris topology* $\mathcal{T}_\mathcal{V}$ on $\mathcal{P}_0(X)$ is the supremum $\mathcal{T}_\mathcal{U} \vee \mathcal{T}_\mathcal{L}$ of the upper and lower Vietoris topologies.

Thus, $\mathcal{T}_\mathcal{U}$ has a subbasis

$$\mathcal{S}_\mathcal{U} = \{\{B \in \mathcal{P}_0(X) : B \subseteq U\} : U \in \mathcal{T}\},$$

and $\mathcal{T}_\mathcal{L}$ has a subbasis

$$\mathcal{S}_\mathcal{L} = \{\{B \in \mathcal{P}_0(X) : B \cap U \neq \emptyset\} : U \in \mathcal{T}\}.$$

Since being contained in each of $U_1, U_2, \ldots, U_n \in \mathcal{T}$ is equivalent to being contained in $U_1 \cap \cdots \cap U_n \in \mathcal{T}$, $\mathcal{S}_\mathcal{U}$ is closed under the formation of finite intersections, so $\mathcal{S}_\mathcal{U}$ is in fact already a basis $\mathcal{B}_\mathcal{U}$ for the upper Vietoris topology $\mathcal{T}_\mathcal{U}$. However, $B$ intersecting each of $U_1, \ldots, U_n \in \mathcal{T}$ is a weaker condition than $B$ intersecting $U_1 \cap \cdots \cap U_n$ and is stronger than $B$ intersecting $U_1 \cup \cdots \cup U_n$, so the subbasis $\mathcal{S}_\mathcal{L}$ is not a basis for a topology. A basis for $\mathcal{T}_\mathcal{L}$ is given by

$$\mathcal{B}_\mathcal{L} = \{\{B \in \mathcal{P}_0(X) : B \cap U_i \neq \emptyset \ (i = 1, \ldots, n)\} : U_1, \ldots, U_n \in \mathcal{T}\}.$$

The Vietoris topology $\mathcal{T}_V = \mathcal{T}_U \vee \mathcal{T}_L$ has subbasis $\mathcal{B}_U \cup \mathcal{B}_L$. Because $\mathcal{B}_L$ and $\mathcal{B}_U = \mathcal{S}_U$ are each closed under finite intersections, $\mathcal{T}_V$ has basis elements of the form

$$[U; V_1, V_2, \ldots, V_n] = \{B \in \mathcal{P}_0(X) : B \subseteq U\}$$
$$\cap \{B \in \mathcal{P}_0(X) : B \cap V_i \neq \emptyset \ (i = 1, \ldots, n)\}$$
$$= \{B \in \mathcal{P}_0(X) : B \subseteq U, B \cap V_i \neq \emptyset \ (i = 1, \ldots, n)\},$$

where $U, V_1, \ldots, V_n \in \mathcal{T}$. Thus,

$$\mathcal{B} = \{[U; V_1, \ldots, V_n] : U, V_1, \ldots, V_n \in \mathcal{T}\}$$

is a basis for the Vietoris topology $\mathcal{T}_V$.

Another basis for $\mathcal{T}_V$ is

$$\mathcal{B}' = \{\langle V_1, \ldots, V_n \rangle : V_1, \ldots, V_n \in \mathcal{T}\}$$

where

$$\langle V_1, \ldots, V_n \rangle = \left\{ B \in \mathcal{P}_0(X) : B \subseteq \bigcup_{i=1}^{n} V_n, B \cap V_i \neq \emptyset \ (i = 1, \ldots, n) \right\}.$$

It is left to the exercises to verify that $\mathcal{B}'$ really is a basis for a topology on $\mathcal{P}_0(X)$. To see that $\mathcal{B}$ and $\mathcal{B}'$ generate the same topology, note that

$$\langle V_1, \ldots, V_n \rangle = [X; V_1] \cap \cdots \cap [X; V_n] \cap \left[ \bigcup_{i=1}^{n} V_i; \bigcup_{i=1}^{n} V_i \right] \in \mathcal{T}_{\mathcal{B}}$$

and

$$[U; V_1, \ldots, V_n] = \langle U \rangle \cap \langle X, V_1, \ldots, V_n \rangle \in \mathcal{T}_{\mathcal{B}'}.$$

Thus, $\mathcal{B}' \subseteq \mathcal{T}_{\mathcal{B}}$ and $\mathcal{B} \subseteq \mathcal{T}_{\mathcal{B}'}$, so $\mathcal{T}_{\mathcal{B}} = \mathcal{T}_{\mathcal{B}'}$.

Using the basis $\mathcal{B}'$ for the Vietoris topology $\mathcal{T}_V$, a basic open neighborhood of $A \in \mathcal{P}_0(X)$ is determined by a finite open cover $\mathcal{C} = \{V_1, \ldots, V_n\}$ of $A$ such that each $V_i$ intersects $A$. The basic neighborhood of $A$ is then the collection of all sets $B \in \mathcal{P}_0(X)$ covered by $\mathcal{C}$ such that $B$ intersects each $V_i \in \mathcal{C}$.

The Vietoris topology was introduced in 1922 in one of the first publications of the Austrian mathematician Leopold Vietoris. Vietoris continued to publish papers for over 70 years, to the age of 103. Before his death at the age of 110 in 2002, he was the oldest man in Austria.

Most applications of the Vietoris topology occur when $X$ is a metric space and $\mathcal{P}_0(X)$ is replaced by $\mathcal{F}_0(X)$ or $\mathcal{K}_0(X)$.

We will show that if $(X, d)$ is a metric space, the Vietoris topology on $\mathcal{K}_0(X)$ is metrizable.

**The Hausdorff metric.** In his 1914 text *Grundzüge der Mengenlehre*, Felix Hausdorff introduced a distance function on the collection $\mathcal{F}_0(X)$ of nonempty closed subsets of a metric space $(X, d)$.

**Definition 13.4.3.** Let $(X, d)$ be a metric space.
(a) For $a \in X$ and $B \in \mathcal{F}_0(X)$, $d(a, B) = \inf_{b \in B} d(a, b)$.
(b) For $A, B \in \mathcal{F}_0(X)$, $d(A, B) = \sup_{a \in A} d(a, B)$.
(c) The *Hausdorff metric* on $\mathcal{F}_0(X)$ is the function $h : \mathcal{F}_0(X)^2 \to \mathbb{R} \cup \{\infty\}$ defined by $h(A, B) = \max\{d(A, B), d(B, A)\}$.

The next definition and theorem provide a geometric characterization of the Hausdorff metric.

**Definition 13.4.4.** Let $(X, d)$ be a metric space. For $\varepsilon \geq 0$, the *$\varepsilon$-enlargement* (or the *$\varepsilon$-fattening*, or the *$\varepsilon$-collar*) of $B \in \mathcal{F}_0(X)$ is

$$B_\varepsilon = \{x \in X : \exists b \in B \text{ with } d(x, b) \leq \varepsilon\} = \bigcup_{b \in B} \bar{B}(b, \varepsilon),$$

where $\bar{B}(b, \varepsilon) = \{x \in X : d(b, x) \leq \varepsilon\}$.

**Theorem 13.4.5.** *If $(X, d)$ is a metric space and $A, B \in \mathcal{F}_0(X)$, then $d(A, B) \leq \varepsilon$ if and only if $A \subseteq B_\varepsilon$. Thus,*

$$h(A, B) = \inf\{\varepsilon \geq 0 : A \subseteq B_\varepsilon, B \subseteq A_\varepsilon\}.$$

This result is suggested by a careful consideration of Figure 13.9. The proof is left to the exercises.

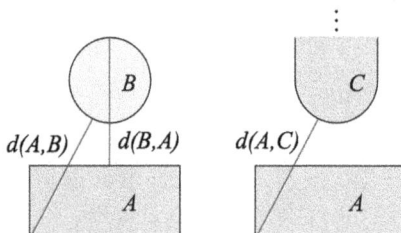

**Figure 13.9:** $h(A, B) = d(A, B); h(A, C) = \infty = d(C, A).$

Since the Hausdorff metric may assume the value $\infty$, it does not officially satisfy the definition of a metric. There are two approaches to remedy this. One could consider the Hausdorff metric only on closed subsets of a bounded metric space so $h(A, B)$ is never $\infty$, or one could allow distances to be infinite. An *extended metric* on $Y$ is a function $h : Y \to \mathbb{R} \cup \{\infty\}$ satisfying all the properties of a metric. Given any extended metric $h$, $\hat{h}(x, y) = \max\{h(x, y), 1\}$ defines a bounded metric which produces the same topology as $h$, so there is little harm in allowing extended metrics.

**Theorem 13.4.6.** *If $(X, d)$ is a metric space, the Hausdorff metric $h$ on $\mathcal{F}_0(X)$ is an extended metric.*

*Proof.* Since the other properties are straightforward, we will only verify the triangle inequality. Suppose $A, B, C \in \mathcal{F}_0(X)$. Given $a \in A, c \in C$, we have

$$d(a, B) = \inf_{b \in B} d(a, b)$$
$$\leq \inf_{b \in B} (d(a, c) + d(c, b))$$
$$= d(a, c) + \inf_{b \in B} d(c, b)$$
$$= d(a, c) + d(c, B)$$
$$\leq d(a, c) + \sup_{c \in C} d(c, B)$$
$$= d(a, c) + d(C, B).$$

Since this holds for every $c \in C$, we have

$$d(a, B) \leq \inf_{c \in C} d(a, c) + d(C, B)$$
$$= d(a, C) + d(C, B)$$
$$\leq \sup_{a \in A} d(a, C) + d(C, B)$$
$$= d(A, C) + d(C, B).$$

Since this holds for every $a \in A$, $d(A, B) = \sup_{a \in A} d(a, B) \leq d(A, C) + d(C, B)$. Now it is easy to show that if $d$ satisfies the triangle inequality, then $h$ defined by $h(A, B) = \max\{d(A, B), d(B, A)\}$ satisfies the triangle inequality. □

**Lemma 13.4.7.** *If $A$ is a compact subset of a metric space $(X, d)$ and $U$ is an open set containing $A$, then there exists $\varepsilon > 0$ such that $A_\varepsilon \subseteq U$.*

*Proof.* For each $a \in A$, there exists $\varepsilon_a > 0$ such that $B(a, \varepsilon_a) \subseteq U$. The open cover $\{B(a, \varepsilon_a/2) : a \in A\}$ has a finite subcover $\{B(a_i, \varepsilon_{a_i}/2) : i = 1, \ldots, n\}$. Let $\varepsilon = \min\{\varepsilon_{a_i}/2 : i = 1, \ldots, n\}$. Now if $b \in A_\varepsilon$, there exists $a \in A$ such that $d(b, a) \leq \varepsilon$. Pick $a_i$ such that $a \in B(a_i, \varepsilon_{a_i}/2)$. Now $d(b, a_i) \leq d(b, a) + d(a, a_i) < \varepsilon + \varepsilon_{a_i}/2 \leq \varepsilon_{a_i}$. Thus, $b \in B(a_i, \varepsilon_{a_i}) \subseteq U$. □

**Theorem 13.4.8.** *On the collection $\mathcal{K}_0(X)$ of nonempty compact subsets of a metric space $(X, d)$, the metric topology $\mathcal{T}_h$ from the Hausdorff metric $h$ is the Vietoris topology $\mathcal{T}_V$.*

*Proof.* First we will show that each $\mathcal{T}_h$ ball $B_h(A, \varepsilon)$ around $A \in \mathcal{K}_0(X)$ contains a $\mathcal{T}_V$ neighborhood of $A$. Suppose $A \in \mathcal{K}_0(X)$ and $\varepsilon > 0$ is given. The open cover $\{B(a, \varepsilon/4) :$

$a \in A\}$ of $A$ has a finite subcover $\{B(a_1, \varepsilon/4), \ldots, B(a_n, \varepsilon/4)\}$, and $\langle B(a_1, \varepsilon/4), \ldots, B(a_n, \varepsilon/4)\rangle \in \mathcal{T}_V$. Now clearly $A \in \langle B(a_1, \varepsilon/4), \ldots, B(a_n, \varepsilon/4)\rangle$. If $B \in \langle B(a_1, \varepsilon/4), \ldots, B(a_n, \varepsilon/4)\rangle$ and $b \in B$, then $b \in B(a_i, \varepsilon/4)$ for some $i \in \{1, 2, \ldots, n\}$, so $d(b, a_i) < \varepsilon/4$ and thus $b \in A_{\varepsilon/4}$. This shows $B \subseteq A_{\varepsilon/4} \subseteq A_{\varepsilon/2}$. Furthermore, if $a \in A$, there exists $j \in \{1, \ldots, n\}$ with $a \in B(a_j, \varepsilon/4)$, and since $B \in \langle B(a_1, \varepsilon/4), \ldots, B(a_n, \varepsilon/4)\rangle$, there exists $b \in B(a_j, \varepsilon/4) \cap B$. Now $\{a, b\} \subseteq B(a_j, \varepsilon/4)$ implies $d(a, b) < \varepsilon/2$, so $A \subseteq B_{\varepsilon/2}$. Thus, $h(A, B) = \inf\{\varepsilon \geq 0 : A \subseteq B_\varepsilon, B \subseteq A_\varepsilon\} \leq \varepsilon/2 < \varepsilon$, so $B \in B_h(A, \varepsilon)$. Thus, $\langle B(a_1, \varepsilon/4), \ldots, B(a_n, \varepsilon/4)\rangle \subseteq B_h(A, \varepsilon)$, so $\mathcal{T}_V$ is finer than $\mathcal{T}_h$.

Now suppose $\langle U_1, \ldots, U_n\rangle \in \mathcal{T}_V$ and $A \in \langle U_1, \ldots, U_n\rangle$. For each $i = 1, \ldots, n$, pick $a_i \in A \cap U_i$ and pick $\varepsilon_i > 0$ such that $B(a_i, \varepsilon_i) \subseteq U_i$. By Lemma 13.4.7, we may choose $\varepsilon_0 > 0$ such that $A_{\varepsilon_0} \subseteq U_1 \cup \cdots \cup U_n$. Let $\varepsilon = \min\{\varepsilon_0, \varepsilon_1/2, \ldots, \varepsilon_n/2\}$. We will show $B_h(A, \varepsilon) \subseteq \langle U_1, \ldots, U_n\rangle$. To this end, suppose $B \in B_h(A, \varepsilon)$. Then $h(A, B) = \inf\{y \geq 0 : A \subseteq B_y, B \subseteq A_y\} < \varepsilon$. In particular, $B \subseteq A_\varepsilon \subseteq A_{\varepsilon_0} \subseteq U_1 \cup \cdots \cup U_n$. Now it only remains to show $B \cap U_i \neq \emptyset$ for each $i = 1, \ldots, n$. Given $i$, $A \subseteq B_\varepsilon \subseteq B_{\varepsilon_i/2}$, so for every $a \in A$, there exists $b \in B$ with $d(a, b) \leq \varepsilon_i/2$. In particular, for $a_i \in A$ there exists $b \in B$ with $d(a_i, b) \leq \varepsilon_i/2 < \varepsilon_i$. Now $b \in B \cap B(a_i, \varepsilon_i) \subseteq B \cap U_i$, showing $B \cap U_i \neq \emptyset$, as needed. $\square$

## Exercises

1. Verify that $\mathcal{B}'$ given after Definition 13.4.2 really is a basis for a topology on $\mathcal{P}_0(X)$.
2. Given $A, B \in \mathcal{F}_0$ and $\varepsilon \geq 0$, verify that $d(A, B) \leq \varepsilon$ if and only if $A \subseteq B_\varepsilon$.
3. If $f : \mathbb{R}^2 \to \mathbb{R}^2$ is continuous, define $\hat{f} : (\mathcal{K}_0(\mathbb{R}^2), \mathcal{T}_V) \to (\mathcal{K}_0(\mathbb{R}^2), \mathcal{T}_V)$ by $\hat{f}(A) = f(A)$. Show that $\hat{f}$ is continuous.
4. In $\mathbb{R}^2$, let $U_1 = B((0, 0), 2)$ and $U_2 = B((2, 0), 2)$. Suppose the rectangle $A = [a, b] \times [c, d]$ is in the $\mathcal{T}_V$-open set $\langle U_1, U_2\rangle$. How many corners of $A$ can lie in $U_1$? More specifically, for which $k \in \{0, 1, 2, 3, 4\}$ is it possible to find a rectangle $A = [a, b] \times [c, d]$ in $\langle U_1, U_2\rangle$ with exactly $k$ of its corners in $U_1$? Justify your answers.
5. Suppose $A$ is a closed set in a metric space and $\varepsilon, \delta > 0$. Show that $(A_\varepsilon)_\delta \subseteq A_{\varepsilon + \delta}$ and that equality need not hold.
6. Under what conditions does $\bigcap_{\varepsilon > 0} A_\varepsilon = A_0$?
7. (a) If $A$ is a compact subset of a metric space $(X, d)$ and $\mathcal{C}$ is an open cover of $A$, prove that there exists a number $\delta > 0$ such that, for any $a \in A$, the ball $B(a, \delta)$ is contained in some $U \in \mathcal{C}$. The number $\delta$ is called the *Lebesgue number* of the cover.
   (b) Show that (a) implies Lemma 13.4.7.
8. Let $h$ be the Hausdorff metric on $\mathcal{F}_0(\mathbb{R}^2)$. For the pairs of sets $A$ and $B$ given below, find $d(A, B)$, $d(B, A)$, and $h(A, B)$.
   (a) $A = [0, 1]^2, B = [4, 1] \times [5, 3]$
   (b) $A = \{(x, y) \in \mathbb{R}^2 : x^2 + (y - 2)^2 = 1\}, B = \{(x, y) \in \mathbb{R}^2 : x^2 + y^2 = 16\}$
   (c) $A = \{(x, y) \in \mathbb{R}^2 : x^2 + y^2 \leq 1\}, B = ([-4, -3] \times [-1, 1]) \cup ([1, 2] \times [-1, 1])$
9. Complete the following characterizations. Justify your answers.

(a) The Hausdorff distance between two lines $L$ and $M$ in the Euclidean plane $\mathbb{R}^2$ is finite if and only if ....

(b) The Hausdorff distance between two closed half-planes $A$ and $B$ in the Euclidean plane $\mathbb{R}^2$ is finite if and only if ....

10. If $(X, d)$ is a metric space, show that $f : (X, d) \to (\mathcal{F}_0(X), h)$ defined by $f(x) = \{x\}$ is an isometry, so (by Exercise 17 of Section 6.1), $X$ is homeomorphic to the subspace $\{\{x\} : x \in X\}$ of $(\mathcal{F}_0(X), h)$.

11. If $(X, d)$ is a metric space, the Hausdorff metric on $\mathcal{F}_0(X)$ is defined by $h(A, B) = \max\{d(A, B), d(B, A)\}$, where $d(A, B) = \sup_{a \in A} d(a, B)$. Show that $d : \mathcal{F}_0(X) \times \mathcal{F}_0(X) \to \mathbb{R}$ is a hemimetric. (Hint: See the proof of Theorem 13.4.6.)

12. If $(X, d)$ is a metric space and $h : \mathcal{F}_0(X) \times \mathcal{F}_0(X) \to \mathbb{R}$ is the Hausdorff metric $h(A, B) = \max\{d(A, B), d(B, A)\}$, the proof of Theorem 13.4.6 showed that $d$ satisfies the triangle inequality. Complete the missing steps there to show that $h$ satisfies the triangle inequality.

# Index

# Bibliography

[1]  Fatemah Ayatollah Zadeh Shirazi and Nasser Golestani, Functional Alexandroff spaces, Hacet. J. Math. Stat. 40 (4) (2011) 515–522.

[2]  Stefan Banach and Alfred Tarski, Sur la décomposition des ensembles de points en parties respectivement congruentes, Fundam. Math. 6 (1924) 244–277.

[3]  Robert G. Bartle, The Elements of Real Analysis, 2nd ed., John Wiley & Sons, New York, 1976.

[4]  Kathleen Bell and Tom Richmond, Transitions between 4-intersection values of planar regions, Appl. Gen. Topol. 18 (1) (2017) 183–202.

[5]  Gerald Beer, Topologies on Closed and Closed Convex Sets, Mathematics and Its Applications Series, Kluwer Academic Publishers, Boston, 1993.

[6]  Garrett Birkhoff, Lattice Theory, 3rd ed. American Mathematical Society Colloq. Publications, Vol. 25, 1967.

[7]  M. P. Berri, Minimal topological spaces, Trans. Am. Math. Soc. 108 (1963) 97–105.

[8]  D. Bushaw, Elements of General Topology, J. Wiley, New York, 1963.

[9]  D. E. Cameron, Maximal and minimal topologies, Trans. Am. Math. Soc. 160 (1971) 229–248.

[10]  D. E. Cameron, A survey of maximal topological spaces, Topol. Proc. 2 (1977) 11–60.

[11]  Stefan Cobzaş, Functional Analysis in Asymmetric Normed Spaces, Birkhäuser Frontiers in Mathematics, Springer, Basel, 2013.

[12]  Fred J. Damerau, A technique for computer detection and correction of spelling errors, Commun. ACM 7 (3) (1964) 171–176, doi:10.1145/363958.363994.

[13]  B. A. Davey and H. A. Priestley, Introduction to Lattices and Order, 2nd ed., Cambridge University Press, 2002.

[14]  Othman Echi, The categories of flows of Set and Top, Topol. Appl. 159 (9) (2012) 2357–2366.

[15]  M. Egenhofer and R. Franzosa, Point-set topological spatial relations, Int. J. Geogr. Inf. Syst. 5 (2) (1991) 161–174.

[16]  R. Engelking, Outline of General Topology, North-Holland Publishing Co., Amsterdam, 1968.

[17]  Peter Fletcher and William F. Lindgren, Quasi-Uniform Spaces, Lecture Notes in Pure and Applied Mathematics, Vol. 77, Marcel Dekker, New York, 1982.

[18]  H. Fürstenberg, On the infinitude of primes, Am. Math. Mon. 62 (1955) 353.

[19]  Y. U. Gaba and H.-P. A. Künzi, Splitting metrics by $T_0$-quasi-metrics, Topol. Appl. 193 (2015) 84–96.

[20]  Y. U. Gaba and H.-P. A. Künzi, Partially ordered metric spaces produced by $T_0$-quasi-metrics, Topol. Appl. 202 (2016) 366–383.

[21]  Jean Goubault-Larrecq, Non-Hausdorff Topology and Domain Theory, New Mathematical Monographs, Vol. 22, Cambridge University Press, 2013.

[22]  Asli Güldürdek and Tom Richmond, Every finite topology is generated by a partial pseudometric, Order 22 (4) (2005) 415–421.

[23]  Ting Hong, Note on minimal bicompact spaces, Bull. Am. Math. Soc. 54 (1948) 478–479.

[24]  J. E. Joseph, Continuous functions and spaces in which compact sets are closed, Am. Math. Mon. 76 (10) (1969) 1125–1126.

[25]  John L. Kelley, The Tychonoff product theorem implies the axiom of choice, Fundam. Math. 37 (1950) 75–76.

[26]  Erwin Klein and Anthony C. Thompson, Theory of Correspondences, Canadian Mathematical Society Series of Monographs and Advanced Texts, John Wiley and Sons, New York, 1984, pp. 196–215.

[27]  Ralph Kopperman, All topologies come from generalized metrics, Am. Math. Mon. 95 (2) (1988) 89–97.

https://doi.org/10.1515/9783110686579-015

[28] H.-P. Künzi, A. E. McCluskey and T. A. Richmond, Ordered separation axioms and the Wallman ordered compactification, Publ. Math. (Debr.) 73 (3–4) (2008) 361–377.

[29] H.-P. Künzi and F. Yildiz, Extensions of $T_0$-quasi-metrics, Acta Math. Hung. 153 (1) (2017).

[30] R. Lalitha, A note on the lattice structure of $T_1$ topologies on an infinite set, Math. Stud. 35 (1967) 29–33.

[31] R. E. Larson and S. J. Andima, The lattice of topologies: a survey, Rocky Mt. J. Math. 5 (2) (1975) 177–198.

[32] R. E. Larson and W. J. Thron, Covering relations in the lattice of $T_1$-topologies, Trans. Am. Math. Soc. 168 (1972) 101–111.

[33] Vladimir I. Levenshtein, Binary codes capable of correcting deletions, insertions, and reversals, Sov. Phys. Dokl. 10 (8) (1966) 707–710.

[34] N. Levine, When are compact and closed equivalent, Am. Math. Mon. 72 (1) (1965) 41–44.

[35] Rezsö L. Lovas and István Mezö, Some observations on the Furstenberg topological space, Elem. Math. 70 (3) (2015) 103–116.

[36] S. G. Matthews, Partial Metric Topology, in Papers on General Topology and Applications, Eighth Summer Conference at Queens College. Eds. S. Andima, et al., Annals of the New York Academy of Sciences, Vol. 728, 1994, pp. 183–197.

[37] Jacob Menix and Tom Richmond, The lattice of functional Alexandroff topologies, Order (2020), doi:10.1007/s11083-020-09523-6.

[38] L. Nachbin, Topology and Order, Vol. 4, Van Nostrand Mathematical, Princeton, N.J., 1965.

[39] Sam Nadler, Jr., Hyperspaces of Sets: A Text with Research Questions, Marcel Dekker, Inc., New York, 1978.

[40] B. Richmond and T. Richmond, Metric spaces in which all triangles are degenerate, Am. Math. Mon. 104 (8) (1997) 713–719.

[41] B. Richmond and T. Richmond, A Discrete Transition to Advanced Mathematics, American Mathematical Society Pure and Applied Undergraduate Texts, Vol. 3, 2004.

[42] Tom Richmond, Ball transitive ordered metric spaces, in Proceedings of the 24th National Conference of Geometry and Topology, Timisoara, Romania, July 5–9, 1994, Part one, Lectures. Eds. Adrian C. Albu, Mircea Craioveanu and Mirton Timisoara, 1996, pp. 137–143.

[43] Tom Richmond, Quasiorders, principal topologies, and partially ordered partitions, Int. J. Math. Math. Sci. 21 (2) (1998) 221–234.

[44] N. Smythe and C. A. Wilkins, Minimal Hausdorff and maximal compact spaces, J. Aust. Math. Soc. 3 (1963) 167–171.

[45] Lynn Arthur Steen and J. Arthur Seebach, Jr., Counterexamples in Topology, 2nd ed., Springer Verlag, New York, 1978.

[46] J. P. Thomas, Maximal connected topologies, J. Aust. Math. Soc. 8 (1968) 700–705.

[47] Wolfgang J. Thron, Topological Structures, Holt, Rinehart and Winston, New York, 1966.

[48] Andrey N. Tychonoff, Über die topologische Erweiterung von Räumen, Math. Ann. 102 (1) (1930) 544–561, doi:10.1007/BF01782364.

[49] Daniel Velleman and Gregory Call, Permutations and combination locks, Math. Mag. 68 (4) (1995) 243–253.

www.ingramcontent.com/pod-product-compliance
Lightning Source LLC
Chambersburg PA
CBHW080926220326
41598CB00034B/5689